T0305989

Introduction to Dual Polarization Weather Radar

An interdisciplinary and easy-to-understand introduction to the subject, covering fundamental theory and practical applications supported by numerous operational examples. This balanced text will allow you to begin from how the radar makes observations and move deeper through electromagnetic scattering theory and cloud microphysics to understand and interpret data as they appear on the display. It uses illustrations and images of real radar observations to convey the concepts and theory of atmospheric processes typically observed with weather radar, as well as presenting a working knowledge of the radar system itself. In addition to covering the fundamentals of scattering and atmospheric physics, topics include system hardware, signal processing, and radar networks. This is the perfect tool for scientists and engineers working on weather radars or using their data, as well as senior undergraduate and graduate students studying weather radars for the first time.

V. Chandrasekar (Chandra) is a University Distinguished Professor at Colorado State University. He is a Fellow of the Institute of Electrical and Electronics Engineers (IEEE), the International Union of Radio Science (URSI), and the American Meteorological Society.

Robert M. Beauchamp is a systems engineer at the Jet Propulsion Laboratory, California Institute of Technology. His expertise is in radar systems and signal processing.

Renzo Bechini is a weather radar expert with 25 years of professional experience in the operation of dual polarization systems. He is currently a weather radar specialist at the regional agency for environmental protection, Arpa Piemonte, in Turin, Italy.

Introduction to Dual Polarization Weather Radar

Fundamentals, Applications, and Networks

V. CHANDRASEKAR

ROBERT M. BEAUCHAMP

RENZO BECHINI

University Printing House, Cambridge CB2 8BS, United Kingdom

One Liberty Plaza, 20th Floor, New York, NY 10006, USA

477 Williamstown Road, Port Melbourne, VIC 3207, Australia

314–321, 3rd Floor, Plot 3, Splendor Forum, Jasola District Centre, New Delhi – 110025, India

103 Penang Road, #05–06/07, Visioncrest Commercial, Singapore 238467

Cambridge University Press is part of the University of Cambridge.

It furthers the University's mission by disseminating knowledge in the pursuit of education, learning, and research at the highest international levels of excellence.

www.cambridge.org
Information on this title: www.cambridge.org/9781108423175
DOI: 10.1017/9781108772266

First published 2023

A catalogue record for this publication is available from the British Library.

Library of Congress Cataloging-in-Publication Data
Names: Chandrasekar, V., author.
Title: Introduction to dual polarization Doppler weather radar :
    fundamentals, applications, and networks / V. Chandrasekar,
    Robert M. Beauchamp, Renzo Bechini.
Description: New York : Cambridge University Press, 2023. |
    Includes bibliographical references and index.
Identifiers: LCCN 2021037838 (print) | LCCN 2021037839 (ebook) |
    ISBN 9781108423175 (hardback) | ISBN 9781108772266 (epub)
Subjects: LCSH: Radar meteorology. | Doppler radar. | Polariscope. |
    Radio waves–Polarization–Measurement. | BISAC: SCIENCE /
    Physics / Electromagnetism
Classification: LCC QC973.5 .C43 2023 (print) | LCC QC973.5 (ebook) |
    DDC 551.63/53–dc23
LC record available at https://lccn.loc.gov/2021037838
LC ebook record available at https://lccn.loc.gov/2021037839

ISBN 978-1-108-42317-5 Hardback

Additional resources for this publication at [insert URL here].

Contents

Preface

Historically, new ideas and hypotheses start in specific research areas fostered by a small community through journal papers and monographs. As their appreciation and practicality toward applications grow through the community of researchers, the work is standardized. It is added to the curriculum and textbooks, becoming part of the education at universities. There are numerous examples in history, such as special topics in applied mathematics and statistics being transformed into modern signal processing, or advanced semiconductor physics becoming the electronics and computing boom of the second half of the twentieth century. Weather radars, and more specifically for our purposes, polarimetric weather radars, have undergone the same evolutionary process. Dual polarization radar is part of the standard radar meteorology or remote-sensing curriculum in universities. Through its evolution from a fledgling field of research in the late 1970s to standard weather-observing equipment today, many advanced books have become available in the literature on the topic, starting in 2001 with the first book on polarimetric Doppler weather radar from Cambridge University Press. This book continues this goal: to distill and disseminate fundamental science and practical engineering knowledge about dual polarization weather radar and its applications in an introductory style.

Radar meteorology is a unique subject. The title itself indicates its interdisciplinary nature, combining the fields of radar and meteorology. Radar systems are the workhorse of weather services and atmospheric researchers around the world. They are used to observe the big picture of atmospheric systems as well as the detailed structure of storms. Weather radars are routinely used in multiple lines of business, including national, regional, and private weather services; hydrologic and agricultural industries; and the aviation sector. The adoption of dual polarization for radar systems beyond the research community coincided with a number of other technological revolutions that made it possible. Examples include the leapfrogging of technology in microwave and radio-frequency devices driven by the revolution in communication and digital signal processing, as well as the exponential growth of affordable computing power and data storage. The conversion of weather radar systems all over the world to the use of dual polarization technology began in the 1990s. Today, dual polarization is standard equipment for most operational weather radars.

The application of radars to meteorology has widespread appeal and has become a core topic in the curricula of meteorology, atmospheric science, atmospheric physics, electrical engineering, and geoscience programs. Fundamentally, radar meteorology is an interdisciplinary topic covering the engineering of the radar system, the physics of wave propagation and scattering, the science of atmospheric processes, and the applications of weather observation and prediction. There is a real need for an introductory-level textbook devoted to the interdisciplinary aspects of "radar" and "meteorology" comprising "radar meteorology." This book aims to serve that goal. With all worth-while endeavors come challenges, and the task of writing an introductory textbook for dual polarization radar meteorology is no different. We recognize that students in these curricula follow different paths and academic interests, from their core classes through their junior year's undergraduate curricula. This book takes on the challenge of developing a cogent introductory textbook for an audience from diverse science and engineering backgrounds.

The book consists of 10 chapters, with the early chapters focused on fundamentals. The first chapter introduces dual polarization weather radar and the book in general. At the chapter's end, a route map through this book's 10 chapters is presented as a guide for students, teachers, and practitioners as they take a journey through this book. Pure science students can skip a subset of the engineering-oriented chapters, and similarly, pure engineering students can skip a subset of the physical science chapters. For professionals pursuing a goal of "lifelong learning," or to refresh their knowledge on new techniques and technologies, an accelerated pathway is presented that is focused on applications. The material in this book is intended for students and practitioners who have completed junior-level science, technology, engineering, and mathematics (STEM) coursework. Each chapter has a set of assignments and online supplements.

We had to make another important decision as part of writing this book: whether to present the illustrations in color. After extensive deliberation, we decided to keep the print version of this book in black and white (B&W), for multiple reasons, including (a) B&W books leave a smaller footprint on the environment, (b) many publications are resorting to color rendering in electronic form to make room for future growth in color illustrations, and (c) this book's focus is on introductory concepts.

After the introduction, Chapter 2 presents a basic introduction to polarization and radar systems for readers of all backgrounds. Chapters 3 and 4 are science-oriented chapters and present the essential precipitation physics and introductory scattering principles for precipitation. Chapters 5 and 6 shift the focus toward engineering, introducing radar signals, followed by weather signal processing. Starting with Chapter 7 on data quality and data science, the book merges the science and engineering branches, focusing on application. This trend continues for radar observations (Chapter 8) and rainfall estimation (Chapter 9) and ends with Chapter 10 on radar networks. The last four chapters are applications directly suited for radar meteorology practitioners.

We would like to thank the funding agencies for supporting the research that has resulted in the topics presented here. We want to acknowledge and thank those who donated their valuable time to provide feedback throughout the evolution of this book (in alphabetical order): Chad Baldi, Robert Cifelli, Brenda Dolan, Nicholas Kedzuf, Patrick Kennedy, Eric Perez, and Richard Roy. To thank all the people who have contributed directly or indirectly toward this work would run many pages. It's to this group we dedicate this book: to our families for supporting us, and to all our teachers, students, and colleagues who have taught us everything presented here.

V. Chandrasekar, Robert M. Beauchamp, and Renzo Bechini

Notations and Acronyms

$[F_{rx}]$	receive antenna's dual polarization matrix
$[F_{tx}]$	transmit antenna's dual polarization matrix
$[M]$	transmitter's dual polarization matrix
$[R]$	radar receiver's dual polarization matrix
$[S]$	dual polarization scattering matrix
$[T]$	propagation channel's dual polarization matrix
$[T_{rx}]$	receive channel's dual polarization matrix
$[T_{tx}]$	transmit channel's dual polarization matrix
α	Erlang distribution's shape parameter
α	Tukey window's taper fraction
α	attenuation constant
$\bar{\eta}$	reflectivity
$\bar{b}_h$	mean minimum beam height within the network's cell
$\bar{b}_s$	mean radar beam size within the network cell
$\bar{E}$	average collection efficiency of raindrops
$\bar{r}_i$	ice average radius
$\bar{r}_m$	mass-weighted mean axis ratio
$\bar{r}_z$	reflectivity-weighted axis ratio
$\bar{S}$	averaged power spectrum
$\bar{v}$	mean Doppler velocity
$\bar{v}_p$	precipitation's mean velocity
$\bar{Z}_{\min}$	mean detection sensitivity within the network cell
β	canting angle
β	phase constant
β	slope parameter for linear drop shape model
β_e	equivalent slope for linear drop shape model
$\Delta\omega$	angular frequency resolution
Δf	frequency resolution
δk	weighted difference of specific attenuation
Δr	range resolution
δr	Doppler frequency induced range migration
Δt	elapsed time
$\Delta t_{2\text{way}}$	round-trip time delay
Δv	Doppler velocity resolution

δ	exponent of snowfall velocity-size relation
δ_{co}	backscatter differential phase shift
$\dot{\psi}_{ant}$	antenna angular scan rate
ϵ	electric permittivity
ϵ	measurement error
ϵ_0	electric permittivity of free-space ($\approx 8.854 \cdot 10^{-12}$ Farad m^{-1})
ϵ_{eff}	effective permittivity
ϵ_e	environment's permittivity
ϵ_i	inclusion's permittivity
ϵ_r''	dielectric constant's imaginary part
ϵ_r'	dielectric constant's real part
ϵ_r	dielectric constant, relative permittivity
ϵ_s	static, low-frequency relative permittivity of water
η	characteristic impedance
Γ	gamma function
Γ	reflection coefficient
γ	Barnes analysis annealing parameter
γ	propagation constant, $\gamma = jk$
$\hat{\Psi}_{dp}$	measured differential phase shift
$\hat{h}$	horizontal polarization's unit vector
$\hat{i}$	incident direction's unit vector
$\hat{k}$	propagation direction's unit vector
$\hat{K}_{dp}$	estimated specific differential phase shift
$\hat{r}_{xy}$	biased covariance estimate
$\hat{s}$	scattering direction's unit vector
$\hat{v}$	vertical polarization's unit vector
κ	velocity extension ratio
Λ	particle size distribution's size (slope) parameter
λ	smoothing factor for spline fitting
λ	wavelength
$\mathbf{A}$	multi-Doppler geometry matrix
$\mathbf{a}$	unit vector along the radar's radial direction
$\mathbf{C}$	covariance matrix
$\mathbf{E}$	electric field vector
$\mathbf{H}$	magnetic field vector
$\mathbf{h}$	filter vector
$\mathbf{J}$	current density vector
$\mathbf{m}$	waveform vector
$\mathbf{P}$	polarization current vector
$\mathbf{P}$	polarization sampling vector
$\mathbf{S}$	Poynting vector
$\mathbf{S}$	dual polarization scattering matrix
$\mathbf{S}$	scattering matrix
$\mathbf{v}$	particle velocity vector

$\mathbf{X}$	signal matrix
$\mathbf{X}_f$	waveform's sidelobe matrix
$\mathbf{y}$	convolution output vector
ATAR	alternate transmit, alternate receive
ATSR	alternate transmit, simultaneous receive
CPI	coherent processing interval
DSD	raindrop size distribution
HPBW	half-power beamwidth
ICPR_2	two-way integrated cross-polar ratio
$\text{ICPR}_2^{(ub)}$	two-way integrated cross-polar ratio's upper bound
ISL	integrated sidelobe level
IWC	ice water content
LDR	linear depolarization ratio
$\text{LDR}_{\text{limit}}$	antenna's minimum LDR
LDR_{hv}	vertical linear depolarization ratio
LDR_{vh}	horizontal linear depolarization ratio
MBF	fuzzy logic membership function
NCP	normalized coherent power
$\text{NCP}_{\text{thres}}$	normalized coherent power's threshold for signal detection
NF	noise figure
PDF	probability density function
PIA	path-integrated attenuation
$\text{PIA}_{1\text{way}}$	one-way path-integrated attenuation
PPI	plan position indicator
PRF	pulse repetition frequency
PRT	pulse repetition time
PSD	particle size distribution
PSL	peak sidelobe level
RCA	relative calibration adjustment
RCS	radar cross-section
RHI	range height indicator
RH	relative humidity
SNR	signal-to-noise ratio
SQI	signal quality index
STSR	simultaneous transmit, simultaneous receive
TBP	time-bandwidth product
μ	magnetic permeability
μ	mean
μ	particle size distributions' shape parameter
μ^0	magnetic permeability of free-space ($4\pi \cdot 10^{-7}\ \text{Henries}\,\text{m}^{-1}$)
μ_m	m^{th} spectral moment
ν	kinematic viscosity of air
Ω	solid angle
ω	angular frequency

ω_0	radar carrier's angular frequency
ω_s	pulse repetition angular frequency
ϕ	azimuth angle
$\dot{\phi}$	rate of change of the drop size
ϕ_1	antenna's half-power beamwidth in azimuth
Φ_{dp}	propagation differential phase shift
Φ_{dp}^{sys}	system differential phase
ϕ_{pc}	phase code
ϕ_{sys}	system's phase shift
ϕ_{vel}	velocity-induced phase shift
ψ_{ant}	angular scanning extent of a CPI
Ψ_{dp}	total differential phase shift
ρ	correlation coefficient
ρ	density
ρ_{co}	complex-valued copolar correlation coefficient
ρ_{snow}	snowflake density
ρ_s	spatial correlation coefficient
ρ_s	water vapor density at water saturation
ρ_v	water vapor density
ρ_w	water density ($997\,\mathrm{kg\,m^{-3}} \approx 1\,\mathrm{g\,cm^{-3}}$)
ρ_{cx}	cross-polarization correlation coefficient
ρ_{hv}	copolar correlation coefficient, linear polarization
$\rho_{hv}^{(m)}$	measured copolar correlation with a reflectivity gradient
ρ_{si}	water vapor density at ice saturation
ρ_{vs}	water vapor density at the drop's surface
ρ_{xx}	autocorrelation of x
ρ_{xy}	cross-correlation of x and y
σ	electrical conductivity
σ	radar cross-section
σ	standard deviation
σ^2	variance
σ_a	absorption cross-section
σ_b	backscatter radar cross-section
σ_e	extinction cross-section
σ_h	clutter's horizontal polarization scattering amplitude
σ_s	total scattering cross-section
σ_v	Doppler spectrum width
σ_v	clutter's vertical polarization scattering amplitude
σ_w	surface tension of water ($0.07275\,\mathrm{J\,m^{-2}}$)
σ_w^2	white-noise signal's variance
σ_{bi}	bistatic radar cross-section
σ_{vc}	clutter's spectrum width
σ_{vn}	normalized Doppler spectrum width

σ_{vp}	precipitation's spectrum width
$\tan\delta$	loss tangent
τ	time delay
$\mathbf{s}$	far-field scattering amplitude vector
θ	complex-valued angular differential phase shift
θ	elevation angle
θ	incidence angle
θ	wave's incident angle
θ_1	antenna's half-power beamwidth in elevation
θ_{err}	antenna's pointing error
θ_{HPBW}	antenna's half-power beamwidth
θ_{sep}	network's angular separation between radars
θ_{ik}	azimuth difference for Barnes analysis
$\tilde{\mathbf{h}}$	minimum integrated sidelobe level filter's vector
$\tilde{p}$	signal's power estimate
$\tilde{r}_{xy}$	unbiased covariance estimate
φ	radar signal's phase
φ_{frdp}	receive antenna's differential phase
φ_{ftdp}	transmit antenna's differential phase
φ_{rdp}	receiver's differential phase
φ_{txh}	horizontal polarization's transmit phase
φ_{txv}	vertical polarization's transmit phase
φ_{tx}	relative transmit phase
$\widehat{\text{LHC}}$	left-hand circular polarization's unit vector
$\widehat{\text{RHC}}$	right-hand circular polarization's unit vector
$\widehat{S}$	estimated power spectrum
$\widehat{X}$	estimated signal spectrum
A	cross-sectional area
A	normalized coherent power threshold's performance constant
A	specific attenuation
a	K_{dp}-Z scale coefficient
a	antenna gain's model coefficient
a	beta MBF's spread parameter
a	radar signal's amplitude
a	radius of Earth
a	spheroid's equatorial radius
a	transmitter's gain coefficient
$A_{1\text{way}}$	one-way specific attenuation
$A_{2\text{way}}$	two-way specific attenuation
$A_{\text{beam}}^{(\text{weighted})}$	V_6 cross-sectional area
A_{dp}	specific differential attenuation
A_a	antenna collecting area
A_e	antenna effective area

a_e	effective radius of Earth
B	bandwidth
b	K_{dp}-Z exponential coefficient
b	beta MBF's slope parameter
b	observation volume's diameter with a circular antenna
b	spheroid's radius along the axis of symmetry
b_ϕ	observation volume's width in azimuth
b_θ	observation volume's height in elevation
b_h	minimum beam height
b_s	radar beam size
C	$\mathrm{Re}(p_h - p_v)$ vs. $(1 - r)$ slope
C	capacitance of the ice crystal
C	weather radar constant
c	IWC estimator's scale coefficient
c	electromagnetic wave's propagation speed, speed of light
c	speed of light
c_0	speed of light in free space ($299{,}792{,}458 \text{ m s}^{-1} \approx 3 \cdot 10^8 \text{ m s}^{-1}$)
C_{gas}	two-way atmospheric gas specific attenuation
C_h	horizontal polarization's weather radar constant
C_v	vertical polarization's weather radar constant
CN	complex-valued Gaussian (normal) distribution's PDF
D	drop's equivalent spherical diameter
D	equivalent spherical diameter of ice particles
D	raindrop's equivalent-volume spherical diameter
d	Euclidean distance
D_0	median equivolume diameter
D_{ev}	drop's diameter that completely evaporates after falling h distance
D_{max}	maximum drop diameter of the DSD
D_{min}	minimum drop diameter of the DSD
d_a	antenna diameter
D_e	volume-equivalent spherical diameter
D_m	mass-weighted mean diameter
D_p	DSD's p^{th} moment
D_s	maximum snowflake dimension
D_v	diffusion coefficient for water vapor in air
D_w	melted ice equivalent drop diameter
D_z	reflectivity-weighted mean diameter
df	frequency offset
E	collision-coalescence collection efficiency of raindrops
e	electric field component
e	prolate shape factor
e	water-vapor pressure
e_a	antenna aperture's efficiency
E_h	horizontal polarization's electric field

e_h	horizontal polarization's received signal
e_s	saturation vapor pressure with respect to water
E_v	vertical polarization's electric field
e_v	vertical polarization's received signal
e_{si}	saturation vapor pressure with respect to ice
E_{tx}	transmitted energy
$f(\mu)$	normalized PDS's shape function
$f(Z_{\mathrm{dr}})$	rain/hail partition function
F	antenna feed's forward distance
F	antenna pattern's complex amplitude
F	noise factor
f	complex-valued forward-scattering amplitude
f	frequency
f	normalized antenna pattern's complex amplitude
f	oblate shape factor
f	observation field
f	volume fraction
f_0	radar carrier's frequency
f_1	pulse repetition frequency for PRT 1
f_2	pulse repetition frequency for PRT 2
f_e	environment's volume fraction
F_{ice}	model for IWC or snowfall rate
f_{IF}	intermediate frequency
f_i	inclusion's volume fraction
f_{LO}	local oscillator's frequency
F_d	vapor diffusion term
f_D	probability density function
f_d	Doppler frequency shift
F_d^i	vapor diffusion term for ice
f_E	Erlang distribution's probability density function
F_h	horizontal force acting on the raindrop
F_k	heat-conductivity term
F_k^i	heat-conductivity term for ice
f_s	pulse repetition frequency
F_v	vertical force acting on the raindrop
f_v	ventilation factor
f_φ	phase distribution's probability density function
f_a	amplitude distribution's probability density function
F_{co}	antenna pattern's copolar complex amplitude
f_{co}	normalized antenna copolar pattern's complex amplitude
F_{cx}	antenna pattern's cross-polar complex amplitude
f_{cx}	normalized antenna pattern's cross-polar complex amplitude
F_{Em}	Power estimator's cumulative density function
f_{Em}	Power estimator's probability density function

F_{hh}	antenna pattern's horizontal copolar complex amplitude
f_{hh}	normalized antenna pattern's horizontal copolar complex amplitude
F_{hv}	antenna pattern's vertical cross-polar complex amplitude
f_{hv}	normalized antenna pattern's vertical cross-polar complex amplitude
f_N	Gaussian (normal) distribution's probability density function
f_p	power distribution's probability density function
f_{rcx}	receive antenna's cross-polar complex-valued gain
f_{rhh}	receive antenna's horizontal copolar complex-valued gain
f_{rvv}	receive antenna's vertical copolar complex-valued gain
f_{tcx}	transmit antenna's cross-polar complex-valued gain
f_{thh}	transmit antenna's horizontal copolar complex-valued gain
f_{tvv}	transmit antenna's vertical copolar complex-valued gain
F_{vh}	antenna pattern's horizontal cross-polar complex amplitude
f_{vh}	normalized antenna pattern's horizontal cross-polar complex amplitude
F_{vv}	antenna pattern's vertical copolar complex amplitude
f_{vv}	normalized antenna pattern's vertical copolar complex amplitude
G	Gaussian-shaped spectrum
G	antenna gain
G	antenna's one-way gain, including radar calibration correction factor
G	component gain
g	Gaussian-shaped time-domain function
g	coefficient for K_{dp}/Z_h relation
g	gravitational acceleration constant (9.81 m s^{-2} on Earth)
G_0	antenna boresight gain
G_0	antenna's one-way gain
G_n	antenna pattern's normalized gain model
G_{h0}	antenna horizontal polarization's boresight gain
G_{rx}	receiver's gain
G_r	receiving antenna's gain
G_t	transmitting antenna's gain
G_{v0}	antenna vertical polarization's boresight gain
h	fall distance
h	filter's impulse response
h	height
h	window function
h_0	height of the radar's antenna
H_{dr}	hail differential reflectivity parameter
H_h	horizontal polarization's magnetic field
h_t	triangle window function
H_v	vertical polarization's magnetic field
I	scatterer power
I_Z	network cell's sensitivity improvement factor
j	imaginary number, $\sqrt{-1}$
K	velocity extension factor

k	angular wavenumber or phase constant
k	frequency sample index
k	propagation constant, $k = -j\gamma$
k_0	propagation constant of free-space
K_{dp}	specific differential phase shift
k_{eff}	effective propagation constant for a mixture
K_{opt}	window constant for calculating optimal number of samples
k_θ	Barnes analysis azimuth smoothing parameter
K_a	thermal conductivity of air
k_a	antenna beamwidth factor
k_B	Boltzmann constant ($1.380649 \cdot 10^{-23}$ J K^{-1})
k_d	Barnes analysis fall-off range parameter
k_L	attenuation coefficient
k_r	Barnes analysis radial smoothing parameter
L	depolarization factor
L	latent heat of vaporization
l	attenuation through precipitation
l_{isl}	mainlobe's maximum sample lag
l_{radome}	one-way radome loss
l_{rx}	receiver loss
l_{tx}	transmitter loss
L_D	filter loss from Doppler frequency shift
L_f	filter loss factor
L_s	latent heat of sublimation
L_w	waveform loss factor
M	K_{dp} estimator's sample length
M	network's overlap ratio
M	number of countered particles
M	number of spectrum to average
M	range-time signal's length
m	raindrop mass
m	transmit waveform
m_{isl}	mainlobe's sample width
m_h	horizontal polarization's transmitted waveform
m_v	vertical polarization's transmitted waveform
$N(D)$	particle size distribution
$N(D)$	raindrop-size distribution
N	Gaussian (normal) distribution's probability density function
N	number of radars in the network
N	number of samples in the integration time
N	number of samples in the waveform
n	an integer index
n	index of refraction
n	noise power

N_0	particle size distribution's concentration (intercept) parameter
N_{norm}	normalized particle-size distribution
N_{opt}	optimal number of samples for maximum spectrum SNR
N_c	number of hydrometeor classes
n_c	number concentration
N_I	number of independent samples
N_i	ice particle number concentration
N_s	snowflake particle size distribution just above the melting layer
N_v	number of classifier input variables
N_w	normalized PSD's intercept parameter
N_w	raindrop size distribution just below the melting layer
P	polar coordinate domain
p	polarization factor
p	total source power
P_c	clutter's power
P_i	intrinsic (unattenuated) echo power
P_m	measured (with attenuation) echo power
P_n	noise's power
P_p	precipitation's power
P_r	power at the receiver
P_t	transmitter's power
P_s	scattered wave's power
P_s	signal's power
P_{th}	horizontal polarization's transmit power
P_{tv}	vertical polarization's transmit power
q	K_{dp} estimator error function weight
R	particle or drop radius
R	rain rate
r	covariance
r	particle or drop radius
r	raindrop's axis ratio
r	range
R_{ATSR}	ATSR samples' covariance matrix
r_{cell}	network cell range
r_{max}	radar's maximum range
r_{sep}	network's separation range between radars
R_{STSR}	STSR samples' covariance matrix
r_a	unambiguous range
r_m	radius of influence
r_r	range from the scatterer to receiver
r_s	calibration sphere's radius
r_t	range from the transmitter to scatterer
R_v	gas constant for water vapor ($461.5 \ \mathrm{J \ kg^{-1} \ K^{-1}}$)

r_{ff}	Fraunhofer distance, the antenna far-field's starting range
r_{ik}	radial difference for Barnes analysis
R_{SS}	scatterer's linear polarization covariance matrix
r_{xx}	autocovariance of x
r_{xy}	cross-covariance of x and y
S	power density
S	power spectrum
S	water-equivalent snowfall rate
s	echo signal
s	ice crystal surface
s	path segment
s	spline solution
s	vertical wind shear
S_a	azimuth's arc distance
S_c	clutter's Doppler spectrum model
s_e	elevation's arc distance
S_i	incident power density
S_i	water vapor saturation ratio with respect to ice
S_k	fuzzy logic class's confidence
S_n	noise's Doppler spectrum model
S_p	precipitation's Doppler spectrum model
S_r	power density at the receiving antenna
S_s	scattered power density
S_w	water vapor saturation ratio
S_x	received signal's Doppler spectrum model
$S_{\Phi_{dp}}$	spectral differential propagation phase
$S_{\rho_{hv}}$	spectral copolar correlation
s_{hh}	horizontal polarization's copolar complex scattering amplitude
s_{vv}	vertical polarization's copolar complex scattering amplitude
$S_{Z_{dr}}$	spectral differential reflectivity
T	temperature
T	total observation time
t	time
T_0	noise factor's reference temperature
T_D	decorrelation time
T_n	noise temperature
T_s	pulse repetition interval
T_u	PRT for equivalent unfolded unambiguous velocity
T_{sys}	system's noise temperature
T_{tx}	transmitted pulse's duration
U	horizontal wind velocity
u	particle velocity in the x direction
V	sphere's volume
V	volume

v	particle velocity in the y direction
v	scatterer's velocity
v	wave's propagation speed
V_6	radar observation volume at -6 dB extent
v_a	unambiguous velocity
v_r	radial Doppler velocity
V_t	terminal velocity
v_u	unfolded unambiguous velocity
V_{ts}	terminal velocity of snow and ice precipitation particles
w'	modified weighting parameter
W	liquid water content
w	K_{dp} estimator error function weight
w	Doppler spectrum width
w	Tukey window function
w	fuzzy logic membership weight
w	particle velocity in the z direction
w	updraft velocity
w	white-noise signal
w^*	Bergeron process's threshold for vertical velocity
X	frequency spectrum of a radar signal
x	Cartesian position in the west-to-east direction
x	MBF input variable
x	radar signal
x_{2way}	signal after two-way phase shift
x_{coded}	phase-coded signal
$x_{decoded}$	signal demodulated by the phase code
x_E	relative position in the west-to-east direction
x_I	radar signal's in-phase (real) component
x_Q	radar signal's quadrature (imaginary) component
x_{bb}	complex-valued signal at base-band frequency
x_{rf}	complex-valued signal at the radio frequency
x_{rf}	radio frequency signal
y	Cartesian position in the south-to-north direction
y_E	relative position in the south-to-north direction
Z	reflectivity factor
z	Cartesian position in the nadir-to-zenith direction
z	altitude
Z^{95}	clutter's 95th-percentile measured reflectivity
Z_{dr}	differential reflectivity
$Z_{dr}^{(bis)}$	bistatic differential reflectivity
$Z_{intrinsic}$	scattering volume's true equivalent reflectivity factor
Z_{max}	minimum sensitivity at maximum range
z_{top}	initial altitude
Z_e	equivalent reflectivity factor

z_E	relative position in the nadir-to-zenith direction
$Z_e^{\min}$	minimum detectable reflectivity factor
Z_h	horizontal polarization's equivalent reflectivity factor
Z_n	noise's equivalent reflectivity factor
Z_v	vertical polarization's equivalent reflectivity factor

1 Introduction

Radar was developed for military use leading up to and during World War II. The first direct application of the use of radio waves to detect objects was by the German inventor Christian Hülsmeyer in 1904. It wasn't until June 17, 1935, that radio detection and ranging were demonstrated in Britain by Robert Watson-Watt and Edward George Bowen, who are generally credited with the invention of what was later termed "radar." *Radar* is an acronym for RAdio Detection And Ranging (or sometimes RAdio Direction and Ranging) and was first used by the U.S. Navy in 1940 [1]. Since its inception, the applications of radar have grown and diversified in numerous ways, including applications in the military, geophysics, astronomy, manufacturing, meteorology, automobile navigation, and more. During World War II, it was first observed that weather was causing unwanted echoes on the radar display. This led to the beginning of meteorological applications, thanks to the abundance of radar systems that were made available after the war for civilian research and the study of precipitation. Weather radar has since evolved independently. The first weather radar networks during the 1960s and 1970s could only measure reflectivity (these are known as reflectivity radar or *noncoherent radar*), providing information on the position and intensity of precipitation.

In May 1973, for the first time, a Doppler radar was used to document the life cycle of a tornado in Oklahoma. The radar observations showed the appearance of a cyclonic circulation aloft before the tornado touched ground [2]. This offered a glimpse of the forecasting potential deriving from additional radial velocity observations using coherent radar systems, marking the inception of the first key technological innovation in the history of weather radar, the Doppler weather radar. By the late 1980s, networks of Doppler radars started to be deployed worldwide, allowing forecasters not only to see the position and intensity of precipitation but also to get an unprecedented picture of the storm's dynamics and detect possible rotation in supercell storms (mesocyclones) and tornadoes. Between 1980 and 2000, Doppler weather radar networks became an essential component of the meteorological observation infrastructure in North America, Europe, Japan, and other developed countries. During this time, the observations from these radars started to be intensively used by national weather services, research institutions, and television stations.

The second major key innovation in weather radar technology has its roots in the research initiated during the 1970s, but eventually fed into operational systems and national weather networks around the turn of the millennium. In addition to coherency to exploit phase measurements in Doppler radar systems, dual polarization weather radar added a completely new dimension to the radar observation's capability, introducing a change comparable to going from black-and-white to color pictures in photography. These new colors, the polarimetric variables are rigorously defined and discussed in Chapter 4. Dual polarization technology had been a research topic for many years before becoming mainstream for radar operations. Nowadays, products based on polarimetric observations are routinely used to improve fundamental radar applications, such as flood forecasting and hail detection.

A third wave of innovation that benefited this field is the concept of small radar networks. At the time of writing this book this wave is still in its early stages, but the fundamental deployment strategy of weather radars has been changed. Historically, the national radar networks were considered expensive and were part of the national infrastructure for weather and aviation. However, the lower-cost small radar networks spearheaded by the Collaborative Adaptive Sensing of the Atmosphere (CASA) program [3] have changed the strategy. Now, local and state governments, and even private entities, are deploying small and dense weather radar networks. Of course, all these waves of innovative developments in radar meteorology have been aided by rapid technological advances in signal processing, computing, and communications.

The continuous evolution of weather radar technology and science has led to the new discipline of radar meteorology, which is taught in many universities around the world. Radar meteorology is often part of atmospheric sciences programs, but it's not a pure atmospheric discipline like most other course topics in these programs. The concept of weather radar is, rather, a combination of four disciplines: microphysics, dynamics, signals, and electromagnetics. Therefore, radar meteorology is fundamentally a multidisciplinary topic simultaneously housed in programs of atmospheric sciences, physics, and electrical engineering. The radar meteorology field has therefore enjoyed immense innovation afforded by the varied cross-disciplinary thinking over the decades of growth. A person taking a course taught from this book should be able to develop a basic understanding for practice in radar meteorology, as well as conduct research and operations with dual polarized weather radar. This book provides the necessary foundation for students to interpret dual polarized weather radar observations for both research and operational purposes. This book is intended for students who have completed basic college-level math and science prerequisites, such as physics, algebra, and calculus. Upon completion of the course, the student should be able to do the following:

- Understand the basic principles of a dual polarization weather radar system and be able to identify the radar system elements.
- Have the basic precipitation physics background needed to understand and interpret dual polarization radar observations.
- Be able to understand the principles of electromagnetic scattering from precipitation particles, including scattering of dual polarized electromagnetic waves.

- Understand the basic theory of weather radar signals.
- Develop a good understanding of the signal-processing system associated with a modern dual polarized weather radar.
- Understand the basic principles and applications of hydrometeor classification used with all modern operational and research dual polarized weather radar systems.
- Understand and apply rainfall estimation from weather radars.
- In addition to all these, the chapter on data quality provides a scientific basis for data quality and provides an understanding of limitations on interpretations. In the modern era of data science, the data quality chapter prepares the student to develop confidence in the dual polarization radar data set.
- Once a student has mastered all the essential fundamentals and applications, this book also prepares the student for the future of the field. The student will not only be able to analyze and interpret the individual radar observations but also to design an appropriate network and redefine network operations.

1.1   Radar in Meteorology

Earth's atmosphere has a layered structure that is based on how the temperature changes with height above the surface (Fig. 1.1). The troposphere is the lowest layer of the atmosphere, extending from the ground level up to about a global average of 13 km in altitude (locally, the height of the troposphere varies between 9 km over the polar regions and 17 km above the equatorial regions). The temperature decreases upward in the troposphere, reaching an average lowest value of $-57°C$ on the top, and then starts to rise in the adjacent layer above, the stratosphere. The upper boundary of the troposphere is marked by the tropopause, a relatively thin layer (typically a few hundred meters deep) where the temperature inversion keeps the humid air of the troposphere separated from the dry stratosphere.

The troposphere is the portion of the atmosphere where weather occurs because it contains 99 percent of all the water vapor, the fundamental constituent of clouds. When an air mass in the troposphere becomes saturated with water vapor (reaching 100 percent relative humidity), condensation produces clouds, which may then form precipitation as soon as cloud particles (either liquid droplets or ice crystals) attain a size large enough to fall as a result of gravity. Precipitation has been measured historically by means of rain gauges, which provide the amounts of accumulated water at given locations over a time interval. Clouds and precipitation can also be remotely observed by exploiting the electromagnetic waves in a range of frequencies that goes from the visible portion of the spectrum to the radio wave, using airborne, spaceborne, and ground-based instruments. Active remote sensors like radar and lidar (*lidar* is an acronym for Laser Imaging, Detection, And Ranging) emit their own source of energy to illuminate a target (a cloud) and then analyze the returned signal. Conversely, passive sensors require some energy to be naturally available, either reflected (e.g., the

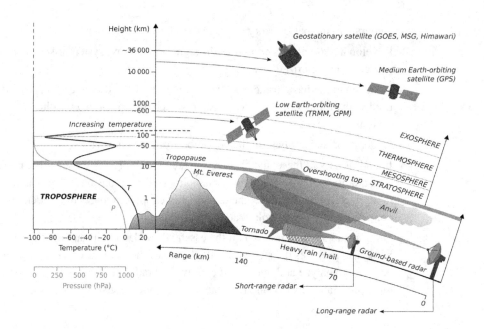

Figure 1.1 Vertical structure of the atmosphere, with height in logarithmic scale. On the left, the average profiles of pressure (gray line) and temperature (black line) drive the partition of the atmosphere in successive layers. The tropopause is the boundary between the troposphere and the stratosphere, located at an average height of 13 km. On the right, short-range and long-range ground-based radars are depicted as they observe a thunderstorm penetrating into the stratosphere (overshooting top). Weather systems can also be observed from space by low Earth-orbiting satellites, such as the Tropical Rainfall Measuring Mission (TRMM) and the Global Precipitation Measurement (GPM) mission, and by geostationary satellites flying at altitudes of $\sim 36{,}000$ km, such as the European Meteosat Second Generation (MSG), the Geostationary Operational Environmental Satellite (GOES) operated by the United States, and Japan's Geostationary Meteorological Satellite "Himawari" (GMS).

sun's light reflected by clouds and detected in the visible channels of meteorological satellites) or thermally emitted (the infrared channels).

Ground-based weather radars have become an essential component of meteorological weather and aviation services around the world, with operational networks deployed in most countries and continuously providing updated observations every few minutes. Weather radar has evolved to the point that it plays a critical role in both developing our understanding of weather and also protecting people and property during severe weather events and supporting the transportation sector. The advent of dual polarization systems brought a paramount contribution to our understanding of precipitation microphysics. Since the early days of radar polarimetry during the 1970s, research focused primarily on rainfall, contributing to revealing the inner structure of precipitation down to the level of the individual raindrop's shape. During the last couple of decades, the ice phase of precipitation has received increasing interest as a result of the advances in modeling the electromagnetic response of complex ice-crystal shapes. Dual polarization observations have revealed a number of significant recurring signatures associated with different types of weather systems, from columns of differential reflectivity in convective storms to signatures of specific differential phase aloft in connection with enhanced snow growth (these and other characteristic signatures are discussed in Chapter 8).

Recent advances in radar technology include the development of lower-cost systems operating at shorter wavelengths. In fact, whereas operational weather radar networks have mainly been using the S band (10-cm wavelength, e.g., the WSR-88D network in the United States) or the C band (5-cm wavelength, e.g., in Europe and Japan), smaller networks of X-band radars (3-cm wavelength) are being deployed in many regions. The lower cost of shorter-wavelength systems directly derives from the smaller antenna size and shorter-range design, thus requiring lower-cost transmitters (Chapter 2), and allows for denser networks of short-range radars (as opposed to sparser networks of long-range systems), thus coping with the observation gap induced by the Earth's curvature (Chapter 10).

The radar observations are reported in a spherical coordinate system using azimuth, elevation, and range to spatially locate the observation volume, as shown in Figure 1.2. The observation volume (denoted by V) has a finite size in azimuth, elevation, and range, which are all determined by the radar's design and operating parameters. The azimuth follows the cardinal directions, where north is $0°$, and the azimuth increases as the antenna rotates clockwise toward east ($90°$), south ($180°$), and west ($270°$). As the azimuth continues to increase, it "aliases" when crossing over north so that the azimuth is always from $0° \leq$ azimuth $< 360°$ (where $0°$ and $360°$ are the same position). The radar's elevation is typically reported from $0°$ to $90°$, with $0°$ pointing parallel to the surface at the radar's location (see Section 2.4.5 for beam-height discussion). At $90°$ elevation, the radar is pointing vertically (this is the mode of operation for vertical profilers).

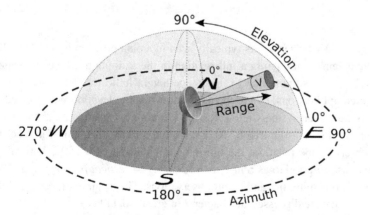

Figure 1.2 A diagram of the radar's spherical scanning volume whose coordinate system uses range, azimuth, and elevation. The radar measurements integrate a volume (V) defined by the antenna's beamwidths in the azimuth and elevation and pulses extent in the range (refer to Section 2.4.5 for a detailed discussion of the radar's geometry).

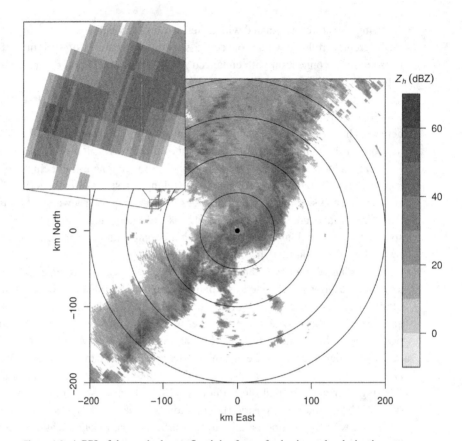

Figure 1.3 A PPI of the equivalent reflectivity factor for horizontal polarization at 1.5° elevation. The expanded view highlights the polar geometry of the radar volume, as well as the detail of the information (refer to Section 2.4.5 for discussion of the radar observation geometry).

Operational weather radars typically scan in a plan position indicator (PPI) mode, with the antenna completing a full rotation in the azimuth at a fixed elevation angle (Fig. 1.3). It is also common to find range-height indicator (RHI) scans, especially in research applications, where the antenna moves in elevation with a fixed azimuth angle. Some radar systems allow for the elevation to scan from 0° to 180°, which provides a "horizon-to-horizon" RHI scan, but strictly speaking, for describing the radar observation's position, the elevation is limited to 0° – 90°. Vertical pointing systems (also referred to as *zenith radar systems* or *vertical profilers*) are common in air motion, precipitation, and cloud observations. The scanning modes and the radar view geometry are discussed in Chapter 2 in greater detail.

Weather radar can also be carried by low-orbiting satellites, such as the Tropical Rainfall Measuring Mission (TRMM) and the Global Precipitation Measurement (GPM), represented at an altitude of ~400 km in Figure 1.1. Although the operating principle of satellite-borne and ground-based radars is basically the same, the "art" of processing the collected observations requires specific expertise, which is essentially related to the different vantage points of the observing sensors.

1.2 Weather Radar Frequencies

The selection of the radar's operating frequency influences every aspect of the radar system design and performance, including physical size, system cost, and radar detection sensitivity. In addition, the coexistence of radar with other systems used by society, such as communication devices, also plays a role in the available choices. The various application groups, such as weather and climate, communication, and aviation, compete for access to a common resource: the frequency spectrum. The actual allocation of the frequency for the radars is managed by the frequency-allocation agencies of national and international regulations [4]. Using a higher operating frequency is one of the basic design decisions to achieve enhanced sensitivity at a lower cost. The radar's frequency selection is often an economic decision, and in many cases, the other "factors" that weigh on the radar performance also vary as the radar's frequency is varied. For certain applications, there are clear advantages to the use of some frequencies rather than others. In Chapter 4, we'll see that higher frequencies also attenuate more, which can reduce the sensitivity over long distances or in heavy rain. The scattering properties of hydrometeors are also frequency dependent. These frequency-dependent scattering characteristics can be informative, especially when comparing observations at two or more frequencies, allowing us to measure the microphysical characteristics of precipitation and clouds.

During World War II, radar frequency bands were referred to by letters for secrecy. Afterward, these designations were declassified, and the naming convention is still commonly followed, with a few accommodations. Previously, what is now divided into the Ku, K, and Ka bands was referred to as the K band. The subdivision was a result of the water-vapor absorption line at approximately 22 GHz, which hindered radar and communication system performance. This resulted in the Ku and Ka bands being identified as their own frequency bands (Ku being "K-under" and Ka being "K-above"). Other standards with different designations (or modified frequency ranges for the same letter designation) can be found. Here, we refer to the Institute of Electrical and Electronics Engineers (IEEE) standard, which was first issued in 1976. In 1984, the millimeter wave region was also revised to include letter designations, which are now the V and W bands. In Table 1.1, the 2002 revision [5] of the IEEE radar frequencies bands are listed. The sub-bands that may typically be used (but are not limited to) for weather radar applications are also shown.

As a reminder, f, the frequency (in hertz) and wavelength, λ, (in meters) are related through the speed of light, c, as $f = c/\lambda$. The speed of light in free space, $c = 299,792,458 \, \mathrm{m \, s^{-1}}$, is typically used as a standard value for the speed of light in Earth's atmosphere.

Although radars operating at frequencies below the S band can detect precipitation, they are not typically used for such applications. Compared with the size and

Table 1.1 IEEE radar frequency band designations with typical weather radar frequency ranges.

IEEE Radar Band	IEEE Frequency Range	Typical Weather Radar Band
HF	3–30 MHz	
VHF	30–300 MHz	
UHF	300–1000 MHz	
L	1–2 GHz	
S	2–4 GHz	2.7–2.9 GHz
C	4–8 GHz	5.6–5.65 GHz (in the U.S.)
X	8–12 GHz	9.3–9.5 GHz
Ku	12–18 GHz	13.4–13.75 GHz
K	18–27 GHz	24.05–24.25 GHz
Ka	27–40 GHz	35.5–36.0 GHz
V	40–75 GHz	
W	75–110 GHz	94.0–94.1 GHz

performance possible with S-band and higher-frequency systems, radars in the VHF, UHF, and L bands are typically limited to characterizing winds, turbulence, and air motion. Vertically pointing wind profiles, operating at frequencies as low as 50 MHz, are used to measure air motion, but the vertical motion of precipitation is also measured. Vertical profilers for air motion (and precipitation) also include the 404-MHz National Oceanic and Atmospheric Administration (NOAA) profiler network and the 449-MHz and 915-MHz profilers as examples.

S-band weather radar systems are commonly used around the world in places where long operating ranges are needed. The WSR-88D (commonly known as NEXRAD) weather radar network is composed of S-band radars observing to ranges of 300 km and more. S band is ideal for observing precipitation, especially severe weather, because of the negligible attenuation in rain or ice. For all but the most severe weather (e.g., large hail), the scatterers are in the Rayleigh regime, and the S-band equivalent reflectivity factor closely matches the reflectivity factor (see Section 2.3.1). S-band weather radars are large systems (with 8.5-m-diameter antenna). The S-band frequency spectrum is crowded, and for radars located near populated areas, interference sources can degrade performance.

Like S band, C band is commonly used for operational weather radars around the world. C band has become more popular since the introduction of dual polarization, which can be used to mitigate the effects of attenuation due to precipitation (see Chapter 9), and the radars are lower cost compared with S band. In the United States, the terminal Doppler weather radars utilize the C band for monitoring weather around major metropolitan airports. For the same antenna beamwidth, a C-band antenna's diameter is approximately half that of an S-band antenna. The C-band radar is typically a more compact system, largely due to the reduced size of the antenna and pointing system. C-band weather radars experience low attenuation, except for the heaviest convection with the largest raindrops or hail, for which dual polarization methodology can mitigate the effects of attenuation. Because of the combination of higher sensitivity with frequency and lower system cost (a result of a smaller antenna),

C-band radars have been extremely popular. As with the S band, the frequency spectrum at the C band is also crowded and subject to radio-frequency interference from communication infrastructure that can hamper radar operations.

X-band systems provide a relatively compact solution for mobile weather radars and networked short-range radar systems. One major advantage of X-band radar systems is that they are small and compact enough that they can be deployed on existing infrastructure, such as building rooftops, communication towers, and trailers. X-band radars are more adversely affected by attenuation in rain, and as a result, this can limit the X-band radar's effective range during extreme weather if the attenuation effect is not compensated by an increased transmitter power. To keep costs low, X-band radar systems are typically operated with observation ranges on the order of 50 km. As a result of the compact size and relatively low costs of short-range X-band radars, dense networks of overlapping systems are being deployed to provide higher spatial and temporal resolution. This solution has found rapid adoption in urban systems, where it may be cost prohibitive to find real estate to deploy large radar systems because their social footprint is fairly high. Because of the small size of X-band radars, they are easily deployed on trucks and used for mobile deployment to track severe storms or for temporary/seasonal deployments.

Ku-band performance is similar to that of the X band. For the same beamwidth, the Ku-band radar has a slightly smaller antenna but also suffers from slightly more attenuation in rain. Ku-band weather radar systems are found on precipitation-observing satellites and are also used in atmospheric research. K-band radars are subject to enhanced attenuation as a result of increased water-vapor absorption around 22.24 GHz. Millimeter-wavelength radars (i.e., 30 GHz and above) used for precipitation observations primarily fall into the Ka and W bands. Because of their high attenuation rates in rain, these wavelengths focus on observing clouds, drizzle, and light rain at short ranges from the ground. Similarly, the differences in the scattering and attenuation at these frequencies make their observations important for characterizing the properties of the clouds and precipitation, especially when combined with observations at other frequencies. As a result of the short wavelength, these radars can be implemented in compact form factors while still achieving high sensitivities and fine spatial resolution.

1.3   Book Content

The book is organized as visually illustrated in Figure 1.4. Chapter 2 introduces the fundamental concepts of radar systems, including a description of the radar-centric observation geometry, the weather radar equation, and typical system configurations, along with a discussion of the main radar components. Then, the book presents three main themes:

- Essential background topics, namely, precipitation physics, are described in Chapter 3, whereas the necessary background electromagnetics are described in Chapter 4. These chapters provide essential background for the interpretation of

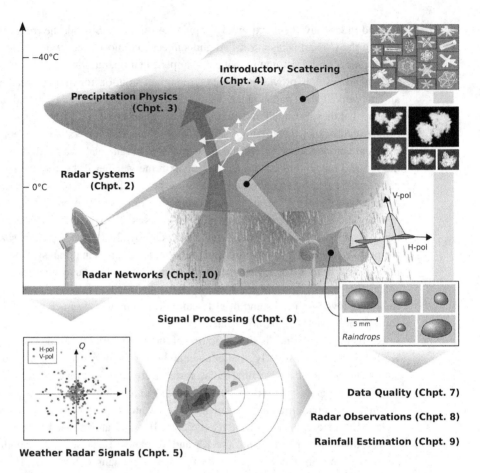

Figure 1.4 A network of three radars looking in different regions of the cloud provides a graphical overview of the book content. Crystal images in the top-right corner courtesy of SPEC$_{inc}$

the radar polarimetric observations, in terms of both the microphysical processes leading to precipitation in its various forms and the electromagnetic response of precipitation particles to the transmitted polarized wave.

- The next theme is the topic of radar signals, where the basics and origin of weather radar signal theory and properties are described in Chapter 5, and signal processing is explained in Chapter 6. In these two chapters, the theory of weather radar signals is described to provide the basis for the estimation of the radar variables from received signals. Some advanced signal-processing concepts, such as the pulse compression used in modern radars with solid-state transmitters, are also included in Chapter 6.
- The third theme is weather radar applications and includes chapters on data quality (Chapter 7), radar observations (Chapter 8), rainfall estimation (Chapter 9), and radar networks (Chapter 10). Data quality is discussed in Chapter 7 to provide guidance on the characterization of measurement accuracy and the interpretation

of radar measurements. Although example radar observations are provided throughout the book in support of the relevant topics being discussed, Chapter 8 is specifically devoted to the illustration of salient features of dual polarization observations in different meteorological contexts. Chapter 9 focuses on one of the important applications of weather radar, namely, for hydrological applications. Rainfall estimation and attenuation correction are discussed from the fundamental theoretical basis of rain microphysics and the capability of dual polarization measurements, leading to estimation algorithms. Finally, Chapter 10 is devoted to a discussion of radar networks, their design, the different types of networks, implications for the basic radar equation, and specific network-based applications, such as the retrieval of three-dimensional wind fields.

Students and researchers often take up the study of weather radar from either an atmospheric science, atmospheric physics, or engineering perspective, depending on their background or specific application needs. In this book, we made an effort to support learners of all disciplines in their efforts toward understanding this multi-disciplinary field. A sequential reading of the book should give the reader a comprehensive picture of the multidisciplinary essence of radar meteorology (Fig. 1.4). However, knowing that some readers may be more interested in either the scientific or the technological side of weather radar, the book also offers possible alternative reading pathways, as illustrated in Figure 1.5. One pathway goes from Chapter 2 through the physics section to explore the processes taking place in the atmosphere (the formation and evolution of precipitation particles and their scattering response to the radar-transmitted wave). An alternative pathway goes from Chapter 2 directly into the signals section, following the path of a radar signal entering the radar systems on the left in Fig. 1.5 and then going into the signal processor to eventually produce the polarimetric moments we routinely see on radar displays. Finally, meteorological practitioners may use Chapters 3 – 6 as reference material and proceed from Chapter 2 directly to Chapter 7, where all pathways eventually reconnect.

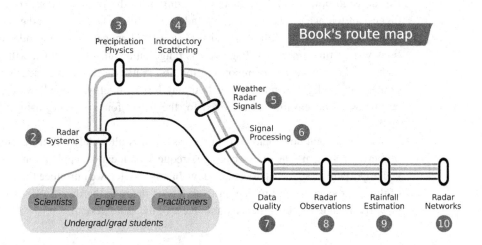

Figure 1.5 Alternative book's reading pathways.

2 Elements of Dual Polarization Radar Systems

A radar system is the central element in the study of radar meteorology. In order to better interpret weather radar observations of precipitation, it is useful to understand the basics of the radar system itself. The fundamentals of the radar system, its performance characteristics, and its operations are introduced in this chapter. We introduce the subsystems that commonly make up a weather radar system, and we discuss the features that determine radar performance, in particular, those relating to its sensitivity for detecting clouds and precipitation. Precipitation radars and cloud radars are generally delineated by their intended use. The microphysics that define clouds and precipitation are covered in Chapter 3. Although it is convenient to partition weather as clouds and precipitation, as a radar observable, these exist as a continuum with overlap. Cloud reflectivities can be well below -50 dBZ, whereas precipitation reflectivities can exceed 70 dBZ with large hail (reflectivity is defined in Section 2.3.1).

The exact choices of components and subsystems and the details of weather radar system designs continue to evolve with advances in technology. Even so, the basic subsystem concepts have not changed significantly since the inception of radar itself. For a radar to operate, it needs a source of power (a transmitter); a means to focus and direct the power into the target of interest and, similarly, to receive the echoes signals (an antenna); and finally, a receiver and data system to convert and visualize the echo power in a way that is meaningful to the radar's users. Arguably, the biggest difference between the early radars and the radars in use today is in how we capture, store, and utilize/visualize the received echoes. The first radar observations were a time trace of voltage versus range on a cathode-ray tube, with photography as the only means of recording the observations. Today, the radar's received echoes can be digitally recorded without any loss of fidelity. This allows the data to be processed in real time or archived for the future for processing using advanced algorithms.

The design and development of a radar system is ultimately a multifaceted trade-off between system size, radar coverage, frequency allocation, application, sensitivity, and importantly, radar cost, to name a few. In this chapter, we introduce these elements of the radar system and provide an overview of the radar hardware. We also discuss how to estimate the performance parameters of the radar system.

2.1   Polarization of an Electromagnetic Wave

The radar transmits and receives electromagnetic waves. The wave's electric field is generated (for transmitting) or measured (for receiving) by the radar. Time-varying electric and magnetic fields are linked, and these coupled oscillating fields form the electromagnetic waves. In the 1870s, James Clark Maxwell developed the theory to explain the electromagnetic waves, which is summarized by a set of equations called *Maxwell's equations*. Although the topic of electromagnetic waves and wave polarization is an in-depth field in itself, we will attempt to provide a simple explanation to introduce it here. (Additional details are provided in Chapter 4.)

Physicist Heinrich Hertz showed that electromagnetic waves can leave the wire and propagate in free space. The wave's polarization is related to the orientation of the electromagnetic wave's electric field. When transmitting, the wave propagates radially outward from the antenna in the form of an expanding sphere, as shown in Figure 2.1. As the range increases, the curvature of the sphere becomes negligible for small regions on the spherical surface, and the wave can be approximated locally as a plane wave (consider, e.g. a region around a raindrop positioned many kilometers from the radar). Following this, the plane-wave approximation is frequently used.

The polarization of electromagnetic waves is the topic of interest here. The electromagnetic wave propagates in three-dimensional space. One dimension is the direction of propagation, and the other two dimensions form the polarization plane. Polarization is a basic property of an electromagnetic wave. The polarization describes the

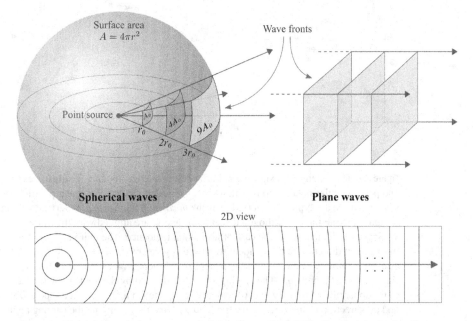

Figure 2.1 Spherical and plane waves. At large distances, spherical waves can be locally approximated by plane waves.

direction of the wave's electric field, and the polarization is defined in a plane that is perpendicular to the direction in which the wave is propagating. For dual polarization radar, we precisely control and observe the polarizations of the wave. For weather radar, linear polarizations are the most commonly used, and the electric-field vectors are the horizontal and vertical polarizations. The horizontal and vertical orientations are determined by the antenna, as illustrated in Figure 2.2, where the polarization bases are shown together with the reference frame of a raindrop represented by an oblate spheroid. The unit vectors $\hat{h}_i$ and $\hat{v}_i$ denote the directions of the incident wave's

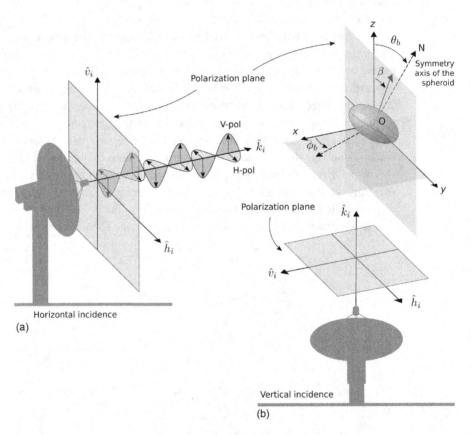

Figure 2.2 The antenna transmitting at horizontal incidence in panel (a) illustrates the parallel and perpendicular polarizations with respect to Earth's surface. For linear polarization systems, these are typically referred to as the *horizontal* and *vertical* polarizations (unit vectors $\hat{h}_i$ and $\hat{v}_i$). Note that the orthogonal vectors $\hat{h}_i$ and $\hat{v}_i$ are aligned with the true horizontal and vertical directions only for the case of horizontal incidence. The unit vector $\hat{k}_i$ represents the propagation direction of the incident wave and is given by $\hat{k}_i = \hat{v}_i \times \hat{h}_i$. Because of the circular symmetry of the spheroid in panel (b), its orientation can be described by only two Euler angles (θ_b and ϕ_b). The canting angle β is defined by the angle between the projection of the spheroid's symmetry axis (oriented along a line from O to N) on the polarization plane and the direction of the vertical electric field $\hat{v}_i$. Note that changing the canting angle of the particle is equivalent to rotating the polarization plane around the propagation direction.

"horizontal" and "vertical" components. As illustrated by the two different antenna positions (horizontal and vertical incidence), it is worth remarking that the terms *horizontal* and *vertical* keep their intuitive meaning (with respect to Earth's surface) only when the antenna is pointing at the 0° elevation angle.

When discussing polarimetric radar, it is implied that the polarizations of the radar's signals are controlled so that the precipitation's polarization characteristics can be measured consistently. For weather radar applications, linear horizontal and vertical polarizations are the most commonly used (although in the early days, there were circularly polarized weather radars).

There are two interesting properties of electromagnetic waves that can be combined to describe the concept of polarization. First, the electric field is a vector and therefore has a magnitude and direction. Second, the oscillatory fields that produce waves can be considered as phasors, such as a sinusoid. A sinusoid voltage (or field) at any given time is described not only by its amplitude but also by its phase. The electric-field component that is the horizontal polarization can be described by a time-varying sinusoid as

$$E_h = |E_h| \cos(\omega t + \phi_h), \tag{2.1}$$

where ω is the angular frequency ($\omega = 2\pi c/\lambda$). Similarly, a vertically oriented electric-field component can also be described by a sinusoid as

$$E_v = |E_v| \cos(\omega t + \phi_v), \tag{2.2}$$

where the relative phasing between the two related sinusoids is $\phi = \phi_h - \phi_v$. The peak amplitudes of the horizontal and vertically oriented electric fields are $|E_h|$ and $|E_v|$, respectively.

A wave is a propagating disturbance that varies with time. Consider a ripple on a pond or a note on a string instrument; the amplitude of the wave depends on the time and location at which it is observed. The instantaneous field of a propagating plane wave is similarly dependent on the time and range at which it is observed. In eq. (2.1), the amplitude of the sinusoidal electric field is considered as a function of time for a fixed position (i.e., a range of zero). For a wave that propagates at the speed of light, c, we can calculate the time it takes to travel a range r as $\tau = r/c$. With a propagating wave, because it takes additional time to travel a distance r, the plane wave's electric field as a function of time and range are captured together:

$$E = E_0 \cos(\omega(t - \tau)) = E_0 \cos\left(\omega t - \frac{\omega}{c}r\right) = E_0 \cos(\omega t - k_0 r), \tag{2.3}$$

where k_0 is the free-space wave number ($k_0 = \omega/c = 2\pi/\lambda$).

If the range is held constant, the signal oscillates with time. Similarly, if the time is held fixed, the signal varies sinusoidally with range. This principle of eq. (2.3) is shown in panel (a) of Figure 2.3, where a vertically oriented electric field oscillates as the range from the radar changes. This also holds true for the horizontal polarization. In each case, the direction of the electric field is the polarization vector. If the

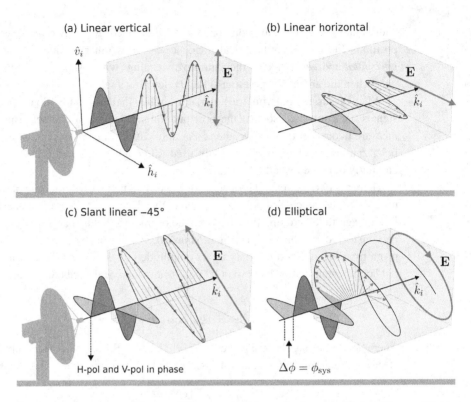

(a) Linear vertical   (b) Linear horizontal

(c) Slant linear −45°   (d) Elliptical

H-pol and V-pol in phase $\Delta\phi = \phi_{\text{sys}}$

Figure 2.3 Propagation of the electric field for vertical (a), horizontal (b), slant linear −45° (c), and elliptical polarization (d).

electromagnetic wave has only a vertically oriented electric field, it is called a *vertically polarized wave*, whereas if it has only a horizontally oriented electric field, it is called a *horizontally polarized wave*.

When we combine these two sinusoids, which represent the horizontal and vertical polarization of the plane wave, and vary their amplitudes $|E_h|$ and $|E_v|$ and their relative phase (ϕ), we can produce widely varying patterns in the polarization planes as a function of time. If the amplitudes are the same for the horizontal and vertical polarizations ($|E_h| = |E_v|$) and the phases are the same ($\phi = 0°$) or opposite ($\phi = 180°$), the resulting electric field is polarized at a either +45° or −45° angle, as shown in panel (c) of Figure 2.3 ($\phi = 180°$) (Note, Slant −45° can also be referred to as Slant 135°.) If there is amplitude imbalance or a nonzero phase shift (i.e., $|E_h| \neq |E_v|$ or $\phi \neq 0$), the polarization will trace out an elliptical shape with a range and time like that shown in panel (d) of Figure 2.3. The general vector electric field of the wave (which includes the components of both polarizations) can be written as

$$\mathbf{E} = (|E_h|\hat{h} + |E_v|e^{j\phi}\hat{v})e^{j(\omega t - k_0 r)}, \tag{2.4}$$

where $\hat{h}$ and $\hat{v}$ are the horizontal and vertical unit vectors, and ϕ is the phase of the vertical polarization relative to the horizontal polarization. (Note the change to a complex phase notation, rather than the cosine, which is adopted here to represent the forward-propagating wave and is consistent with the discussions in Chapters 4 and 5.) The term $\exp(j\omega t)$ describes the time-harmonic (or oscillatory) nature of the electric field. The combination $\exp[j(\omega t - k_0 r)]$ describes a traveling electromagnetic wave that propagates in the $\hat{k}$ direction (where $\hat{k} = \hat{v} \times \hat{h}$). The sinusoidal nature of $E_h = \mathbf{E} \cdot \hat{h}$ and $E_v = \mathbf{E} \cdot \hat{v}$, with corresponding arbitrary phases, can be described as

$$\mathbf{E} = (|E_h|e^{j\phi_h}\hat{h} + |E_v|e^{j\phi_v}\hat{v})\,e^{j(\omega t - k_0 r)}. \tag{2.5}$$

Consider the different resulting shapes of the transmitted dual polarization signal in Figure 2.4, where we plot the instantaneous polarization in the plane defined by the $\hat{h}$ and $\hat{v}$ unit vectors. As time advances, the curve will trace out the polarization's behavior, whether it be linear, circular, or something in between. Any variation of these shapes can be created with control over both polarizations. Special cases include the following:

- $|E_h| = |E_v|$, and $\phi_h = \phi_v$: The wave is linearly polarized in the "slant 45°" configuration (panel [b] of Fig. 2.4).

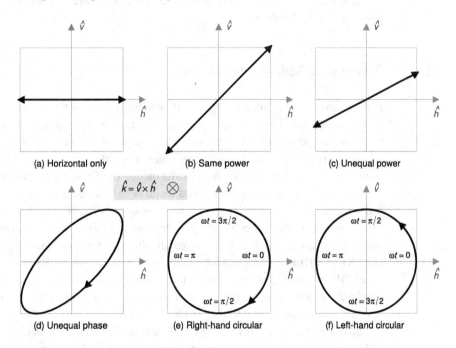

(a) Horizontal only   (b) Same power   (c) Unequal power

(d) Unequal phase   (e) Right-hand circular   (f) Left-hand circular

Figure 2.4 The relationship between the dual polarization signals in the linear polarization coordinates (horizontal and vertical). The amplitude and phase of the linear dual polarization signals are modified, giving the curves shown here (see eq. [2.5]). The direction of propagation is perpendicular and into the page ($\hat{k} = \hat{v} \times \hat{h}$). The arrow indicates the direction in which the signals vary with time. (a) Horizontal only: $E_v = 0$, and ϕ is any value. (b) Slant 45°: $|E_h| = |E_v|$, and $\phi = 0°$. (c) Unequal power: $|E_v| = |E_h|/2$, and $\phi = 0°$. (d) Unequal phase: $|E_h| = |E_v|$, and $\phi = 45°$. (e) Right-hand circular: $|E_h| = |E_v|$ and $\phi = +90°$. (f) Left-hand circular: $|E_h| = |E_v|$, and $\phi = -90°$.

- $|E_h| = |E_v|$, and $\phi_h - \phi_v = \pm 90°$: The wave is circularly polarized. For $\phi_h - \phi_v = +90°$, it is right-hand circularly polarized (panel [e] of Fig. 2.4), and for $\phi_h - \phi_v = -90°$, it is left-hand circularly polarized (panel [f] of Fig. 2.4).
- $E_h = 0$: The wave is vertically polarized.
- $E_v = 0$: The wave is horizontally polarized (panel [a] of Fig. 2.4).

An arbitrary combination of $|E_h|$, $|E_v|$, ϕ_h, and ϕ_v results in elliptical polarization. Thus, depending on the type of polarization state needed to probe the precipitation, the radar antenna transmits the corresponding type of polarization wave. The antenna is the primary device that determines the type of polarization transmitted and received by the radar. The antenna is the subsystem that transmits the electromagnetic waves and also receives the waves coming back into the radar. The antenna determines the polarization state. Because the polarization is essentially the direction of the electric field, one can think of a unit polarization vector that represents the polarization state. Any polarization state can be written as the composition of two orthogonal components. The most common orthogonal components are the horizontal and vertical polarization states, where the term *horizontal* refers to the horizontal direction when the antenna is pointing at a 0° elevation.

2.2   Polarimetric Radar Architectures

Polarimetric radars are defined by how they implement transmitting and receiving the two polarizations. Three operating modes are used for dual polarization radar. If a polarimetric radar transmits two polarization states simultaneously (e.g., horizontal and vertical) and receives the same two states, that radar is operating in *simultaneous transmit and simultaneous receive* (STSR) mode. The STSR mode also has a variety of other names, including simultaneous mode or SIM mode, simultaneous horizontal and vertical (SHV) mode, simultaneous transmit and receive (STAR), and hybrid mode. Similarly, two other modes of operation are radars that selectively transmit only one polarization at a time and either alternate transmit and alternate receive (ATAR) or alternate transmit and simultaneously receive (ATSR) both polarization states. The rationale for operating in the different modes is discussed in Chapter 4. (A simultaneous-transmit, alternate-receive radar is not considered because it does not have advantages over the other modes.)

For the STSR dual polarization mode, the ideal polarization state is slant 45° or 135°. The STSR mode is sometimes referred to as a *slant 45°* polarization. In practice, these systems typically transmit elliptical polarization because the phase between the horizontal and vertical polarizations are not exactly 0° (for 45° slant polarization) or not exactly 180° (for 135° slant polarization), and the amplitude between the horizontal and vertical polarizations is not exactly equal.

STSR mode requires a high degree of polarimetric isolation to minimize bias in the dual polarization variables. This also implies that the scatterers themselves must have negligible cross-polarization scattering, or else the scatterer's cross-polar signals will bias the measurements. The cross-polarization bias can be significantly reduced in ATSR mode, and the scatterer's cross-polarization characteristics can be estimated. The downside is that the horizontal and vertical polarizations' copolar measurements are not made at the same time. As we'll discuss in Chapter 5, the motions of the scatterers result in a loss of correlation between the horizontal and vertical measurements, which can increase the estimation error for the polarimetric variables in some instances (ATSR estimators are discussed in Chapter 6).

To implement the three modes, various radar architectures have been developed. The selection of transmit/receive polarizations is either fixed by the radar manufacturer or, for some radar designs, it is an option that can be configured during operation. Based on the polarization states that the radar can transmit and receive, dual polarization weather radars are broadly divided into four system architectures. As with most engineering endeavors, linear dual polarization radar systems can be implemented in a number of ways. Not all radar systems support all modes of operation, and in fact, the majority of operational dual polarization weather radars only operate in STSR mode. (As an example, the WSR-88D operates in STSR mode.) Some research radar systems allow the selection of either the ATSR and STSR sampling modes depending on the science application (e.g., the Colorado State University–Chicago Illinois radar [CSU-CHILL] or the National Aeronautics and Space Administration [NASA] D3R radars).

Figure 2.5 illustrates four common radar architectures for dual-polarization weather radar systems. The architecture in panel (a) of Figure 2.5 uses a single transmitter and receiver with a switch to select the antenna's polarization. Before the introduction of polarimetric radars, systems used one transmitter and one receiver. Dual polarization could be adapted to these systems with a radio-frequency (RF) switch and a suitable antenna. The early dual polarization observations used a mechanical switch to change polarizations. Today, advanced dual transmitter systems are available that can enable full polarimetric mode to sample the four elements of the scattering matrix.

The most common weather radar system architecture that implements STSR mode is shown in panel (b) of Figure 2.5. This architecture uses a signal transmitter whose power is evenly split between the two polarizations using a power divider (e.g., a "Magic T"). Both polarizations have their own duplexer (shown as a circulator) and their own dedicated receiver. Panel (c) of Figure 2.5 shows another single-transmitter, dual receiver architecture. This architecture only implements ATSR mode (in contrast to the architecture in panel [b]). The architecture in panel (b) splits the transmitted power between the two polarizations. Because the architecture in panel (c) doesn't split the power between the two polarizations, it has 3-dB-higher transmit power for the selected polarization and therefore has 3-dB-better sensitivity.

The final architecture that is more commonly found in advanced research radar is shown in panel (d) of Figure 2.5 and has two transmitters and two receivers. This architecture provides flexibility for the radar operators, allowing STSR, ATSR, or variants of these operating modes to be used depending on the application.

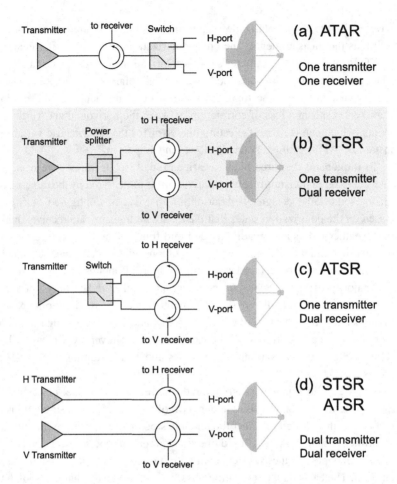

Figure 2.5 Example architectures for linear dual polarization. (Variations on these architectures can be found.) The majority of operational dual polarization weather radar systems operate in STSR using the mode option shown in panel (b). Note that STSR could be implemented with the option shown in panel (d) by using the same transmitted signal and timing for both the H and V transmitters. Similarly, the option shown in panel (d) can also implement ATSR mode with proper selection of the H and V transmitter for each pulse.

2.3   Polarimetric Doppler Weather Radar Measurements

A Doppler weather radar measures the amplitude and phase of echoes for one or more polarizations (i.e., the horizontal and vertical polarization for a dual polarization radar). These measurements are sampled at different ranges in the direction in which the antenna is pointing. The amplitude is proportional to the size, composition, and number of scatterers. The phase is related to the distance to the scatterers, their velocity, and scattering properties. Three common Doppler radar observations are estimated: equivalent reflectivity factor (Z_e), mean Doppler velocity ($\bar{v}$), and Doppler spectrum width (σ_v). With dual polarization (using ATSR and STSR modes), three additional variables are estimated: differential reflectivity (Z_{dr}), differential phase shift (Φ_{dp}), and copolar correlation (ρ_{hv}). In ATSR mode, the linear depolarization ratio (LDR) and cross-polar correlation are also available.

2.3.1 Reflectivity (η), Reflectivity Factor (Z), and Equivalent Reflectivity Factor (Z_e)

Radar reflectivity has entered everyday language, just like other weather-related terms. This term (or simply *reflectivity*) is commonly used to designate the intensity of the precipitation in weather radar maps, and it will also be largely used in this book to indicate the reflectivity observed by the radar. However, the term *reflectivity* can be used to reference several different quantities, all of which are related but have subtle differences. The following discussion highlights the various "reflectivities" and, more importantly, the convention that will be followed through the remainder of this book.

There are commonly used terms that are quite distinct, namely, *radar reflectivity*, *radar reflectivity factor*, and *equivalent reflectivity factor*. There is also a difference between the radar's *observed reflectivity* and the precipitation's *intrinsic reflectivity*. The radar reflectivity factor is an attribute of the precipitation within the radar's observation volume. The radar's actual observation is only an estimate of the radar reflectivity factor of this volume, whose value may vary because of system calibration, finite observation times, and attenuation effects resulting from the radar's signal propagating through the atmosphere and precipitation.

Observations of clouds may produce values below -50 dBZ, whereas -20 dBZ provides a lower bound for the lightest drizzle [6], and 10 dBZ is typically associated with very light precipitation (approximately 0.1 mm h^{-1} or less). Heavy rain in convective systems is in general less than 55–60 dBZ. Higher reflectivities such as 60–70 dBZ could be from hail, whereas values well above 70 or 75 dBZ could be from nonmeteorological objects, such as buildings or mountains, typically termed *clutter*.

The radar reflectivity factor (Z) for raindrops can be expressed as

$$Z = \frac{\sum_n D_n^6}{V}, \tag{2.6}$$

where D is the raindrop's equivolume diameter, and the summation is taken over all drops within a volume V. To express reflectivity in the common units of mm^6 m^{-3}, we can rewrite eq. (2.6) as

$$Z = 10^{18} \frac{\sum_n D_n^6}{V}, \tag{2.7}$$

where D is in mm, V is in m^3, and the factor 10^{18} is used to convert from m^6 m^{-3} to mm^6 m^{-3}. The basic concept of the D^6 relationship with the reflectivity is shown in Chapter 4.

The reflectivity factor has units of $\text{mm}^6\,\text{m}^{-3}$ but is typically referred to using a logarithmic scale with units of dBZ:

$$Z\,(\text{dBZ}) = 10\log_{10}(Z). \tag{2.8}$$

Note that there are no radar-dependent parameters used to calculate the reflectivity factor. The radar cross-section (σ) and its wavelength (λ) are not found in the calculation. The reflectivity factor has only two assumptions: the particles are spherical, and the particle sizes are small with respect to a radar's wavelength (the Rayleigh-scattering assumption). The radar reflectivity factor is a measurable property of rain within a volume and can be calculated just like the liquid water content:

$$W \propto \frac{\displaystyle\sum_n^N D^3}{V}, \tag{2.9}$$

which is a measure of the amount of water in a unit volume. Compare this to the radar reflectivity factor, which is proportional to the sixth power of the diameter. This is to say that the radar reflectivity is generally related (although not uniquely; see, e.g., the discussion of Fig. 3.3) to the liquid water content (i.e., the amount of rain).

The radar reflectivity, $\bar{\eta}$, is defined as the radar cross-section (RCS), σ, per unit volume. Intuitively, the concept of the RCS of an object refers to how big the object appears, "electromagnetically" (the concept of the RCS is introduced formally in Chapter 4). The radar reflectivity is defined as the mean RCS per unit volume for all of the scatterers within the volume, V:

$$\boxed{\bar{\eta} = \frac{\displaystyle\sum_n^N \sigma_n}{V}.} \tag{2.10}$$

For a weather radar, when the precipitation particles are much smaller than the wavelength, λ, the RCS of a spherical particle with diameter D is as follows:

$$\sigma(D) = \frac{\pi^5}{\lambda^4}|K|^2 D^6, \tag{2.11}$$

where $|K|^2$ is the dielectric factor, which is a function of the complex-valued relative permittivity (or complex-valued refractive index) of the scatterer (see Section 4.2.1).

The equivalent reflectivity factor (Z_e) is a (nearly) frequency-independent measurement (as opposed to $\bar{\eta}$), which makes it suitable for comparing observations at different frequencies and also for relating the physical properties of the precipitation. For raindrops that are small compared with the radar wavelength (Rayleigh-scattering regime), the radar's equivalent reflectivity factor is the same as the reflectivity factor (for Rayleigh scattering, $Z = Z_e$).

It is important to note that the value measured by weather radar systems, neglecting attenuation effects, is by convention named the *equivalent reflectivity factor Z_e*, which is related to the radar reflectivity $\bar{\eta}$ following

$$Z_e = \frac{\lambda^4}{|K_w|^2 \pi^5} \bar{\eta},$$

(2.12)

where $|K_w|^2$ is the dielectric factor of water. Typically, $|K_w|^2 = 0.93$ for precipitation radars (refer to Section 4.2.1 for details on the dielectric factor for ice and water). Equation (2.12) reduces to eq. (2.6) for Rayleigh scatterers (use eq. [2.10] with the RCS from eq. [2.11]).

In a typical radar, the received power is proportional to $\bar{\eta}$, and therefore we need to make an assumption to convert that to reflectivity, and the convention is to use the dielectric factor of water (independent of what type of scatters are present; e.g., the signals could be coming from ice crystals, buildings, or even insects). The intrinsic reflectivity of a precipitation particle with arbitrary composition (i.e., water, ice, or mixtures) using the corresponding dielectric factor is

$$Z_{\text{intrinsic}} = \frac{\lambda^4}{|K|^2 \pi^5} \bar{\eta},$$

(2.13)

where $|K|^2$ varies according to the hydrometeors, composition.

From the foregoing discussion, the following should be clear:

- The radar reflectivity factor Z is a quantity defined in terms of the particle size distribution, which is the fundamental microphysical property of precipitation (Section 3.1).
- Radar reflectivity $\bar{\eta}$, the radar cross-section per unit volume, is directly related to what the radar measure via the electromagnetic response of the scatterers (Chapter 4).
- The equivalent radar reflectivity factor Z_e is directly related to the radar reflectivity $\bar{\eta}$, but it is converted to a microphysically significant quantity similar to Z using the dielectric factor of water. It is an estimate of the radar reflectivity factor. It is the "reflectivity" variable ordinarily shown on radar displays.

All of the three different "reflectivities" are often used in the literature related to weather radar, and all can be referred to as just "reflectivity." Most frequently when discussing radar observations, we use *reflectivity* to specifically discuss the equivalent reflectivity factor (because this is what the radar is actually measuring, neglecting attenuation effects), but care should be taken to interpret the context of the discussion.

For the remainder of this book, when referring to reflectivity observations, we are specifically talking about the radar's measured estimate of the equivalent reflectivity factor for the horizontal polarization. This estimate often includes the effects of path-integration signal attenuation. This may also be referred to as *observed reflectivity* or *measured reflectivity*. If techniques to estimate and remove the effects of signal attenuation are used for the reported estimates, they are referred to as *corrected reflectivity*.

The corrected reflectivity is the estimate of the equivalent reflectivity factor for the horizontal polarization.

2.3.2   Doppler Velocity ($\bar{v}$) and Spectrum Width (σ_v)

For coherent radars (systems for which there is a known relationship between the phase of the transmitted and received pulses; see Section 2.6), the mean Doppler velocity and the Doppler spectrum width can also be measured. These additional measurements complement the microphysics information provided by the radar reflectivity with the capability of sensing the atmospheric dynamics.

The mean Doppler velocity ($\bar{v}$) is determined by measuring the Doppler frequency shift of the scatterers (Section 5.3.2) that results from a weighted average of the scatterers' velocity within the radar resolution volume. It is important to note that the average is weighted by the reflectivity and the observation volume, which takes into account both the antenna's illumination pattern and the transmitted pulse's "shape" (Section 2.4). Therefore, the resulting measured Doppler velocity is primarily influenced by the velocity of the larger particles within the sampled volume. For radars scanning at low-elevation angles, the Doppler velocity gives the radial component of the wind velocity vector. Observations at the vertical incidence provide information about the falling velocity of particles and can be useful, for example, to discriminate between rain and snow (Section 3.6.1).

The Doppler spectrum width (σ_v or sometimes w) is a measure of the dispersion of velocities within the resolution volume and can provide information about wind shear and turbulence. This is used in particular to help in the decision process for severe thunderstorms and tornado warnings.

2.3.3   Polarimetric Radar Measurements: Z_{dr}, Φ_{dp}, ρ_{hv}

Similar to Doppler radar observations, polarimetric radar observations rely on measurements of the amplitude and phase of the signal, although in this case from two orthogonal (horizontal and vertical) polarizations. The terms *observations* and *measurements* are used interchangeably. Both the propagating and the scattered waves can be described by their complex amplitudes. Several radar variables can be derived based on the magnitude of the backscattered wave at the horizontal and vertical polarizations, their phase, or both.

The most straightforward additional measurement that a polarimetric radar can provide is the differential reflectivity, Z_{dr}. It is the ratio of the equivalent reflectivities at horizontal and vertical polarizations (Z_h and Z_v, respectively). Note that Z_{dr}, just like reflectivity, is often expressed in logarithmic scale (dB):

$$Z_{dr}\,(dB) = 10\log_{10}(Z_{dr}) = Z_h\,(dBZ) - Z_v\,(dBZ).$$ (2.14)

Throughout this book, the symbol Z_{dr}, just like Z_h, represents the variable in linear units, whereas when we refer to, for example, Z_{dr} in dB, this will be explicitly noted in the text, or the corresponding log transform (dB scale) will be added in the figure axis.

Z_{dr} reveals information about the shape, composition, and density of the hydrometeors. For any given particle shape, Z_{dr} may vary depending on the particular composition (i.e., water, ice, or mixed phase), as a consequence of the different relative permittivity of the material (Section 4.2). For example, oblate raindrops will have a larger Z_{dr} than solid ice particles with the same shape (see, e.g., Fig. 4.10). Low-density aggregates, such as snowflakes, will have even smaller Z_{dr} values (typically close to 0 dB) because of the small effective relative permittivity of diluted ice–air mixtures (Fig. 4.21).

Z_{dr} reveals properties of the precipitation medium based on the ratio of the backscattered signal strength at the two orthogonal polarization channels. Analogously, the differential phase shift Φ_{dp} (i.e., the phase difference between the received signals at horizontal and vertical polarization) can be highly informative about the composition of the particles and concentration along the propagation path.

The copolar correlation coefficient between the horizontal and vertical polarization signals (ρ_{hv}) provides a measure of how similar the horizontal and vertical scattering cross-sections of the objects are. A higher value of ρ_{hv} (close to 1) is typically observed in rain, as a result of the consistency of the raindrops' shape and fall behavior. Lower values generally indicate a greater variability in the scatterers' population; for example, lower values are normally observed in the melting layer (coexistence of liquid, solid, and mixed-phase particles with varying shape) or with non-meteorological echoes, such as tornado debris, insects, and birds (Section 7.3.5).

Finally, for radars that transmit one polarization at a time and receive both the copolar and cross-polar signals (e.g., ATSR systems; panel [c] of 2.5), an additional variable can be estimated. Similar to Z_{dr}, the magnitude of two received polarization signals are compared, although in this case the ratio is between the cross-polar and the copolar power. For example, if the radar transmits a horizontal polarization, the LDR is given by the ratio of the power received on the vertical polarization (cross-polar signal) to the power received on the horizontal polarization (copolar signal). Although the measurement is realized in the same way as for Z_{dr} (two power levels are measured by means of two different receivers connected to the antenna), the significance of the LDR observations is different. Similar to ρ_{hv}, LDR can be a good indicator of regions where a mixture of precipitation types occurs (typically when the radar resolution volume is populated with randomly oriented particles). In other instances, LDR can be used to detect a preferential orientation of the hydrometeors, such as in the presence of strong electric fields in thunderstorms (see discussion in Sections 4.9 and 8.5.3).

2.4   Observation Geometry

The radar's observation geometry is most naturally described using a modified form of spherical coordinates: range (r), azimuth (ϕ), and elevation (θ). The radar is located

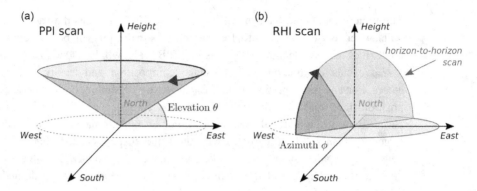

Figure 2.6 Illustration of weather radar elemental scan modes. The elevation angle, θ, is measured above the horizontal plane. The azimuth, ϕ, is measured clockwise from north (north is $\phi = 0°$; east is $\phi = 90°$). In PPI mode (a), the antenna completes a full rotation in the azimuth, ϕ, at a fixed elevation angle, θ. The PPI scan makes observations on a conical surface. In RHI mode (b), the antenna moves in elevation, θ, with the azimuth angle, ϕ, fixed. The RHI collects data on a vertical plane that may extend across the zenith to obtain a "horizon-to-horizon" scan.

at the center of the sphere and observes outward. For scanning radar systems, the radar can steer the radar's beam in both the azimuth and the elevation. The radar's transmitted power travels radially away from the radar's antenna, and the echoes from precipitation are received along the same radials. For a fixed-pointing radar system, there are no scanning capabilities; the radar is pointed along a single radial, typically pointing vertically, in the zenith direction.

For scanning radar systems, different scanning strategies have been adopted, but all are typically composed of the range-height indicator (RHI) and plan position indicator (PPI) scans, shown in Figure 2.6. Multiple RHI or PPI scans can be combined to scan a volume. Similarly, the extent of the RHI or PPI scan can be limited to cover only a limited sector of interest.

For radar observations, the cardinal directions and Cartesian coordinates are related as follows:

- The positive x-axis is east, and the negative x-axis is west.
- The positive y-axis is north, and the negative y-axis is south.
- The positive z-axis is the zenith, and the negative z-axis is the nadir (which is not encountered for ground-based radar).

The radar's azimuth angle starts pointing north ($\phi = 0°$) and increases as it rotates clockwise with east ($\phi = 90°$), south ($\phi = 180°$), and west ($\phi = 270°$). The radar's elevation angle starts pointing parallel to the ground ($\theta = 0°$) and increases in angle toward its zenith ($\theta = 90°$). For radars with the capability of scanning from horizon to horizon, the elevation angles are sometimes represented as $0° < \theta \leq 180°$, where elevation angles $\theta > 90°$ correspond to azimuth angles $\phi + 180°$. Section 2.4.5 discusses the conversion between the radar and Cartesian coordinates in more detail.

2.4.1 Radar Ranging, r

In free space, the speed of light is $c = 299,792,458$ m s^{-1}. When determining the distance that a radar's signal travels, the speed of light in Earth's atmosphere is assumed as the speed of light in a vacuum. At this speed, it takes light 33.356 μ s for the radar's energy to travel 1 km.

Consider a monostatic radar detecting a target at a distance of $r = 1$ km. The transmitted pulse takes 33.356 μ s to travel from the radar to the target. A portion of the transmitted pulse's energy is scattered by the target, and some of this scattered energy is directed back at the radar (the energy is said to be *backscattered*). The backscattered energy then takes another 33.356 μ s to travel from the target back to the radar, where it is received. The radar's signal travels $2r$. The total round-trip travel time of the radar's signal is

$$\Delta t = \frac{2r}{c}.$$ (2.15)

In practice, the radar system measures the time delay between the transmitted signal and the received signal, Δt, and therefore the radar's range to the target can be calculated:

$$r = \frac{\Delta t \cdot c}{2}.$$ (2.16)

A simple relationship to remember is that every 1 μ s of elapsed time is equivalent to 150 m of radar range. Since the introduction of the radar, the technology for accurate time measurements (and therefore accurate range measurements) has improved significantly and is an important technical aspect of modern radars.

2.4.2 Range Resolution, Δr

In pulsed radar (the most common weather radar), the radar signal is transmitted for a short period of time. The ranges covered by the pulse can be calculated from the relative difference in the starting and ending times of the transmitted pulse (see eq. [2.16]). For a pulsed radar whose pulse duration is T_{tx}, the range resolution is calculated as

$$\Delta r = \frac{c T_{tx}}{2}.$$ (2.17)

For pulsed radar systems, the pulse width and bandwidth are inversely related as $T_{tx} = 1/B$.

The radar's range resolution can be rewritten using the radar signal's bandwidth as

$$\boxed{\Delta r = \frac{c}{2B}.}$$ (2.18)

For radars that utilize pulse compression techniques, the range resolution is still determined by the radar signal's total bandwidth (but the pulse length T_{tx} is no longer related to range resolution because $B \neq 1/T_{tx}$). Pulse compression is an advanced topic and is discussed in Section 6.5. Equation (2.18) is generally more appropriate for relating the radar's operating parameters to the range resolution.

When determining the range resolution, the bandwidth, B, is the effective bandwidth of the transmitted signal convolution with the receiver's filter. The pulse-shaping effects of the transmitter and receiver set the resolution of the radar's signal. After convolution, the $-6\,\text{dB}$ power threshold determines the range resolution for estimation of the V_6 observation volume. This may be approximated as the half-power level (i.e., $-3\,\text{dB}$) of the transmitter's pulse (which assumes a receiver filter with similar bandwidth is used).

2.4.3   Observation Volume, V_6

Although the first radars were introduced to detect hard targets such as aircraft, the applications for radar have vastly extended beyond detecting single targets, which are often referred as *point targets*. The applications for radars have expanded to look at surfaces from airborne and spaceborne radars, as well as volumes of scatterers that include our particular application: precipitation and weather. Weather radars observe a class of scatterers broadly referred to as *volume targets*. For weather radar, the formulation of the radar equation must be updated to accommodate a volume of scatterers.

Atmospheric systems of clouds and precipitation can extend over large areas and heights, whereas at the smallest microphysical scales, these are individual ice crystals or water drops that exist among a vast number of other particles and fill the volume. The radar's antenna pattern and the range resolution of the radar's signal, Δr, and the range from the radar, r, all determine the shape and size of the radar's observation volume, V_6 (often referred to simply as V), that connects to the received signal at a given time. Figure 2.7 illustrates the range resolution volume and its dependencies.

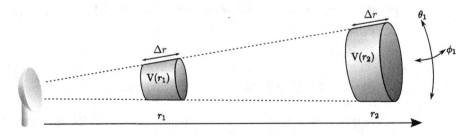

Figure 2.7 The radar's observation volume is a function of the range resolution, Δr; the radar range, r; and the antenna pattern. The one-way antenna pattern's half-power beamwidth for elevation, θ_1, and azimuth, ϕ_1, are used. The radar range is from the radar to the center of the range resolution volume. The range resolution remains constant for all ranges but, as the range from the radar increases, so does the distance covered in the azimuth and elevation directions.

The transmitted pulse's amplitude with respect to time determines how the radar volume is illuminated in the range dimension. The orthogonal dimensions of the volume (in the spherical coordinate system) are the azimuth and elevation angles, which are determined by the transmitting antenna's beam pattern. Within the weather radar equation for a monostatic system, the transmitting and receiving antennas' 3-dB, one-way azimuth and elevation beamwidths are used along with the radar's range to determine the observation volume [7]:

$$V_6 = (\Delta r)\frac{\pi\theta_1\phi_1}{8\ln2}r^2, \tag{2.19}$$

where θ_1 and ϕ_1 are the one-way, -3-dB antenna beamwidths in the elevation and azimuth, respectively. The observation volume, denoted V_6, is a disk, assuming a circularly symmetric beam, whose angular and range extents represent the -6-dB two-way beamwidth of the monostatic antenna (the -3-dB beamwidth of the transmitting antenna and the -3-dB beamwidth of the receiving antenna). The V_6 volume represents the volume where the majority of the radar signal's power is focused at an instance in time, using the traditional 3-dB convention commonly used in electrical engineering to represent beamwidths and filter bandwidths. As shown in Figure 2.7, as the range from the radar increases, the size of the radar volume increases as a result of the increasing width and height of the disk in the azimuth and elevation dimensions. The range extent of the volume is determined by the range resolution, which remains fixed as a function of range.

2.4.4 The Beamwidth

It is clear from Figure 2.7 that as the range increases, the width and height of the observation volume increase. The angular beamwidths are determined by the antenna and therefore constant (for most mechanical antennas), but the arc distance spanned by the radar volume scales with the radar's range. In the elevation direction, the distance spanned by the beam is

$$b_\theta = 2r\sin\left(\frac{\theta_1}{2}\right) \approx r\theta_1, \tag{2.20a}$$

and the beam's width in the azimuth direction is

$$b_\phi = 2r\sin\left(\frac{\phi_1}{2}\right) \approx r\phi_1. \tag{2.20b}$$

The approximations shown adopt the small-angle assumption, $\sin x \approx x$. The beamwidths b_θ and b_ϕ both have the same units as r (typically meters). The beamwidths only consider the size of the beam as a result of the antenna and do not take into consideration the broadening effects when the radar is scanning.

For a Gaussian approximation of the antenna pattern, which is common for weather radars with reflector antennas, the cross-sectional area can be approximated as an ellipse:

$$A = \pi \frac{b_\theta b_\phi}{4} = \pi \frac{r^2 \theta_1 \phi_1}{4}. \tag{2.21a}$$

From eq. (2.19), the pattern-weighted antenna beam area, after removing the range resolution contribution from V_6, is

$$A_{\text{beam}}^{(\text{weighted})} = \frac{\pi \theta_1 \phi_1}{8 \ln 2} r^2. \tag{2.21b}$$

Comparing the weighted pattern's area to the uniform cross-section area reveals that:

$$A_{\text{beam}}^{(\text{weighted})} = \frac{A}{2 \ln 2}. \tag{2.21c}$$

The Gaussian antenna pattern's weighting scales the total echo power from the observation volume by a factor of $1/(2\ln 2)$, which is found in the formulation of the weather radar equation.

2.4.5   The Beam Height and Ground Distance

The weather radar observations are given in the spherical coordinates of azimuth (ϕ), elevation (θ), and range (r). Although the spherical coordinates are natural from the perspective of how the radar observes, it is not typically intuitive for users when trying to place the location of the observation relative to different towns or cities. (Rectilinear grids are also natural for comparing multiple sensors, numerical weather models, and some processing algorithms.) Conversion to Cartesian coordinates for other applications or users is done through geometric relations:

$$\begin{bmatrix} x \\ y \\ z \end{bmatrix} = \begin{bmatrix} r \cos \left(\pi/2 - \phi \right) \cos \left(\theta \right) \\ r \sin \left(\pi/2 - \phi \right) \cos \left(\theta \right) \\ r \sin \left(\theta \right) \end{bmatrix}. \tag{2.22}$$

The zenith direction is $+z$, west is $-x$, and east is $+x$; south is $-y$, and north is $+y$ (refer to Fig. 2.6). For observations close to the radar, the surface can initially be assumed as a flat surface with little error. As the range from the radar increases, however, the error in the estimated height above the surface can become significant if Earth's curvature and atmospheric refraction are neglected.

In the troposphere, the index of refraction (n) decreases with increasing altitude (h) [8]. Gradients in temperature, pressure, and water vapor within the atmosphere all result in variations in the index of refraction. From Snell's law,

$$n_1 \sin \theta_1 = n_2 \sin \theta_2, \tag{2.23}$$

the angle of propagation of the electromagnetic wave changes as a result of differences in the layer's index of refraction. The beam-height model assumes a nominal gradient in Earth's troposphere, $\frac{dn}{dh} = -\frac{1}{4a}$, where a is the radius of Earth. The implies that the radar beam does not travel along a straight line but is bending toward the surface as it propagates.

For estimating the beam height, it is useful to adopt a modified Earth's radius such that the ray path is straight, as illustrated in Figure 2.8. Earth is slightly oblate, and its

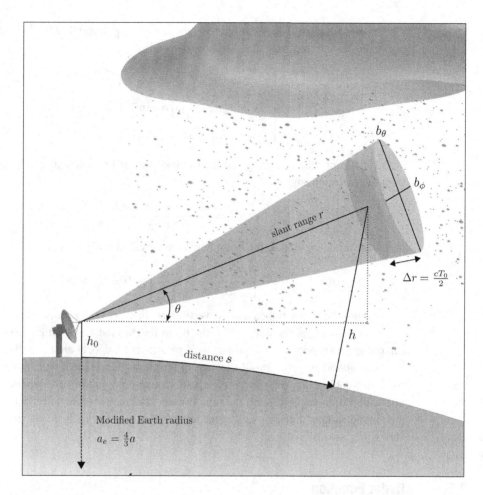

Figure 2.8 Schematic illustration of the propagation path and radar resolution volume.

radius varies from approximately 6357 km at the poles to 6378 km at the equator. For the purposes of estimating the beam's height, the variation of Earth's radius is small. The spherical Earth model with Earth's nominal radius of $a = 6371$ km is frequently assumed. Using the average gradient of the refractive index, $\frac{dn}{dh} = -\frac{1}{4a}$, leads to an effective radius of $a_e = \frac{4}{3}a$.

With simple trigonometric considerations, the equation for the height of the radar beam in standard atmosphere can be written as follows (see, e.g., Doviak et al. [9]):

$$h(r, \theta) = h_0 + \left(r^2 + a_e^2 + 2r\,a_e\,\sin\theta\right)^{\frac{1}{2}} - a_e, \qquad (2.24)$$

where h_0 denotes the height of the radar antenna above the surface. The great-circle distance s (the distance along Earth's surface) is obtained with the law of sines, as follows:

$$s(h, \theta) = a_e \sin^{-1}\left(\frac{r\cos\theta}{a_e + h}\right). \qquad (2.25)$$

The ground relative distance and height above a spherical Earth's surface (i.e., neglecting topography) can then be approximated as

$$
\begin{bmatrix} x_E \\ y_E \\ z_E \end{bmatrix} = \begin{bmatrix} \cos\left(\pi/2 - \phi\right) a_e \sin^{-1}\left(\frac{r\cos\theta}{a_e+h}\right) \\ \sin\left(\pi/2 - \phi\right) a_e \sin^{-1}\left(\frac{r\cos\theta}{a_e+h}\right) \\ h = h_0 + \left(r^2 + a_e^2 + 2r\,a_e\,\sin\theta\right)^{\frac{1}{2}} - a_e \end{bmatrix}.
\tag{2.26}
$$

Other useful relations are given by Doviak et al. [9] to calculate the slant range from the elevation angle:

$$
r(s, \theta) = \frac{(a_e + h_0)\sin(s/a_e)}{\cos\theta}.
\tag{2.27}
$$

or the elevation angle from the beam height and great-circle distance:

$$
\theta(h, s) = \tan^{-1}\left[\frac{H\cos(s/a_e) - a_e}{H\sin(s/a_e)}\right]; \quad H = a_e + (h - h_0).
\tag{2.28}
$$

As atmospheric conditions and the vertical gradients of the index of refraction vary from the nominal values assumed by the beam-height model, the actual beam height can change. Deviations in the propagating path from the model are typically referred to as *anomalous propagation*. Refraction of the atmosphere can vary the radar beam's path. A notable example in radar observations can be found in time lapses of scans at low-elevation angles around the time the sun sets: the presence of reflections from the ground and building can vary as the temperature gradient in the lower troposphere varies.

2.5   Radar Equation

The weather radar system transmits a signal and receives the echoes from scatterers in the troposphere. The radar equation is used to calculate the received power for a given radar system's and scatterer's parameters. The radar equation is an important tool to estimate the radar system's performance.

First, consider the objects that the radar will observe: the scatterers, such as raindrops, hail, or ice crystals. Generally speaking, the larger the size of the scatterer, the easier it is for the radar system to detect it. For smaller scatterers, a higher-sensitivity radar system is required to detect its weaker echo. With this simple concept of "smaller" and "larger" scatterers, when considered in the context of a radar, the size of the scatterer is formally referred as its RCS. Specifically, the RCS is directly related to the ratio of scattered power to the incident power, which is a measure of how much of the incident power (from the transmitter) is scattered back toward the receiving antenna. The RCS is given by

$$
\sigma = \lim_{r\to\infty} 4\pi r^2 \frac{S_s}{S_i}
\tag{2.29}
$$

where r is range, and S_i and S_s are the incident and scattered power densities, respectively (in watts per square meter). In the case of a monostatic radar system, the scattered direction is opposite the incident direction (i.e., $\hat{s} = -\hat{i}$). (As an example, in Fig. 2.2, the direction of the incident wave's propagation, $\hat{k}_i$, would be $\hat{i}$.) In eq. (2.29), the limit as r approaches infinity is meant only to imply that the scatterer and the observing antennas are far apart (more accurately, the range is greater than the antenna's far-field range). The RCS is not dependent on the range.

The RCS, σ, is an area with units of meters squared and is discussed in greater detail in Chapter 4. The RCS corresponds to the apparent area of the target, assuming the scattered power is isotropic (scattered equally in all directions). The RCS's area and the target's physical area are not necessarily the same. The RCS can also depend on the radar's frequency. Typically, the larger the physical size of the object, the larger the RCS. For example, hail has a larger RCS compared with cloud water droplets. For more complex targets, the RCS can change significantly as the incidence and scattering angles vary.

2.5.1   Single-Target Radar Equation

The single-target radar equation (also called the *point-target radar equation*) is a simple concept used to calculate the echo power at the radar's receiving antenna for a target with a known RCS. For any geometry, the scatterer can be treated as a single point in space with a fixed RCS, σ. The radar equation is used to calculate the received power for a monostatic radar system, where the transmitter's and receiver's antenna are the same. Similarly, it can also be used for bistatic radar systems where the transmitter's and receiver's antennas are in different locations (and therefore at different ranges).

For a given transmit power (P_t), to determine the received power density (S_r), eq. (2.29) can be rearranged to calculate the scattered power density (at the scatterer). Assuming the transmitting antenna is an isotropic source (i.e., the transmit antenna's gain $G_t = 1$), the wave's power density incident on the scatterer at range r_t is

$$S_i = \frac{P_t}{4\pi r_t^2},\tag{2.30}$$

where $1/(4\pi r_t^2)$ is the surface area of a sphere with radius r_t. The transmitting antenna focuses the transmitted power, which effectively increases the power density incident on a scatterer. If the transmitting antenna is not isotropic, the total transmitted power is confined to a smaller surface area, which increases the power density. The increased incident power density due to the transmit antenna's gain (G_t) is included as

$$S_i = \frac{P_t G_t}{4\pi r_t^2}.\tag{2.31}$$

The fraction of the incident power that is reflected (or scattered) is determined by the RCS. The scattered wave's power (at the scatterer) is

$$P_s = S_i \sigma = \frac{P_t G_t}{4\pi r_t^2}\sigma,$$ (2.32)

where P_s has the unit of watts (recall that S_i has the unit of watts m^{-2}, and σ has the unit of m^2). The scatterer, in effect, becomes a second transmitting source whose power, P_s, propagates away from the scatterer. At the radar receiver, located at range r_r from the scatterer, the scattered wave's power density is

$$S_r = \frac{P_s}{4\pi r_r^2} = \frac{P_t G_t}{4\pi r_t^2}\sigma\frac{1}{4\pi r_r^2}.$$ (2.33)

Similarly, the receiver antenna acts to collect the scattered signal using its large surface area. The received power is increased as a result of the scattered power density being integrated over the antenna's effective area. To calculate the received power, multiply by the antenna's effective area (A_e; see eq. [2.42]) in square meters to give

$$P_r = S_r A_e = \frac{P_t G_t}{4\pi r_t^2}\sigma\frac{A_e}{4\pi r_r^2}.$$ (2.34)

The receiving antenna's gain is related to the antenna's effective area, A_e, following

$$G_r = \frac{4\pi A_e}{\lambda^2}.$$ (2.35)

Substituting G_r into eq. (2.34) gives the point-target radar equation:

$$P_r = \frac{P_t G_t}{4\pi r_t^2}\sigma\frac{\lambda^2 G_r}{4\pi}\frac{1}{4\pi r_r^2}$$ (2.36a)

$$= \frac{\lambda^2 P_t G_r G_t}{(4\pi)^3 r_r^2 r_t^2}\sigma.$$ (2.36b)

For a monostatic radar system, where the same antenna is used for both transmitting and receiving (i.e., $G = G_r = G_t$ and $r = r_r = r_t$), the point-target radar equation simplifies to the well-known form:

$$\boxed{P_r = \frac{\lambda^2 P_t G^2}{(4\pi)^3 r^4}\sigma.}$$ (2.36c)

It should also be noted that the antenna gain used for the point-target radar equation is the gain in the direction of the scatterer being observed. The antenna's gain varies as a function of angle, and therefore the actual antenna gain for the observed scatterer is not necessarily the antenna's boresight gain (the boresight gain is its peak gain). Note that, approximations for a parabolic antenna's boresight gain as a function of its diameter and wavelength, or its beamwidth, are given later by eqs. (2.43) and (2.44), respectively.

The signal's powers encountered in a radar system can span many orders of magnitude, from femtowatts (10^{-15}) for weak echoes to megawatts (10^6) for radar transmitter power levels. As such, the radar signal's power is frequently reported using a logarithmic scale with units of dBm or dBW. The power in dBm is the power ratio with respect to 1 milliwatt, in decibels. Similarly, dBW is the power ratio with respect to 1 watt. (Note that 1 dBW = 30 dBm.)

2.5.2 Weather Radar Equation

Precipitation is a collection of hydrometeors, which can each be treated as point targets from the radar's perspective. A typical approximation assumes that because of the low-volume fraction in space, there are no scattering interactions between hydrometeors (e.g., raindrops or snowflakes), and therefore the received power from each individual particle can simply be added together. This implies that the received power from a precipitation-filled radar volume is the summation of all particles' point-target responses within the radar volume. For the monostatic radar system, this is

$$P_r = \sum_V P_t(r) \frac{\lambda^2 G^2(\theta, \phi)}{(4\pi)^3 r^4} \sigma_n, \tag{2.37}$$

where σ_n indicates the RCS of the nth particle, and V is the full spatial volume illuminated by the pulse's range resolution around range r (typically represented in spherical coordinates that correspond to the radar's azimuth, elevation, and range). The transmit power, P_t, is shown as a function of range to select the extent in the range of the radar volume (see Fig. 2.7). In practice, the extent over which the summation is carried out is the V_6 radar volume (see eq. [2.19]). It is also typical to characterize the scatterers within the volume as uniformly distributed (the particle sizes may vary according to a distribution, but the same distribution applies throughout the volume). With this, the total RCS of the radar's resolution volume (with units m^2) can be described by its reflectivity, $\bar{\eta}$, which is the mean volumetric RCS with units m^2 m^3. The total RCS and radar reflectivity are then related as $\sigma = \bar{\eta} V_6$. The concept of volumetric reflectivity is used to convert the discrete summation over a large number of precipitation particles to an integral. Using the volume-averaged mean reflectivity, the received power can be written as an integral:

$$P_r = \iiint_V P_t(r) \frac{\lambda^2 G^2(\theta, \phi)}{(4\pi)^3 r^4} \bar{\eta} \, dV. \tag{2.38a}$$

The antenna's pattern and the radar's range resolution act to spatially select the volume defined by eq. (2.19). The monostatic weather radar equation gives the received power:

$$P_r = P_t \frac{\lambda^2 G^2}{(4\pi)^3 r^4} \left(\frac{c T_{tx}}{2} \frac{\pi \theta_1 \phi_1}{8 \ln 2} r^2 \right) \bar{\eta}. \tag{2.38b}$$

The radar's observation volume is centered at range r. The transmit pulse's duration, T_{tx}, is used to define the pulse's range extent for integrating the reflectivity $\bar{\eta}$, and G is the antenna's boresight (peak) gain. Comparing the monostatic radar equations for the point-target and volume weather targets (eq. [2.36c] and [2.38b], respectively), the most significant difference is the received power's dependence on range. The point-target radar equation has $1/r^4$ dependence, whereas the weather radar equation has $1/r^2$ dependence.

The signal P_r is processed by the radar receiver's electronics. The receiver filter's output power is related to the observed volume's equivalent reflectivity factor by

$$P_o = G_{rx} \left(\frac{cT_{tx}}{2} \right) \left(\frac{\lambda^2 P_t G^2}{(4\pi)^3} \right) \left(\frac{\pi \theta_1 \phi_1}{8 \ln 2} \right) \left(\frac{\pi^5 |K_w|^2}{\lambda^4 r^2} \right) \frac{Z_e}{10^{18}}. \tag{2.39a}$$

The 10^{18} factor converts from m^6 to mm^6 to give the appropriate unit for Z_e, which is $mm^6 m^{-3}$. The receiver gain, G_{rx}, includes the gains (or losses) in the receiver and signal-processing paths. In practice, the majority of the terms are constant for a radar system, and as such, the equation for the equivalent reflectivity factor is simplified to

$$P_o = Z_e \frac{C}{r^2}. \tag{2.39b}$$

The equivalent reflectivity factor can be calculated from the received power and range as

$$\boxed{Z_e = \frac{P_o r^2}{C},} \tag{2.39c}$$

where the weather radar constant is

$$\boxed{C = \frac{G_{rx} P_t G^2 \pi^3 |K_w|^2 c T_{tx} \theta_1 \phi_1}{10^{18} \lambda^2 1024 \ln 2}.} \tag{2.39d}$$

Note that C assumes r is in meters, but some formulations may use range in kilometers. If the range in eq. (2.39c) is kilometers instead of meters, an appropriate scale factor (i.e., 1 km = 1000 m) must be included. If the range is presented in kilometers instead of meters, 10^{18} becomes 10^{24} in eq. (2.39d).

For dual polarization weather radar, the radar systems typically are designed to provide nearly identical performance at both polarizations. This would imply that the same radar constant, C, could be used for both the horizontal and vertical polarizations. In practice, there can be slight differences in performance between the two polarizations which can bias some of the radar measurements. While these differences are small, this results in different radar constants for each polarization state (i.e., C_h and C_v for the horizontal and vertical polarizations, respectively). As we'll discuss in Chapter 7, these differences can sometimes be included as calibration factors to apply to an otherwise-common radar constant.

2.6   Radar System Components

A block diagram of a dual polarization weather radar system is shown in Figure 2.9. This is an example of a weather radar that operates in the STSR mode, using one transmitter whose power is split between the horizontal and vertical

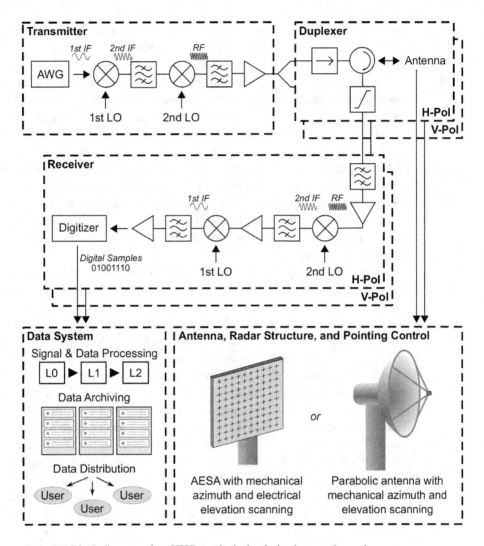

Figure 2.9 Block diagram of an STSR-mode dual polarization weather radar system (see the architecture in panel [b] of Fig. 2.5). Two antenna and scanning systems are shown: a mechanically scanning parabolic antenna and an active electronically scanned array (AESA), electronically scanning in the elevation and mechanically scanning in azimuth. Other antenna and scanning configurations could also be used for this block diagram. The relative frequencies of the radar's signal as it propagates through the radar are shown before and after the respective mixer stages. A two-stage up- and down-conversion is shown in this example.

polarizations (h-pol and v-pol, respectively). For the system shown, each polarization has a dedicated duplexer and receiver. The duplexer provides an interface between the radar electronics and the antenna. The antenna and pointing-control system could be mechanical scanning, electrical scanning, or a combination of both. The digitized data are communicated to the data system, where they are processed into data products (as reflectivity, velocity, rainfall rate, etc.). The processed data are archived for future use but are also distributed to radar operators, meteorologists, and other data systems.

The local oscillators (LOs) shown in Figure 2.9 are used to convert from the low-frequency signal of the transmitter's arbitrary waveform generator (AWG) to the radar's RF signal. An AWG is one of many techniques for generating the transmitter's signal. To ensure that only the frequency of interest is selected, it is sometimes necessary to use intermediate frequencies (IF), generated with multiple mixer stages and appropriate filters, to implement the frequency up- and down-conversion in the transmitter and receiver, respectively. With recent advances in high-speed analog-to-digital and digital-to-analog converters, directly sampling the RF signal, or only a single stage of analog frequency conversion, may be considered. This diagram and the explanation are meant to provide a high-level description of how the radar's RF signals are generated.

To estimate the Doppler velocity and the covariance between the dual polarization echo signals, the system's phase relationship between the transmitter and receiver must be known or measured so that it can be removed from the observations. When the phase of the received signal is known, or is accurately measured with respect to the transmitted signal, the radar system is said to be "coherent." The transmitter and receiver can be made coherent by phase locking all of the frequency sources to a common clock (e.g., a stable local oscillator [STALO]). When the transmitter and receiver are phase synchronized by design, this is known as a *coherent-on-transmit* design. Alternatively, instead of controlling the phase of the transmitter, the receiver can be used to sample the transmitted signal and estimate its phase for each pulse. This estimated phase can be used to correct the received signal on a pulse-by-pulse basis. This is known as *a coherent-on-receive design*. For magnetron-based Doppler radar systems, the transmitter's phase cannot be controlled, and therefore a coherent-on-receive design must be used. This has also become a standard technique for use with coherent transmitters to correct for second-order phase drifts in the radar system.

2.6.1   Electronically Scanned Arrays

The active electronically scanned array (AESA) antenna is typically designed as an integral system element rather than an interchangeable antenna component. The AESA radar is also commonly referred to as a *phased-array radar*. AESAs can provide a new paradigm for weather radar operations. With these new operational capabilities come additional considerations for system designs. Radar systems using an AESA antenna are highlighted as electronically scanned arrays or phased arrays to differentiate their enhanced capabilities compared with more traditional radar systems using reflector antennas.

Mechanically scanning systems are limited by the inertia of the antenna and the pointing system's solid structure. The radar's observation volume is set by the antenna pattern (which is fixed) and the direction in which the antenna is pointing. The pointing is mechanically controlled and therefore limited by the speed at which the antenna can change positions. For large S-band and C-band radar systems, the inertia of the antenna systems requires significant structures to withstand the large forces to move the antenna. Generally, mechanically scanning radars move continuously in the azimuth direction and step through multiple elevations (sometimes with research radars, they move continuously in elevation and step through the azimuth). This presents a trade-off between the time spent observing a specific radar volume and the total coverage that can be achieved within a given time period.

With electronically scanned systems, the observation volume is selected by electronically "steering" the antenna's pattern. The antenna's pattern is no longer fixed but is a controllable parameter in the radar's operation. The antenna's pattern and the pointing control are now combined together as one system. Because the pointing of the antenna beam is electronically controlled, the AESA does not have mechanical restrictions on how long it takes to position the antenna beam in the desired location, and the antenna's beam can remain fixed on the same volume (as opposed to scanning across the volume, which results in a changing volume over time).

The AESA is composed of a number of smaller antenna elements whose signals are combined together to synthesize the desired antenna pattern. Numerous system designs are possible, and these continue to evolve with the availability of new technologies. Electronically scanning radar systems can be implemented in a number of ways with a combination of electrical and mechanical scanning or only electrical scanning. The AESA can be designed to scan in one direction only (e.g., in elevation) or both azimuth and elevation. More complex AESAs and receiver systems can simultaneously sample multiple beams at one time. With individually digitized elements (or subarrays of elements), the phased array can beamform on receive to sample multiple discrete radar volumes at once. Simpler forms of the AESA provide a single beam that more closely approximates the mechanical antenna. Figure 2.10 shows

Figure 2.10 An X-band radar system using a planar electronically scanned antenna. Image courtesy of the Raytheon Technologies.

an example of an X-band AESA, which is implemented as a flat array of antenna elements, each of which can be electronically controlled to steer the antenna's pattern on transit and receive.

2.6.2   Transmitter

Radars are "active" instruments, meaning they provide their own signal to detect echoes. (In contrast, passive instruments only listen for other signal sources that may either be naturally occurring or are generated by other sources of opportunity.) As such, the transmitter is an essential component of the radar, providing the source signal to be used for "detection and ranging."

The key element of the transmitter subsystem is the source for the high-power RF signal. High-power amplifiers (e.g., klystrons, traveling-wave tubes, solid-state electronics) and high-power oscillators (magnetrons) are shown in Figure 2.11. The transmitter subsystem's coordination and timing are tightly controlled between the transmitter and receiver subsystems. For transmitters using power amplifiers, in addition to the high-power RF amplifier, the transmitter subsystem typically also includes the low-power RF signal-generation and frequency-conversion electronics. The power oscillator generates a high-power RF signal at its resonant frequency and does not use an input signal other than an on/off control. The most common transmitter types for weather radar systems are the magnetrons and solid-state power amplifiers (SSPAs). For some applications, the klystron and traveling-wave tube amplifiers (TWTAs), which use vacuum tube technology, provide the most effective solution.

Magnetrons (panel [b] of Fig. 2.11) were the first available techniques for generating microwave frequencies with sufficiently high power for use in radar systems. The most common magnetron design for radar transmitters is the resonant-cavity magnetron. The basic magnetron structure consists of a cathode located in the center of an anode ring. A permanent magnet is aligned so that the electrons interact with the magnetic field as they move through the cavity. As the electrons travel from the cathode to the anode, the magnetic field modifies their trajectory, forcing them to circle within the cavity. As they circle, a small percentage will eventually make it to the anode and generate a high-frequency electrical current. When correctly tuned, the frequency and stability of the electronic currents are suitable for generating RF or microwave signals. The operating frequency of a magnetron is largely a function of the mechanical dimensions of its cavity. A high-voltage modulator is used to energize the cathode, creating a voltage between the anode and cathode, which in turn generates the RF signal. The modulator typically has voltages in the tens of kilovolts and can generate peak RF powers in the megawatts.

The magnetron and its associated modulator circuit are a cost-effective method of generating high-power RF for radar systems as a result of the benefits from the legacy of large-scale manufacturing. The magnetron's resonance frequency is determined by its cavity dimensions, and therefore the magnetron's frequency can drift over temperature. Modern magnetrons typically allow the operating frequency to be tuned during

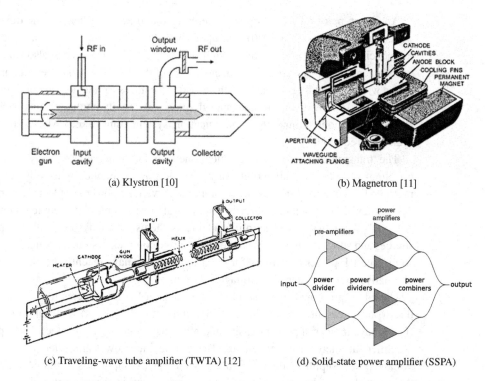

(a) Klystron [10]

(b) Magnetron [11]

(c) Traveling-wave tube amplifier (TWTA) [12]

(d) Solid-state power amplifier (SSPA)

Figure 2.11 Diagrams illustrating the concepts of operation for common technologies used as high-power transmitter sources or amplifiers in weather radar systems. The klystron (a) and traveling-wave tube (c) use vacuum tube amplifier technology. The solid-state power amplifiers (d) combine the power from multiple integrated circuits to achieve the desired transmitter power levels. These amplifiers reproduce the phase and frequency of an input reference signal while increasing the amplitude. The magnetron (b) is a power oscillator. The magnetron is both the transmitter's oscillator and a high-power source.

operation to conform to the radar's assigned frequency band. Similarly, because the magnetron is a power oscillator, the signal's starting phase is random. The frequency drift and random starting phase impose additional requirements on the receiver system to track the magnetron's RF frequency and phase via "automatic frequency control" circuits and coherent-on-receive signal-processing techniques.

The extended interaction klystron amplifier (EIKA; panel [a] of Fig. 2.11) is a power amplifier that uses vacuum tube technology. Unlike the magnetron, which oscillates at its resonant frequency, the klystron is a power amplifier that requires an RF signal to be provided as an input. The klystron tube amplifier takes a low-power input RF signal and amplifies it. Because the klystron is a power amplifier, the amplitude and phase of the output signal can be controlled via the low-power RF input signal. The klystron uses resonant cavities, which limits the range of frequencies for which it can operate to a relatively narrow frequency band. Klystron amplifiers are capable of generating high peak powers (>1 MW) but are typically limited to relatively low duty cycles ($<10\%$).

Another type of tube-based power amplifier found in RF systems is the traveling-wave tube amplifier or TWTA (panel [c] of Fig. 2.11). TWTA designs are available for frequencies from 300 MHz to over 50 GHz. TWTA output powers can also vary from watts to tens of kilowatts or more. An advantage of the TWTAs is that they can operate over a wide bandwidth (the bandwidth can extend octaves for the helical configuration and may be 10–20 percent of the center frequency for cavity configurations). TWTAs are also commonly found in communication system transmitters where high-duty cycles are required.

The transmitter power sources mentioned so far – magnetrons, klystrons and traveling-wave tubes – all require high-voltage power supplies with voltages from multiple kilovolts to 100 kilovolts or more. High-voltage power supplies are complex devices and have potential hazards for the staff responsible for their maintenance. Although the actual power-amplifier elements themselves may be small, the power sources and support electronics can result in large, complex systems. As an alternative, solid-state RF power amplifiers for radar applications have recently become a viable alternative to these other high-voltage solutions.

Useful output power from solid-state systems at RF frequencies has largely been due to advances in semiconductors. Gallium arsenide (GaAs) and gallium nitride (GaN) have increased power density and efficiency to enable cost-effective RF power amplifiers at microwave frequencies. These solid-state power amplifiers rely on low-voltage power supplies commonly used by the other components of the radar system. An example of solid-state RF amplifiers power combined to increase the total RF power is shown in panel (d) of Figure 2.11.

The peak powers achieved with solid-state amplifiers are generally an order of magnitude lower that those achieved with the high-voltage counterparts previously discussed, but solid-state amplifiers can be operated at high duty cycles. From a review of the radar equation, the radar's sensitivity is related to the amount of energy that is transmitted ($E_{tx} = P_t T_{tx}$). To successfully utilize solid-state technologies as power amplifiers for weather radar, pulse-compression techniques have been adopted [13–15]. (Pulse compression is discussed in Section 6.5.) By using pulse-compression techniques and operating the solid-state transmitters at higher duty cycles, the solid-state power amplifiers (SSPAs) can transmit sufficient energy without sacrificing the radar's range resolution so that the radar systems achieve the required sensitivity. Although solid-state amplifiers may have a lower peak power than the other transmitters discussed, they can achieve similar transmitted energy with higher operating duty cycles and advanced signal processing.

SSPA transmitters provide characteristics that can be advantageous for numerous applications. They do not require high-voltage power systems, and their designs can be lightweight and compact. The solid-state integrated circuits that make up the power amplifier also enable modular configurations. These integrated circuits can be combined to increase the total available transmission power. For phased-array systems, where a physically distributed array of transmitting elements is desired, solid-state power amplifiers provide a cost-effective and compact solution. The combination of multiple lower-power devices to make a higher-power system inherently provides fault tolerance (albeit with degraded performance).

2.6.3 Antenna Systems

The antenna is the radar system's mechanism to transition the energy generated in the transmitter to free space (or from free space to the receiver). The antenna provides an electrical "match" from the nominal 50-ohm impedance of the radar system's RF components to the 377-ohm intrinsic impedance of the free space. Just as important for dual polarization weather radar, the antenna determines the radar's polarization characteristics. All antennas inherently have a polarization. In dual polarization radar systems, the polarization characteristics of the antenna are controlled and used toward a specific architecture, such as those presented in Section 2.2. In general, the antenna is the main device that controls the polarization state of the radar.

Antenna theory and technology is a well-developed field, with numerous textbooks and journals devoted to the latest research. The presentation here is a brief introduction in the context of weather radars. Figure 2.12 shows examples of a subset of antennas types (and types of antenna feed horns) that are used in dual polarization weather

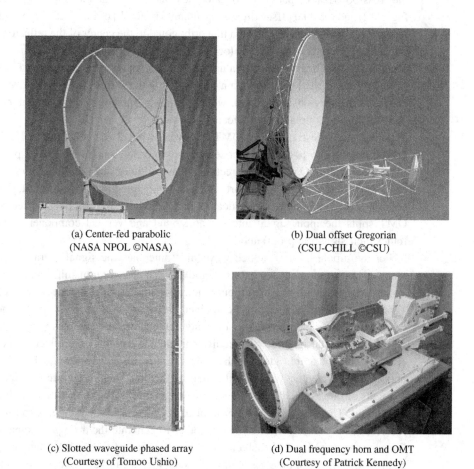

(a) Center-fed parabolic
(NASA NPOL ©NASA)

(b) Dual offset Gregorian
(CSU-CHILL ©CSU)

(c) Slotted waveguide phased array
(Courtesy of Tomoo Ushio)

(d) Dual frequency horn and OMT
(Courtesy of Patrick Kennedy)

Figure 2.12 Common weather radar antenna types and a dual frequency, dual polarization feed horn with OMT.

radar. For a reflector antenna, an electromagnetically reflective surface is used like a mirror to focus and direct the energy between the free space and a feed. The feed horn is the point where the transmitted and received fields of the radar are generated and collected, respectively. The reflector dish acts to collimate and focus the antenna beam, ultimately forming the antenna's pattern and determining its beamwidth. Reflector antennas are relatively simple and robust designs and provide excellent performance due to their legacy of scale manufacturing. For designs like the dual offset Gregorian antenna, multiple reflectors may be used.

The reflector antenna requires the feed horn (e.g., panel [d] of Fig. 2.12) to be placed in front of the antenna at a focal point; the forward distance F is defined by the parabolic surface curvature. For most parabolic antennas, the ratio for forward distances to diameter (also referred to as "F over D"), F/d_a, is 0.25–0.8 [16]. While D is commonly used to represent the antenna diameter, d_a is used here to differentiate it from the median drop diameter. The structures (or struts) to hold the feed in place and the feed itself can interact with the signal and distort the antenna's pattern (especially the cross-polarization pattern). Consider the center-fed parabolic antenna (panel [a] of Fig. 2.12) and the dual offset Gregorian antenna (panel [b] of Fig. 2.12). The center-fed antenna has its feed placed along a line orthogonal to the center of the reflector, where the peak of the antenna pattern is located, and the struts run across the entire antenna. The dual offset Gregorian antenna uses two reflectors and places the feed focal point outside of the direction of the main beam so that there is no interference within the antenna's primary observing area, which results in improved antenna sidelobe levels and reduces cross-polarization effects.

The orthomode transducer (OMT), shown in panel (d) of Figure 2.12, located behind the feed horn, combines or separates two orthogonal polarizations paths. The linear polarization OMT is a bidirectional waveguide device with three ports: a combined port, a horizontal signal port, and a vertical signal port. On transmit, the OMT combines the horizontal and vertical signals together to be output. On receive, the OMT splits the input signal into the horizontal and vertical components and then routes their signals appropriately.

For the dipole horn or slotted waveguide antennas, the signal is transmitted and received by the antenna element itself; the antenna directly transmits and receives the signal from the environment without requiring interactions with a dedicated reflector. Typically, multiple elements are combined as part of a larger system to generate a desired antenna pattern. An array of many simple antenna elements (e.g., dipoles or patches) can be designed to have equivalent antenna patterns as a reflector antenna.

The phase and amplitude of each element are precisely controlled so that the contributions of each element work together to form the final antenna pattern. The individual elements of the array are typically simple antenna designs, such as a patch, dipole, or waveguide. The amplitude and phase of each of the AESA's elements (on transmit, receive, or both) are then combined together to generate the array's antenna pattern. By adjusting the phase of each element, the direction of the AESA's antenna pattern can be steered. Amplitude weighting of the elements can be employed to control the antenna's side-lobe levels below the desired limits.

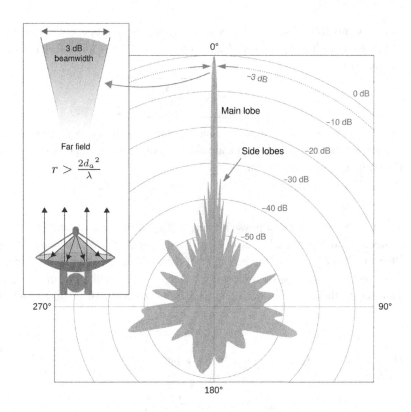

Figure 2.13 A typical two-dimensional normalized pattern of a weather radar antenna in polar coordinates (azimuth plane). The 3-dB beamwidth is $\sim 1°$. The inset diagram on the left illustrates how the reflector collimates and focuses the wave from the feed horn. The resulting antenna pattern on the right is valid for the far field. Close to the antenna, the full area of the reflector contributes to focusing the wave.

The gain of the AESA is generally determined by the peak boresight gain of the individual element multiplied by the total number of elements used to form its beam. The gain for the transmit and receive may be different. As one might expect, AESAs are complex systems compared with a simple reflector antenna. The complexity can greatly increase the radar system's capabilities but also the cost. With multiple active elements in the AESA, calibration of the antenna may be required periodically, and compensation for component performance drifts (e.g., due to temperature variations) may be required.

For most weather radar systems, the transmitting and receiving antennas are the same (i.e., a monostatic configuration). Figure 2.13 is an example cut through an antenna's illumination pattern (here, the cut is across the antenna's azimuth plane). The pattern illustrates the antenna's one-way gain with respect to changes in the azimuth angles from the antenna's boresight direction. For monostatic weather radar systems, the one-way 3-dB beamwidth defines the azimuth and elevation extent of the V_6 radar volume.

The beamwidth of an antenna is generally inversely proportional to the diameter of the antenna. The half-power beamwidth (HPBW; in radians) of the antenna is approximated as

$$\boxed{\theta_{\text{HPBW}} = \frac{k_a \lambda}{d_a},} \tag{2.40}$$

where d_a is the antenna's diameter, and k_a is a factor that depends on the antenna's architecture and efficiency (a typical value for parabolic antennas is $k_a = 1.22$ [17]). The gain of a parabolic antenna is

$$G_0 = \frac{4\pi A_a e_a}{\lambda^2}, \tag{2.41}$$

where A_a is the antenna's physical area, and e_a is the antenna aperture's efficiency between 0 and 1 (a typical value of e_a is 0.6–0.7 for parabolic antennas). The aperture efficiency is a function of the geometry of the antenna, the illumination area, and the shape and roughness of the reflector's surface. For a circular reflector antenna with diameter d_a, the antenna's area is $A_a = \pi d_a^2/4$. The effective antenna area is then

$$A_e = e_a A_a = \frac{e_a \pi d_a^2}{4}. \tag{2.42}$$

The typical boresight antenna gain for a parabolic reflector ($e_a = 0.7$) is approximated as

$$G_0 \approx 0.7 \left(\frac{\pi d_a}{\lambda} \right)^2. \tag{2.43}$$

Combining these approximations for a parabolic reflector, the antenna's gain and beamwidth are related as

$$G_0 \approx 10.3 \left(\frac{1}{\theta_{\text{HPBW}}} \right)^2. \tag{2.44}$$

These approximations are valid for observations in the antenna's far field, that is, at ranges larger than the Fraunhofer distance r_{ff} [17]:

$$r_{ff} = \frac{2 d_a^2}{\lambda}. \tag{2.45}$$

The far field also requires that $r_{ff} \gg \lambda$ and $r_{ff} \gg d_a$. For scatterers at ranges that are closer than the far-field distance, the antenna's pattern and gain can vary as a function of range, and near-field antenna patterns are required for interpreting the received signals.

Consider a 1.0° one-way, HPBW, center-fed parabolic antenna with $k_a = 1.22$ and $e_a = 0.7$. At the S band (2.9 GHz), the antenna is approximately 7.2 m in diameter. To achieve the same one-way beamwidth at the X band (9.4 GHz), the antenna is 2.23 m in diameter. If the antenna with $d_a = 7.2$ m is used at the X band, a one-way beamwidth of approximately 0.31° is achieved. For an S-band radar with a 7.2-m-diameter antenna, the far field typically starts at around $r_{ff} \approx 1000$ m, whereas the X-band 2.23-m antenna's far field starts at $r_{ff} \approx 311$ m.

Dual polarization antennas are specifically designed with two objectives: (1) to provide nearly identical patterns at the two orthogonal polarizations and (2) to isolate orthogonal polarizations from one another. In particular for weather radar, minimizing the cross-polarization leakage and maintaining identical patterns at the two polarizations is important for analyzing weather observations. The differences in the observed signals between two polarizations are expected to be due to the precipitation and not the antenna patterns. If the copolar signal leaks into the cross-polar signal because of the antenna, this limits the radar's ability to accurately measure the cross-polarization signature of the scatterers (for the LDR) and can even bias measurements when comparing the dual polarization copolar signals, in particular, the differential reflectivity. For reflector antennas, an OMT enables the two polarizations. The OMT is a waveguide device that combines or separates two orthogonal polarizations. In antennas such as slotted waveguides, dipoles, or patch arrays, the linear polarizations' directions are defined by mechanically orienting the antenna elements 90° to each other. For radar systems using a reflector antenna, the system's polarization isolation depends on the performance of all elements, including the OMT, the feed horn, and the reflector and its mechanical supports. For the center-fed parabolic antenna, the mechanical struts that hold the feed horn in place are sources of cross-polarization that reduce the polarimetric isolation of the system [18].

The antenna pattern's amplitude and phase (F) are characterized as a function of elevation (θ) and azimuth (ϕ). For dual polarization, this gives us two copolar patterns, F_{hh} and F_{vv}, and two cross-polar patterns, F_{vh} and F_{hv}. F_{hv} is the amplitude and phase of the horizontally polarized radiated electric field when the vertical antenna port is energized, and vice versa for F_{vh}. Ideally for dual polarization antennas, the copolar and cross-polar antenna patterns are the same for the horizontal and vertical polarizations, giving $F_{co} = F_{hh} = F_{vv}$ and $F_{cx} = F_{vh} = F_{hv}$.

Recall that the antenna pattern is proportional to the electric field, and the antenna gain is proportional to power. The antenna's gain, as a function of look angle, is

$$G(\theta, \phi) = |F(\theta, \phi)|^2. \qquad (2.46)$$

A parabolic antenna ideally has the boresight along look angle $\theta = 0$ and $\phi = 0$, with the peak copolar gains being equal for both polarizations, $G_{h0} = G_{v0}$. The gain of the antenna is included in F. It is sometimes useful to define a normalized antenna pattern (so that the peak gain is 1, or 0 dB, as in Fig. 2.13):

$$f(\theta, \phi) = \frac{F(\theta, \phi)}{\sqrt{G_0}}. \qquad (2.47)$$

The copolar and cross-polar patterns are normalized by their respective polarization's copolar peak gains (e.g., $F_{co} = \sqrt{G_0} f_{co}$ and $F_{cx} = \sqrt{G_0} f_{cx}$).

The dual polarization performance of the antenna is typically quantified by its cross-polarization isolation. If the polarimetric antenna patterns are available for multiple planes (or "cuts") through the measurement space, the two-way (including the antenna's effects for both transmit and receive) integrated cross-polar ratio (ICPR$_2$) is calculated as follows [19]:

$$\text{ICPR}_2 = \frac{\iint |f_{hh} f_{vh} + f_{hv} f_{vv}|^2 \, d\Omega}{\iint |f_{hh}^2 + f_{hv}^2|^2 \, d\Omega}, \qquad (2.48a)$$

where $d\Omega$ is the elemental solid angle ($d\Omega = \sin\theta d\theta d\phi$ in spherical coordinates) and the integration if over the sphere's surface. If we assume the copolar and cross-polar patterns are the same for the horizontal and vertical polarizations, the equation simplifies to

$$\text{ICPR}_2 = \frac{4 \iint |f_{co} f_{cx}|^2 \, d\Omega}{\iint |f_{co}^2 + f_{cx}^2|^2 \, d\Omega}. \qquad (2.48b)$$

Although the antenna patterns are complex valued (the ICPR$_2$ is calculated using both the magnitude and phase), often only the magnitudes of the antenna patterns are available. Without the phase information, an upper bound of the ICPR$_2$ can be calculated as

$$\text{ICPR}_2^{(ub)} = \frac{4 \iint (|f_{co}| \, |f_{cx}|)^2 \, d\Omega}{\iint (|f_{co}|^2 - |f_{cx}|^2)^2 \, d\Omega}. \qquad (2.48c)$$

For the CSU-CHILL X-band patterns shown in Figure 2.14, the ICPR$_2$ is calculated to be -42.0 dB. With only the antenna patterns' magnitude, the ICPR$_2^{(ub)}$ is calculated as -39.0 dB for the same antenna. The 3-dB difference is due to the effects of the antenna phase, which result in a slight decorrelation between the copolar and cross-polar patterns. A dual polarization antenna should have ICPR$_2 < -30$ dB to ensure negligible bias in the radar's dual polarization measurements for STSR mode (in particular, Z_{dr} and ρ_{hv}) [20]. The ICPR$_2$ determines the the LDR limit of the antenna for the ATSR mode (the antenna's LDR limit is discussed in Section 7.4.1). For interested readers, an in-depth examination of the polarimetric antenna pattern and its effect on the performance of dual polarization weather radar can be found in [18–21], and the references within.

The majority of radar systems operate with the antenna's surface covered by a radome. The radome protects the antenna and feed from rain, snow, and other debris and shields it from the forces of the wind. Consider the additional requirements placed on the antenna structure and the antenna's positioner systems if it must overcome the forces required to move a large antenna (which acts like a sail) in high winds. Ideally, radomes have a hydrophobic coating to limit the size of the droplets and the amount of water or ice adhering to the radome. Especially for high-frequency radars, the precipitation that adheres to the radome can add additional attenuation to the signal (see discussion in Section 7.4.3), which can also include polarimetric biases (e.g., the attenuation can be different between the horizontal and vertical polarizations as a result of the vertically oriented streams of water running down the radome). The radome's seams and brackets can also introduce biases in the dual polarization observations.

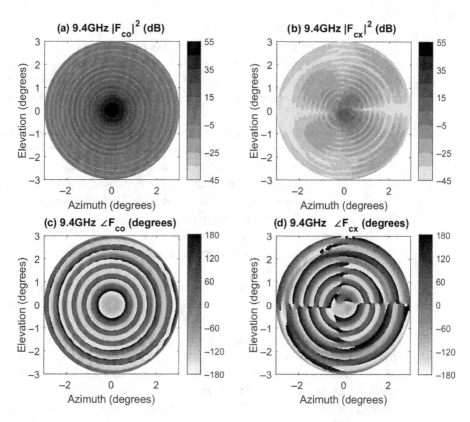

Figure 2.14 Antenna patterns from the CSU-CHILL's 8.5-m dual offset Gregorian measured at the X band (9.4 GHz). The patterns' magnitudes (in decibels) are $|F_{co}|^2$ (a) and $|F_{cx}|^2$ (b). Similarly, the antenna patterns' phases, $\angle F_{co}$ (c) and $\angle F_{cx}$ (d), are shown. Note that the cross-polar magnitude is lowest along the principal planes (i.e., the $0°$ cut, where elevation is zero and azimuth is varied, and the $90°$ cut, when azimuth is zero and elevation is varied). The cross-polar magnitudes exhibit increased antenna side lobes along both the $45°$ and $135°$ cuts. Also note that an odd symmetry about the boresight is observed for the cross-polar phase (d).

2.6.4   Duplexer

The duplexer allows the monostatic system to transmit and receive using a single antenna while providing isolation between the transmitter and receiver. Isolation between the transmitter and receiver is necessary to ensure that the transmitted power is routed to the antenna, the incoming echoes are routed to the receiver, and the amount of transmitter power that "leaks" into the receiver is sufficiently low to ensure that the receiver's electronics are not damaged. The transmitter's peak power levels can be on the order of 30 dBm (1 watt) to 90 dBm (1 megawatt), whereas the power of the received echoes can be as low as -110 dBm (10 femtowatts). The large differences between the peak transmitter's power and the received echoes' power can present a significant practical challenge.

Most of the monostatic pulsed radar systems operate using time-division multiplexing: this means the radar system is in either transmitting or receiving mode (but not both). Transmitting power and receiving echoes at the same time is not possible with time-division multiplexing. This is to ensure there is sufficient isolation between the two paths to prevent damage to the receiver electronics and ensure that maximum power is transmitted.

One of the most common elements of a radar duplexer is a ferrite circulator. The circulator is a nonreciprocal device with three RF ports, which ideally only allows RF energy to travel in one direction: from the transmitter to the antenna and also from the antenna to the receiver (the circulator also allows power from the receiver port to the transmitter port, but there is no RF power source in this direction). In practice, the circulator typically provides a low forward-insertion loss and high reverse isolation of 25 dB or more. When additional isolation is required, additional components are included in the duplexer to increase the isolation between the transmitter and receiver paths.

As part of the duplexer, a power limiter may be used to protect the sensitive electronics in the receiver. The power limiter is designed to prevent higher-than-expected power, which may enter the receiver path, from damaging the radar's receiver. The excess receive power may be due to large targets close to the radar or RF interference from other RF transmitters, such as wireless communication equipment or even other radars. Examples of power limiters include diode semiconductors and gas-discharge tubes. Diodes are used to limit the peak voltage in the receiver but have limited power-handling capability. For high-RF-power applications, ionized gas-discharge tubes are frequently used.

2.6.5   Receiver

The antenna collects the energy from the incoming signal (e.g., backscatter from precipitation) and routes each incoming polarization's signal to its respective antenna port. From the antenna port, the received signal passes through the duplexer and into the radar's receiver. The receiver includes all of the electronics to amplify, frequency convert, filter, and finally digitize the received signals. For radar architectures that simultaneously receive both polarizations, each polarization has its own receiver electronics, which are typically identical copies for each channel.

The receiver's components are selected to optimize the receiver gain so that the expected signal- and noise-power input from the antenna are within the range of powers that the digitizer can accommodate. The digitizer is an electronic device that periodically samples the analog voltage signal and converts it to a digital count that is proportional to the sampled voltage's magnitude. The digitizer is also referred to as an *analog-to-digital converter*. The dynamic range is the ratio of maximum to minimum power that the receiver can operate. The receiver's components, each with its individual gains and losses, are selected to maximize the receiver's dynamic range. As the receiver's input power becomes too high, components of the receiver can start to "compress," and the gain decreases. When the receiver's components are in compression, there is no longer a constant relationship between the change in input power and output power. Operating the radar receiver in the compression region is usually avoided for weather radars, and the gain should ideally remain constant over its dynamic range. Once the signal is digitized, digital signal processing can further increase the dynamic range of the radar system beyond the analog dynamic range of the receiver.

One of the purposes of the radar receiver elements is to convert between intermediate frequencies (IFs) and the radar's operating RF, which contains the signal's information. The radar's IFs are selected to accommodate available hardware resources, such as the digital-to-analog and analog-to-digital converters and signal-processing electronics. LO frequencies are constant frequencies that aid in frequency conversion only (they do not provide any new information about the transmitter's or receiver's signal). Although less common in modern radar designs, multiple frequency conversion stages may be used, which require multiple LOs (as shown in Fig. 2.9). The frequency conversion is typically accomplished using a mixer that effectively multiplies the two signals, one of which is the LO. The transmitter's frequency converter is referred to as an *upconverter* (because the frequency is increased), and the receiver's frequency converter is referred to as a *downconverter*. (Section 5.1.5 considers the mathematics and signal theory of frequency conversion.)

For coherent radar systems, a common reference clock is required between the radar's clocks (e.g., radar timing, sampling clock, LOs). The same LO and sampling clocks are commonly split to provide symmetry in the transmitter and receiver frequency converters. Doppler radars rely on the receiver and transmitter to be phase coherent, so the transmitted pulse's phase is determinstic with respect to the receiver's sampling time. The phase coherence (or phase lock) ensures that the phase difference between the system's transmitter and the receiver is constant, and therefore any measured change in phase is a property of the echo.

2.6.6   Radar Mechanical Structure and Pointing Control

Ground-based weather radar systems are typically designed to scan the hemisphere above the ground (some research radars make fixed, vertical-pointing observations). For operational scanning systems, the azimuth is typically scanned at a fixed speed, and the elevation is stepped in discrete intervals once per azimuth revolution. The specific elevation intervals are determined by the volume coverage pattern (VCP). Some radar systems require additional agility, with capabilities to provide full PPIs, sectors of a PPI, RHIs, fixed pointing, or combinations of all of these with various requirements for timing and scanning speeds.

For mechanically scanning systems, the radar system's pedestal anchors the radar to a solid platform (e.g., roof, tower, or ground), and the pedestal provides the mechanisms for full 360° motion in the azimuth direction. Elevation scanning from 0 to 90° or even 0 to 180°, which allows for horizon-to-horizon vertical slices, is possible. The maximum scanning elevation may be limited to something less than 90° if the radar application does not require this full range of motion.

For scanning radar systems, the mechanical assembly that does the scanning is balanced around the rotation axis (usually with the addition of counterweights) to minimize the stress on the motion-control system in static states. The motion control must still be designed to overcome the inertia of accelerating the mass, which can require substantial electrical power and mechanical strength in the structure.

For unrestricted azimuth operation, no fixed wires can be used, and a slip ring and/or waveguide rotary joint is needed to provide electrical or RF connections (the exact configuration depends on the location of the receiver or transmitter). These let the pedestal rotate continuously without concern for wrapping or binding wires or waveguides between the stationary platform base and the rotating antenna. Depending on the radar's design, one or both may be needed in various combinations to achieve the desired range of motion for the system. Slip rings in the pedestal are used to provide electrical paths for data communications and power. Modern solid-state and low-power radars frequently carry both the transmitter and the receiver on the rotating subsystems to reduce RF losses [15, 22].

The integration time, sometimes referred to as the *coherent processing interval* (CPI), indicates the duration over which N signal samples are collected for estimation of the volume's radar variables (Fig. 2.15). For a mechanically scanning system, the angular extent that is covered (ψ_{ant}) for a given scan rate ($\dot{\psi}_{ant}$) and pulse-repetition period (T_s) is

$$\psi_{ant} = \dot{\psi}_{ant} \cdot N \cdot T_s. \tag{2.49}$$

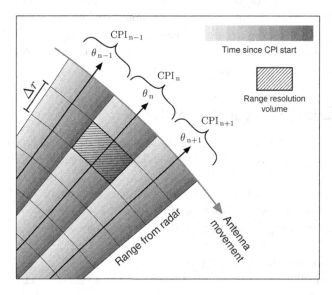

Figure 2.15 The received signal is integrated over multiple samples to estimate the radar variables. The integration time of the CPI is shown here for a scanning antenna. Each CPI includes a number of radar pulses that are averaged, and the estimated values within echo range resolution volume are typically assigned to the midpoint of the volume, both in range and angular positions (e.g., θ_n).

Ideally, the radar antenna's scanning speed and integration time should be selected so that no more than half of the antenna's half-power beamwidth is traversed during the integration time (i.e., $\psi_{ant} < \theta_{HPBW}/2$). This is to limit decorrelation of the observation volume due to the antenna motion (which is discussed in Section 5.7). The antenna's speed can vary during the scan. Radar signal processors typically provide the option to process over fixed angular resolutions (e.g., every 0.5° or 1°, allowing N to vary) or for a fixed number of pulses (e.g., $N = 32$ or $N = 64$ and accepting the resulting variations in angular resolution). As an example, for a radar scanning at $\dot{\psi}_{ant} = 18°$ s^{-1}, $N = 64$ samples, and a pulse-repetition period of $T_s = 1$ ms, the antenna rotates $\psi_{ant} = 1.15°$.

Electronically scanned systems can significantly deviate from their mechanically scanned counterparts, in both operational capacity and the design of the structure and pointing-control system. A fully electronically scanned system does not require any moving parts. The phased array can selectively sample volumes on a pulse-by-pulse basis. The phased-array system enables electronically steerable antennas, increasing the speed over which a volume can be scanned.

Mechanically controlled systems have a fixed antenna aperture (the beamwidth is fixed by the antenna design), and the pointing direction of the antenna is determined by the positioner system. Changing the pointing angle requires physically moving the antenna. The speed at which the antenna can be pointed to a new direction is largely a function of the antenna's mass and size and the capability of the positioner. For weather radars, the acceleration of the scanning system can be between 1°/s^2 and 60°/s^2, and the radar scanning rates can be between 18°/s and 60°/s (or 3–10 RPM).

For fully electronically steered antennas, the beam can be pointed to any direction within its field of view on a pulse-to-pulse basis (or nearly so), enabling an almost instantaneous scanning rate. With the use of beamforming techniques, multiple radar volumes can be observed at the same time. With intelligent scanning algorithms, the AESA radar can also track storms of interest (e.g., mesoscale convective systems) at high temporal rates while still providing surveillance of the entire domain at a lower (or standard) rate. A hybrid compromise, with electronic scan in the elevation and mechanical scan in the azimuth, is evolving as a more cost-effective compromise in weather radars. Although electronically scanned systems have operational advantages, mechanically scanned systems provide the most cost-effective solution for general-purpose weather surveillance.

2.6.7 Data System

The earliest techniques of recording weather radar observations were drawings or tracings by the operator or photographs of a cathode-ray tube's output taken at regular intervals. Modern computers revolutionized radar processing and data archiving, trading analog for digital. As digital data storage became more affordable and computer processing allowed, radar moments could be archived to magnetic tapes on a limited basis. As this trend of decreasing data-storage costs and increasing data-processing power continued, the fidelity of the radar observations similarly improved, and continuous data archiving became standard practice. For some radars (typically, research radars that are only periodically operated), the raw instrument data are archived and available for future reprocessing. This allows for new signal-processing algorithms to be applied to legacy data sets.

Radar data products are typically classified into "processing levels," where the higher the level, the more the data have been refined or processed. The higher-level data typically include calibration and unit conversions applied to the raw instrument data, as well as additional derived results from the lower-level data (e.g., rainfall rate), or placing data on a regular grid. The availability of these data to the general public varies by the desired data level and the radar's operator/owner. For example, the WSR-88D Level 2 and Level 3 data product archives are easily accessible, with data generally available within 1–3 days, depending on the data product. Chapters 6 – 10 cover the various aspects of signal and data processing used for weather radar.

Table 2.1 Example description of the levels of data processing. Each instrument may adopt variations of these descriptions.

Level 0	Raw, unprocessed instrument data. Calibration and geolocation information may be separate and at different sampling rates.
Level 1	The data from Level 0 are time referenced, with instrument and other ancillary information appended (e.g., radiometric calibration and georeferencing information). The ancillary data include the necessary information to process to higher levels, but no corrections or changes are made to the raw data at this point. The data are at the original resolution, without any data loss between Level 0 and Level 1. For some radar systems, Level 1 data are the lowest-level data available.
Level 2A	The geolocation and calibration information is applied to the Level 1 data at the original spatial resolution. Data are converted to the sensor's units (e.g., Doppler velocity, equivalent reflectivity factor). Data reduction may be performed (i.e., the Level 2A data cannot be reverted to Level 1 data because information, such as the uncalibrated raw data and calibration values, is missing). Data quality control is applied. Some processing systems may classify aspects of this processing as Level 1B. Note that Level 2 data are not always split into 2A and 2B processing.
Level 2B	Derived geophysical parameters in the instrument's spatial resolution and reference frame (e.g., range, azimuth, and elevation). Examples of derived geophysical parameters are attenuation corrected reflectivity, rainfall rate, and specific differential phase. Additional data quality control may be applied if necessary.
Level 3	Geophysical parameters from the instrument (e.g., Level 2 data) that have been spatially and/or temporally resampled (e.g., mapped to a Cartesian grid at specific time intervals). A common spatial and temporal sampling interval is typically used between multiple instruments of various types.
Level 4	Composite results from multiple sensors or model analysis from lower-level data. The data are typically at the same spatial and temporal grid resolution as Level 3 data. These Level 4 derived products are not directly from a single instrument (e.g., multi-Doppler velocity).

A general description of the data-processing levels is provided in Table 2.1. The specific details of the data-processing levels for a given instrument may vary. Additional letters may be used to define additional processing stages (1C, 2A, 2B, etc.). The data-processing levels build on prior processing levels. A single instrument can include process levels from 0 to 3. Multi-instrument data (or model results derived from the observations of a single instrument) are Level 4 data. Not all processing levels are used by all instruments. For some radar systems, the radar's signal and data-processing system may internally process the data so that the first data level archived is equivalent to the Level 1 data. The processed data are streamed to multiple users: radar operators and forecasters for real-time use, data archives for storage and future use, modern artificial intelligence (AI) applications for data discovery, and Level 4 data-processing systems for fusion with multiple sensors and models. The Level 4 data-processing systems can also provide nearly real-time data to applications as fused data products or near-term forecasts.

2.7   Radar System Sensitivity

There are a number of fundamental performance metrics of weather radar that are considered when selecting a system for a particular application. One of the most important considerations is the radar's sensitivity. The radar is typically limited in the dynamic range of the power (and therefore reflectivity) it can observe. These limitations are often due to system noise or transmitted power. In either case, a radar is designed to observed a range of reflectivities. The sensitivity of the radar is the lowest reflectivity the radar can observe and is usually given for a standard distance (e.g., 1 or 10 km).

The sensitivity also assumes different levels of signal-processing gains. As part of the design, the radar's minimum and maximum range and the range resolution must be defined. The selection of the range resolution has an impact on the reflectivities that can be observed. The selection of the maximum range must also be considered, along with the expected range of velocities the radar will need to unambiguously resolve.

These are some of the factors that must be considered when designing or operating a radar. Some of the choices are not limited by physics but by other constraints, such as available frequency bands or economics. This section discusses how the radar subsystems' performance metrics effect the radar's sensitivity.

2.7.1   Noise in Radar Systems

The radar receiver's signal includes noise in addition to the echo power. This noise power, P_n, determines the minimum detectable signal power of the receiver. This noise power comes from the environment, which is received by the antenna, but noise is also generated within the radar receiver's electronics. In the absence of any echo power (i.e., our signal), the power measured at the output of the receiver is noise. Interference from anthropogenic sources is neglected because it does not typically drive the estimates of radar system performance (other than making sure the radar can tolerate/survive it).

The noise power of a noise temperature (T_n) and a system bandwidth B is

$$P_n = k_B T_n B, \tag{2.50}$$

where $k_B = 1.380649 \cdot 10^{-23} \, \text{J} \, \text{K}^{-1}$ is Boltzmann's constant. For determining receiver noise for the minimum detectable signal, B is the receiver filter's bandwidth. The total noise temperature is a function of the radar system and the environment. This noise temperature is largely dominated by the receiver's physical temperature but can be influenced by the background temperature in the direction the antenna is pointing (e.g., Earth's surface, the troposphere, or space).

Each component has its own noise contribution. Some can be negligible, some significant. The noise factor, F, defines the ratio of the input signal-to-noise ratio (SNR) to the output SNR as

$$\boxed{F = \frac{\text{SNR}_{\text{in}}}{\text{SNR}_{\text{out}}},} \tag{2.51}$$

where $F \geq 1$. This means that the SNR of the output is the same as or lower than the SNR at the input of the device (without changing the bandwidth). The noise figure (NF) is typically used to describe a component's effect on the system's SNR and is simply

$$\text{NF} = 10 \log_{10}(F). \tag{2.52}$$

For radar systems, the noise contribution of each component in the receiver's path can be cascaded together. The cascaded noise figure is used to determine the overall impact it has on the SNR between the antenna and the receiver's output (the output is typically sampled by the digitizer). The SNR determines the radar's ability to detect a given signal power, and therefore the radar system's performance is strongly tied to the receiver's noise figure.

The noise figure is defined with an input reference temperature of $T_0 = 290° \, \text{K}$. If the physical temperature of the device is varied, the input noise temperature and therefore noise factor are also varied. The additional noise temperature contribution of the system (T_{sys}) and the noise factor, F, are related as

$$F = \frac{T_{\text{sys}}}{T_0} + 1 \tag{2.53a}$$

or

$$T_{\text{sys}} = T_0(F - 1). \tag{2.53b}$$

The noise power can be calculated using the system and reference temperatures, or equivalently the noise factor, as

$$P_n = k_B(T_{\text{sys}} + T_0)B = k_B T_0 F B \tag{2.54}$$

Note that the radar signal's bandwidth, B, is typically the only parameter available to the radar operator once the radar is designed and built.

A radar's receiver consists of a number of devices where the signal is routed in series from one to the next from the input to the output. For a cascade of devices, starting with the first device ($n = 1$) and moving through all N devices, the cascaded noise factor is

$$F = F_1 + \frac{F_2 - 1}{G_1} + \frac{F_3 - 1}{G_1 G_2} + \cdots + \frac{F_N - 1}{\prod\limits_{n=1}^{N-1} G_n}, \tag{2.55}$$

where G_n denotes the gain (in linear units) of each stage. Note that for some components, such as a filter or attenuator, their gain is less than 1 (or negative in decibels). For these components, a fraction of the input signal power does not make it to the output because it is either absorbed or reflected back toward the input by the device. Also note that for these passive components (no electrical power other than the signal goes into the device), the gain and noise factor are the inverse of one another. The cascaded gain is simply the product of the gains of all components in the series:

$$G = \prod_{n=1}^{N} G_n = G_1 G_2 G_3 \cdots G_N. \tag{2.56}$$

The output signal power through the series of cascaded devices is the input signal power $P_s^{(\text{in})}$ multiplied by the total gain of the entire set of components:

$$P_s^{(\text{out})} = P_s^{(\text{in})} \cdot G. \tag{2.57}$$

With an input noise power of $P_n^{(\text{in})}$, the noise power at the output of the series of cascaded devices is

$$P_n^{(\text{out})} = P_n^{(\text{in})} \cdot F \cdot G. \tag{2.58a}$$

Like the output signal power, the output noise power includes the total gain of the series, but it also includes an increase in noise power due to the cascaded noise factor in the signal path (represented by F). The output SNR ($P_s^{(\text{out})}/P_n^{(\text{out})}$) is reduced compared with the input SNR by the noise factor (following its definition in eq. [2.51]).

A low-noise amplifier (LNA) is selected to provide amplification with a low-noise factor. With careful design, the first LNA dominates the receiver's overall noise figure. Any attenuation or loss before the first gain stage adversely affects the receiver's noise figure. This is why the first LNA is typically placed as close to the antenna feed as possible (on satellite dish receivers, you can find the electronics at the feed). For X-band radars, where losses are higher per length of waveguide, the electronics are again placed close to the antenna ports, usually on the back of the antenna if possible. For S-band radars, the waveguide losses are typically easier to tolerate (and the electronics are large), allowing (necessitating) longer distances between the antenna and receiver electronics.

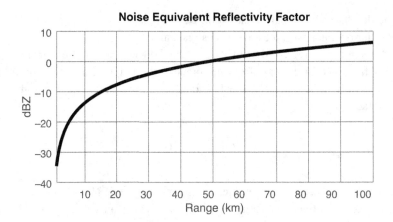

Figure 2.16 An example of the noise-equivalent reflectivity factor, Z_n, as a function of range for the CSU-CHILL S-band radar. Whereas the noise power P_n is constant for all ranges, Z_n varies in proportion to the range-squared (i.e., $Z_n \propto r^2$).

2.7.2   Noise-Equivalent Reflectivity Factor

When describing the radar system's performance, a metric that is used is the minimum detectable signal for a single pulse. This is frequently described as a minimum detectable reflectivity factor at a given range (1- or 10-km ranges are common). With the radar system's noise power P_n, a noise-equivalent reflectivity factor, Z_n, can be calculated from eq. (2.39c) as

$$Z_n = \frac{P_n r^2}{C}, \tag{2.59}$$

where weather radar constant C is calculated from eq. (2.39d). The noise-equivalent reflectivity factor is a measure of the single-pulse-detection performance of the radar and is a function of range, r. (This provides a performance metric to consistently evaluate radar performance for only one pulse. Improved detection can be achieved though signal processing techniques.) Figure 2.16 illustrates the noise-equivalent reflectivity factor for a constant noise power, $P_n = -105$ dBm. To estimate the noise-equivalent reflectivity factor, this example uses $G_{rx} = 1$, $P_t = 87$ dBm, $G = 43$ dB, $|K_w|^2 = 0.934$, $T_{tx} = 1 \ \mu s$, $\theta_1 = \phi_1 = 1.1°$, and $f_0 = 2.725$ GHz. (Note that P_n assumes the signal's nominal bandwidth and includes the receiver's noise figure.)

For the design of weather radar systems, a typical objective is to maximize the sensitivity of the radar so that the radar can detect the weakest signal possible. The radar's sensitivity can be increased by reducing Z_n. Some of the parameters reduce the noise power, P_n, whereas other parameters act to increase the signal power, $P_s = P_o$. It is possible to evaluate how the parameters affect the system's SNR, which is directly related to the radar's sensitivity:

$$\text{SNR} = Z_e \frac{G_{rx} P_t G^2 \pi^3 |K_w|^2 c T_{tx} \theta_1 \phi_1}{10^{18} \lambda^2 1024 \ln 2} \frac{1}{r^2 k_B T_0 B F}. \tag{2.60}$$

The constants are neglected to focus on the radar system parameters that can be affected by the radar's design. From Section 2.6.3, the antenna's gain and beamwidth are related to one another following $G \propto 1/(\theta_1 \phi_1)$. After reduction of the previous equation, the parameters that can be controlled by the radar design or operation and how they affect the radar's sensitivity through consideration of the following relationship:

$$\text{SNR} \propto \frac{P_t G T_{tx}}{\lambda^2 r^2 B F}. \tag{2.61}$$

Some radar systems have constraints on the largest antenna size that can be accommodated. The antenna size and λ directly affect the antenna gain G. From eq. (2.61), it is clear that the SNR increases by increasing the amount of energy transmitted, either through a higher peak transmitter power P_t or through a longer duration of the transmitted pulse T_{tx}. (This is one motivation for using pulse-compression techniques to enhance radar sensitivity, as discussed in Chapter 6.) The SNR can also be increased by reducing the noise factor F, reducing the bandwidth of the radar signal B, reducing the operating range r, or reducing the wavelength λ. (Note that the scattering behavior is wavelength dependent and can be nonlinear. This is covered in Chapter 4.) Although not directly evident in this equation, the implementations and performance of the technologies are interdependent as well (e.g., the relationship between wavelength, antenna size, and antenna gain). If a network of radars can be used, the maximum range from any radar to a given volume can be reduced. (This and other performance characteristics of the radar network are covered in Chapter 10).

The minimum detectable reflectivity may be lower than the noise-equivalent reflectivity factor. The weather radar's minimum detection sensitivity includes additional signal-processing gains that enable the radar to estimate radar variables when SNR < 0 dB. A common technique to improve the minimum detectable reflectivity is to use noise correction (see Chapter 7). With noise correction and an integration time with N pulses, the radar's sensitivity, or its minimum detectable reflectivity, is approximated as

$$Z_e^{\min} = \frac{Z_n}{\sqrt{N}}. \tag{2.62}$$

When designing (or acquiring) a weather radar, the process typically starts with a cost cap and other practical implementation boundaries, such as limits on the available locations to deploying the radar and its social footprint. The availability of dual polarization capabilities for attenuation correction, solid-state transmitter technology, and the concept of networked radar systems have fundamentally altered the solution space for weather radars and have made higher-frequency bands practical (and attractive) options.

2.8   Selected Problems

1. What is the time delay of a radar echo that is received from a scatterer at a range of 250 km from the radar? What is the highest pulse repetition frequency that can be used for an unambiguous range of 250 km?
2. Calculate the altitude of the radar beam at 150-km slant range when the antenna is pointing at $1.0°$ elevation angle using eq. (2.24). Compare the result with the altitude obtained using a instead of a_e, that is, neglecting the vertical variation of the refractive index.
3. Assuming a simple pencil beam cone model, how wide will the antenna beam be at ranges of 40 km, 60 km, 100 km, 200 km and 300 km? For a range resolution of 150 meters, what is the radar observation volume in cubic kilometers?
4. A radar with a $1°$ beamwidth is scanning at a low elevation of $1°$ (to avoid local obstruction). How high will the beam center be at a distance of 40 km, 60 km, 100 km, 200 km and 300 km? To account for refractive index variation over the surface of the Earth, use a 4/3 Earth radius model.
5. For a bandwidth of 10 MHz and a noise temperature of 290 K, what is the noise power at the input of the antenna in dBm? If the bandwidth is changed to 1 MHz, what is the input noise power in dBm?
6. The CSU CHILL S-band radar operates at a frequency $f = 2.725$ GHz. It has a 1 MW transmitter and uses a 1 μs pulse width. Its dual-offset antenna has an antenna gain of approximately 43 dB and a half-power beamwidth of $1.0°$. Assume $G_{rx} = 1$ (the power at the antenna port can be converted to reflectivity) and the dielectric factor of water is $|K_w|^2 = 0.93$:
   a) calculate the weather radar constant, C.

   b) Using CSU-CHILL's S-band weather radar constant and a receiver noise power of -110 dBm, what are the noise-equivalent reflectivity factor at 1 km, 10 km, and 100 km? (Note that dBm is the power ratio with respect to 1 milliwatt.)
7. Calculate the cascaded noise figure for three cascaded amplifiers, each with 20 dB gain and 3 dB noise figure. What is the system's noise temperature? Repeat the calculation, but with a 2 dB attenuator before each amplifier (for a 2 dB attenuator, its gain is -2 dB with a 2 dB noise figure). What is the contribution to the total noise figure of the first attenuator?
8. With a noise figure of 5.5 dB and gain of 24 dB, what is the noise power and signal power at the output of the receiver for an input noise power of -108 dBm and an input signal power of -70 dBm? What are the input and output SNRs for this receiver?

9. If a the radar volume is uniformly filled with a density of one 2-mm raindrop per cubic meter, what is the radar reflectivity factor in dBZ? For a radar with a minimum sensitivity of -10 dBZ at 1 km, what is the maximum range that the radar would detect the precipitation echo? If the radar's minimum sensitivity was -30 dBZ at 2 km, what is the maximum detection range?

10. Metal spheres are commonly used as radar calibration targets. For spheres whose diameters are large with respect to the radar wavelength, and therefore in the optical scattering regime, the radar cross-section is the sphere's cross-sectional area: $\sigma_b = \pi r_s^2$. Assuming a wavelength of $\lambda = 3$ cm and the metal sphere has a dielectric factor $|K|^2 = 1$:

   a) What is the RCS of a 30 cm diameter metal sphere?

   b) For a radar observation with $V_6 = 25890$ m^3 (the approximate radar volume for $\theta_1 = \phi_1 = 1°$ HPBW, $\Delta r = 150$ m, and r = 1000 m), what is the equivalent mean reflectivity $\bar{\eta}$ for the same RCS as the sphere?

   c) What is the metal sphere's reflectivity factor Z with $|K_w|^2 = 0.93$?

   d) If the radar range increases from 1 km to 10 km, what reflectivity factor would the radar measure?

   e) If the backscatter is calculated assuming the Rayleigh scattering regime, where particles must be much smaller than the wavelength, what is the metal sphere's RCS?

   f) If the Rayleigh approximation for the large calibration sphere's RCS is used instead of the appropriate optical regime approximation, what is the error in the reflectivity factor's calibration in decibels?

3 Essential Precipitation Physics for Dual Polarization Radar

After observing a radar display for a sufficient amount of time, it becomes obvious to identify specific patterns and features in the dual polarization parameter space that are recurring among different precipitation events. The most notable pattern is likely the so-called bright band, a circular band of enhanced reflectivity often appearing in high-elevation plan position indicators (PPIs) during stratiform precipitation. The bright band is the result of the layered structure of the atmosphere during widespread precipitation, in combination with the change of the dielectric factor arising from the solid-to-liquid phase transition of precipitation particles and the geometry of the radar scan, with observations being collected on a conical surface intersecting the melting layer. This particular example illustrates the need for a comprehensive understanding that includes geometric, electromagnetic, and physical factors. To aid in this endeavor, this chapter is devoted to the introduction of fundamental concepts of cloud physics relevant to the interpretation of radar observations.

3.1 The Microscale Structure of Precipitation

Clouds are a collection of particles suspended in air. The collection of particles (water droplets or ice crystals) remains stable as long as the particles are suspended. When precipitation particles form, they tend to fall outside of the cloud medium, and the system becomes unstable.

In order to describe the time evolution of cloud and precipitation systems, their microscale structure can be characterized by specifying the phase, shape, size, and number concentration of the constituent particles. A convenient way to synthesize this information is by means of the particle size distribution (PSD). Given the well-established size–shape relation of liquid drops (Section 3.3), the spectra of clouds and raindrops can be easily represented by mathematical expressions considering only the size of the drops. In the case of ice particles, additional information regarding the shape and density of the particles is generally required to provide an appropriate description.

3.1.1 The Particle Size Distribution

The PSD is the probability density function of particle sizes. For spherical particles, the size is represented by the diameter (D), although for more complex particles, different definitions of size may be adopted. For ice particles, the maximum dimension is used in some instances, or the equivalent melted-drop diameter (the equivalent diameter of the water drop to which the ice particle melts). The use of the equivalent melted diameter requires assumptions about the density–size relationship. The PSD is conventionally indicated by $N(D)$, the distribution of particle sizes per unit volume per unit size interval. The number concentration may be expressed in different units, which usually vary depending on the context. Widely used units are $cm^{-3} \mu m^{-1}$ (cloud droplets), $m^{-3} mm^{-1}$ (raindrops), and cm^{-4} (snow particles).

The PSD represents the physical link between the cloud and precipitation parameters and the remote sensing observations within the sampling resolution volume. In addition to remote sensing, knowledge of the PSD is also important for applications in cloud physics, agriculture, numerical models, and hydrology.

Figure 3.1 shows some characteristic size distributions of liquid and frozen precipitation particles, spanning different size and concentration ranges. The PSD of rain (panel a) is conventionally referred to as the *drop size distribution* (DSD). The two DSDs representative of continental and maritime rain in panel (a) of Figure 3.1 are characterized by similar water content but different shapes. In general, the particular shape and time evolution of the size distribution can help elucidate important traits of the underlying precipitation mechanisms. For example, the longer right tail of the continental DSD reveals the dominance of the ice-phase processes in precipitation formation. In continental stratiform precipitation, the depositional growth of ice crystals (Bergeron process) followed by aggregation and subsequent melting below the freezing level may lead to larger raindrops, in comparison with the typical sizes attainable through warm rain processes in tropical regions (Section 3.5). In continental convective precipitation, most raindrops also originate from the melting of frozen particles, which in this case mainly consist of graupel and hail.

One of the most popular analytical expressions for the PSD in remote-sensing applications is the two-parameter exponential distribution:

$$N(D) = N_0 e^{-\Lambda D},$$ (3.1)

where N_0 is the concentration (intercept) parameter, and Λ is the size (slope) parameter. For rain, Marshall and Palmer [23] fitted an exponential model to raindrop size distributions observed in Canada. The model uses a fixed intercept parameter ($N_0 = 8 \times 10^3 \ mm^{-1} \ m^{-3}$) and a slope parameter expressed as a function of the rain rate R ($\Lambda = 4.1 \ R^{-0.21} \ mm^{-1}$), with D in mm and R in $mm \ h^{-1}$. One of the limitations of the exponential form is the poor representation of the smallest drop sizes. However, given the high moments of the DSD involved in radar applications, this can be generally regarded as a minor inconvenience. As compared with raindrops, ice particles span a wide range of sizes, from tens of microns to a few centimeters. For ice aggregates larger than a few tenths of a millimeter, it has been shown that the PSD can also be well represented by the exponential function [24, 25], as depicted in panel (b) of Figure 3.1. Field and Heymsfield [26] have shown how the growth of ice particles through aggregation translates into a typical scaling of the $N_0 - \Lambda$ relation.

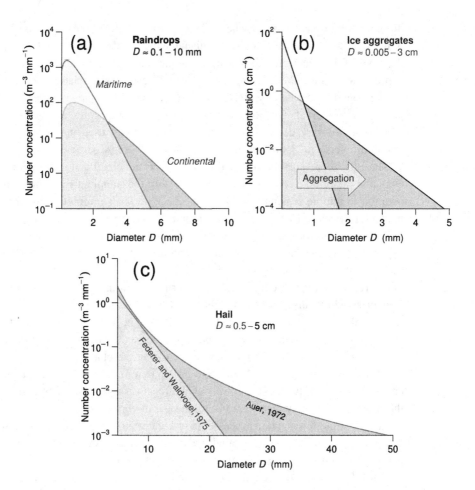

Figure 3.1 Characteristic PSDs for rain (a), ice aggregates (b), and hail (c). The two distributions in panel (a) have the same liquid water content of $\sim$1.5 g m^{-3} but different shapes as a result of distinct precipitation-formation mechanisms. The light- and dark-gray ice particle distributions in panel (b) represent two successive stages of evolution through aggregation, showing the strong coupling of $N_0 - \Lambda$ (both parameters of the exponential PSD decrease with increasing aggregation).

A notable advantage of the exponential form of the PSD is that it allows for simple calculation of any moment of the distribution:

$$\int_0^\infty D^n N(D) dD = N_0 \frac{\Gamma(n+1)}{\Lambda^{n+1}}, \tag{3.2}$$

where Γ is the gamma function ($\Gamma(n+1) = n!$ for n, an integer). This analytical formulation for the nth moment makes the exponential form especially convenient for use in theoretical work in general. Although eq. (3.2) requires an infinite upper limit of

Table 3.1 Typical ranges of size and terminal fall velocity of precipitation particles.

Precipitation Particle	Size (mm)	Velocity (m s^{-1})
Small raindrops	0.5–2	2–6
Large raindrops	2–8	6–9
Pristine ice crystals	0.1–5	0.1–1
Snow	1–30	0.5–2
Graupel	1–20	2–20
Hail	10–100	10–50

integration, it is generally a good approximation for real cases with a finite maximum particle size because of the fast-decreasing tail of the distribution.

A more accurate representation of the PSD, in particular for the left portion of the size range (smallest particles), is attained using the three-parameter gamma distribution:

$$N(D) = N_0 D^\mu e^{-\Lambda D}. \tag{3.3}$$

With respect to the exponential form, the additional multiplicative term D^μ provides better control of the distribution "shape," with the parameter μ being called the *shape parameter*. The gamma form in eq. (3.3) is widely used to represent the size distribution of both cloud droplets and raindrops (panel [a] of Fig. 3.1), providing a simple tool to represent a wide range of rainfall conditions. The exponential form in eq. (3.1) can be regarded as a special case of the more general three-parameter gamma distribution, obtained by setting $\mu = 0$ in eq. (3.3). Another special case of the gamma distribution is the power-law distribution ($\Lambda = 0$ in eq. [3.3]). Power-law relations have been used, for example, to represent the hail-size distribution by Auer [27] (panel [c] of fig. 3.1).

Depending on the composition, size, shape, and orientation, the precipitation particles may fall with a characteristic velocity. The terminal fall velocity is defined as the constant speed reached by precipitation particles when the gravity force (downward) is balanced by the aerodynamic drag force (upward) exerted by the surrounding air. Table 3.1 lists the values of terminal fall velocity for common precipitation particles, showing a wide range of values going from approximately 0.1 m s^{-1} for ice crystals to approximately 50 m s^{-1} for the largest hailstones.

3.1.2   Measuring the Size Distribution: The Disdrometer

Disdrometers are measuring devices specifically meant for the direct observation of the PSD. Different measuring systems have been developed, resulting in a variety of disdrometer types, including acoustic, displacement (Joss–Waldvogel disdrometer), optical, and image (2DVD) disdrometers. Some disdrometers, in addition to the size spectrum, can also measure the velocity of the particles, and some can measure the shape too. As an example, Figure 3.2 illustrates the principle of an optical laser disdrometer.

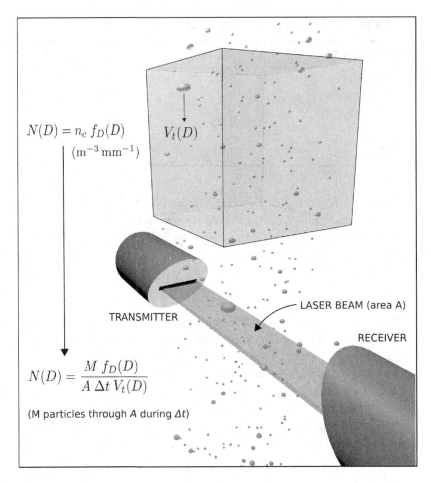

$$N(D) = n_c \, f_D(D)$$

$$(\mathrm{m}^{-3} \, \mathrm{mm}^{-1})$$

$$V_t(D)$$

LASER BEAM (area A)

TRANSMITTER

RECEIVER

$$N(D) = \frac{M \, f_D(D)}{A \, \Delta t \, V_t(D)}$$

(M particles through *A* during *Δt*)

Figure 3.2 Measuring principle of a laser disdrometer. The PSD is defined as the number of particles per unit volume per unit size interval, $N(D) = n_c f_D(D)$, where n_c is the number concentration (m^{-3}), and $f_D(D)$ is the probability density function (mm^{-1}). Note that the computation of the PSD from the observed disdrometer counts involves knowledge of the particles' fall velocity $V_t(D)$.

Although the PSD is defined as the number of particles per unit volume per unit size interval (i.e., the drops in the cubic volume in Fig. 3.2), disdrometers often measure the flux of particles during a given time interval (e.g., the number of particles falling through the laser beam during time Δt in Fig. 3.2) and then convert to drops concentration using assumed or measured fall velocities. This suggests how the actual measuring principle should be carefully considered when interpreting measured data. For example, in windy conditions may lead to unreliable estimation of the real drop concentration.

3.2   The Size Distribution of Raindrops

The DSD is of great relevance for radar rainfall applications, as illustrated in Figure 3.3, which shows four different DSDs with the same radar reflectivity of 37 dBZ but rainfall rates ranging between 1 and 36 $mm\,h^{-1}$. In fact, depending on how the total mass in the volume is distributed over different drop sizes, the rainfall rate may greatly vary and attain the largest values for the distributions dominated by small drops. This is a direct consequence of Z and R representing different moments of the DSD, which will be thoroughly discussed in Chapter 9.

For a generic raindrop size distribution $N(D)$, it is useful to define some special diameters. The median volume diameter D_0 is the diameter such that half the total liquid water content W is contributed by diameters smaller than D_0 and the other half by diameters larger than D_0. Given the total water content resulting from the third moment of the DSD,

$$W = \frac{\pi}{6} \times 10^{-3} \rho_w \int_{D_{\min}}^{D_{\max}} D^3 N(D) dD \quad [\mathrm{g\,m^{-3}}], \tag{3.4}$$

where $\rho_w \approx 1\,\mathrm{g\,cm^{-3}}$ is the water density and the drop's diameter, D, is in mm, the median volume diameter is such that

$$\frac{\pi}{6} \times 10^{-3} \rho_w \int_{D_{\min}}^{D_0} D^3 N(D) dD = \frac{1}{2} W. \tag{3.5}$$

In addition to the median volume diameter, various mean diameters can be calculated from the expression:

$$D_p = \frac{\int_{D_{\min}}^{D_{\max}} D^{p+1} N(D) dD}{\int_{D_{\min}}^{D_{\max}} D^p N(D) dD}. \tag{3.6}$$

In particular, the mass-weighted mean diameter,

$$D_m \equiv D_3, \tag{3.7}$$

and the reflectivity-weighted mean diameter,

$$D_z \equiv D_6. \tag{3.8}$$

When integrating a moment of the DSD as in the previous equations, the integral is evaluated between the lower and upper limits $D_{\min}$ and $D_{\max}$. In general, there is uncertainty about these limits, which is related to the particular way the DSD is measured. Any measuring device can evaluate the drop diameter within a specific limited range. For most disdrometers, the typical useful measurement range is $\sim$0.2–10 mm. Droplets smaller than 0.2 mm are difficult to measure with disdrometers

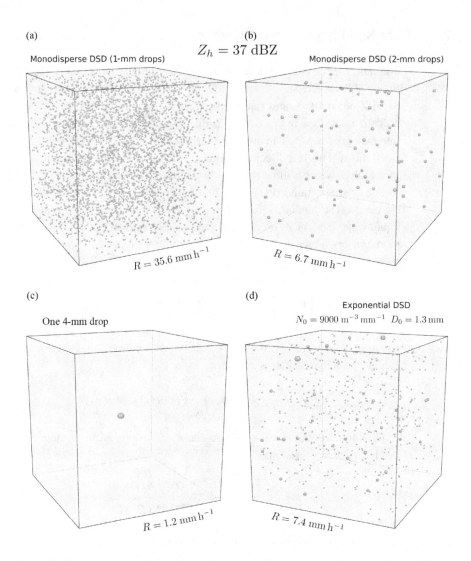

Figure 3.3 The same radar reflectivity can be observed from a unit volume containing different DSDs. All four represented DSDs show the same reflectivity but very different rainfall-rate values. Panels (a) and (b) represent monodisperse DSDs with ~5000 1-mm drops (a) and 74 2-mm drops (b). A single 4-mm drop (c) can also produce the same 37-dBZ reflectivity but a greatly reduced rainfall rate. A typical exponential DSD is also shown for comparison (d). All drops are scaled 20:1 with respect to the 1-m^3 volume.

[28, 29]. However, the choice of $D_{\min}$ has a negligible impact on the evaluation of the typically high-order moments considered in radar applications, and in general, $D_{\min}$ is simply set to 0 mm. A proper selection of $D_{\max}$ is more relevant for the evaluation of precipitation-related moments like W (third-order moment), R (3.67th-order moment), and Z (sixth-order moment). In many applications, it is quite common

to use a fixed maximum diameter in the 6- to 10-mm range (e.g., $D_{\max} = 8$ mm). An accurate estimation of the maximum diameter is related to the sample size of the available measurements. In fact, the true distribution of the drop diameters cannot be known because of the limited time-space physical sampling of any measuring device. In general, the estimation of the radar and precipitation variables is a statistical problem of estimating the characteristics of a population from a sample drawn from that population [30]. The larger the sample size, the more accurate the estimate of the population characteristics will be. When dealing with actual disdrometer measurements having a limited sample size, a generally good approximation is to relate the maximum diameter to the mass-weighted mean diameter D_m. A widely used approximation is $D_{\max} = 2.5 D_m$ [21] because drops larger than $2.5 D_m$ are rarely observed in nature.

It is also worth noting that for a gamma distribution with positive shape parameter μ (Section 3.2.2), the impact of the uncertainty on $D_{\max}$ is greatly reduced as a result of the faster-decreasing tail of the distribution, with respect to the exponential form (Fig. 3.5). For greater ease of reading, the integral limits will be dropped from the integral symbols from now on, except where explicitly noted.

3.2.1 The Exponential Model for Raindrops

For the exponential size distribution (eq. [3.1]) in rain, the slope parameter Λ is related to the median volume diameter D_0 and to the mass-weighted mean diameter D_m [21]. For a large enough maximum diameter $(D_{\max}/D_0 \geq 2.5)$, it can be shown that

$$\Lambda D_0 = 3.67 \tag{3.9}$$

$$\Lambda D_m = 4, \tag{3.10}$$

allowing us to rewrite eq. (3.1) as

$$\boxed{N(D) = N_0 e^{-3.67 \frac{D}{D_0}} \ [\text{m}^{-3}\,\text{mm}^{-1}]} \tag{3.11}$$

or

$$N(D) = N_0 e^{-4 \frac{D}{D_m}} \ [\text{m}^{-3}\,\text{mm}^{-1}]. \tag{3.12}$$

3.2.2 The Gamma Model for Raindrops

Although the exponential model represents a good approximation for space-time-averaged size distributions, "instantaneous" DSDs (i.e., DSDs measured over a small time interval) can be better approximated by a three-parameter gamma model [31]. It is common and useful to represent the gamma DSD in different forms to take advantage of specific properties, depending on the application. The most frequently used expression directly follows as an extension of the exponential form in eq. (3.1) and was presented in eq. (3.3).

For the gamma distribution, the median volume diameter and the mass-weighted mean diameter are related to both Λ and μ:

$$\Lambda D_0 = 3.67 + \mu \tag{3.13}$$

$$\Lambda D_m = 4 + \mu. \tag{3.14}$$

Another form of the DSD is the classic probability density function:

$$N(D) = n_c f_D(D); \quad [\text{m}^{-3}\,\text{mm}^{-1}], \tag{3.15}$$

where n_c is the number concentration (units of m^{-3}), and $f_D(D)$ is a probability density function with unit integral. For the special case of a gamma distriibution, $f_D(D)$ is expressed as (Fig. 3.4):

$$f_D(D) = \frac{\Lambda^{\mu+1}}{\Gamma(\mu+1)} e^{-\Lambda D} D^{\mu}; \quad [\text{mm}^{-1}]. \tag{3.16}$$

Using this form, any nth moment of the DSD can be calculated with the following formula:

$$\int_0^\infty D^n f_D(D) dD = \frac{\Gamma(\mu+1+n)}{\Gamma(\mu+1)} \frac{1}{\Lambda^n}. \tag{3.17}$$

Although eq. (3.17) requires the integral limits to be $(0, \infty)$, it still provides an approximation for the integral parameter of the DSD with real limits $(D_{\min}, D_{\max})$ whenever $D_{\min}/D_0$ is small and $D_{\max}/D_0$ is large. Figure 3.5 shows example empirical fits on real disdrometer observations using the gamma model (solid lines) and the exponential model (dashed lines).

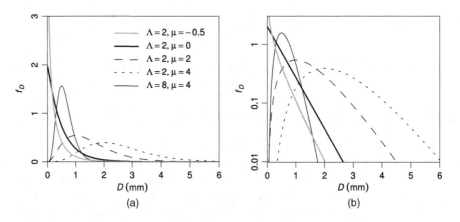

Figure 3.4 (a) gamma density function for different values of the slope parameter Λ and shape parameter μ. When $\mu = 0$, the distribution reduces to the exponential distribution (thick solid black line). (b) Same as panel (a), with y-axis in logarithmic scale.

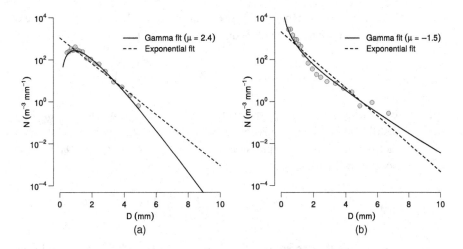

Figure 3.5 $N(D)$ versus D for 1-minute DSD measured by a laser disdrometer. Panel (a) shows a gamma fit with a positive shape parameter μ (the most frequent occurrence in real rainfall), whereas panel (b) shows a distribution fitted by a gamma with a negative μ. Note the overestimation of the large-drop concentration with the use of the exponential approximation (dashed line) for the distribution in panel (a) ($\mu > 0$).

3.2.3 The Normalized Drop Size Distribution

When comparing different distributions with varying water content, it is useful to introduce the concept of the normalized distribution. Equations (3.11) and (3.12) provide expressions where the drop diameter is normalized by either the characteristic median drop diameter or the mass-weighted mean diameter. In order to complete the normalization, the particle concentration needs to be considered next. Sekhon and Srivastava [32] and Willis [33] introduced the concept of the normalized distribution relying on the total water content W (eq. [3.4]) and the median volume diameter D_0. Using eq. (3.17) to calculate the third moment of the DSD results in the following:

$$
\begin{aligned}
W &= \frac{\pi}{6} \times 10^{-3} \rho_w \, n_c \, \frac{\Gamma(\mu + 4)}{\Gamma(\mu + 1)} \frac{1}{\Lambda^3} \\
&= \frac{\pi}{6} \times 10^{-3} \rho_w \, \frac{N_0 \Gamma(\mu + 4)}{\Lambda^{\mu+4}} \\
&= \frac{\pi}{6} \times 10^{-3} \rho_w \, \frac{N_0 \Gamma(\mu + 4)}{(3.67 + \mu)^{\mu+4}} D_0^{\mu+4},
\end{aligned}
\tag{3.18}
$$

where eq. (3.13) has been used, in addition to the relation between N_0 and n_c (eqs. [3.3], [3.15], [3.16]):

$$
N_0 = n_c \frac{\Lambda^{\mu+1}}{\Gamma(\mu + 1)}.
\tag{3.19}
$$

For the case of an exponential ($\mu = 0$), from eq. (3.18), N_0 can be expressed in terms of the water content W and median volume diameter D_0 as

$$N_0 = 10^3 \frac{(3.67)^4}{\pi \rho_w} \left(\frac{W}{D_0^4} \right). \tag{3.20}$$

Expressing the intercept parameter N_0 in terms of W and D_0 in the exponential distribution suggests a possible way to normalize the more general gamma DSD. In fact, by dividing the expression in eq. (3.3) by $(10^3 W)/(\rho_w D_0^4)$, we can define a normalized distribution:

$$N_{norm}(D) = \left(\frac{\rho_w D_0^4}{10^3 W} \right) N_0 D^\mu e^{-\Lambda D}. \tag{3.21}$$

Substituting the expression of W from eq. (3.18) to eliminate N_0 and using eq. (3.13) to normalize the particle diameter gives

$$N_{norm}(D) = \frac{6}{\pi} \frac{(3.67 + \mu)^{\mu+4}}{\Gamma(\mu + 4)} \left(\frac{D}{D_0} \right)^\mu e^{-(3.67+\mu)\frac{D}{D_0}}. \tag{3.22}$$

Finally, again multiplying eq. (3.22) by $(10^3 W)/(\rho_w D_0^4)$ leads to a third expression of the gamma DSD, that is, the normalized DSD:

$$N(D) = N_w f(\mu) \left(\frac{D}{D_0} \right)^\mu e^{-(3.67+\mu)\frac{D}{D_0}}, \tag{3.23}$$

where:

$$N_w = 10^3 \frac{(3.67)^4}{\pi \rho_w} \left(\frac{W}{D_0^4} \right) \tag{3.24}$$

is the normalized intercept parameter, and

$$f(\mu) = \frac{6}{(3.67)^4} \frac{(3.67 + \mu)^{\mu+4}}{\Gamma(\mu + 4)}. \tag{3.25}$$

Comparing eqs. (3.24) and (3.20), it is clear that N_w is equal to the intercept parameter N_0 of an equivalent exponential DSD with the same water content and the same median volume diameter as the gamma DSD.

Using eq. (3.14) in eq. (3.18) to express the water content in terms of the mass-weighted mean diameter D_m, instead of D_0, an alternative formulation of the normalized DSD is given by Testud [34] as

$$N(D) = N_w f(\mu) \left(\frac{D}{D_m} \right)^\mu e^{-(4+\mu)\frac{D}{D_m}}, \tag{3.26}$$

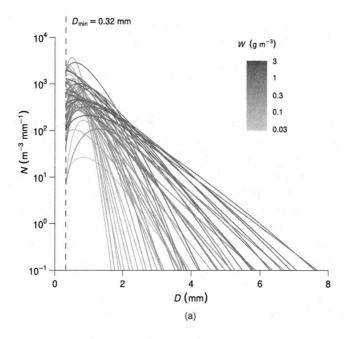

(a)

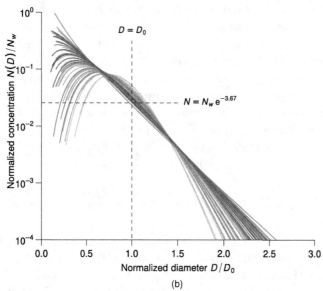

(b)

Figure 3.6 (a) Fifty gamma distributions, randomly selected after fitting over $40,000$ 1-minute DSDs measured by 14 laser disdrometers in Iowa. The plotted fits are different shades of gray based on the value of the water content (the corresponding rain rate associated with the observed DSDs varies between 1 and 80 mm h^{-1}). (b) Same DSD fits as in panel (a) but plotted over the normalized axes. Note the "compression" toward $D = D_0$ and $N = N_w e^{-3.67}$, which corresponds to the exponential distribution ($\mu = 0$ in eqs. [3.23] and [3.25]). Note also the typical dominance of fitted distributions with a positive shape parameter μ.

where

$$N_w = 10^3 \frac{4^4}{\pi \rho_w} \left(\frac{W}{D_m^4} \right) \tag{3.27}$$

$$f(\mu) = \frac{6}{4^4} \frac{(4+\mu)^{\mu+4}}{\Gamma(\mu+4)}. \tag{3.28}$$

Using the illustrated normalization approach, the DSD is expressed as a function of three independent and physically meaningful parameters. The median volume diameter D_0 (or the mass-weighted mean diameter D_m) and N_w have, respectively, the physical units of length and concentration, whereas the unitless parameter μ describes the shape of the DSD. On the other hand, in the standard form of the gamma DSD (eq. [3.3]), there is an implicit relation between μ and N_0, which has no intuitive physical meaning (N_0 being expressed in units of $m^{-4-\mu}$). Figure 3.6 shows an example of raindrop size distributions obtained after fitting a gamma distribution to disdrometer observations, with the corresponding normalized distributions.

3.3   Drop-Shape Models

It has long been recognized on the basis of theoretical and experimental studies that raindrops have shapes resembling an oblate spheroid, with their symmetry axis aligned close to the vertical. The shape of a raindrop is primarily the result of the balance between surface tension and gravitational forces. Green [35] proposed a simple model for approximating the shape of raindrops in terminal fall equilibrium, based on the balance between surface tension and hydrostatic forces. According to this model, the shape of a raindrop can be represented by an oblate spheroid; that is, it can be completely specified by the value of its axis ratio b/a, where b is the minor axis radius, and a is the major axis radius (Fig. 3.7). The volume-equivalent spherical diameter $D_e = 2\,(b)^{1/3}(a)^{2/3}$ is given in millimeters as follows [35]:

$$D_e = 10^3 \times 2 \left(\frac{\sigma_w}{g \rho_w} \right) \frac{\left[(b/a)^{-2} - 2(b/a)^{-1/3} + 1 \right]^{1/2}}{(b/a)^{1/6}}, \tag{3.29}$$

where $\sigma_w = 0.07275\,\mathrm{J\,m^{-2}}$ is the surface tension of water, $g = 9.81\,\mathrm{m\,s^{-2}}$ is the gravitational acceleration, and $\rho_w = 997\,\mathrm{kg\,m^{-3}}$ is the water density. The axis ratio relation of the Green model is shown in Figure 3.7 as a thin solid black line:

For ease of notation, the e subscript is dropped from D_e and D is used to represent the volume-equivalent diameter of raindrops.

For applications not requiring very high accuracy, the axis ratio relation can be approximated by a linear relation. The commonly used linear formulation is the one given by Pruppacher and Beard [37], which is valid for diameter sizes ranging between 1 and 9 mm:

$$\frac{b}{a} = 1.03 - 0.062D; \quad 1 \le D \le 9\,\mathrm{mm}. \tag{3.30}$$

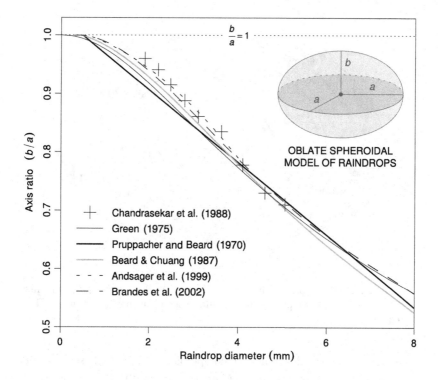

Figure 3.7 Axis ratio of raindrops as a function of the diameter. Some of the most relevant relations are illustrated, in addition to the aircraft observations of Chandrasekar et al. [36].

Beard and Chuang [38] elaborated an accurate numerical model of the raindrop shape that considers the effects of the aerodynamic pressure arising from the flow around the drop, in addition to the surface tension and the hydrostatic pressure as in the Green model [35]. Figure 3.8 shows a representation of the Beard and Chuang model [38], evidencing the flattened base of the larger raindrops with respect to the simpler oblate spheroidal models. For most radar applications, it is generally sufficient to use a simplified spheroidal model with an axis ratio relation given by a polynomial fit to the Beard and Chuang model:

$$\frac{b}{a} = 1.0048 + 0.00057D - 0.02628D^2 + 0.003682D^3 - 0.0001677D^4, \quad (3.31)$$

which is valid for D between 0.5 and 7 mm.

Many studies have shown that mean raindrop shapes are more spherical than the equilibrium values given by either the Green model or the Beard and Chuang model. The reason is related to the fact that raindrops falling in the free atmosphere are subject to both axisymmetric (oblate-prolate mode) and transverse-mode oscillations [36, 39, 40]. These oscillations have frequencies ranging from a few hundred hertz for the smaller raindrops ($D \approx 1$ mm) to a few tens of hertz for the larger drops ($D > 5$ mm) [41].

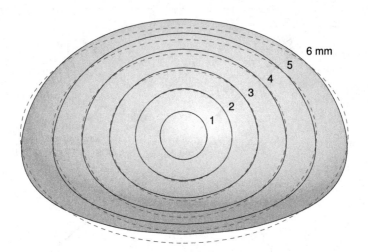

Figure 3.8 Representation of the Beard–Chuang drop-shape model [38] (solid lines). The dashed contours represent the simplified spheroidal approximation given by eq. (3.31).

Although the axisymmetric mode produces a symmetric scatter around the equilibrium axis ratio (two-sided variation about unity), with negligible effects on the resulting mean shape, the transverse-mode oscillations result in a one-sided scatter. This leads to an upward shift in the mean axis ratio relative to the equilibrium shape. In particular, for drops smaller than 4 mm, Andsager et al. [42] provided an experimental fit to laboratory measurements in the form of a second-order polynomial (gray thin solid line in Fig. 3.7):

$$\frac{b}{a} = 1.012 - 0.0144D - 0.0103D^2. \tag{3.32}$$

According to this relation, drops with diameters smaller than 4 mm exhibit a mean axis ratio larger than that obtained at equilibrium. The laboratory results of Andsager et al. are in agreement with the aircraft observations previously studied by Chandrasekar et al. [36], who reported mean axis ratios higher than the equilibrium value for drops with diameters between 2 and 3.6 mm (+ symbol in Fig. 3.7). A frequent approach adopted in radar studies is the combination of eqs. (3.31) and (3.32), applying the Andsager et al. fit (eq. [3.32]) for $D < 4$ mm and the Beard–Chuang relation (eq. [3.31]) for $D > 4$ mm. More recently, Brandes et al. [43] introduced a polynomial approximation that combines the observations of Pruppacher and Pitter [44], Chandrasekar et al. [36], Beard and Kubesh [45], and Andsager et al. [42]:

$$\frac{b}{a} = 0.9951 + 0.0251D - 0.03644D^2 + 0.005030D^3 - 0.0002492D^4. \tag{3.33}$$

This relation agrees well with the fit provided by Andsager et al. [42] for $D < 4$ mm, and it provides a good match between the observed and predicted radar variables obtained from scattering simulations on real DSDs [43].

3.3.1   Terminal Velocity of Raindrops

Similar to the shape of the drops, the fall speed of raindrops is another quantity related to the drop size. Gunn and Kinzer [46] performed measurements of the terminal fall velocity of drops in stagnant air, which have been the basis for several empirical formulations. Figure 3.9 shows the original measurements with two widely used fitting relations. Atlas and Ulbrich [47] derived a power-law expression based on the measurement of Gunn and Kinzer [46]:

$$V_t(D) = 3.78D^{0.67},\qquad(3.34)$$

where the diameter D is expressed in millimeters and the fall velocity in meters per second. This widely used formulation provides a good approximation for drops up to ~4 mm. The power-law relation (eq. [3.34]) has the notable advantage of facilitating the incorporation of the drops' fall velocity into the expression for the rainfall rate (see Section 9.1 for more details):

$$R = 7.12 \times 10^{-3} \int D^{3.67} N(D) dD,\qquad(3.35)$$

where R is expressed in millimeters per hour.

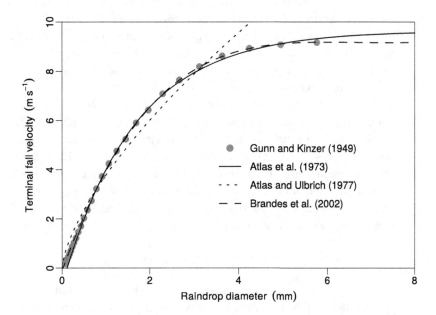

Figure 3.9 Terminal fall velocity of raindrops at sea-level. Measurements by Gunn and Kinzer [46] (circles) and fitting models by Atlas et al. [49], Atlas and Ulbrich [47], and Brandes et al. [43].

A more accurate fit to the measurements of Gunn and Kinzer [46] is another widely used expression given by Atlas et al. [49]:

$$V_t(D) = 9.65 - 10.3\,e^{-0.6D}, \tag{3.36}$$

which is valid for $D > 0.1$ mm. Even closer fits to the Gunn and Kinzer [46] data can be obtained using polynomial approximations, like the one given by Brandes et al. [43] (dashed line in Fig. 3.9):

$$V_t(D) = -0.1021 + 4.932D - 0.9551D^2 + 0.07934D^3 - 0.002362D^4. \tag{3.37}$$

The power-law approximation (eq. [3.34]) is a convenient formulation for developing analytical relations involving the drops' fall velocity. However, for real DSDs that include large drops (i.e., $D_{max} > 4$ mm), eq. (3.36) or (3.37) are a more appropriate representation of the raindrops' terminal velocities. Note that all above formulations for the drop's fall velocity are valid in the standard atmosphere at the sea level. To get the drop's fall velocity at higher altitudes, the right-hand side of the equations should be scaled by the correction factor $(\rho_0/\rho)^m$, where ρ_0 is the air density at sea level, ρ is the air density aloft and $m \approx 0.4$.

3.3.2   Canting of Raindrops

So far, we have assumed raindrops falling in still air with their symmetry axis along the vertical. Radio-wave-propagation studies revealed that raindrops, in addition to forming an anisotropic medium composed of highly oriented particles with their axis of symmetry close to the vertical [50], also produce cross-polarization between the orthogonal horizontal and vertical linear polarizations [51]. This phenomenon led to the inference that the main symmetry axis of the raindrop may not be perfectly aligned with the vertical. The canting angle is defined as the angle between the projection of the drop's symmetry axis on the polarization plane and the projection of the local vertical direction on the same plane.

A uniform horizontal wind without a vertical gradient would not affect the orientation (canting) of the drop, which could still be regarded as falling within an inertial frame of reference; that is, the drop is moving at a constant speed (the terminal fall velocity) with zero net force acting upon it. Brussaard [52] speculated that the observed canting angle of raindrops could be explained as the result of wind shear in the lower portion of the troposphere, where windspeed decreases towards the surface due to friction. In fact, approximating the horizontal wind U by a laminar flow varying with height and with fixed direction, he expressed the canting angle β as follows [52]:

$$\beta = \tan^{-1}\left(\frac{s\,V_t}{g}\right), \tag{3.38}$$

where $s = dU/dh$ is the vertical wind shear, and g is the gravitational acceleration.

According to this simple model, the combination of two different forces acting on a raindrop of mass m ($F_h = V_t s m$ and $F_v = gm$) leads the drop to assume a slanted orientation, with the rotational symmetry axis of the drop parallel to the relative airflow. The canting angle is also independent of the drop size because the mass m cancels out when taking the ratio F_h/F_v (eq. [3.38]). This implies that raindrops would take a preferential orientation with respect to the vertical within a given atmospheric volume which is subject to wind shear. However, Brussaard [53] noted that the wind gradient is generally small at altitudes above 100 m, implying that the average canting angle should also be small ($\beta < 3°$) at the heights typically sampled by a weather radar. A preferential average orientation in a raindrop population is actually rarely observed. Most measurements in rain can be explained by a symmetric distribution of the canting angles centered around zero.

In general, regardless of the specific physical forcing, the observed canting-angle distribution of raindrops can be interpreted as a manifestation of the oscillatory behavior of the drops. As previously mentioned, drops tend to oscillate in two preferential modes [45]: the axisymmetric mode oscillations between oblate and prolate shapes, and the asymmetric transverse mode (see, e.g., Fig. 3 in [40]). Although it has been shown that a strong coupling may exist between oscillation modes and eddy shedding in the drop's wake [54], to this date, the physical mechanisms responsible for raindrop oscillations are still not completely understood. Whether these are related to extrinsic forcing, such as drops collisions, wind shear, and turbulence, or promoted by intrinsic aerodynamic forces is still to be clarified. In particular, further progress on the understanding of the nature of drop oscillation requires a clarification of the interactions between eddy shedding in the wake of the drop and the drop itself [41, 55].

More importantly for radar applications, it has been widely shown that regardless of the underlying physics of oscillations or canting, the use of canting-angle distributions allows modeling the dual polarization observations. Numerous studies have been carried out assuming a narrow Gaussian distribution with zero mean and standard deviation in the range of 5–10°. For example, Huang et al. [56] analyzed the data set of artificially generated drops falling 80 m from a bridge during low-wind conditions described by Thurai and Bringi [57]. They found that the distributions of the canting angle from the two disdrometer cameras were nearly symmetric about a mean of 0°, with a standard deviation of 7–8°. A substantial agreement with these estimates is provided by the radar observations discussed by Bringi et al. [58]. For applications such as electromagnetic scattering simulations of radar moments, it then appears reasonable to represent the canting-angle distribution assuming a Gaussian model with 0° mean and a standard deviation between 5° and 10°.

3.4 Precipitation Phases

Weather radars operating at 3–10 GHz (S, C, and X band) are generally referred to as *precipitation radars*, as opposed to the cloud radars operating at higher frequencies. The name originates from the sensitivity of the specific frequency range (corresponding to the 3- to 10-cm wavelength) to precipitation particles. Although the primary scope of weather radar is largely associated with the observation and quantification of rainfall, other nonliquid particles are routinely observed, such as ice crystals, dry and melting aggregates (snow), graupel, and hail.

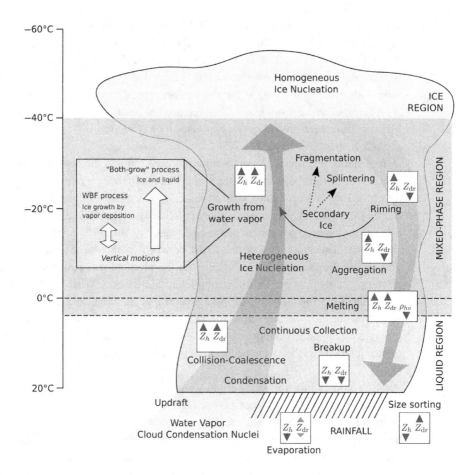

Figure 3.10 Schematic illustration of possible paths to rain formation. The major physical interactions between cloud particles are represented with the corresponding typical trend of radar variables Z_h and Z_{dr}. The Wegener-Bergeron-Findeisen (WBF) process is a primary mechanism for ice growth in mixed-phase clouds, whereas the "both-grow" process indicates a coexistence regime (see Section 3.5.6).

Clouds can be classified as liquid, glaciated (exclusively composed of ice particles), or mixed phase (containing both ice particles and supercooled water). Figure 3.10 shows a schematic representation of the main precipitation processes within a cumulus cloud. For each of the main processes, the typical trend of the polarimetric radar variables is indicated. A review of the main processes of precipitation formation and growth is provided in Section 3.5, with special emphasis on the impact on dual polarization radar moments.

In addition to the fundamental role of temperature, the water-vapor pressure is the other important factor behind the processes of precipitation formation and growth. The water-vapor pressure (e) is the partial pressure exerted by the water vapor in thermodynamic equilibrium with its liquid (or solid) condensed phase. The amount of water vapor that can exist in a cloud decreases with temperature. A relative humidity of 100 percent is reached when the amount of water vapor is in equilibrium above a flat water surface. This is called the *saturation vapor pressure with respect to water* (e_s). The water vapor saturation ratio is defined as

$$S_w = \frac{e}{e_s} = \frac{\text{RH}}{100},\tag{3.39}$$

where RH is the relative humidity.

Analogously, the water-vapor saturation ratio with respect to ice is defined as

$$S_i = \frac{e}{e_{si}} = S_w \frac{e_s}{e_{si}},\tag{3.40}$$

where e_{si} is the saturation vapor pressure with respect to ice. The saturation vapor pressure with respect to ice is always lower than the saturation vapor pressure with respect to water ($e_{si} < e_s$; see Fig. 3.11) because of the stronger bonds between the adjacent molecules of the ice particles. As a consequence, the water vapor in a cloud may be saturated, at 100 percent relative humidity ($S_w = 1$), with respect to a water droplet, but at the same time, it would be supersaturated with respect to an ice particle ($S_i > 1$).

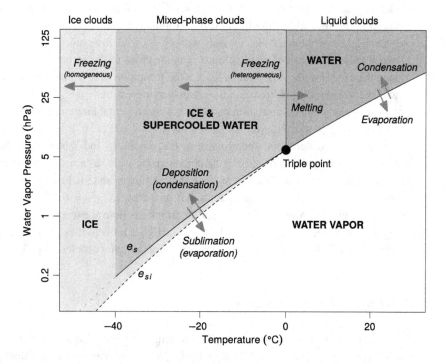

Figure 3.11 Phase diagram for water in the atmosphere. The solid (dashed) line indicates the saturation vapor pressure with respect to water (ice).

Both the saturation vapor pressure with respect to water (e_s) and the saturation vapor pressure with respect to ice (e_{si}) increase with temperature as a result of the release of water vapor through evaporation and sublimation (Fig. 3.11). At subfreezing temperatures, for a given water vapor pressure e, when $e_{si} < e < e_s$, the water vapor will attempt to return to equilibrium, through deposition onto the ice surface of the crystals. Deposition occurs whenever ice crystals exist in a cloud supersaturated with respect to ice, by means of water vapor diffusion. If a mixed-phase cloud is supersaturated also with respect to water ($e > e_s$), then both ice crystals and supercooled drops may grow at the same time (Section 3.5.6).

The phase diagram of water in Figure 3.11 provides an indication of the fundamental phase transitions. When cloud droplets rise above the freezing level, they do not readily freeze, as confirmed by the common observation of liquid water droplets in the clouds well above the freezing level (supercooled liquid drops). In fact, the common knowledge of water freezing below $0°C$ cannot be readily applied to water in the atmosphere. The experience of water freezing below $0°C$ is valid for bulk water, whereas water in the clouds is distributed over many small droplets. This particular fragmentation of the available water in tiny spherical particles increases the resistance to freezing [59]. Small droplets of pure water freeze spontaneously at temperatures of around $-40°C$ (homogeneous freezing). However, the presence in the atmosphere of particles acting as ice nuclei provides a substrate on which water molecules can stick, thus favoring freezing at much warmer temperatures (heterogeneous freezing). Heterogeneous freezing occurs when a solid aerosol particle acts as an ice nucleus, favoring the freezing of the liquid drops at temperatures as warm as $-10°C$. Below $-15°C$, a large number of ice crystals will typically be observed. In general, the water vapor pressure in mixed-phase clouds remains close to the saturation over water. In fact, the excess water vapor would lead to the formation of new precipitation particles, with a corresponding decrease in the water-vapor pressure.

The phase of the precipitation particles is a fundamental factor influencing the radar backscattered power. Figure 3.12 shows the normalized radar cross-section for water and ice at two widely used operating radar frequencies (S and X bands). Radar echoes from the upper portion of ice clouds are generally weak as a result of the combined effect of the lower water-vapor content at high altitudes and the lower dielectric factor of ice ($|K_i|^2 = 0.19$) compared with that of water ($|K_w|^2 = 0.93$). Typical values of reflectivity from pristine ice crystal distributions do not exceed 15 dBZ, whereas crystal aggregates may reach $\sim$30–35 dBZ. Higher reflectivities at subfreezing temperature normally require the presence of graupel or hail particles (Fig. 3.13).

3.5   Precipitation Processes

Precipitation-formation mechanisms are traditionally classified as either warm-rain or cold-rain (snow) processes. *Warm rain* refers to clouds where all constituent particles are liquid drops. Cold-rain processes take place below $0°C$ and may involve both glaciated and liquid (supercooled) particles. The region of the cloud where liquid and

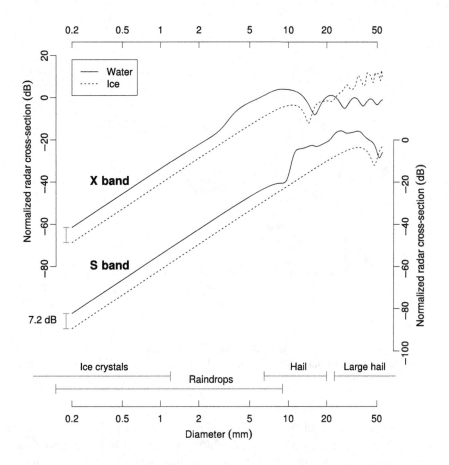

Figure 3.12 Normalized radar cross-section of spherical water (solid line) and ice particles (dashed line) for X-band (10-GHz, left y-axis) and S-band (3-GHz, right y-axis) radar frequencies. In the Rayleigh region (small diameters), the backscattered power from ice particles is approximately 7 dB lower than the power scattered from water drops with the same dimension. This may affect the detectability of cloud regions dominated by small pristine ice crystals. Note the different behavior between the X band and S band for large diameters. At the X band, ice particles with a diameter above $\sim$2 cm (e.g., large hail) may produce a stronger return with respect to an equivalent water particle.

ice particles are present simultaneously is called the *mixed-phase region* and may extend between the freezing level ($0°C$) and $-40°C$, which represents the lower limit where water can exist in a supercooled state. Warm rain is prevalent in the tropical region, mainly over the oceans, whereas cold rain is most frequent at middle and high latitudes.

Condensation of water vapor in the atmosphere generally occurs in a rising air mass as a consequence of adiabatic cooling. Water drops are characterized by a relatively high surface tension, which is reflected in the characteristic spherical shape of cloud droplets and small raindrops. For a droplet to form from condensation of the water

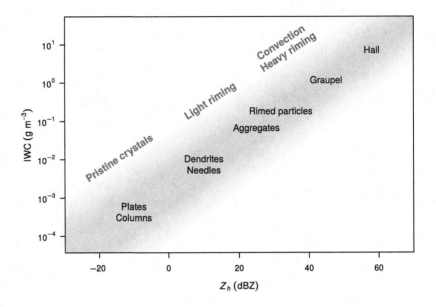

Figure 3.13 Illustration of the dominant ice particle types for varying ranges of reflectivity (Z_h) and ice water content (IWC). (The estimation of the IWC and snowfall rate is discussed later in Section 8.5.2.).

vapor, a rather high vapor pressure is necessary to prevail over the surface-tension force. Smaller drops have greater curvature and require larger vapor pressure to keep water molecules from evaporating (curvature effect).

In a pure water-vapor atmosphere, a relative humidity of several hundred percent would be needed for the water vapor to condense into a cloud drop (homogeneous nucleation). Such large values of relative humidity never occur in the atmosphere. In fact, the abundant presence of cloud condensation nuclei (CCN) makes it much easier for cloud droplets to form (heterogeneous nucleation). The CCN are suspended wettable aerosol particles, with a typical size of ∼1 μm, thus lowering the equilibrium vapor pressure of the liquid drops by decreasing the curvature effect (the particles are larger compared with droplet embryos) and through molecular-scale effects (in the case of water-soluble particles). The presence of CCN allows the formation of cloud droplets when the relative humidity barely exceeds 100 percent. Given the role of the aerosol particles in the equilibrium supersaturation ratio, it is clear how the size and composition of the CCN have a crucial role in the composition of the cloud DSD and the subsequent development of rainfall [60].

Significant differences exist between maritime and continental clouds as a consequence of the different air masses where they develop. Continental air masses are generally richer in CCN with respect to marine air masses. Marine aerosols have a number concentration of the order of 10^3 cm^{-3}, whereas the concentration of continental aerosols can reach ∼10^5 cm^{-3}. However, a higher percentage of maritime aerosols (10–20 percent) can serve as CCN because of the large presence of water-soluble particles like sea salts, in contrast with only $\sim$1 percent for continental aerosols. This peculiar difference, considering the observed similar amount of liquid water content in maritime and continental clouds, implies that maritime clouds are generally composed of fewer but larger cloud droplets in comparison with continental clouds (see panel a in fig. 3.1).

Once formed, the cloud droplets can grow by diffusion of the surrounding water vapor (growth by condensation). The rate of growth is linearly dependent on the ambient supersaturation and is inversely proportional to the drops' radius (see later eq. 3.53). The slower growth of larger droplets leads to a narrowing of the DSD, hindering the development of precipitation (fewer collisions and therefore fewer chances of coalescence).

Analogously to the role of the CCN in the condensation of water droplets, ice-crystal formation in cold clouds is favored by the presence of foreign particles acting as ice nuclei, as mentioned in the Section 3.4. For ice particles, however, the formation process is more efficient than the corresponding process of cloud drops as a result of the supersaturation of water vapor with respect to ice (Fig. 3.11). Heterogeneous freezing generally does not take place until the temperature drops below $-10°C$, so it is not uncommon for warm clouds to have tops extending to subfreezing temperatures as cold as $-15°C$.

Upon the formation of water droplets and ice crystals in the cloud, other processes come into play for the subsequent development of precipitation. The main mechanism for the onset of precipitation is the aggregation of particles of different sizes and fall speeds. The aggregation of water drops is referred to as the *collision-coalescence process* (warm rain). At subfreezing temperatures, in addition to the aggregation of different ice particles, accretion plays a major role, especially in convective systems, and refers to the capture of supercooled drops by an ice-phase precipitation particle (cold-rain process). The aggregation processes do not affect the overall water content in the cloud (unlike condensation/evaporation and deposition/sublimation) but may lead to precipitation through a redistribution of the mass between the smaller and the larger particles.

3.5.1   Collision-Coalescence

In the warm-rain process, the precipitation formation occurs through the collision-coalescence of liquid drops. As reported in Table 3.1, cloud drops have size diameters up to approximately 0.1 mm and a negligible fall speed as a result of the equilibrium between the gravitational force and the frictional resistance of the air. Larger drops are called *drizzle* (0.1–0.25 mm) and *raindrops* (D > 0.25 mm). These drops have a non-negligible terminal velocity that increases with size, thus contributing to precipitation. When drops with different fall speeds are present inside a cloud, collision between pairs of drops becomes possible. Upon collision, drops may bounce off each other (no change to the DSD), break up (Section 3.5.2), or merge, leading to the formation of a new, larger drop (coalescence).

Collision-coalescence is the rain-formation mechanism in warm rain, prevalent in the tropics, whereas it is generally of minor importance in middle and higher latitudes. However, even if the primary rain-formation processes at higher latitudes rely on the crucial role of ice crystals, collision-coalescence still takes place in the lower portion of the cloud and can significantly reshape the raindrop size distribution below the melting layer. This reflects in a modulation of the vertical profile of precipitation and dual polarization radar variables (see examples in Section 3.5.4).

The simplest method to compute the growth rate of liquid drops by collision-coalescence is based on the continuous-growth model [59]. Considering a drop with radius R falling within a population of smaller droplets with radius r ($r < R$), the collision volume dV is given by the collisional area of both the large and the smaller drop, multiplied by the difference in terminal fall speed:

$$dV = \pi(R+r)^2 [V_t(R) - V_t(r)], \qquad (3.41)$$

where V_t is the drop's terminal fall speed. For a continuous population of drops with number concentration $N(r)$, the volume growth rate is obtained by the integration with upper limit R:

$$\frac{dV}{dt} = \frac{4}{3}\pi \int_0^R (R+r)^2 r^3 N(r) E(R,r) [V_t(R) - V_t(r)] \, dr, \qquad (3.42)$$

where $E(R,r)$ is the collection efficiency, that is, the product of collision efficiency (the fraction of the water drops in the path of a falling larger drop that collide with the larger drop) and coalescence efficiency (the fraction of all collisions that results in the actual merging into a single larger drop).

In terms of the drop's radius, eq. (3.42) can be rewritten as

$$\frac{dR}{dt} = \frac{\pi}{3} \int_0^R \left(\frac{R+r}{R}\right)^2 r^3 N(r) E(R,r) [V_t(R) - V_t(r)] \, dr. \qquad (3.43)$$

If $R \gg r$ and $V_t(R) \gg V_t(r)$, eq. (3.43) can be approximated by

$$\frac{dR}{dt} = \frac{\overline{E}W}{4\rho_w} V_t(R), \qquad (3.44)$$

where $\overline{E}$ is the average collection efficiency, ρ_w is the water density, and W is the liquid water content (mass per unit volume) of all the droplets with radius $r < R$. The variation of the drop's radius with height is given by

$$\frac{dR}{dz} = \frac{dR}{dt}\frac{dt}{dz} = \frac{dR}{dt}\frac{1}{w - V_t(R)}, \qquad (3.45)$$

where w is the updraft velocity. If w is negligible, then

$$\frac{dR}{dz} = -\frac{\overline{E}W}{4\rho_w}, \qquad (3.46)$$

which suggests that the drop's radius should decrease with height (i.e., increase approaching the surface).

Although the continuous-growth model can reasonably reproduce the average raindrop-spectrum broadening, it is not able to explain the rapid development of precipitation from the cloud stage. Using eq. (3.45), Bowen [61] showed the impact of the collection efficiency, cloud water content, and updraft intensity on the growth of cloud droplets by coalescence. From Bowen's calculations, the time required for the cloud droplets to reach a millimeter-size raindrop is approximately 1 hour. This time frame is at least a factor of 2 longer than common observations of rain developing from the cumulus clouds, which takes about 15–20 minutes. This discrepancy led researchers to address the growth by coalescence as a statistical problem involving a large ensemble of discrete particles subject to random fluctuations and local variations in droplet concentration. Stochastic collision-coalescence models provide a better understanding of the process by which warm cloud precipitation develops in relatively short times [59]. In the stochastic models (e.g., that by Telford [62]) the onset of precipitation is due to the discrete collisions between drops, as opposed to the continuous-growth models, and in particular to the extraordinary growth of few large "fortunate" drops that fall in a region with a high concentration of smaller droplets. Turbulent fluctuations may also increase the effective fall path of some large drops, contributing to a further broadening of the cloud DSD.

The collision-coalescence process is particularly effective for raindrop size distributions with a large liquid water content mainly contributed by small drops. In this case, the large drops are able to grow very rapidly while falling through the population of smaller drops.

3.5.2 Drop Breakup

Breakup and coalescence are the two possible outcomes of a collision between drops (drops may also bounce off upon collision, although this has no effect on the DSD). For warm rain, the combination of breakup and coalescence is the main physical mechanism driving the evolution of the DSD. Being tightly related, the two processes are in general considered together in numerical modeling, as part of the stochastic collection-breakup equation (see, e.g., Prat and Barros [63]).

Just like collision-coalescence, breakup is a mass-redistribution process that affects the drop-size spectrum and determines the maximum size of raindrops. There are two main mechanisms for drop breakup: (1) aerodynamic breakup arising from hydrodynamic instability due to the flow around large drops and (2) collisional breakup. Collisional breakup has been recognized to be the most important mechanism, whereas aerodynamic breakup becomes relevant only for drops with a diameter larger than $\sim$3 mm. When $D > 6$ mm, raindrops are unstable and tend to have a short life, although undisturbed water drops can survive falling in air and reach a maximum size of $\sim$8 mm before breakup due to aerodynamical deformation.

Collisional breakup between one large drop and one smaller drop occurs every time the collisional kinetic energy is able to overcome the surface tension, as opposed to being dissipated by the viscous motion of water molecules inside the coalesced drop [64]. From laboratory studies, Low and List [65] identified three main geometric shapes assumed by the drops after initial contact (filaments, sheets, and disks), which are associated with specific statistical distributions of the fragment sizes. These results led to a widely used parameterization of the collision, coalescence, and breakup process in numerical cloud modeling [66, 67].

3.5.3   Evaporation of Raindrops

When the environmental air is below water-vapor saturation (RH < 100 percent), for example, when dry air is present below the cloud base, the raindrops start to evaporate by the process of water-vapor diffusion. This is the opposite of condensation, which is water-vapor diffusion toward the raindrops. Recall, condensation allows for continuous growth in a saturated environment (RH ≥ 100 percent). From a radar point of view, evaporation is of great interest because of the well-recognizable dual polarization signatures that may show up in rain. Depending on the level of environmental subsaturation, very dry atmosphere below the cloud base may have RH as low as ~30 percent. On the other hand, the relative humidity in rain may barely exceed 100 percent, and there are typically other growth processes going on, such as collision-coalescence, that have a larger impact on the raindrop size distribution. For this reason, condensation does not lead to distinct radar signatures in rain.

Besides the direct impact on the DSD and precipitation evolution, it is worth noting that evaporation is also very important for its influence on the storm's dynamics. In fact, evaporational cooling below the cloud base may enhance downdrafts, producing damaging winds, and develop a cold pool at the surface, which plays a fundamental role in the storm's development and propagation.

The rate of diffusion, that is, the rate of mass increase (condensation) or decrease (evaporation), can be approximated by the gradient of the water-vapor density around the drop of radius R [68]:

$$\frac{dm}{dt} = 4\pi R f_v D_v \left[\rho_v(\infty) - \rho_v(R) \right], \tag{3.47}$$

where m is the mass of the drop, D_v is the diffusion coefficient for water vapor in air, f_v is a ventilation factor, and ρ_v is the water-vapor density (g m^{-3}) in the environment (∞) and at the drop's surface (R).

In the condensation of water vapor on a drop, latent heat is released to the environment, whereas during evaporation, the latent heat is absorbed by the drop. In both cases, it is assumed that the latent heat is transported by diffusion away from the droplet. This results in a balanced equation that is analogous to eq. (3.47):

$$L\frac{dm}{dt} = 4\pi R f_v K_a \left[T(R) - T(\infty) \right], \tag{3.48}$$

where L is the latent heat of vaporization, K_a is the coefficient of thermal conductivity of air, $T(\infty)$ is the ambient temperature, and $T(R)$ is the temperature at the drop's surface. The ventilation factor f_v in eqs. (3.47) and (3.48) is required to account for the altered diffusion of water vapor and heat for a drop falling through the surrounding air

[59]. Combining eqs. (3.47) and (3.48) results in the expression for the two unknowns $\rho_v(R)$ and $T(R)$:

$$\frac{\rho_v(\infty) - \rho_v(R)}{T(R) - T(\infty)} = \frac{K_a}{D_v L}, \tag{3.49}$$

where the fraction on the right-hand side varies weakly with temperature and pressure.

Considering the equation of state for an ideal gas ($e_s = \rho_{vs} R_v T$, with R_v being the gas constant for water vapor), and assuming that the vapor density at the drop's surface can be approximated by the saturation vapor density ($\rho_v(R) \approx \rho_{vs}$), these combine to give a second expression relating $\rho_v(R)$ and $T(R)$:

$$\rho_v(R) \approx \frac{e_s(T)}{R_v T(R)}. \tag{3.50}$$

This approximation neglects the effects of the drop's curvature and the solution (the water drop is assumed to be sufficiently diluted), both of which are only relevant for small cloud droplets. A numerical solution for $\rho_v(R)$ and $T(R)$ from eqs. (3.49) and (3.50) eventually allows us to calculate the change in the drop's mass by either condensation or evaporation.

An alternative analytical approximation for the growth by condensation is as follows [59]:

$$R\frac{dR}{dt} = f_v \frac{S_w - 1}{\rho_w(F_k + F_d)}, \tag{3.51}$$

where $S_w = e/e_s$ is the ambient saturation ratio, and F_k and F_d are, respectively, the heat-conductivity term and the vapor-diffusion term, defined as

$$F_k = \frac{L}{K_a T}\left(\frac{L}{R_v T} - 1\right); \quad F_d = \frac{R_v T}{D_v e_s(T)}, \tag{3.52}$$

where ρ_w is the liquid water density.

Considering precipitation in calm air, the change in drop diameter for a given fall distance is $dD/dH = (dD/dt)(1/V_t)$, where H is the distance of fall below a given altitude. Equation (3.51) can then be written in terms of the vertical variation of the drop diameter:

$$V_t D\frac{dD}{dH} = 4 f_v \frac{S_w - 1}{\rho_w(F_k + F_d)}. \tag{3.53}$$

Equation (3.53) shows that the evaporation is particularly important for small drops, given the inverse relation between the derivative dD/dH and the drop size. For the same reason, the impact of evaporation increases as the process progresses, and the size of the drop is further reduced. The opposite happens with condensation, which is more relevant at the beginning of the process, for small cloud droplets. As the drops start to grow by condensation, the process progressively loses importance, in favor of the collision-coalescence mechanism, which is needed to produce precipitation-size drops.

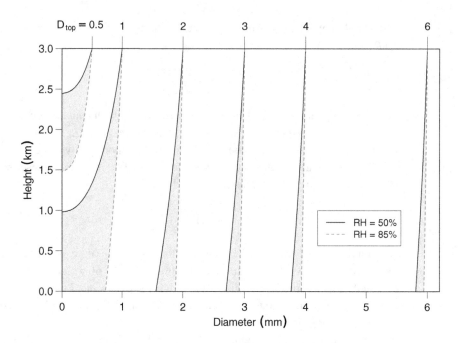

Figure 3.14 Modification of the drop size with height by evaporation for different initial diameters D_{top}. A uniform relative humidity of 50 percent (black solid line) and 85 percent (gray dashed line) is assumed through the 3-km layer. The temperature is set to $0°C$ at the top of the layer (3000 m) and increases downward, following a dry adiabatic lapse rate $(9.8°C \ km^{-1})$.

The impact of evaporation is illustrated in Figure 3.14, where the modification of the drop size with height is represented for several initial drop diameters and two saturation ratios ($S_w = 0.5$ and $S_w = 0.85$). The smaller drops evaporate much faster than the large drops. For the 3-km-depth atmospheric layer, drops with an initial diameter of 0.5 mm evaporate before reaching the surface, even with a relative humidity barely below 100 percent. On the other hand, 1-mm-sized drops may or may not reach the surface, depending on the environmental subsaturation conditions, whereas larger drops are, in general, marginally affected by evaporation. The application example in Section 3.5.4 examines in more detail how evaporation affects different initial DSDs and the associated impact on the radar variables.

3.5.4   Simplified Examples of Warm-Rain Processes Associated with Dual Polarization Radar Signatures

The basic equations governing the warm-rain processes described in the previous sections can be used as a basis for simplistic models to help understand how individual rain processes may affect the radar variables. The following examples are intended to show the characteristic radar response to individual processes, rather than to simulate radar observations under realistic atmospheric conditions where multiple processes and interactions are taking place at the same time.

The radar variables are evaluated according to the Rayleigh–Gans approximation (eqs. [9.4] and [9.5]) and with the use of the Beard–Chuang axis ratio approximation for the shape of the raindrops (eq. [3.31]). The Rayleigh–Gans approximation is useful to simulate observations at the S band. For higher radar frequencies, in addition to the simplified physical approximations about the rain processes, differences with respect to real observations may arise as a result of Mie scattering effects.

Radar Signatures of Collision-Coalescence Growth in Calm Air

For the sake of simplicity, assume an idealized one-dimensional setting where collision-coalescence is the only process affecting the evolution of the raindrop-size distribution along the vertical within a 3000-m-depth precipitation layer. The continuous-growth model (eq. [3.43]) demonstrates the broadening of the raindrop size distribution resulting from collision-coalescence for a given DSD at the top of the precipitation layer.

Continuous rain in a widespread stratiform system may be approximated as a steady-state process, in which the precipitation quantities are constant with time at any given altitude [59]. In such an ideal setting, in calm air, the rain mass is continuously replaced at an altitude z_0 below the melting layer. A conservation of the volume flux of water is assumed to assess the qualitative effect of the collision-coalescence of raindrops on the vertical profiles of the radar variables.

Consider the differential expression for the conservation of the rainwater flux:

$$N(D,z)D^3 V_t(D)dD = N(D_{top}, z_{top})D_{top}^3 V_t(D_{top})dD_{top} \qquad (3.54)$$

where D_{top} and $N(D_{top}, z_{top})$ are respectively the diameter and inital DSD at altitude z_{top} (the top of the rain layer). We want to see how the DSD is being modified toward the surface in the absence of vertical air motions ($w = 0\,\mathrm{m\,s^{-1}}$). Recalling the rate of change of the drop size with altitude (eq. [3.45]), we define:

$$\phi(D) \equiv \frac{dD}{dz} = -\frac{dD}{dt}\frac{1}{V_t(D)}, \qquad (3.55)$$

from which:

$$\begin{aligned} D_{top} &= D + H\phi(D) \\ dD_{top} &= dD + H\phi'(D)dD, \end{aligned} \qquad (3.56)$$

where $H = z_{top} - z$ is the distance of fall below z_{top}.

Substituting in eq. (3.54), we get:

$$N(D,H) = N(D + H\phi(D), z_{top}) \left[\frac{D + h\phi(D)}{D}\right]^3 \frac{V_t(D + H\phi(D))}{V_t(D)}[1 + H\phi'(D)]. \qquad (3.57)$$

Using the expression (3.36) for the terminal velocity of raindrops, eqs. (3.43) and (3.57) can be easily solved numerically to obtain the evolution of the DSD below the initial altitude z_{top}. Let's assume an exponential Marshall–Palmer DSD with a rain rate of $R = 15\ \mathrm{mm\,h^{-1}}$ at the top of the layer ($N_0 = 8 \times 10^3\ \mathrm{mm^{-1}\,m^{-3}}$,

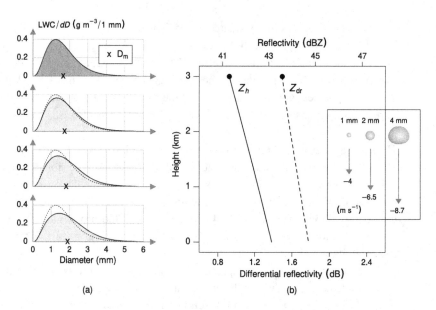

(a)   (b)

Figure 3.15 (a) A DSD with a constant rain rate ($R = 15 \, \mathrm{mm \, h^{-1}}$), evolving by collision-coalescence in calm air, given an exponential Marshall–Palmer DSD located at 3000 m (darker shading). The total liquid water content (LWC) is proportional to the area under the DSD curve and is approximately constant ($W \approx 0.8 \, \mathrm{g \, m^{-3}}$) along the vertical profile. The dotted lines reproduce the initial DSD at $z_{top} = 3000$ m, helping to visualize the mass redistribution across the size spectrum as the collision-coalescence proceeds to lower altitudes ($z =$ 3, 2, 1 and 0 km are illustrated, top to bottom). The 'x' at each altitude are the DSDs' mass-weighted mean diameters, D_m. (b) Corresponding profile of radar variables Z_h and Z_{dr}, calculated under the Rayleigh–Gans approximation (eqs. [9.4] and [9.5]), and using the Beard–Chuang axis ratio approximation for the shape of the raindrops (eq. [3.31]). In the absence of vertical air motions, the magnitude of the drop's vertical velocity is simply given by the terminal fall velocity (inset with example raindrops). The radar variables show increasing values at lower altitudes, as a consequence of the broadening DSD.

$\Lambda = 2.4 \, \mathrm{mm^{-1}}$). Figure 3.15 shows the resulting raindrop-size spectra at different heights, evolving by collision-coalescence from the initial Marshall–Palmer DSD at an altitude of $z_{top} = 3000$ m. The DSD is represented as mass density (liquid water content per unit diameter) as opposed to number concentration to highlight the growth of the larger drops at the expense of the smaller ones. In this representation, the area under the DSD curve is proportional to the total water content, which is nearly constant through the depth of the precipitation layer. The mass-weighted mean diameter D_m increases as the collision-coalescence process continues to broaden the raindrop-size spectrum at lower altitudes. As a consequence, both radar variables Z_h and Z_{dr} also increase by virtue of their particular sensitivity to the larger particles.

For this example, the overall increase in Z_h and Z_{dr} is, respectively, $\sim$2 dB and $\sim$0.3 dB. Values of similar magnitude can be frequently observed below the melting layer in continental precipitation (refer, e.g., to the orographic precipitation case in Fig. 8.5). However, the amount of the increase may significantly vary depending on the shape of the DSD aloft and the total water content. Given the same amount of liquid water content, DSDs characterized by a greater contribution of small drops (larger shape parameter μ) tend to produce a larger enhancement of the radar variables along the vertical profile. For tropical convective rain during the DYNAMO/ARM-AMIE campaign in the central Indian Ocean, Kumjian and Prat [69] reported observed increases of Z_h and Z_{dr} up to ~20 dB and ~1 dB, respectively, over a 3-km layer. They noted, however, that these large dual polarization signatures ascribed to collision-coalescence may have been enhanced by additional raindrop growth due to accretion of cloud water and by the role of vertical air motions.

The contribution of collision-coalescence may be enhanced in an ascending air column, where the larger drops are favored in the collision process because of their longer stay in the cloud. Considering a uniform updraft with magnitude $w = 10\,\text{ms}^{-1}$, a small 1-mm drop will have an upward velocity of ~$6\,\text{m s}^{-1}$, whereas a 4-mm drop will rise at just about $1\,\text{m s}^{-1}$. This leads to an increasingly longer right tail of the DSD. The role of collision-coalescence within an updraft is especially relevant in explaining the often-observed appearance of a vertical intensification of Z_{dr} in convective storms, as discussed in detail in Section 3.6.2.

Radar Signatures of Evaporation Below the Cloud Base

Although the collision-coalescence mechanism is a mass-conservative process where the DSD is reshaped as illustrated in the previous examples, the evaporation of raindrops always involves an overall loss of the water mass in addition to a modification of the DSD shape.

Li and Srivastava [70] proposed an intuitive and simple way to represent the evaporation process and its impact on the radar variables. Following a similar approach, consider the approximate analytical solution in eq. (3.53) to represent the modification of the drop's diameter with height. Rearranging the terms in such a way to separate the left- and right-hand sides to be, respectively, a function of D and H only, results in

$$\frac{V_t D}{f_v}\,dD = 4\frac{S_w - 1}{\rho_w(F_k + F_d)}\,dH. \tag{3.58}$$

In eq. (3.58) the terminal fall velocity V_t is assumed invariant with altitude, neglecting the effect of the reduced air density aloft, and the ventilation factor f_v is an increasing function of the drop diameter (e.g., as in Pruppacher and Klett [60]):

$$f_v = 0.78 + 0.308 \left(\frac{\nu}{D_v}\right)^{1/3} \left(\frac{V_t D}{\nu}\right)^{1/2}, \tag{3.59}$$

where ν is the kinematic viscosity of air.

The integration of the left-hand side of eq. (3.58) between 0 and D can be fitted by a quadratic form:

$$\int_0^D \frac{V_t D}{f_v}\,dD \approx c_1 D + c_2 D^2, \tag{3.60}$$

where $c_1 = 2.008 \text{ cm}^2 \text{ s}^{-1}$ and $c_2 = 30.146 \text{ cm s}^{-1}$ for a midlevel at 800 hPa and $T = 10°C$ [70]. Considering two altitude levels z_1 and z_2 within a dry layer, integration of the right-hand side of eq. (3.58) results in

$$(c_1 D_1 + c_2 D_1{}^2) - (c_1 D_2 + c_2 D_2{}^2) = 4 \int_{z_1}^{z_2} \frac{S_w - 1}{\rho_w (F_k + F_d)} dz \qquad (3.61)$$

where D_1 and D_2 are the drop diameters at altitudes of z_1 and z_2.

If we set $D_1 = 0$ mm in eq. (3.61), a diameter D_{ev} is then defined as the diameter of a drop that, falling from an initial level $z_2 = z_{top}$, would completely evaporate after a fall distance H, at altitude $z_1 = z_{top} - H$:

$$\left[c_1 D_{ev}(H) + c_2 D_{ev}(H)^2 \right] = -4 \int_{z_{top}-H}^{z_{top}} \frac{S_w - 1}{\rho_w (F_k + F_d)} dz \qquad (3.62)$$

The integral on the right-hand side can be solved by considering the temperature dependency of F_k and F_d (consider the example coefficients values for eq. [3.52] included in problem 4 at the end of this Chapter). This establishes a unique relation between D_{ev} and the the fall distance H. As an example, consider a layer of constant relative humidity (RH $= 50\%$) and a dry adiabatic lapse rate (i.e., $-dT/dz$) of $9.8°C \text{ km}^{-1}$. Under these simplified environmental conditions, $D_{ev} = 0.5$ mm would correspond to a fall through a 1-km-depth atmospheric layer, whereas a 1-mm drop ($D_{ev} = 1.0$ mm) would need to fall through a 2.3-km-depth layer to completely evaporate.

The idealized environment with constant relative humidity and uniform adiabatic lapse rate is a rough assumption that is, however, suitable to show the qualitative response of the radar variables to the evaporation for a typical raindrop size distribution. In a real sub-cloud environment, the evaporation of raindrops would cause an increase in the water-vapor pressure and a decrease in temperature as a consequence of the evaporative cooling, leading to a feedback loop between the DSD evolution and the environmental thermodynamic parameters.

To calculate the decrease in size of a raindrop falling from the top of a dry layer subject to evaporation, eqs. (3.61) and (3.62) are used to obtain D at a generic fall distance H from the following quadratic expression:

$$\left[c_1 D(H) + c_2 D(H)^2 \right] = (c_1 D_{top} + c_2 D_{top}{}^2) - \left[c_1 D_{ev}(H) + c_2 D_{ev}(H)^2 \right] \quad (3.63)$$

where D_{top} is the drop diameter at altitude z_{top} and $D_{top} \geq D_{ev}$.

To evaluate the changes in a DSD subject to collision-coalescence, a steady-state process was assumed by imposing the conservation of the mass flux (eq. [3.54]). When evaporation is the only process affecting the DSD, mass is not conserved, but a steady-state scenario can be effectively simulated by assuming conservation of the number flux of raindrops:

$$N(D, z)V_t(D)dD = N(D_{top}, z_{top})V_t(D_{top})dD_{top} \qquad (3.64)$$

Figure 3.16 shows the results of this simple evaporation model for two different initial raindrop size distributions characterized by the same rain rate of $R = 5 \text{ mm h}^{-1}$

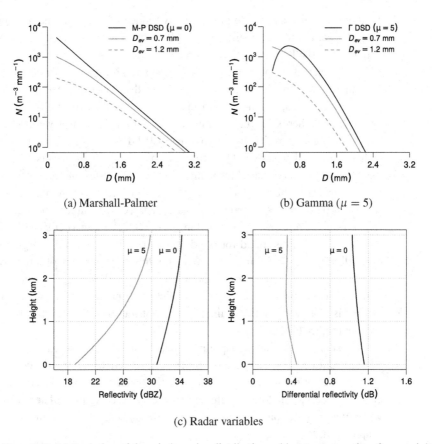

Figure 3.16 (a) Evolution of the raindrop size distribution subject to evaporation, from an initial Marshall–Palmer DSD. $D_{ev} = 1.2$ mm corresponds to a fall of about 3 km. (b) Same as panel (a) but for an initial gamma distribution with $\mu = 5$. (c) Corresponding vertical profiles of the radar variables Z_h (left) and Z_{dr} (right) below $z_0 = 3000$ m. See text for details.

but a different shape. The DSD in panel (a) is a Marshall–Palmer distribution ($\mu = 0$), whereas in panel (b), a gamma distribution is considered with $\mu = 5$. The most notable feature in the modification of the DSD with increasing D_{ev} values (larger evaporation) is the more pronounced decrease in the concentration of the small raindrops compared with the larger ones for both cases. This is expected, considering the inverse relation between dD/dH and the drop diameter (eq. [3.58]).

The shape of the DSD at the top of the dry layer has an impact on the subsequent evolution through the dry layer and specifically on the radar variables Z_h and Z_{dr} (panel [c] in Fig. 3.16). The radar reflectivity Z_h always decreases because of evaporation, although the amount of the decrease depends on the shape of the initial DSD. In particular, the decrease can be much larger for a gamma DSD with a positive μ as compared with an exponential distribution ($\mu = 0$). In fact, the larger contribution of the small drops to the total mass for the gamma distribution (panel [b] of Fig. 3.16) makes evaporation especially efficient with a larger reduction of the overall water content. The rain rate is also more heavily reduced for the gamma distribution (from 5 mm h^{-1} at the initial level $z_{top} = 3000$ m to 0.2 mm h^{-1} at $z = 0$ m) as compared with the Marshall–Palmer distribution (from 5 to 1.3 mm h^{-1}).

The impact of evaporation on Z_{dr} is less straightforward. The larger decrease of the smaller drops often implies an increase of the mass-weighted mean diameter and Z_{dr}. In Figure 3.16, Z_{dr} shows an overall increase of $\sim$0.2 dB through the 3-km-depth layer. On the other hand, for the gamma distribution, Z_{dr} is nearly constant for the first 2 km of fall and increases by approximately 0.1 dB through the last 1000 m.

Kumjian and Ryzhkov [71] performed simulations of the effects of evaporation on the polarimetric radar variables using a bin microphysics one-dimensional rain shaft model. Similar to the simplistic example presented here, they found that Z_{dr} generally shows subtle increases, whereas Z_h and K_{dp} decrease significantly. (The specific differential phase shift, K_{dp}, is defined in Section 4.7.1.) In particular, the largest reduction of Z_h and K_{dp} is found for raindrop size distributions with a large concentration of small drops ($D < 2$ mm) and a relatively low concentration of large drops ($D > 4$ mm). The relative impact of evaporation on Z_h and K_{dp} is also higher for weaker rain rates as opposed to the collision-coalescence process. The collision-coalescence process is more important for raindrop size distributions characterized by a large water content (eq. [3.44]).

For radar applications, evaporation below the cloud base can be detected when a significant decrease in the reflectivity is observed with a nearly constant Z_{dr}. For high-frequency radar, such as X band or Ku band, a significant decrease in K_{dp} can also be observed at lower altitudes. In fact, K_{dp} is tightly related with the liquid water content (Section 9.1.1) and is consequently sensitive to the mass loss caused by evaporation.

The different response of reflectivity and differential reflectivity to the evaporation process is also exploited by Kumjian and Ryzhkov [71] to improve the estimation of the rainfall rate (R) in regions of limited radar visibility (beam blockage as an example), where the low-level atmospheric layer cannot be adequately sampled. In particular, it is demonstrated that the relative error $\Delta R/R$ depends on the value of Z_{dr} at the top of the evaporating layer (just below the melting layer), whereas it is independent of the value of Z_h aloft. Coupling the radar observations aloft with thermodynamic information from sounding or numerical models may then allow for the introduction of a physically based correction for the estimation of the rain rate at the surface.

Sedimentation of Raindrops and Size Sorting

The previous examples provided a simplified representation of two fundamental warm-rain processes (collision-coalescence and evaporation), relying on a steady-state approximation (eqs. [3.54] and [3.64] respectively). This is a useful way to represent the effect of a rain process on the vertical profile of the radar moments. In fact, under the steady-state approximation, the DSD is invariant with time but shows a characteristic modulation with height, which is a signature of the specific dominant microphysical process.

However, there are processes that cannot be represented by the steady-state approximations because they are inherently transient. A notable example is size sorting. Size sorting indicates a modification of the size distribution, which occurs as a consequence of the relative movement of precipitation particles. The most simple and fundamental example of size sorting is the differential sedimentation of raindrops. Differential sedimentation happens all the time because of the variation of the raindrops' vertical fall velocity as a function of its size (e.g., eq. [3.37]). Size sorting is normally a transient effect. In fact, after precipitation starts below the cloud base, the largest drops fall faster and are the first to reach the surface (hence the usual early appearance of large drops at the ground during the early stage of a convective shower). However, after some time, the smaller drops also reach the surface. If rain production at the cloud base is stationary, then the DSD becomes uniform throughout all the underlying atmospheric layers as well.

This can be easily shown by considering the advection equation for the number concentration of raindrops:

$$\frac{dN}{dt} = -V_t \frac{dN}{dz}. \tag{3.65}$$

The time evolution of the DSD below the cloud base can then be easily simulated by solving eq. (3.65) with a finite-difference method and assuming a continuous replacement of the rainwater aloft, that is, a steady-state DSD with $N(D, t, z_{\text{top}}) = N(D, t_0, z_{\text{top}})$.

The results for an initial Marshall–Palmer DSD with $R = 15$ mm h^{-1} are shown in Fig. 3.17. This clearly demonstrate that a steady-state is reached, and the effects of size sorting disappear approximately 10–15 minutes after the onset of precipitation at the cloud base. During the transition period, however, the radar signatures may be very pronounced. In particular, the mass loss on the left tail of the distribution causes a decrease of Z_h and an increase of Z_{dr} because the distribution becomes more skewed toward larger drops (the mass-weighted mean diameter increases). This is represented in panels (c) and (d) of Figure 3.17, which shows the vertical profiles of Z_h (panel [c]) and Z_{dr} (panel [d]) below the cloud base for several simulation times after the onset of precipitation aloft ($t = 0$). In particular, the profiles of Z_h and Z_{dr} appear almost specular, with a marked decrease of Z_h accompanied by an increase of Z_{dr} up to approximately 2 dB above the steady-state value.

Like the evaporation process, size sorting does not imply any interaction between the drops. Evaporation causes a decrease of the drops' mass for all sizes, with the size reduction being more emphasized for the smaller diameters (eq. [3.58]). This causes a decreasing reflectivity and an almost neutral variation or a weak increase of Z_{dr} (Fig. 3.16). On the other hand, in size sorting, the small drops are physically removed from the distribution (they are sorted out) while the largest drops remain unaltered in a resulting narrowed distribution. As a consequence, the impact on differential reflectivity can be much stronger for size sorting compared with evaporation, with Z_{dr} increasing up to several decibels, as illustrated in Figure 3.17.

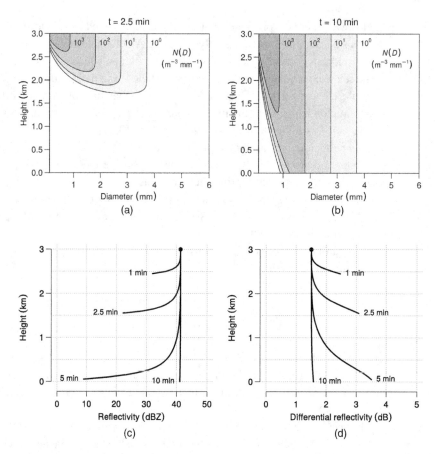

Figure 3.17 (a) and (b) DSD below the cloud base, showing the narrowing due to differential sedimentation for two simulation times (2.5 and 10 minutes) after the onset of precipitation at a 3-km height. (c) and (d) The simulated radar variables at several simulation times. After 10 minutes from the start of rain production aloft, the vertical profile of the radar variables appears almost uniform, indicating the establishment of a steady-state regime.

In other situations, such as in the presence of an updraft or vertical wind shear, size sorting can be enhanced and last longer, producing distinct signatures of the radar variables. One such special signature is the vertical column of Z_{dr}, a columnar enhancement of differential reflectivity often observed in convective storms and resulting from the combined effect of collision-coalescence and size sorting. These columnar regions may extend up to 2–3 km above the environmental freezing level (see Section 3.6.2). The distinct raindrop size distributions arising from size sorting have also been considered in the construction of hydrometeor classification algorithms, which include a specific "large drops" category [72] (refer to Chapter 8).

3.5.5   Cold Rain: Ice-Crystal Growth by Vapor Deposition

In cold clouds, ice formation generally takes place by heterogeneous freezing over ice nuclei below $-10°C$. Once the crystals are formed by nucleation, further growth in the atmosphere may take place by three different processes: deposition, riming, and aggregation. All these processes result in the vertical profile that we observe with radar in ice and mixed-phase clouds (see, e.g., Fig. 3.20).

Deposition occurs whenever ice crystals exist in a water-saturated cloud (super-saturated with respect to ice), by means of water-vapor diffusion. As illustrated in Figure 3.11, both the saturation vapor pressure with respect to ice, and with respect to water, increase with temperature as a result of the release of water vapor through evaporation and sublimation. The difference between the saturation vapor pressures for water and ice $(e_s - e_{si})$ has a maximum at around $-12°C$. This is not readily visible in Figure 3.11 because of the logarithmic plot, but it is shown in the crystal habit diagram in Figure 3.18 for the excess water saturation (solid thick line), expressed as vapor density $(g\,m^{-3})$:

$$\rho_s - \rho_{si} = \frac{e_s - e_{si}}{R_v T}, \tag{3.66}$$

where $R_v = 461.5\,J\,kg^{-1}\,K^{-1}$ is the gas constant for water vapor.

In the cold-rain process, the precipitation formation in the ice phase is mainly attributed to the Wegener–Bergeron–Findeisen (WBF) process, often simply referred to as the Bergeron process, which takes place when ice crystals and supercooled water droplets coexist. In this case, because of the difference in saturation vapor pressure with respect to water and ice, the ice crystals grow by deposition ($S_i > 1$) at the expense of the evaporating water droplets ($S_w < 1$). The growth of ice crystals by diffusion is governed by an equation analogous to the diffusional growth and evaporation of water drops (eq. [3.47]):

$$\frac{dm}{dt} = D_v \int_s \nabla \rho_v ds = 4\pi C D_v \left[\rho_v(\infty) - \rho_v(s)\right], \tag{3.67}$$

where ρ_v is the density of water vapor around the ice crystal (∞) and at the crystal surface (s), D_v is the diffusion coefficient, and C is the capacitance of the ice crystal in length units. The analytical approximation (eq. [3.51]) can also be used for the ice crystals' growth:

$$\frac{dm}{dt} = 4\pi C f_v \frac{S_i - 1}{F_k^i + F_d^i}, \tag{3.68}$$

where in this case, $S_i = e/e_{si}$ is the ambient water-vapor saturation ratio with respect to ice, and F_k^i and F_d^i are the conductivity and vapor-diffusion terms defined following eq. (3.52) but replacing e_s by e_{si} and L by the latent heat of sublimation L_s. In a water-saturated environment, $S_i \approx e_s/e_{si}$ and shows a marked temperature dependence, which is illustrated by the difference between e_s and e_{si} in logarithmic scale in Figure 3.11. Considering the combined variation with temperature and pressure of the term $1/(F_k^i + F_d^i)$ in eq. (3.68), it can be shown that the crystal growth rate by diffusion is at a maximum near $-15°C$ for a wide range of pressures [59].

Table 3.2 Capacitance of simple geometrical shapes. In the following relations, r represents the radius of the sphere and disk shapes, and a (b) are the major (minor) semiaxes of the prolate and oblate shapes. Adapted from Westbrook et al. [75].

Shape	Capacitance
Sphere	$C = r$
Disk	$C = 2r/\pi$
Prolate spheroid	$C = A/\ln[(a + A)/b], A = \sqrt{a^2 - b^2}$
Oblate spheroid	$C = ae/\arcsin(e), e = \sqrt{1 - b^2/a^2}$

Equation (3.67) was derived by Maxwell in analogy with electrostatics (see, e.g., [59]), with C representing the capacitance of the ice crystal expressed with length units (i.e., the electrical capacitance in farads, normalized by $4\pi\epsilon$, where ϵ is the permittivity of the surrounding medium). The capacitance is a function of the size and shape of the ice particle. For a sphere, the capacitance $C = r$, and eq. (3.68) reduces to the growth equation of the water drops (eq. [3.51]). However, the solid phase of water admits a variety of shapes for the ice crystals as a consequence of the internal molecular structure and the environmental parameters of temperature and humidity. The complex shape of ice crystals, as opposed to the spherically symmetric cloud droplets, affects the distribution of the water-vapor density around the particle. The resulting vapor density gradients may be further enhanced by ventilation near the edges of the ice crystals, influencing the formation and growth of the ice crystals in a range of shapes denoted as *crystal habits* [73]. The capacitance of the simple geometrical shapes in Table 3.2 can be used as a first approximation to estimate the growth rate of ice crystals (e.g., the thin disk shape for plate-like crystals or the prolate shape for needles). For an accurate estimation of the capacitance of real ice crystals, more complex approaches have recently been proposed, including finite-element methods to solve Gauss's law (eq. [3.67]) numerically [74] or using Monte Carlo methods [75].

In clouds containing both ice crystals and supercooled drops (mixed-phase clouds), the WBF process is especially efficient between approximately $-12°C$ and $-16°C$, where the difference between the saturation vapor pressure over water and the saturation vapor pressure over ice is the greatest. In this region, dendritic crystals present the most effective shape for the deposition of the ambient water vapor by virtue of the high surface-to-volume ratio [68], as depicted in the ice-crystal diagram in Figure 3.18.

The ice-crystal diagram is adapted according to the recent revision of Bailey and Hallet [76] and shows the approximate range of differential reflectivity values associated with different crystal shapes. The ice-crystal habit is a function of both temperature and supersaturation and presents a succession of plate-like and columnar regimes with decreasing temperatures. With the exception of electrically active thunderstorms,

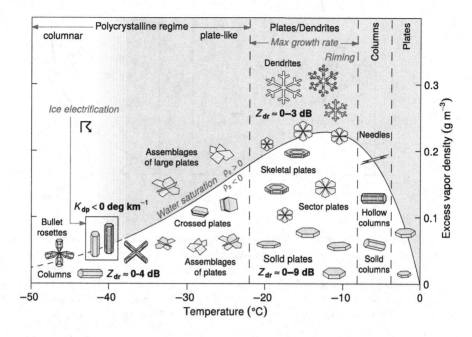

Figure 3.18 Main crystal habits as a function of ambient temperature and excess vapor density following Magono and Lee [77], recently revised by Bailey and Hallet [76]. The vertical dashed lines indicate the primary habit transitions. The solid thick line represents the saturation vapor pressure over water (eq. [3.66]). In the region above (gray shading), the air is supersaturated with respect to water and ice; the region below is only supersaturated with respect to ice. Riming over single crystals generally has the effect of reducing the positive Z_{dr} signal as a result of a decrease in the particles' aspect ratio and a wider distribution of canting angles [78]. Refer also to Figure 4.21 for typical values of Z_{dr} for oblate particles (approximating solid plates and dendrites) and prolate particles (columns and needles) with varying densities. In the glaciated portion of the cloud (leftmost side of the diagram), ice electrification in thunderstorms may induce vertical alignment of ice crystals, typically resulting in a negative specific differential phase shift K_{dp} (Section 8.5.3). Apart from this exception, ice crystals are drawn with their major axis mainly horizontal to represent the standard fall behavior due to aerodynamic forces. The schematic crystal drawings follow the global classification of snow crystals proposed by Kikuchi et al. [79].

where crystals may be oriented vertically, the preferential orientation of ice crystals, including columns, is horizontal with positive differential reflectivity.

3.5.6   Coexistence Regime

Ice crystals and liquid drops may grow simultaneously in mixed-phase clouds under appropriate microphysical and dynamical conditions. When the vapor density is below the water-saturation level (white area under the water-saturation line in Fig. 3.18), ice crystals grow at the expense of the liquid drops, which evaporate in the subsaturated (with respect to water) air. On the contrary, when the level of supersaturation is above the liquid water saturation (gray-shaded area in Fig. 3.18), both ice crystals and liquid

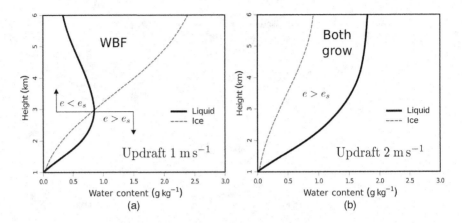

Figure 3.19 Modeled water content change with height for the liquid (solid thick line) and ice phase (gray dashed line) for an adiabatic parcel ascending with a velocity of 1 m s^{-1} (a) and 2 m s^{-1} (b) in a mixed-phase cloud. The different behavior highlights the role of the updraft in the prevalence of the WBF process versus the coexistence regime ("both-grow" process), where ice particles and liquid drops grow simultaneously. Adapted from Korolev [81], ©American Meteorological Society. Used with permission

drops may grow at the same time. However, this is not a stable condition because the depletion of the available water vapor will soon act to decrease the level of saturation below the water saturation. The water supersaturation condition may be maintained if the air mass is undergoing an upward displacement. In this case, the adiabatic cooling would determine a supersaturation growth, counterbalancing the vapor-pressure decrease as a result of the water-vapor depletion by ice particles. The needed updraft may be as small as ~0.1 m s^{-1} and could be provided by turbulence, weak mesoscale ascent in stratiform systems, or convective motions [80].

Figure 3.19 shows the evolution of the water content with height according to the numerical simulation illustrated by Korolev in his critical revision of the WBF process in mixed-phase clouds [81]. The liquid and ice water content show a distinct variation with height, depending on the updraft strength, evidencing a larger amount of supercooled water at higher altitudes for the stronger updraft. In particular, the numerical simulation by Korolev [81] shows that for the weaker updraft, the WBF process is enabled at altitudes above ~3 km, where the ice particles grow at the expense of the evaporating liquid droplets. On the other hand, a 2-m s^{-1} updraft is sufficient to allow the simultaneous growth of ice particles and liquid droplets throughout the vertical column, showing that even a relatively weak updraft can turn off the WBF process. Similar to the simultaneous growth in the updraft, a simultaneous evaporation of liquid drops and ice particles may happen in the downdraft region of the cloud.

The relative abundance of supercooled water in stronger updrafts has an important impact on the subsequent precipitation growth processes and, by extension, on the radar observations of reflectivity and dual polarization variables. In particular, riming through accretion of supercooled water droplets may be relevant even with relatively small updraft velocities.

3.5.7 Further Growth by Aggregation and Riming

Aggregation and riming are two different collection processes. Aggregation only involves ice particles, whereas riming requires the presence of supercooled liquid drops. Similar to the collection of liquid drops (collision-coalescence process), the prerequisite condition for riming and aggregation is a sufficient breadth of the particles' fall speed spectrum, which implies differential sedimentation and increasing chances of collisions. Weiss and Hobbs [82] have shown that the ice-particle growth mode is related to the vertical gradient of the mean particle fall speed. Small gradients are associated with growth by vapor deposition, whereas larger gradients favor the growth by aggregation and riming.

In addition to the differential fall speed, aggregation depends on the particular shape of the ice crystals and the ambient temperature. In general, the increasing stickiness of the ice particles with temperature makes aggregation more likely to occur at warmer temperatures ($> -5°C$) compared to depositional growth, which is most efficient around $-15°C$.

The aggregation of ice crystals results in ice particles with larger diameters and lower bulk density. While the augmented size causes an increase of the radar reflectivity, the lower density has the opposite effect as a result of the decreasing magnitude of the refractive index. Given the strong dependence of the reflectivity on the particles' dimension (D^6), the size effect is dominant, and the reflectivity increases in response to aggregation. As a matter of fact, the vertical profile of reflectivity is often characterized by an increased height derivative as a result of the transition from vapor deposition to aggregation processes [24] (see, e.g., the stratiform profile in Fig. 3.20). Although aggregation normally occurs at temperatures warmer than $-5°C$, a secondary maximum between $-10°C$ and $-16°C$ may exist when the arms of the dendritic crystals become entangled [68]. The particular shape of dendritic crystals with open structures is believed to provide a favorable condition for the crystals to stick upon collisions. This phenomena is illustrated by the inflection point around $-15°C$ in the reflectivity profile in panel (a) of Figure 3.20, which is almost coincident with the increased differential reflectivity associated with dendritic growth by vapor deposition [83].

The decreasing magnitude of the refractive index combined with the random falling behavior of aggregates tends to reduce the value of both Z_{dr} and K_{dp}. In the example profiles of Figure 3.20, Z_{dr} decreases below the $-15°C$ level, where there is a coexistence of dendrite crystals and aggregates. As the temperature increases further when approaching the freezing level, aggregation becomes the dominant process, and Z_{dr} slowly decreases with increasing Z_h.

Aggregation is most important in widespread stratiform precipitation systems with a relatively small amount of supercooled water drops. On the other hand, riming is a dominant precipitation growth process in convective storms, where it may even-

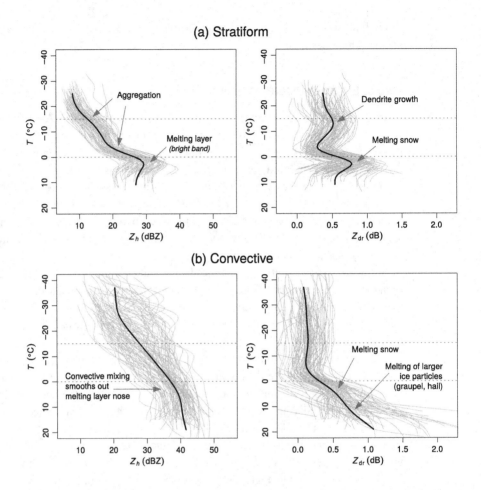

Figure 3.20 Hourly average vertical profiles (gray thin lines) of reflectivity (left) and differential reflectivity (right) from C-band observations during 54 precipitation events in northern Italy [83]. The stratiform (a) and convective (b) profiles are normalized with respect to the environmental temperature in order to compare observations from different events encompassing more than 1 year. The thick black lines represent the median over all hourly profiles.

tually lead to the formation of hail. Riming is also often referred to as *accretion* and involves the capture of supercooled drops by ice-phase particles, which may be either individual crystals or larger aggregate particles. Riming is favored by the intrinsic different fall speeds of liquid drops and ice particles (Table 3.1). Just like aggregation, riming tends to increase the aspect ratio of the ice particles, making them more spherical. Unlike aggregation, the density of the rimed particle increases (up to a value of $\sim$0.3 $\mathrm{g\,cm^{-3}}$ for rimed snow or $\sim$0.8 $\mathrm{g\,cm^{-3}}$ for graupel). Riming is associated with enhanced reflectivity, but generally a low Z_{dr} because the shape effect tends to dominate over the increased refractive index.

Heavy riming in strong convective updrafts may lead to large graupel and hail, which are typically characterized by very high reflectivity (> 50 dBZ) and nearly zero

Z_{dr}. On the other hand, when graupel and hail become wet at warmer temperatures, the dual polarization radar signatures may assume distinctive values (large Z_{dr} and low ρ_{hv}), as discussed in Section 3.6.3.

In the observations of cold clouds, there is a notable discrepancy in the ratio of the number of ice crystals in the cloud and the number of ice nuclei in the ambient air, which can reach a factor of 10^4 [59]. The fact that there is no one-to-one relation between ice particles and ice nuclei implies the existence of secondary processes for ice formation. The main mechanism for ice multiplication is the Hallet–Mossop process [84] (also called *rime splintering*), by which secondary ice particles are produced as graupel grows by accretion in the temperature range between $-3°C$ and $-8°C$. In fact, when a large enough ice particle collides with supercooled drops, the accretion of rime ice is often accompanied by the shedding of small ice splinters. These small ice particles are mostly columnar crystals or needles and may continue to grow by vapor deposition. Therefore, despite the expected low magnitude of Z_{dr} associated with riming, when secondary ice production by the Hallet–Mossop process takes place, Z_{dr} may be enhanced as a result of the presence of smaller ice crystals with preferential horizontal orientation. This was shown by Vogel and Fabry [85] in a study combining observations from a vertical pointing X-band radar and a dual polarization S-band scanning radar. They showed that a higher degree of riming is generally accompanied by a decrease in the observed differential reflectivity. However, when secondary ice production is inferred from the bimodal spectra of the vertical pointing radar, the differential reflectivity may be higher than that for nonriming cases because of the presence of oriented crystals resulting from rime splintering.

The specific differential phase shift K_{dp} is not sensitive to nearly spherical particles such as aggregates, so it may show a marked increase when oriented ice crystals produced by ice splintering appear in the temperature range of $-8°C$ to $-3°C$ [86]. In addition to rime splintering, other possible secondary ice-production mechanisms include collision fragmentation, droplet shattering, and sublimation fragmentation. Field et al. [87] give a comprehensive overview of the secondary ice-production processes supported by in situ observations, remote sensing (including dual polarization radar observations), and modeling studies.

3.6 Stratiform and Convective Precipitation

Another way to classify precipitation is the widely used partition between stratiform and convective systems. This partition is very common in weather radar applications because of the specific distinguishing signatures arising from the different microphysical processes discussed in the previous sections.

The vertical air motion influences the processes controlling the formation of precipitation particles [88]. As such, the precipitation can be broadly classified as stratiform when

$$|w| \ll V_{ts}, \tag{3.69}$$

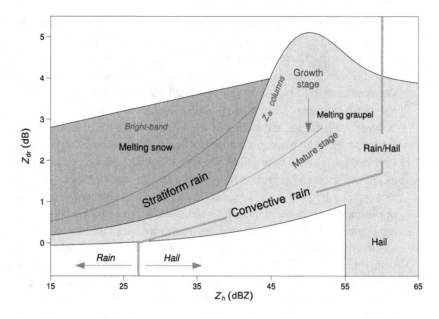

Figure 3.21 Schematic partition of the S-band polarimetric Z_h–Z_{dr} space, showing the salient characteristics of convective (light gray) and stratiform (dark gray) precipitation. The rain-hail partition was originally introduced by Aydin et al. [113] in the eighties and is reported later in eq. 3.76.

where w is the vertical air velocity, and V_{ts} is the terminal fall speed of ice-precipitation particles (~ 0.5–$1\ \mathrm{m\,s^{-1}}$). Conversely, the precipitation type is defined to be convective precipitation when eq. (3.69) does not hold.

Early weather radar observations greatly contributed to characterizing these two types of precipitation, given the specific morphology of the reflectivity field in the horizontal and vertical planes. The advent of dual polarization expanded the benefits of morphological studies with additional radar measurements, leading to the partitioning of the dual polarization parameter space, as shown in the example Z_h-Z_{dr} diagram in Figure 3.21. This type of diagram provides the basis for the hydrometeor classification methods discussed in Chapter 8. Figure 3.21 highlights a common trait in the polarimetric observations of reflectivity and differential reflectivity. For a given weak to moderate Z_h, convective rain shows on average lower Z_{dr} than stratiform rain. This is explained by considering the different dominant rain-formation processes. In particular, in stratiform precipitation, rain originates from ice aggregates (snowflakes), which can lead to the formation of large drops upon melting (Section 3.6.1).

As shown in Figure 3.21, the stratiform rain region is confined to a limited portion of the Z_h–Z_{dr} parameter space. For convective precipitation, the presence of partially melted ice particles or transient DSDs, such as in the leading edge of storm cells, may enlarge the spectrum of differential reflectivity observations considerably. In particular, the Z_h–Z_{dr} pairs show a different behavior during the growth and mature phases of convection. The growth stage is generally characterized by large values of

Z_{dr} (Section 3.6.2). During the subsequent evolution to the mature stage, the DSD evolves from a low concentration of large drops to an exponential shape as a result of increasing collisions that lead to drop breakup. As a consequence, the radar observations in the $Z_{dr}-Z_h$ parameter space shift to the lower Z_{dr} values characteristic of mature convective precipitation. The schematic partition in Figure 3.21 is an idealized representation to highlight the salient dual polarization signatures associated with different precipitation types. In reality, as explained in the discussion of hydrometeor classification methods (Chapter 8), there can be a significant overlapping between the hard-boundary regions depicted in the diagram. For example, observations of moderate Z_h and large Z_{dr} during the growth stage in convective storms may present similar values to observations collected during stratiform precipitation. For this reason, the analysis of pointwise observations may not be sufficient to identify the characteristic precipitation type. The specific morphology of the dual polarization observations ("signatures") provides additional information about the microphysical and dynamical structure of precipitation.

In the following sections, relevant phenomena associated with the characteristic radar signatures in stratiform and convective precipitation are analyzed in more detail. In particular, for stratiform systems, the discussion focuses on the melting layer. For convective precipitation, the radar signatures of strong updraft (Z_{dr} columns) and hail are elaborated.

3.6.1   Precipitation across the Melting Layer

Besides the meteorological definition of stratiform precipitation (eq. [3.69]), from the radar's point of view, stratiform rain systems are easily identified when the characteristic signature of the melting layer appears on the radar display. Examples of PPIs showing the circular signature of melting snow are shown in Chapter 8 (Fig. 8.2). Figure 3.22 shows a conceptual model illustrating the main physical mechanism affecting the radar observations during the transition from the solid phase to the liquid phase. The key physical factors determining the vertical profiles of dual polarization variables within the melting layer are (1) the increase of the dielectric factor due to the phase transition from ice to water, (2) the reduced size of the particles as a consequence of the increased density, and (3) the lower number concentration caused by the larger terminal fall speed of raindrops as compared to ice particles.

During the initial stage of melting, small water droplets begin to form on the tips of the dendrite crystals (see insets in Fig. 3.22). Because of the higher permittivity of water with respect to ice (see Section 4.2), the reflectivity starts to increase. However, at this stage, the initial shape of the aggregates, whether they are irregular or nearly spherical, is maintained because the frame of the ice particles is still almost intact. As a consequence, the differential reflectivity is barely affected by the change in the dielectric constant. As melting proceeds, a larger amount of water affects the particle shape, which becomes increasingly oblate, thus resulting in a rapid increase of Z_{dr}. The meltwater's mass is held together by a few crystal branches. As a consequence, the asymmetrical mass distribution may lead to the formation of individual droplets

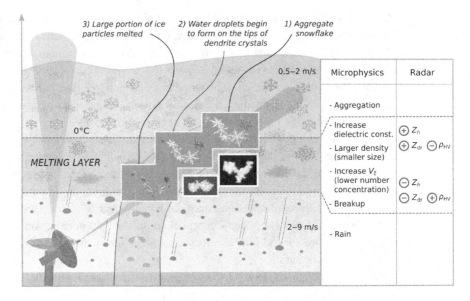

Figure 3.22 Conceptual model of the melting layer. Snowflake images from the smoothed particle hydrodynamics model of Leinonen and von Lerber [89] show the process of melting below the freezing level (1–3), along with actual pictures of melting snowflakes. The main microphysical processes and corresponding changes in the radar variables are synthesized in the right columns.

that are shed from the edges of the flakes. At any given height within the melting layer, because of the distribution of the ice-particle diameters, there is a coexistence of smaller particles that are completely melted and larger particles that are only partially melted. This mixing of hydrometeors with different shapes and dielectric factors causes a drop in the correlation coefficient.

This simplified melting sequence may vary depending on the snowflake size and density, air temperature, and relative humidity. Using a smoothed particle hydrodynamics model, Leinonen and von Lerber [89] have shown that unrimed or lightly rimed aggregates (low-density snow) show a significant amount of water on the surface at the first stage of melting, whereas heavily rimed aggregates (higher-density snow) accumulate water inside as a result of the porous crystal structure. Low-density snow, as opposed to highly rimed aggregates, is also more prone to breakup during the last stage of melting as a consequence of the thinner crystal connections.

Figure 3.23 shows an example of the X-band profiles of Doppler velocity spectrum at vertical incidence (panel [a]) and dual polarization variables Z_h (panel [b]), Z_{dr} (panel [c]), and ρ_{hv} (panel [d]) at 15°elevation. It can be seen that the peak in Z_{dr} is located approximately 100 m below the peak in Z_h, whereas the minimum of the correlation coefficient is somewhere in between. The Doppler spectrum also highlights the phase transition from ice to raindrops from 3.4 to 2 where the Doppler spectrum shifts more negative due to the increased terminal velocity. The gap between the Z_h and Z_{dr} peaks is normally clearly seen at higher elevations (shorter ranges),

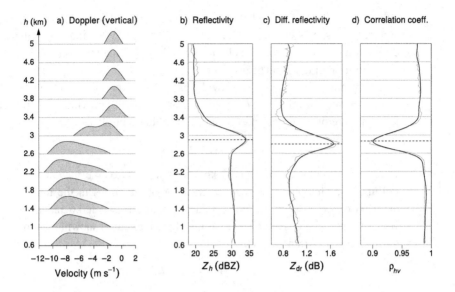

Figure 3.23 Observations through the melting layer in uniform stratiform precipitation with an X-band radar in northern Italy. Normalized Doppler power spectrum from a vertically pointing scan (a). Reflectivity (b), differential reflectivity (c), and correlation coefficient (d) average profiles from a 15° elevation scan taken right after the vertical scan. The thick solid lines highlight the salient features of the dual polarization variables, through a spline fit on the 360° average profiles (gray thin line). The dashed lines mark the height of the peak (Z_h and Z_{dr}) and minimum value (ρ_{hv}) corresponding to the melting of ice particles.

whereas it is typically not perceivable at lower elevations (longer ranges) because of the smoothing effect from beam-broadening. From the microphysics point of view, the height gap between the Z_h and Z_{dr} peaks is related to the snowflakes' distribution aloft and the degree of aggregation above the freezing level. In fact, when aggregation is not significant, melting of small oriented pristine ice crystals (either plate-like or columnar) will result in a small or negligible increase in reflectivity but a well-defined increase of Z_{dr}, typically confined to a thin layer (small particles melt faster than larger aggregates). For this reason, the dual polarization observations of Z_{dr} and ρ_{hv} can help discriminate the melting layer even when there is not a clear "bright band" on the reflectivity map.

Below the Z_h and Z_{dr} peaks, the increasing terminal fall velocity of the melted particles assumes a dominant role in shaping the vertical profiles. In fact, let's consider a simple model where the ice particles evolve subject only to melting, without any other physical interactions. In this case, the conservation of the number concentration across the melting layer is expressed as

$$N_s(D_w)V_{ts}(D_w)dD_w = N_w(D_w)V_t(D_w)dD_w, \tag{3.70}$$

where $N_s(D_w)$ and $N_w(D_w)$ denote, respectively the number concentration of snowflakes (in terms of the melted diameter D_w) and raindrops just above and below the melting layer, and V_{ts} and V_t are the terminal fall velocities of ice particles and raindrops. Because $V_{ts} < V_t$ (Table 3.1), eq. (3.70) implies that the number concentration of raindrops (melted snowflakes) is smaller than the corresponding concentration of dry snowflakes aloft.

The terminal fall velocity of snowflakes is often expressed as a function of the size using a power-law relation. For example, Langleben [90] provided an expression valid for a mixture of dendrites and irregular assemblages of plates:

$$V_{ts} = 0.90\, D_w^{0.61}, \tag{3.71}$$

where the velocity (in $\mathrm{m\,s^{-1}}$) is expressed in terms of the melted diameter D_w in millimeters. This above fit was obtained for observations in the range of a melted diameter of 0.6–2.4 mm, resulting in fall velocities between 0.6 and 1.5 $\mathrm{m\,s^{-1}}$. Other velocity–size relations are given in terms of the maximum snowflake dimension D_s in millimeters, for example [91]:

$$V_{ts} = 0.96\, D_s^{0.12}. \tag{3.72}$$

Considering eq. (3.71) and the analogous formulation for the terminal fall velocity of raindrops (eq. [3.34]), eq. (3.70) is simplified to:

$$N_s(D_w) = \frac{V_t(D_w)}{V_{ts}(D_w)} N_w(D_w) \approx 4 N_w(D_w). \tag{3.73}$$

The concentration of raindrops below the melting layer is then about four times smaller (or -6 dB lower) than the concentration of snowflakes aloft (in terms of the melted diameter).

The snowflake density ρ_{snow} is generally a decreasing function of size. A wide range of empirical density–size or mass–size relations for ice particles can be found in the literature (see, e.g., Heymsfield et al. [92] and references therein) of the form:

$$\rho_{\text{snow}}(D_s) = \alpha\, D_s^{\beta}. \tag{3.74}$$

As an example, in the relation used by Fabry and Szyrmer [93], the coefficients have values $\alpha = 0.15$ and $\beta = -1$; that is, the density is inversely proportional to the size of the ice particle. Assuming spherical snowflakes and enforcing mass conservation ($\rho_w D_w^3 = \rho_{\text{snow}} D_s^3$),

$$D_s = \left(\frac{\rho_w}{0.15}\right)^{1/2} D_w^{3/2}, \tag{3.75}$$

where both D_s and D_w are in millimeters, whereas ρ_w and ρ_{snow} are in grams per cubic centimeter. Under the simplified hypotheses represented by eqs. (3.70)–(3.75), and assuming a Marshall–Palmer DSD below the melting layer corresponding to $R = 5\ \mathrm{mm\,h^{-1}}$ and $Z_h = 35$ dBZ, it is possible to show the corresponding distribution of dry snowflakes above the melting layer (Fig. 3.24). Although the distribution of snowflakes in terms of the melted diameter $N_s(D_w)$ corresponds to the same exponential distribution of raindrops shifted by a factor of 4, the distribution in terms of the snowflake size $N_s(D_s)$ has a modified shape as a consequence of the density–size relation (eq. [3.74]).

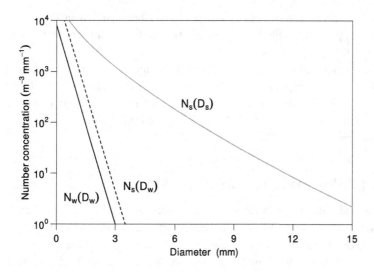

Figure 3.24 PSDs above and below the melting layer. $N_w(D_w)$ is the Marshall–Palmer exponential distribution of raindrops below the melting layer ($N_0 = 8 \times 10^3$ mm^{-1} m^{-3}, $\Lambda = 3.0$ mm^{-1}), and $N_s(D_w)$ is the corresponding snowflake distribution above, expressed in terms of the melted diameter. $N_s(D_s)$ is the snowflake distribution expressed in terms of the snowflake diameter D_s.

The lower concentration of raindrops is one of the factors causing the decrease in the radar reflectivity in the lower portion of the melting layer. Because Z_{dr} is insensitive to the particles' concentration, its maximum value is generally observed closer to the bottom of the melting layer.

The simplified assumption in eq. (3.70) is the basis of model simulations that assume spheroidal shapes for melting snowflakes in the absence of accretion and aggregation (e.g., Szyrmer and Zawadzki [94]). This is generally a good approximation to explain the observed vertical profiles of dual polarization variables. The conservation of the mass flux across the melting layer (i.e., the similarity of the snowfall rate just above the melting layer and the rain rate immediately below) is also confirmed by in situ observations, as reported by Heymsfield [95]. Barthazy et al. [96] have shown strong observational evidence of aggregation of partially melted snowflakes within the upper part of the melting layer and breakup within the lower part. The net result is still a correspondence between the number flux density of snowflakes just above the melting layer and the number flux density of raindrops below, implying that on average, one snowflake yields one raindrop.

Accretion may play an important role in the microphysical evolution of precipitation within the melting layer (see also Section 8.1.1). Troemel et al. [97], in their study of the backscatter differential phase shift δ_{co} within the melting layer, developed a simple model for the representation of accretion that is able to explain the observed signatures of δ_{co} at the S and C bands. (See Section 4.7.2 for an introduction to δ_{co}.) The smaller-size snowflakes are the first to completely melt near the top of the melting layer, resulting in faster-falling liquid drops. The accretion of these small droplets by larger, partially melted snowflakes may lead to a significant increase in the water content of larger particles.

The breakup of partially melted snowflakes and raindrops in the lower part of the melting layer is responsible for the decrease in differential reflectivity (panel [c] of Fig. 3.23). During this stage, as the largest snowflakes are being completely melted, the correlation coefficient (panel [d]) rises to near unity, reflecting the coherent structure of rainfall.

3.6.2   Strong Updraft and Convection – Z_{dr} Columns

In the dual polarization parameter space represented in Figure 3.21, there is a notable departure from the stratiform/convective partition corresponding to the leading edge of convective storms, where the most vigorous growth phase is commonly taking place. In this region of the storm, the updraft contributes to depleting the smaller drops through size sorting (Section 3.5.4), whereas large raindrops may grow through enhanced collision-coalescence (Section 3.5.1), leading to a more skewed DSD and higher differential reflectivity. Z_{dr} in this region can reach up to approximately 5 dB at the S band and 7 dB at the C band as a result of non-Rayleigh scattering. These features can be easily detected in PPI scans as very high-Z_{dr} spots near the storm's edge. Despite their limited spatial extension, these regions with anomalously high Z_{dr} values may reveal important microphysical and dynamical processes of convective storms. When the positive Z_{dr} regions extend upward to considerable altitudes above the environmental freezing level, they are named Z_{dr} *columns*. These columnar enhancements of Z_{dr} observed since the 1980s [98, 99] have been attributed to the lofting of large supercooled raindrops above the freezing level. More recent studies [100, 101] confirmed the tight connection with the strength of the updraft and revealed a lagged correlation with the intensity of surface precipitation and the occurrence of hail in deep convection.

However, the occurrence of Z_{dr} columns is not an exclusive trait of deep convective storms; it is commonly observed in shallow convection as well [102]. It can be regarded as a signature indicating a particular relevance of the warm-rain processes for the growth of precipitation in convective systems. Plummer et al. [102] observed that shallow convective cells in England are associated with higher rainfall intensities compared with the general population of convective cells, suggesting that this particular signature reveals the activation of an efficient rain-producing mechanism.

The recirculation of raindrops formed aloft back into the updraft, as illustrated in Figure 3.25, may favor the collisions between drops, making coalescence effective. In fact, when the updraft intensity matches the fall speed of the large raindrops, these may remain suspended and increase their chances of collision with smaller drops ascending from the lower levels, which are in turn greatly sorted out by the updraft because of their slower terminal velocity. The combined effect of collision-coalescence and size sorting (refer to the simplified examples in Section 3.5.4) explains the typical morphological evolution of the Z_{dr} column. When the coalescence-grown large drops

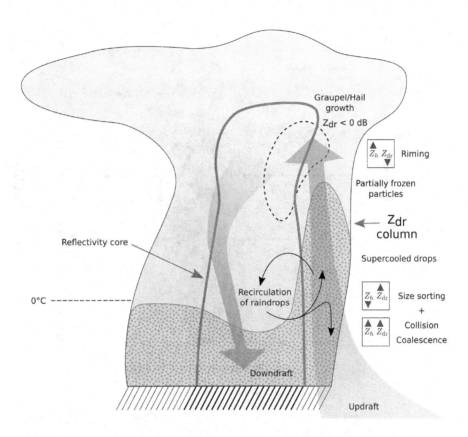

Figure 3.25 Conceptual diagram of a Z_{dr} column. The offset between the column and the reflectivity core is related to the updraft/downdraft circulation and may favor longer-lived storms.

are lofted upward above the freezing level, they do not immediately freeze [60] and may remain in the liquid phase for several minutes. During this time, the particles may typically ascend 1–2 km within the updraft. As the drops get partially frozen, the correlation coefficient decreases, and a linear depolarization ratio (LDR) signature may appear to cap the Z_{dr} column [99, 103], indicating the development of mixed-phase conditions initiated by the freezing of supercooled raindrops. With enhanced cooling aloft, the frozen particles may produce graupel and become embryos for the growth of hail [103]. Finally, depending on the vertical wind shear inside the storm, the falling graupel and hail particles may rapidly deplete the Z_{dr} column signature (in the case of ordinary thunderstorms with negligible wind shear) or develop an adjacent hail signature with nearly zero or negative differential reflectivity (Fig. 3.25). In this case, the Z_{dr} column may persist for a longer time.

Figure 3.26 shows an example of an ordinary convective storm with an extended Z_{dr} column centered within the peak updraft, as inferred from multiple-Doppler analysis [21]. The 1-dB contour of Z_{dr} reaches an altitude of approximately 7 km (around $-10°C$), with highest Z_{dr} values in the lower portion of the convective cell

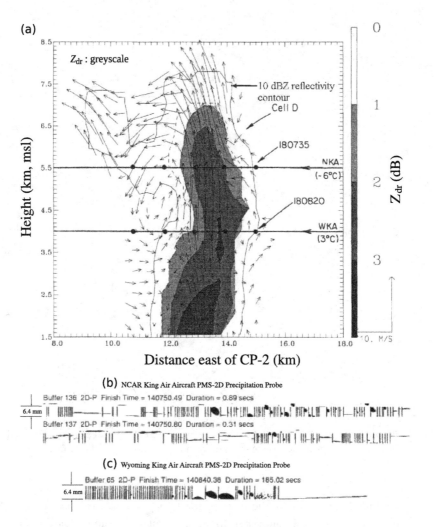

Figure 3.26 (a) Vertical section of radar data (from the NCAR CP-2 radar) in a vigorous growing cell in Florida. Gray scales depict Z_{dr} values (from 0 to 3 dB), with darker shades representing larger values. An outer 10-dBZ reflectivity contour is also shown. Arrows depict triple-Doppler-derived wind vectors in the vertical plane. Two aircraft penetration tracks are shown (NKA: NCAR King Air; WKA: University of Wyoming King Air), with solid dots spaced at 10-second increments. (b) Drop images from an imaging probe on the NKA. (c) Drop images from an imaging probe on the WKA. From Bringi et al. [99], ©American Meteorological Society. Used with permission

(Z_{dr} reaches 3.6 dB at a 1.7-km altitude). The peak reflectivity of 45 dBZ is located at an altitude of 2 km. The two horizontal lines with filled circles indicate the tracks of two aircrafts (NKA: NCAR King Air; WKA: University of Wyoming King Air) at altitudes of 5.5 and 4.0 km, respectively. Along these tracks, the peak reflectivity is 37 dBZ, whereas Z_{dr} reaches 2.7 dB. The updraft magnitude within the central portion of the NKA penetration is around 12 m s^{-1}, whereas it is $6\text{--}9 \text{ m s}^{-1}$ along the WKA penetration. Particle images recorded by imaging probes on both aircraft are shown in panels (b) and (c) of Figure 3.26, with smooth elliptical images indicating liquid drops with a maximum diameter of up to 8 mm. These observations demonstrate the dominance of warm-rain processes involving collision-coalescence of drops within the positive Z_{dr} column. The presence of large raindrops but in very low concentrations typically results in moderate Z_h values and large Z_{dr} values.

In general, the appearance of the Z_{dr} column is an indication of the storm's intensification. In particular, the evident physical relation between the column and the updraft suggests the possible use of this signature for prognostic applications. Good correlation has been observed between the height of the Z_{dr} column and the hail mass at the surface. In particular, Picca et al. [104] reported peak positive correlations between the Z_{dr} columns and the hail-core intensity (defined as the ratio of the 60-dBZ reflectivity volume to the 40-dBZ volume below the environmental freezing level) at a lag of 20–25 minutes.

The analysis of the Z_{dr} columns appears particularly promising for operational applications when multiple Doppler observations are not available to perform a wind analysis and identify the location of the updrafts. Adaptive scanning strategies, with dedicated volume or elevation radar scans focused on active convective cells, may allow for better use of the information associated with the vertical structure of Z_{dr}.

3.6.3   Hail

Hail is one of the most impactful weather phenomena worldwide, causing large amounts of economic damage every year. The essential ingredient for hail formation is a broad and intense updraft to carry raindrops aloft in the subfreezing areas of the cloud, where hailstones grow by collision and accretion of supercooled drops. The growth may continue until the updraft is no longer able to sustain the particles, which then fall to the ground. As a consequence, the stronger the updraft, the larger the hailstones that can be produced by the storm.

Weather radars have greatly contributed to revealing the dynamics and microphysical behavior of the processes of hail formation and growth. In particular, the analysis of reflectivity and Doppler observations in hailstorms (e.g., [106, 108, 109]) led to the development of the basic conceptual models of hail growth, such as the one illustrated in Figure 3.27. The recirculation of hail embryos (typically frozen drops or graupel particles), schematically represented by the white numbered arrow in Figure 3.27, has been widely recognized as the cause of the commonly observed layered structure of hailstones at the ground. Using aircraft observations and multiple Doppler radar measurements, Kennedy and Detwiler [110] confirmed the recirculation mechanism by direct observation of embryos reentering the base of a mature updraft in a multicell thunderstorm.

The onion-like appearance of the hailstone image in Figure 3.29 is the result of alternate wet and dry growth of the hailstone as it goes through portions of the cloud

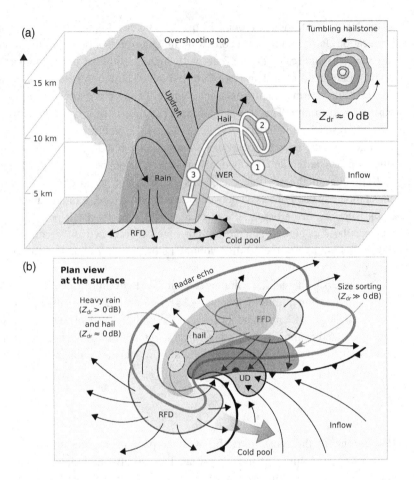

Figure 3.27 Schematic representation of hail formation and growth within a supercell, based on the observations and conceptual models of Browning and Foote [105], Foote and Wade [106], and Lemon and Doswell [107]. In panel (a), the white numbered arrow represents a possible trajectory of hail embryos growing to large hailstones, often showing a typical layered structure arising from alternate wet and dry growth processes (see text). In supercell storms, strong updrafts create a rain-free area called the *weak echo region* (WER), which is bounded on one side and above by a very intense reflectivity. The plan view at the surface in panel (b) illustrates the typical supercell circulation near the surface, with the main updraft (UD) and downdraft regions (FFD is the forward flank downdraft, and RFD is the rear flank downdraft). Differential reflectivity may show a wide range of values, ranging from ~0 dB near the storm's core when dry hail reaches the ground to several decibels arising from size sorting near the storm's boundary in the inflow region (see example in Section 8.2.1).

with either mild or very cold subfreezing temperatures. When the hailstone grows through the accretion of supercooled droplets, the release of latent heat of fusion increases its surface temperature. On the other hand, the hailstone may lose heat by conduction with the surrounding air or sublimation. The resulting heat balance on the hail surface determines the equilibrium temperature [68]. When this temperature remains below $0°C$, the hailstone goes through a dry growth phase, and the super-cooled drops quickly freeze on the surface. During this rapid freezing, air bubbles remain trapped, and a layer of white (opaque) ice is formed. When heavy riming occurs, the large amount of latent heat released during freezing may keep the hailstone surface close to $0°C$, that is, several degrees warmer than the surrounding air. In this case, the supercooled drops do not immediately freeze, resulting in a clear ice layer (air bubbles escape while water spreads across the tumbling hailstone). During this wet growth phase, part of the water may penetrate into the hailstone, favoring the formation of spongy hail, while some may be lost by shedding. Recirculation of hail embryos, as depicted in Figure 3.27, is not always necessary for hail growth. Hail embryos may also grow in one pass through the sloping updraft [111], with the final layered structure of the hailstones being a result of the particles going through areas of varying temperature and liquid water content.

The advent of dual polarization in the 1980s fostered the development of techniques for the detection of hail within thunderstorms. The early works focused on the use of differential reflectivity to identify the hailshaft, relying on the characteristic scattering behavior of tumbling hailstones. Because of their nearly spherical shape and large size compared with raindrops, hail produces a radar signature often denoted by large Z_h and low Z_{dr}, which was named the "Z_{dr} hole" by Wakimoto and Bringi [112]. Based on the early polarimetric radar observations, Aydin [113] introduced a partition of the Z_h–Z_{dr} space to discriminate between rain and hail:

$$f(Z_{dr}) = \begin{cases} 27 & Z_{dr} \leq 0 \text{ dB} \\ 19Z_{dr} + 27 & 0 \leq Z_{dr} \leq 1.74 \text{ dB} \\ 60 & Z_{dr} > 1.74 \text{ dB}. \end{cases} \tag{3.76}$$

The domain below the $f(Z_{dr})$ line in the Z_{dr}–Z_h space (the left part of the transposed Z_h–Z_{dr} space in Fig. 3.21) encompasses values typically observed in rain, whereas the domain above $f(Z_{dr})$ (right part of the Z_h–Z_{dr} space in Fig. 3.21) includes values characteristic of hail. Based on the partitioning function (eq. [3.76]), the likelihood of hail below the melting layer was then defined through a hail differential reflectivity parameter $H_{dr} = Z_h - f(Z_{dr})$, measuring the departure of the observed reflectivity from the boundary line $f(Z_{dr})$. Despite the application of this "hail signal" providing useful quantitative information for hail remote sensing over the years (e.g., Depue et al. [114] reported thresholds of H_{dr} of 21 and 30 dB, respectively, to identify regions where large and structurally damaging hail was observed), the simple approach was later replaced by more sophisticated techniques for hydrometeor classification involv-ing additional input variables, as discussed in Chapter 8.

Dual polarization observations of hail may show significant departures from the simple "Z_{dr} hole" behavior, in relation to the actual shape and the varying degree of wetting of the hailstones. On one side, wetting changes the effective value of the particle's dielectric factor. On the other hand, large particles such as hail typically

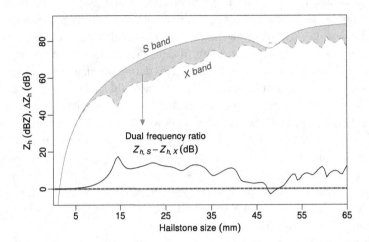

Figure 3.28 Reflectivity at the S band (solid gray line) and X band (dashed gray line) for a solid ice spherical hailstone. The positive ratio of the two reflectivity values (the difference in decibels) is highlighted by the shaded area and represented by the solid black line in the lower part of the plot. It is worth noting that the reported reflectivity for a single particle assumes one particle per cubic meter, whereas in real size distributions, large hailstones are much sparser (e.g., 0.2 particles per cubic meter for a 1-cm hailstone and 0.02 for a 2-cm hailstone [60]). The sparsity of very large hailstones poses a limit on the actual values of reflectivity, which rarely exceed 70 dBZ in the strongest hailstorms.

fall in the Mie scattering region for common radar frequencies in the 3- to 10-GHz range (Chapter 4). This implies that radars operating at different frequencies may provide measurements showing substantial differences. During the 1970s, the different scattering behavior at the S band ($\lambda = 10$ cm) and X band ($\lambda = 3$ cm) was exploited in the study of hailstorms with dual frequency radar systems. The approach relied on the simple idea that medium to large-size hailstones (larger than 1 cm) fall in the Rayleigh scattering regime at the S band ($\pi D/\lambda \approx 0.3$), whereas they fall in the Mie scattering regime at the X band ($\pi D/\lambda \approx 1$). The ratio of the reflectivity factors at these two frequencies can then be used to detect large hail. Figure 3.28 shows the reflectivity of a single spherical dry hailstone (solid ice) with increasing hail size. The shaded area highlights the difference in decibels between the reflectivity at the S band and the X band (dual frequency ratio), which is used to detect hail with a size of > 10 mm. The difference is on the order of 10 dB for hailstones with a size between 10 and 45 mm. However, practical difficulties related to not having perfectly matched antenna beams and the presence of attenuation along the propagation path for the X-band measurements limited the applicability of this methodology [115], in favor of the emerging dual polarization technologies.

The Colorado State University–Chicago Illinois (CSU-CHILL) radar supports simultaneous dual wavelength (S and X bands), dual polarization (linear horizontal and vertical) operation [116] allowing studies of precipitation that unambiguously revealed the impact of Mie scattering by hail. In particular, Mie scattering is detected where the dual wavelength ratio departs significantly from the path-integrated attenuation [117]. In addition to this "classical" approach to combining two far-apart wavelengths (S and X bands) for which scatterers either fall in the Rayleigh regime or in the Mie regime, new investigations also include the possible use of two closely spaced wavelengths for which scatterers fall in the Mie regime. For the purpose of hail detection and sizing, Kumjian et al. [118] have shown that observed differences in reflectivity and differential reflectivity at two close frequencies may be related to hail size, resulting from resonance scattering effects for large particles occurring at slightly different sizes for the two wavelengths. The combination of dual frequency and dual polarization operation, however, is still a research topic with no direct impact on dual polarization operational systems that use a single frequency.

Analyzing data sets collected in North America, Knight [119] observed that small hailstones are nearly spherical, whereas larger hailstones show a tendency to become more oblate with increasing size. The bigger hailstones analyzed (with size of 5–6 cm) may reach a shape factor, defined as the ratio between the shortest and the longest axes, of ∼0.6. The varying size–shape relation among samples collected in different regions may vary in relation to the dominant embryo type and the degree of wet or dry growth. On average, the shape factor of hailstones is approximately 0.8. This is the case for the hailstone collected during a severe storm in Colorado depicted in Figure 3.29. The image shows an often-observed characteristic of hailstones, with the

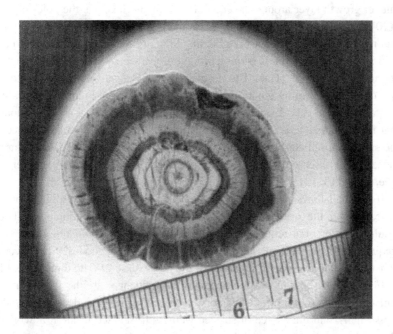

Figure 3.29 Photomicrograph through a large hailstone collected during a severe hailstorm in northern Colorado, showing the different growth layers resulting from alternate wet and dry growth.

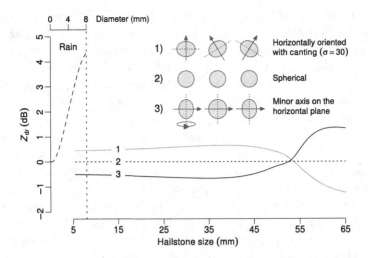

Figure 3.30 Differential reflectivity at the S band for different hail shapes and fall behaviors. For an X-band radar, because of the frequency scaling, the Z_{dr} sign change would show up at a smaller hail size (~15 mm). For comparison, the Z_{dr} corresponding to a single raindrop is also shown ($D < 8$ mm). Hail size is the length of the major axis, and the shape factor of nonspherical hailstones (cases 1 and 3) is assumed to be 0.8.

thicker growth layer around the equator and thinner layers at the poles. Kry and List [120] elaborated a theoretical analysis of the falling behavior of spheroidal hailstones that may explain this behavior. According to their theory, the oblate spheroids tend to spin about their minor axis (symmetric gyration), which traces out a cone symmetric about the horizontal plane. This implies a small negative Z_{dr} for hailstones up to approximately 5 cm in size at the S band, as illustrated in Figure 3.30. In this figure, the fall mode in case 3 represents a simplification of the Kry and List model [120]; that is, the hailstone is assumed to be spinning like a coin on a flat surface, without the superimposed conical motion of the short axis. A negative Z_{dr} from large hail is indeed often observed at the S band, with a negative bound of approximately −1.5 dB (see, e.g., [113, 121]).

If the hailstones are falling with their major axis mainly aligned horizontally instead, with the orientation following a Gaussian distribution (e.g., [114]), the resulting differential reflectivity would be positive (case 1 in Fig. 3.30). In reality, frequent tumbling is generally superimposed on these modes of fall, causing the hailstones to appear nearly spherical to the radar (case 2). The three cases illustrated in Figure 3.30 show how scattering may be influenced by the fall behavior of the hailstones, in addition to other factors, such as the hail size, density, irregular shape (spiky hail), and the degree of surface wetness (discussed in the following section). However, and more importantly, irrespective of the actual fall orientation of dry hailstones, Z_{dr} shows a markedly different behavior with respect to raindrops, with typical values for hail in the range of −1 to +1 dB at the S band.

In addition to influencing the backscattered power (exploited in the dual frequency approach), Mie scattering may produce a distinct Z_{dr} signature for nonspherical hail, depending on the hailstone size. For the S band, the nearly constant Z_{dr} trend is interrupted around a size of 5 cm (Fig. 3.30). For these very large hailstones, Z_{dr} may reverse sign, becoming negative for horizontally oriented hailstones (case 1) and positive for hailstones with its minor axis on the horizontal plane (case 3). It is important to note that theses Mie scattering effects would show up at a much smaller hail size ($\sim$1.5 cm) at the X band as a result of the scaling with frequency, as discussed in detail in Chapter 4. At the same time, it should be considered that the results illustrated in Figures 3.28 and 3.30 are for a single particle. The actual distribution of hailstones (often modeled as exponential) may significantly smooth out the variability associated with Mie scattering, resulting in typically small Z_{dr} values.

Melting Hail

When hailstones fall through the warm lower atmospheric layers, melting starts to form a water coat around the ice particle, increasing its dielectric factor. Wet hailstones can be simply modeled as possessing the dielectric constant of water or as a two-layer body with an inner solid ice core surrounded by an oblate coat of water [21]. As melting progresses, the hailstone becomes increasingly oblate and approaches the shape of a raindrop. Unlike snowflakes, which are characterized by low density ($\rho = 0.01$–0.1 g cm^{-3}) and small fall velocity (0.5–2 m s^{-1}), hailstones are made of high-density ice ($\rho = 0.5$–0.92 g cm^{-3}, with the lower value representing "soft hail" or graupel) and may attain terminal velocities in the range of 10–50 m s^{-1}. For this reason, the melting process is not confined over a thin atmospheric layer like in stratiform precipitation and does not imply a bright-band signature.

To gain insight into the radar's dual polarization variables associated with melting hail, consider the Rasmussen–Heymsfield melting model [122] illustrated in Figure 3.31 (on the left). When a hailstone melts, a thin water coating forms on the ice particle, and a water torus takes shape around the equator as a result of the tangential stress on the lower surface and the airflow separation near the equator of the particle [123]. There is a maximum amount of water that can be retained in this ring-like form on the surface of the ice-core mass. When melting proceeds, small droplets start to shed, determining an overall mass loss and a decrease in terminal velocity. With further melting, the ice core becomes completely embedded in the meltwater, and the particle gradually assumes the shape of a raindrop, as illustrated by the lower-left drawing in Figure 3.31. At this stage, perturbations of the meltwater are damped out by the ice core [123], allowing the particle to reach a larger size in comparison with raindrops, which would become critically unstable and break up around $D \approx 8$ mm. These large particles appear to the radar as "giant" oblate raindrops, causing Z_{dr} to increase substantially relative to a fully melted particle of the same axis ratio.

Realistic profiles of differential reflectivity in melting hail are shown in the right part of Figure 3.31 for large hail at the S band (black solid line) and small hail (gray lines, solid for S band and dashed for C band). Because of the more spherical shape

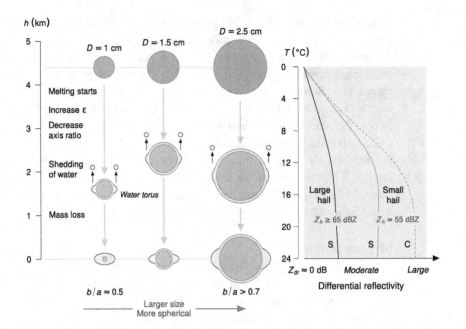

Figure 3.31 Schematic illustration of the transition of high-density hail during melting, according to wind tunnel experiments [123] and theoretical model of Rasmussen and Heymsfield [122, 124]. The vertical profiles of Z_{dr} on the right are approximations to the simulation results of Ryzhkov et al. [125] for exponential distributions of hail with a maximum size of 1.4 cm (small hail) and 3.5 cm (large hail). The Z_{dr} for small melting hail at the C band may be greatly enhanced as a result of Mie scattering (gray dashed line).

of the larger hailstones, for which the fraction of meltwater is lower, the differential reflectivity of the large hail profile is also lower. In fact, because Z_{dr} is an indicator of the reflectivity-weighted mean shape of the particles in the resolution volume, it is mainly affected by the larger hailstones. On the other hand, precipitation with a prevalence of small hail aloft ($D \leq 1.5$ cm) may result in extremely large values of Z_{dr} in the lower levels, especially at the C band (up to 6–7 dB).

Because Z_{dr} increases as hail progressively melts, the detection of melting hail becomes more difficult and should rely on the additional information provided by the correlation coefficient. In contrast to pure rain, which is denoted by a very high correlation coefficient (typically $\rho_{hv} > 0.98$), melting hail generally shows a lower correlation (down to $\rho_{hv} \approx 0.8$). In hydrometeor classification schemes (Chapter 8), it is common to assume a lower correlation for larger hail and for a higher degree of melting [126]. At the S band, giant hail with sizes exceeding 5 cm may produce especially low correlation coefficient values as a result of Mie scattering effects [127]. The observation of an enhanced depression in ρ_{hv} may then provide an indication of the presence of very large hail.

In general, the presence of hail and melting hail complicates the interpretation of radar observations and the quantification of precipitation characteristics because it affects both scattering and propagation through enhanced attenuation and differential attenuation. In particular, the estimation of hail size has been long pursued by the weather radar community but is still an active topic of research.

For the purpose of rainfall estimation in the presence of hail, measurement of the specific differential phase shift K_{dp} (the half-range derivative of the differential phase shift Φ_{dp}) can overcome the limitations affecting the conventional power measurements of Z_h and Z_{dr}. Although the backscattering power is mostly affected by the largest particles within the radar resolution volume (D^6 weighting), the differential phase is only influenced by those particles that generate a difference between the forward-scattering amplitudes of the two polarizations (refer, e.g., to the integral definition in eq. [9.3]). A mixture of hail (large spherical particles) and rain (highly oblate but smaller particles) will then produce the same phase shift as if the resolution volume were exclusively filled with the rain component. For this reason, K_{dp}-based rainfall-rate estimators are especially suitable for applications in heavy convection with potential hail contamination of the radar measurements (Section 9.5).

3.6.4 Closing the Loop: DSD and Precipitation Processes

With the aim of providing the essential background for the proper interpretation of the dual polarization radar variables, the main precipitation processes were discussed with regard to to how they affect the microscale structure of precipitation in terms of the phase, shape, size, and number concentration of the constituent particles. Although the last part of the chapter focuses more on the distinctive traits and the morphology of stratiform and convective precipitation (the macroscale), it is important to highlight how the precipitation appearance (to the radar) is closely related to its microscale composition, that is, the particle size distribution.

The schematic partition of the radar parametric space Z_h–Z_{dr} in Figure 3.21 can be enhanced by considering the microphysical processes leading to rain formation in stratiform and convective systems, as presented in Figure 3.32. The clusters represented by gray shaded areas (convective rain) and hatched areas (stratiform rain) correspond to the rain-formation mechanisms identified by Dolan et al. [128] in their analysis of a large data set of disdrometer measurements spanning a broad range of precipitation regimes at different latitudes. The study of the disdrometer observations, in conjunction with the analysis of the corresponding radar observations, allowed the researchers to relate the rain-formation mechanisms to the characteristics of the DSD observed at the surface and objectively identify six groups with unique characteristics. The work resulted in a conceptual model (Fig. 3.33) where the six groups are represented in the microphysical space defined by the median drop diameter D_0 and the normalized intercept parameter N_w [128]. The N_w–D_0 parameter space is partitioned into two main domains, the lower-left part, which is associated with stratiform precipitation, and the upper-right part, representative of convection. Groups 1, 3, 5, and 6 (filled areas) are related to convective precipitation, whereas groups 2 and 4 (hatched areas) are characterized by stratiform precipitation.

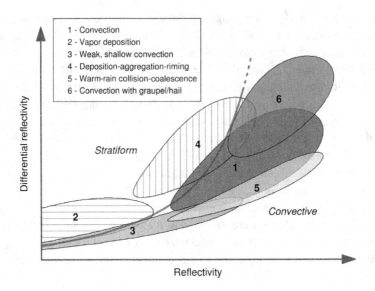

Figure 3.32 Conceptual diagram showing the distribution of differential reflectivity versus reflectivity, according to the main microphysical processes leading to rain formation in stratiform and convective systems.

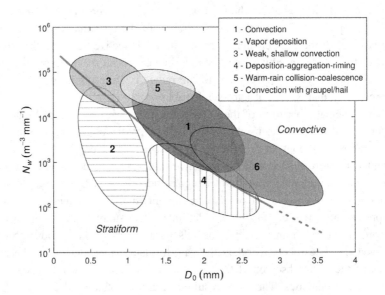

Figure 3.33 Partition of the microphysical space D_0–N_w into six groups according to the dominant precipitation mechanisms identified through the analysis of surface disdrometer observations by Dolan et al. [128].

Following the convective-stratiform separation line from left to right (increasing D_0 and decreasing N_w), there is a progression from warm-rain to ice-phase processes for convective precipitation, which corresponds to the otherwise-known transition from maritime to continental regimes (see, e.g., [129]). In the maritime rain regime, DSDs are characterized by a high concentration of smaller-size drops, whereas the continental rain regime can be described by rain DSDs with a low concentration of larger-size drops, as illustrated in panel (a) of Figure 3.1.

Group 3 is characteristic of warm-rain showers, with large N_w and small D_0. Although this group in general denotes a precipitation regime typical of tropical shallow convection, similar DSD characteristics can also be found in forced orographic precipitation. Group 5 has a high N_w like group 3, but the larger D_0 indicates a major role of collision-coalescence for rain development, which is a characteristic of deeper warm clouds. On the other end of the convection spectrum, group 6 has the largest D_0 and a smaller N_w. This type of DSD results from ice-phase processes in strong convection, where intense vertical motions support the formation of graupel and hail aloft. Melting of these large ice particles as they fall to the surface leads to a raindrop size distribution with relatively few but very large drops. In between, the DSDs clustering in group 1 may result from the combined effect of warm-rain and cold-rain processes and can attain very large values of liquid water content.

On the lower-left side of the parameter space, groups 2 and 4 represent DSDs resulting from stratiform precipitation processes. For group 2, the small drop diameters implicate ice-crystal growth by vapor deposition as the main precipitation process in the cold portion of the cloud. The typical radar signature for this type of precipitation is a weak bright band as a result of the lack of large ice particles aloft. On the other hand, the onset of aggregation (typically around $-5°C$) produces larger ice particles that will turn into larger raindrops below the melting layer. Group 4 is then characterized by DSDs originating from ice-crystal distribution shaped by aggregation, in addition to vapor deposition. The relative abundance of large aggregates reflects in a well-marked bright-band signature. Moving farther to the right in the N_w–D_0 space, group 4 partly overlaps with group 6, indicating a transition to convective precipitation when riming becomes more relevant for the growth of precipitation. With heavier riming, the bright-band signature is increasingly smoothed, and the vertical profiles of the radar variables lose the characteristic behavior of stratiform rain depicted in Figure 3.23.

Although a relation between the DSD parameters and the radar variables cannot be uniquely established (more discussion on this in Chapter 9), mapping the six DSD groups into the Z_h–Z_{dr} radar space in Figure 3.32 (under the simplified hypothesis of exponential DSD) serves to illustrate the general qualitative behavior of radar observations in different precipitation regimes. In comparison with Figure 3.21, drawn in the context of convective and stratiform precipitation, this figure shows how dual polarization radars can identify more than the dominant particle type in the resolution volume (rain, hail, etc.), a goal pursued since the early development of weather radar polarimetry. When considering the observations' arrangement in the space and time domains, dual polarization radars can be used to infer valuable information about microphysical processes in the cloud and how precipitation develops. In the end, the polarimetric radar signatures ultimately resides in the size distribution of the precipitation particles, which is in turn the result of specific precipitation processes.

In Chapter 9, dual polarization will be used to quantitatively account for the different forms rainfall may take through a more in-depth analysis of the relation between the DSD and the radar variables for rainfall estimation.

3.7   Selected Problems

1. For an exponential DSD with $N_0 = 2000 \text{ m}^{-3}\text{mm}^{-1}$, D_0=1.5 mm, and $D_{max} = 8$ mm:
   a) find the volume fraction of rain,
   b) calculate the reflectivity factor,
   c) calculate the rainfall rate.
2. Consider a uniform DSD composed of 2-mm raindrops.
   a) How many raindrops per cubic meter are necessary to make a reflectivity of 30 dBZ?
   b) For 30 dBZ reflectivity, how many drops will be there in the radar volume with range resolution $\Delta r = 0.1$ km and beamwidth of $1°$, at the distances of 40, 60, 100, 200, 300 km? How much water (in liters) will there be?
3. Consider the growth by collision-coalescence example in Section 3.5, but instead with an updraft intensity of $w = 10 \text{ m s}^{-1}$. Given the strength of the assumed updraft which is higher than the terminal fall velocity of the largest drops, all raindrops are subject to an upward displacement. The same initial DSD as in fig. 3.15 ($N_0 = 8 \times 10^3 \text{ mm}^{-1} \text{m}^{-3}$, $\Lambda = 2.4 \text{ mm}^{-1}$) is now representative of the precipitation at the bottom of the layer.
   a) What would be the qualitative behavior of Z_h and Z_{dr} in the precipitation layer as a function of height?
   b) Compute the simple growth model defined by eqs. (3.43) and (3.57), with $w = 10 \text{ m s}^{-1}$, and calculate the vertical profile of the radar parameters Z_h and Z_{dr} over 3 km height rise. How much is the total upward variation of Z_h and Z_{dr} ?
4. Consider a monodisperse DSD with 10 3-mm raindrops per cubic meter falling from $z_{top} = 3000$ m. Assume uniform relative humidity RH=50% (RH = $100 \times (e/e_s)$), T=4°C at z_{top} with standard atmosphere lapse rate (9.8 °C km^{-1}), and the values in the table below for the other relevant constants and properties:

Description	Symbol	Value	Units
Water-vapor gas constant	R_v	461.5	$\text{J kg}^{-1}\text{K}^{-1}$
Coefficient of diffusion*	D_v	2.52×10^{-5}	m^2s^{-2}
Thermal conductivity*	K_a	2.55×10^{-2}	$\text{J m}^{-1}\text{s}^{-1}\text{K}^{-1}$
Dynamic viscosity*	μ	1.815×10^{-5}	$\text{kg m}^{-1}\text{s}^{-1}$

*Values at 20°C [59].

Note that the kinematic viscosity v in eq. (3.59) is defined as $v = \mu/\rho$, where μ is the dynamic viscosity and ρ is the air density (for simplicity, assume $\rho = 1 \text{ kg m}^{-3}$). The following approximate expressions for the saturation vapor pressure e_s [130], the latent heat of condensation L [59], and the ventilation coefficient f_v [60] may be used in eqs. (3.52) and (3.53):

$$e_s \text{ (Pa)} = 610.78 \exp(17.27\, T/(T + 237.3))$$

$$L \text{ (J kg}^{-1}) = 2.501 \times 10^{-6} - 2340\, T$$

$$f_v = 0.78 + 0.308(v/D_v)^{1/3}(V_t\, D/v)^{1/2}$$

where T is in Celsius, D is the drop diameter in meters, and V_t is the terminal fall velocity in m s^{-1} (see, e.g., eq. [3.37]).
  a) Calculate the height $z_1 < z_{top}$ where the drops will have lost 20% of their mass by evaporation.
  b) Calculate the reflectivity of the monodisperse raindrops distribution at z_{top} and z_1. What is the average vertical gradient of reflectivity?

5. Consider the same thermodynamic environment as in problem 4, but in this case imagine a monodisperse DSD with N 1.5-mm raindrops per cubic meter falling from $z_{top} = 3000$ m.
  a) How many 1.5-mm raindrops (N) are needed to get the same reflectivity as in problem 4 at z_{top}?
  b) Considering the new 1.5-mm monodisperse DSD from 5, repeat the calculation of evaporation and estimate the reflectivity at height z_1 (where the drops will have lost 20 percent of their mass). What is the average reflectivity gradient in this case?

4 Basic Scattering Theory and Principles for Radar Meteorology

This chapter considers how the radar's signal, in the form of an electromagnetic wave, propagates and interacts with precipitation. The wave's interaction with precipitation is determined by the scattering properties of the hydrometeors. These properties are fundamental to the interpretation of weather radar measurements. If the scattering characteristics of the precipitation that are observed are known, the radar parameters can be calculated for a specified hydrometeor type. The problem can also be reversed; if the radar parameters are measured, the scattering properties of the precipitation can be estimated.

Electromagnetic theory is the fundamental basis for how and why a radar system works (or, in some cases, its limitations) for observing weather phenomena. This chapter provides a brief introduction to concepts necessary to describe the physics of how the radar's signal interacts with precipitation. Ultimately, the signal's interactions are used to infer the properties of the clouds and precipitation the radar observes.

Electromagnetic theory and scattering theory are vast fields of research beyond what is covered here. This chapter is by no means a complete development of electromagnetic theory. Extensive development and discussion of electromagnetic theory and scattering physics can be found in references such as [21, 131–133].

It is also important to mention that numerous science and engineering communities are concerned with the topic of electromagnetics, wave propagation, and scattering. The notation and conventions adopted across these communities can vary to best suit their objectives, and as a result, some differences (and confusion) can exist. The following discussion attempts to present some of the common variations, but as one might expect, it is not exhaustive.

The radar's observations are used to measure the scattering characteristics within the radar volume. To that end, the scattering volume (and the scatterers within the volume) are characterized by their scattering cross-sections. The backscatter cross-section, σ_b, is also commonly referred to as the *radar cross-section* for monostatic radar systems. The bistatic radar cross-section, σ_{bi}, is measured when the incident and scattering directions are not the same. For weather radar, we are also interested in the extinction cross-section σ_e, which is a measure of how much of the incident power is lost. The extinction cross-section of the scatterer is $\sigma_e = \sigma_a + \sigma_s$. Loss of the radar wave's power is due to the absorption cross-section, σ_a, like the heating effect in a microwave oven, but power is also lost through scattering. The total scattering cross-section, σ_s, describes the power scattered in all directions. (This total scattering cross-section includes σ_b.)

The inversion of the weather radar observations to scattering properties is an ill-posed problem in the most general sense – only a small number of radar measurements are available to estimate the electromagnetic properties of thousands of scatterers. The microphysical properties of precipitation and the observable dual polarization radar parameters can in fact significantly constrain the potential solutions. For example, the three-parameter gamma model from Section 3.2.2, which represents the drop size distribution (DSD) of rain, can be used to estimate the radar observations through scattering models. Similarly, an inversion can be performed, allowing the parameters of the gamma model to be approximated directly from the radar's observations (as discussed thoroughly in Section 9.2).

Naturally occurring hydrometeors assume many shapes. Until the advent of polarimetric radar, most of these shapes were approximated by spheres. However, radar polarimetry has reinvigorated the computation of scattering from the complex geometries of hydrometeors. The simulation of these scattering properties is in general a difficult task that can be substantially simplified by approximating the particles with simple regular shapes. From Chapter 3, raindrops can be represented quite accurately by oblate spheroids. Analogously, most ice particles can also be approximated by either oblate (plate-like crystals, hailstones) or prolate (columnar crystals) spheroids. The use of regular and rotationally symmetric shapes allows fast methods to be applied for the computation of electromagnetic scattering by homogeneous nonspherical particles, like the T-matrix code [134]. On the other hand, the application of methods for computing the scattering characteristics of particles with arbitrary shapes has been limited in the past because of the huge computational demands. Currently, with the availability of large processing resources, methods like the discrete dipole approximation (DDA) are becoming increasingly used for the simulation of complex ice-phase precipitation particles (see, e.g., [135]). For the purpose of illustrating the scattering properties of precipitation particles, the spheroidal approximation is used. It provides an easy and intuitive means for relating the shape and orientation of the particles with their electromagnetic response while still providing a good approximation.

Like dual polarization, dual frequency systems provide additional independent measurements of the scattering environments. With careful selection of the radar frequencies that are systematically combined, the additional information can be microphysically significant. (For example, consider the relationship of scattering characteristics versus particle size for two different frequencies in Figure 3.12). Mathematically, the number of available independent observations determines the number of unknowns that can be solved for. Dual polarization radars are used to directly measure median drop size (through Z_{dr}; see Section 9.2) and rain rate (through K_{dp}; see Section 9.5.3). Dual frequency techniques have been investigated for decades but have received renewed interest in the weather communities as both our understanding of precipitation and the available radar technology have advanced. Dual frequency systems can provide more information about the properties of precipitation by exploiting the differences in the scattering responses resulting from the sizes of the scatterers being observed. Combining dual polarization and dual frequency capabilities is becoming more common for research radar systems to further our understanding of precipitation.

4.1   Electromagnetic Wave Propagation and Scattering

Radar uses propagating electromagnetic waves to remotely observe objects. The power from the radar's antenna propagates away from the radar as traveling electromagnetic waves. The transmitted wave then scatters, as a result of objects in the wave's path. The scattered waves then propagate in a similar way. Some of the scattered power is received and measured by the radar's receiver. As the wave propagates, some of its energy is lost from scattering, and some is absorbed in the objects. As an illustrative analogy, the radar signals' propagation and scattering are similar to using a flashlight as a transmitter and your eye as the receiver. In fog, the water particles scatter some of the light while letting the rest continue to propagate. Weather radar uses electromagnetic waves in the microwave spectrum (i.e., on the order of 3 mm–30 cm), whereas the flashlight uses wavelengths in the visible light spectrum (i.e., 380–700 μm).

The electromagnetic wave from the radar's antenna propagates and expands spherically. As the range increases, the angular extent of the beam stays constant, but the width of the beam increases. This is also readily observed using the flashlight analogy, where the width of the illuminated spot size gets larger as the range increases (but the angular width of the beam remains constant). For small areas, at faraway distances from the radar, the spherical wave can be approximated by a plane wave (see fig. 2.1). This approximation is commonly used to evaluate how the electromagnetic wave interacts with the media it scatters from and propagates through (the air, rain, ice, ground, buildings, insects, etc.).

Weather radars measure the amplitude and phase of the received signal, which are echoes from scatterers. The amplitude is influenced by the attenuation (energy in the forward propagating that is lost) and the backscattered power (which is proportional to the radar cross-section [RCS]). The phase is a function of the range from the target to the scatterer, the phase constant of the medium the wave propagates through, and the backscattering phase shift resulting from the scattering physics. With the amplitude and phase information, the radar's received signal can be directly linked to the properties of the scatterers (e.g., rain and ice) in the environment that the weather radar is observing.

4.1.1   Wave Equation

The radar transmits and receives *transverse electromagnetic (TEM) waves*. The term transverse electromagnetic indicates that the electric and magnetic fields of the wave are both orthogonal to the direction of propagation. TEM waves propagate with coupled electric (**E**) and magnetic (**H**) fields. The electric field and magnetic field are vectors. For the TEM wave, the electric and magnetic fields are in the plane (as an example, the plane of $\hat{h}$ and $\hat{v}$) that is perpendicular to the direction of propagation ($\hat{k}$). The electric and magnetic fields are also perpendicular to each other (see Figure 2.2 for typical coordinate systems). Electromagnetic waves are a well-developed topic with a firm mathematical foundation and well-known equations in the literature. Here, the discussion of these equations is only an introduction. It is meant to encourage a brief familiarity with the terminology and the physics governing weather radars' observations.

The time-harmonic (frequency-domain) wave equation for the electric field in an isotropic, lossy medium is given by the well-known Helmholtz equation:

$$\boxed{\nabla^2 \mathbf{E} + \omega^2 \mu \epsilon \left(1 - j\frac{\sigma}{\omega\epsilon}\right) \mathbf{E} = 0.} \tag{4.1}$$

The wave equation is derived from the differential form of the foundational equations for electromagnetic theory known as *Maxwell's equations* [21]:

$$\boxed{\nabla \times \mathbf{E} = j\omega\mu\mathbf{H},} \tag{4.2a}$$

and

$$\boxed{\nabla \times \mathbf{H} = j\omega\epsilon\mathbf{E} + \sigma\mathbf{E},} \tag{4.2b}$$

where ω is the wave's frequency, and the media's properties are as follows: μ is the magnetic permeability, ϵ is the electric permittivity, and σ is the electrical conductivity (not to be confused with the RCS). The curl ($\nabla\times$) and Laplacian (∇^2) are standard mathematical operators. For the moment, define the direction of the wave's propagation along the positive z-direction, which then places the electric and magnetic fields in the $x - y$ plane. The component of the electric field that is directed along the x-axis will be coupled to the magnetic field along the y-axis, and the electric field component along the y-axis will be coupled to the magnetic field in the negative x-axis. By definition, the x-axis and y-axis electric fields (and their associated magnetic fields) are orthogonal to each other.

For dielectric materials at weather radar frequencies, the conductivity σ can typically be neglected with $\sigma = 0$. The dielectric's losses, which are represented as the imaginary part of the permittivity (Im(ϵ)), are a frequency-dependent absorption loss within the dielectric. Both of these contributions (σ and Im(ϵ)) have the same effect and are indistinguishable at a fixed frequency. They both result in power being absorbed by the medium, and therefore both may be combined into the single frequency-dependent term for the formulation of the wave equation. Referring back to the two Maxwell's equations [eq. (4.2)], the term $\sigma\mathbf{E}$ is used as the current density $\mathbf{J}$. The polarization current within the media, $\mathbf{P}$, is not included. It is another current source that may also be included with the current density to give $\mathbf{J} = \sigma\mathbf{E} + j\omega\mathbf{P}$. (Note that the polarizability of the material is frequency dependent because of its time derivative.)

4.1.2   Poynting Vector and the Wave's Power

The radar's transmitted power is scattered by the hydrometeors. The fraction of the power that is scattered is determined by the RCS, which is a measure of the scatterer's ability to reflect the transmitted waveform in the direction of the radar receiver. More specifically, the RCS is the ratio of the power scattered per unit solid angle in the direction of the radar receiver to the power density incident on the scatterer.

The scattered power is the echo that the radar will receive. The weather radar's detection of clouds or precipitation depends on the power of the echoes received by the radar, and therefore, detection depends on the RCS of the hydrometeors. From eq. (2.29), the RCS is the ratio of the scattered to incident power density:

$$\sigma = \lim_{r \to \infty} 4\pi r^2 \frac{S_s}{S_i}. \tag{4.3}$$

The power density of the wave is defined as the time-averaged Poynting vector [21]:

$$\mathbf{S} = \frac{1}{2} \mathrm{Re} \left(\mathbf{E} \times \mathbf{H}^* \right), \tag{4.4}$$

where Re() is the real component; the superscript asterisk denotes the complex conjugate; and the "×" symbol refers to the vector cross product, indicating that $\mathbf{S}$ is a vector. The magnitude of the incident and scattered power densities, $|\mathbf{S}_i|$ and $|\mathbf{S}_s|$, respectively, are used to calculate the RCS.

For a TEM wave propagating in the $\hat{k}$ direction with orthogonal electric and magnetic fields in the direction of $\hat{h}$ and $\hat{v}$, there is no electric or magnetic field in the direction of propagation. The cross product of the electric and magnetic field of the propagating TEM wave is

$$\mathbf{E} \times \mathbf{H}^* = \begin{bmatrix} \hat{h} & \hat{v} & \hat{k} \\ E_h & E_v & 0 \\ H_h^* & H_v^* & 0 \end{bmatrix} = \left(E_h H_v^* - E_v H_h^* \right) \hat{k}. \tag{4.5}$$

This result, which is proportional to the Poynting vector, shows that the wave's power only propagates in the direction $\hat{k}$. Note that in eq. (4.5), the power from both polarizations are included. To evaluate the dual-polarization scattering, the RCS of each polarization are considered separately.

4.1.3   Propagation Constant

Now consider how the terms in the wave equation (eq. [2.4]) define the propagation of the wave. If the wave propagates through a precipitation medium (instead of propagating through clear air, which can be treated as vacuum to a first approximation), the real-valued wavenumber k_0 is replaced by a generalized, complex-valued wavenumber k as:

$$E_{h,v}(r) = E_{h,v}(0)e^{-jk_{h,v}r} \tag{4.6a}$$

$$= E_{h,v}(0)e^{-j\beta_{h,v}r}e^{-\alpha_{h,v}r}, \tag{4.6b}$$

where the subscripts indicate the polarization. To simplify the following discussion, the subscripts are dropped and one polarization is assumed. The initial magnitude of the electric field, $E(0)$, is at the range $r = 0$. The attenuation constant α in the component $e^{-\alpha r}$ affects the amplitude of the electric field. The phase constant β in the term $e^{-j\beta r}$ only affects the phase of the electric field and not its amplitude. The magnitude of this term is always $|e^{-j\beta r}| = 1$.

The electric (and magnetic) field interacts with the environment as it propagates. For a propagating TEM wave, the magnetic field and electric field are coupled via the wave's characteristic impedance in the material (see Section 4.1.4), and therefore the electric field and characteristic impedance define the wave. These interactions result in scattering and attenuation of the signal. The following discussion assumes isotropic material properties (the material properties are all scalar values). This simplification is used to discuss the phenomena of the propagating and scattering wave because it helps in the interpretation of the radar observations.

The electric and magnetic fields, as well as the material properties, may be anisotropic, meaning that the values vary with the direction in which the wave is propagating or the polarization of the wave. For this case, material properties should be represented as matrices. A more rigorous development can be found in Bringi and Chandrasekar [21] for the interested reader.

From the wave equation (eq. [4.1]), the wave's propagation constant is defined as

$$\gamma = jk = \alpha + j\beta = j\omega\sqrt{\mu\epsilon}\sqrt{1 - j\frac{\sigma}{\omega\epsilon}}. \tag{4.7}$$

For some conventions within the literature, a strict formal definition of k as the angular wavenumber of the wave is $k = 2\pi/\lambda$ with units of radians. The angular wavenumber is also called the *phase constant* (i.e., k and β are equivalent under this formal definition). In an other formulation of the wave equation, which is the one adopted in this book, the definition of k is slightly altered so that it is related to the complex-valued propagation constant as follows:

$$k = -j\gamma = \beta - j\alpha. \tag{4.8}$$

This variation of k is adopted because of the simplification in representing the propagating plane wave (as shown in eq. [4.6a]), where k can be used as only the phase constant or also used as a modified propagation constant. Note that for a lossless medium, where $\alpha = 0$, $k = \beta = 2\pi/\lambda$, which adheres to the more standard definition.

The propagation media relevant to weather radars refers to mostly free space with water and ice particles sprinkled along the way at a small volume fraction. This is different from waves propagating in water or solid ice blocks. Most of radar meteorology will be concerned with characterizing precipitation media described by sparsely filled hydrometeors or cloud particles.

Three electrical properties of the medium determine its propagation constant: the electrical conductivity (σ), which is not to be confused with the RCS; the dielectric permittivity (ϵ); and the magnetic permeability (μ). In free space, which is also a good approximation for clear air, $\sigma = 0$ Sm^{-1}, $\epsilon_0 \approx 8.854 \cdot 10^{-12}$ F m^{-11}, and

$\mu_0 = 4\pi \cdot 10^{-7}\,\mathrm{H\,m^{-1}}$. With these values, the propagation constant of free space is $k_0 = \omega\sqrt{\mu_0\epsilon_0}$. In free space, k_0 only has a real component (no imaginary part), and therefore its attenuation constant is $\alpha = 0$. A medium with $\alpha = 0$ is lossless because it does not attenuate the wave.

The permittivity characterizes the electrical response of a material to an electric field. The electric permittivity of materials is typically given in terms of a dielectric constant ϵ_r, which is also called the *relative permittivity*. The dielectric constant is a scale factor to be applied to the electric permittivity of free space. The permittivity can be complex-valued, and therefore the dielectric constant can be complex-valued:

$$\epsilon = \epsilon_r \epsilon_0 = \left(\epsilon'_r - j\epsilon''_r\right)\epsilon_0. \tag{4.9}$$

In free space, $\epsilon_r = 1$. For all materials, $\epsilon'_r \geq 1$. The permittivity values of materials are sometimes only defined as $\epsilon_r = \epsilon'_r$, with no imaginary part reported. This would imply that the material is lossless and does not attenuate the signal. For dielectrics, the imaginary component of the permittivity is sometimes defined by the loss tangent at a particular frequency, where

$$\tan\delta = \frac{\epsilon''}{\epsilon'}. \tag{4.10}$$

With a real-valued ϵ_r and the loss-tangent $\tan\delta$, the complex-valued electric permittivity can be calculated. For weather radar, both the real and imaginary part of the dielectric constant of water can be rather high, and both are significant. The dielectric properties of water and ice vary across the radio-frequency (RF) and microwave spectra used by weather radar. The permittivity of water and ice (the main constituents of precipitation) are discussed in detail in Section 4.2.

The speed of the propagating wave is calculated by the distance traveled in one time period of oscillation. Because the wave travels one wavelength per period, the speed is given by

$$v = \lambda f, \tag{4.11}$$

where f is the frequency in Hertz (or cycles per seconds). Recall that $\omega = 2\pi f$ is the frequency in radians per second. The phase constant is the angular wavenumber, and therefore $\beta = 2\pi/\lambda$. With $k_0 = \beta$ for free space, the speed of the electromagnetic wave is:

$$c = \frac{\omega}{\beta} = \frac{1}{\sqrt{\mu_0\epsilon_0}}. \tag{4.12}$$

With changes in the material's electromagnetic properties, variations in c can be calculated. For weather radar applications, the speed of light can always be treated as a constant, equal to the speed of light in free space $c_0 = 299,792,458 \approx 3 \cdot 10^8\,\mathrm{m\,s^{-1}}$, for determining radar's range. Once again, the precipitation medium is almost free space with water drops or ice particles sprinkled around at a very low concentration. Therefore, changes in the phase constant result in negligible changes in c with respect to c_0 for determining radar range. Where there are slight differences in the effective dielectric constants between the horizontal and vertical polarizations, the difference in propagation characteristics between the two polarizations can be measured as a relative phase shift between the radar polarizations' signals. (See Section 4.7 for discussion of the polarization's effective dielectric constants.) This phase difference between the weather radar's two polarizations is the measurement Φ_{dp}.

The complex-valued refractive index of the media is

$$n = \frac{k}{k_0}. \tag{4.13}$$

For nonmagnetic materials, the magnetic permeability is the same as free space (i.e., $\mu = \mu_0$). (All precipitation types are nonmagnetic; therefore, these are the only materials considered in this book.) With $\mu = \mu_0$ and a negligible conductivity, the complex-valued refractive index is determined by the dielectric constant:

$$n = \sqrt{\epsilon_r}. \tag{4.14}$$

4.1.4   Wave Characteristic Impedance

The radar systems' antennas typically measure the electric field as a voltage. For a propagating electromagnetic wave, the electric field, **E**, and magnetic field, **H**, are tightly coupled via the characteristic impedance (η, not to be confused with the radar reflectivity $\bar{\eta}$) of the media:

$$\eta = \frac{E_{h,v}}{H_{v,h}} = \sqrt{\frac{j\omega\mu}{\sigma + j\omega\epsilon}}. \tag{4.15}$$

(Recall that the TEM wave's electric field is coupled with the magnetic field in the orthogonal direction.) The wave's Poynting vector (its time-averaged power density) can then be calculated from measurements of the respective electric field. With this, the magnitude of the Poynting vector is the power of the wave along the propagating direction:

$$|S_{h,v}| = \frac{1}{2}\frac{|E_{h,v}|^2}{\text{Re}(\eta)}. \tag{4.16}$$

With a known characteristic impedance, the electric field then provides the necessary information about the wave's power.

4.1.5   Spherically Propagating Waves

The electric field for a plane wave is given by eq. (4.6a). The plane wave is a good approximation of the radar's propagating wave for faraway distances from the radar, where it is considered locally as a plane wave (see fig. 2.1).

Consider a propagating wave generated by a point source. The point source's field propagates away from the source as a sphere with radius r. For a source transmitting finite power, conservation of energy requires that as the range increases, the total power must remain constant over the sphere. The surface area of a sphere is $4\pi r^2$, and therefore the total power p and power density magnitude $|S|$ of the wave are related following $p = |S| \cdot 4\pi r^2$. The electric field and power density are related as $|S_h| \propto |E_h|^2$ for the horizontal polarization and $|S_v| \propto |E_v|^2$ for the vertical polarization. Combining these, the electric field for the point source is then

$$p_{h,v} \propto |E_{h,v}|^2 4\pi r^2 \tag{4.17a}$$

$$|E_{h,v}|^2 \propto \frac{p_{h,v}}{4\pi r^2}, \tag{4.17b}$$

and $\sqrt{p_{h,v}} = E_{h,v}(0)$ is the magnitude of the electric field proportional to the point source's power. Finally, this yields

$$|E_{h,v}| \propto \frac{E_{h,v}(0)}{\sqrt{4\pi r}}. \tag{4.17c}$$

For a propagating wave that is in the source's far field, the wave's source can be approximated as a point. This approximation is used extensively for radar systems. The intensity of the electric field at a distance r from a point source with initial magnitude $E_{h,v}(0)$ is

$$E_{h,v}(r) = E_{h,v}(0)\frac{e^{-jk_{h,v}r}}{\sqrt{4\pi r}}. \tag{4.18}$$

For a signal propagating in free space, $\alpha_{h,v} = 0$, and then $k_{h,v} = k_0 = 2\pi/\lambda$ for eq. (4.18). This is a good approximation of the radar signal when there is no precipitation along the path of the radar pulse. (This also neglects the small amount of attenuation caused by atmospheric gases and water vapor in the RF and microwave spectra.) When precipitation is present, k in eq. (4.18) is the effective propagation constant, k_{eff}, with the contribution of free space and the included precipitation particles in the volume. Calculation of the effective propagation constant for mixtures, such as raindrops in air, is covered in Section 4.8.

4.1.6   Wave Propagation and Absorption

Consider a plane wave that is propagating toward a dielectric surface. A portion of the wave is reflected, and another portion propagates into the dielectric. The reflection coefficient for the plane wave, which is perpendicularly incident on the planar surface, is a function of the characteristic impedance of the incident wave's medium (η_1) and the surface medium (η_2):

$$\Gamma = \frac{\eta_2 - \eta_1}{\eta_2 + \eta_1}. \tag{4.19}$$

Note that this equation is only exact for perpendicular incidence on a planar surface that is large with respect to the wavelength. Exact solutions for other geometries require variations to the equation. Changes in material properties create discontinuities of the characteristic impedance encountered by the plane wave. As the wave propagates from medium 1 to medium 2, these changes in the characteristic impedance result in a fraction of the power being reflected while the rest continues in the incident wave's propagation direction.

For the plane wave incident on a planar surface, the fraction of power that is scattered is $|\Gamma|^2$, implying that $1 - |\Gamma|^2$ of the power is transmitted beyond the interface. Even for a wave propagating in a lossless medium where there is no absorption, a portion of the propagating wave's power is lost as a result of scattering. When the radar's wave is incident on precipitation, it can be seen as a transition from free space to a different medium, where the first reflection happens at the layer containing a set of precipitation particles. The actual scattering characteristics for a volume of precipitation is discussed later.

Attenuation reduces the total power of the signal. In eq. (4.6b), which defines a forward-propagating plane wave, the attenuation constant's term $e^{-\alpha r}$ reduces the electric field intensity. The wave's power density is $S \propto |\mathbf{E}^2|$. The one-way path-integrated attenuation (PIA) of the propagating plane wave's power, which is due to absorption in the media with attenuation constant α, is then

$$\text{PIA}_{1\text{way}} = \frac{1}{|e^{-\alpha r}|^2} = e^{2\alpha r}, \tag{4.20}$$

where r is range in meters. This equation assumes α is a constant value along the propagation path. The factor of "2" in the exponent is to convert from electric field to power density (or volts to power in the receiver's signal). The PIA is defined as power that is lost from the signal, therefore $\text{PIA}_{1\text{way}} \geq 1$. In decibels, this ensures that the PIA is always a nonnegative number, and therefore PIA is a loss to be subtracted from the initial power (which would also be in decibels). With a known starting power, such as the transmitted power P_t, and a given $\text{PIA}_{1\text{way}}$ at range r, the total signal power after the signal propagates a distance r is simply $P_t/\text{PIA}_{1\text{way}}$. Similarly, the amount of power that was absorbed (or scattered if scattering losses are included in the actual PIA) is then $P_t - P_t/\text{PIA}_{1\text{way}}$.

The power of the forward-propagating wave changes with the distance traveled and the environment in which the wave propagates. Factors that affect the wave's power are the initial transmitted power, the wave's propagated distance, and the accumulated losses. Later, the absorption, total scattering, and extinction cross-sections are discussed in detail. The scattering and absorption losses can be combined into one term, the extinction losses. This mathematical form for the attenuation can be relaxed to include both scattering and absorption losses (i.e., use α as a proxy to represent the extinction losses).

By setting $r = 1000\,\text{m}$, the specific attenuation due to absorption within the medium (with units dB km^{-1}) can be calculated as

$$A_{1\text{way}} = 10 \log_{10}\left(e^{2\alpha \cdot 1000}\right) = 10 \cdot 2000\alpha \log_{10}(e) = 4343 \cdot 2\alpha. \tag{4.21}$$

Again, the attenuation factor applies to the power of the wave. For a homogeneous medium, the one-way PIA (in decibels) is the product of the one-way specific attenuation (in decibels per kilometer) and the range: $\text{PIA}_{1\text{way}} = A_{1\text{way}} \cdot r/1000$. If r is instead the range that the monostatic radar signal must travel, the total distance is then $2r$, which gives the two-way specific attenuation that a radar signal experiences:

$$A_{2\text{way}} = 2 \cdot A_{1\text{way}} = 8686 \cdot 2\alpha. \tag{4.22}$$

In Section 9.3.1, a similar derivation of attenuation is provided using the extinction cross-section, σ_e. Note that the attenuation constant α is defined for the wave's electric field density, whereas the extinction cross-section is defined for the wave's power density.

4.2   The Relative Permittivity of Water and Ice

Given the general definition of the radar variables for weather radar from Section 2.3, now consider the physical origin of the different scattering behaviors in relation to the phase of the precipitation particles (e.g., liquid or solid). For water and ice particles, the electrical response of a material to an electric field is described by the permittivity: $\epsilon = \epsilon_r \epsilon_0$, with $\epsilon_0 = 8.854 \times 10^{-12}$ C V^{-1} m^{-1}. Typically, materials are characterized by their relative permittivity, or dielectric constant, $\epsilon_r = \epsilon_r' - j\epsilon_r''$. For water and ice, over the range of frequencies used by weather radar, the dielectric constant varies with temperature and frequency. It can be modeled as a single or multiple Debye process, depending on the range of frequencies and temperatures of interest. For frequencies up to about 30 GHz and temperatures above 0°C, a single Debye model can provide an accurate fit to the available observations. However, for higher frequencies and lower temperatures, the relative permittivity is better represented by a double-Debye model. A simple model used to describe the real and imaginary part of the complex-valued relative permittivity is obtained from the summations [136, 137]:

$$\epsilon_r' = \epsilon_s - (2\pi f)^2 \sum_{i=1}^{2} A_i \tag{4.23a}$$

$$\epsilon_r'' = (2\pi f) \sum_{i=1}^{2} B_i, \tag{4.23b}$$

where f is the frequency in Hz; ϵ_s is the static, low-frequency relative permittivity of water,

$$\epsilon_s = 87.91 - 4.044 \times 10^{-1}T + 9.587 \times 10^{-4}T^2 - 1.328 \times 10^{-6}T^3; \tag{4.23c}$$

and T is the temperature (°C). The terms in the summations in eqs. (4.23a) and (4.23b) are given by

$$A_i = \frac{\tau_i^2 \Delta_i}{1 + (2\pi f \tau_i)^2} \tag{4.24a}$$

$$B_i = \frac{\tau_i \Delta_i}{1 + (2\pi f \tau_i)^2} \tag{4.24b}$$

with

$$\Delta_i = a_i \exp(-b_i T) \tag{4.24c}$$

and

$$\tau_i = c_i \exp\left(\frac{d_i}{T + 134.2}\right). \tag{4.24d}$$

For water, the coefficients a_i, b_i, c_i, and d_i in eqs. (4.24c) and (4.24d) are reported in Table 4.1. The value of the relative permittivity of water computed with this model is represented in the complex plane where $\text{Re}(\epsilon_r)=\epsilon_r'$ and $\text{Im}(\epsilon_r)=\epsilon_r''$ in Figure 4.1 for the range of microwave frequencies 2–95 GHz. The diagram illustrates how the relative permittivity of water varies significantly as a function of temperature and frequency,

Table 4.1 Coefficients for the water relative permittivity model of Turner et al. [137].

	a_i	b_i	c_i	d_i
$i = 1$	81.11	4.434×10^{-3}	1.302×10^{-13}	$6.627 \times 10^{+2}$
$i = 2$	2.025	1.073×10^{-2}	1.012×10^{-14}	$6.089 \times 10^{+2}$

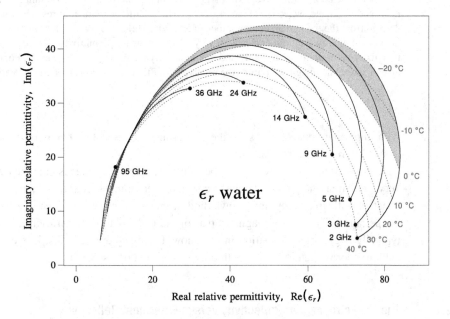

Figure 4.1 The complex relative permittivity of water in the microwave spectral range of weather radar and millimeter-wave cloud radars (2–95 GHz) at different temperatures. Solid lines for frequency (the standard weather radar frequencies are indicated) and dotted lines for temperature. The permittivity values, extending to subfreezing temperatures for application to supercooled drops (gray-shaded area), are calculated according to the Turner et al. [137] fitting model to the historical data set of laboratory measurements between $-18.2°$C and $50°$C. Above ~20 GHz, both the real and imaginary parts decrease monotonically with frequency (refer to Table 4.2 for ϵ_r values at standard frequencies).

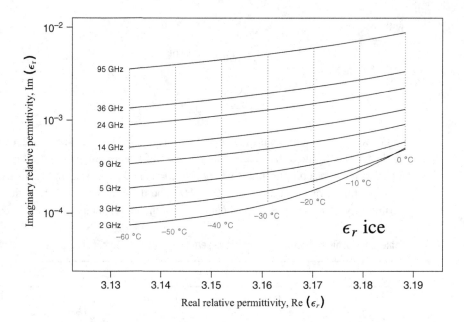

Figure 4.2 The complex relative permittivity of ice in the microwave spectral range of weather radar and millimeter-wave cloud radars (2–95 GHz) at different temperatures. The values are derived from the theoretical computations of Maetzler [138]. Ice is a low-loss dielectric medium at microwave frequencies with very small imaginary permittivity. Above ~3 GHz, the imaginary permittivity is smoothly and monotonically increasing with frequency. For ease of visualization, the imaginary part of the relative permittivity (y-axis) is plotted in logarithmic scale.

with important consequences for the scattering and absorption behavior of raindrops in the different radar bands.

In contrast to water, the relative permittivity of ice (Fig. 4.2) is characterized by a very small imaginary part over the microwave spectrum, whereas the real part is nearly constant with frequency and shows modest change for a wide range of subfreezing temperatures. The small imaginary part of the ice's relative permittivity explains the typically negligible attenuation in dry snow. For many applications, it is often sufficient to approximate ϵ_r of ice with only its real part ($\epsilon_r \approx 3.17$).

4.2.1   The Dielectric Factor: Reflectivity versus Equivalent Reflectivity

In eq. (2.12), the dielectric factor of water $|K_w|^2$ was introduced in the relationship between the radar reflectivity factor and the volume's mean radar cross-section, $\bar{\eta}$. The dielectric factor of dielectric spheres surrounded by air [139] is

$$|K|^2 = \left| \frac{\epsilon_r - 1}{\epsilon_r + 2} \right|^2 . \tag{4.25}$$

Table 4.2 Dielectric constant ϵ_r and dielectric factor of water $|K_w|^2$ for different temperatures and frequencies.

Frequency (GHz)	ϵ_r			$|K_w|^2$		
	0°C	10°C	20°C	0°C	10°C	20°C
2.8	$81 - j23$	$80 - j16$	$78 - j12$	0.934	0.931	0.928
5.6	$64 - j37$	$71 - j29$	$73 - j22$	0.932	0.930	0.928
9.4	$45 - j41$	$55 - j37$	$62 - j32$	0.929	0.929	0.927
13.9	$30 - j37$	$41 - j39$	$50 - j37$	0.923	0.925	0.925
36	$11 - j19$	$14 - j24$	$19 - j28$	0.869	0.894	0.906
95	$7 - j8$	$7 - j11$	$8 - j13$	0.699	0.763	0.811

From Figure 4.1, it is clear that the dielectric properties of water vary with temperature and radar frequencies. Because the composition and temperature of the observed volume are not known, in weather radar, the dielectric constant is fixed to a constant value. Operationally, $|K_w|^2 = 0.93$ is typically used as the dielectric factor in the weather radar constant up to the Ka band. At the W band, 0.70 is more commonly used for operational radar systems. Although there is variation in the dielectric factor, the overall error in the estimated reflectivity factor is small (e.g., 0.64 dB of variation in the estimated equivalent reflectivity at 94 GHz between $|K_w|^2$ at 0°C and 20°C from Table 4.2).

When comparing the RCS between water and ice in Figure 3.12, in the Rayleigh scattering regime, the ice RCS is approximately 7.2 dB lower than that for water. This difference is a function of the ratio of the dielectric factors of ice and water, which is calculated as $10\log_{10}\left(|K_i|^2/|K_w|^2\right) \approx -7.2$ dB (see Figs. 4.1 and 4.2). When estimating reflectivity, the correction for this dielectric factor assumes water ($|K_w|^2$ is used in the radar constant; see eqs. [2.39c] and [2.39d]), and therefore the reflectivity factor in ice is underestimated by approximately 7.2 dB. Although this apparent error in the reflectivity factor may seem large, all radar systems assume the dielectric factor of water. Therefore, simulations, models, and observations by other radars consistently share this error to be able to compare results. It is important to remember these differences when simulating the RCS of ice hydrometeors and then converting to Z_e.

When estimating the Rayleigh reflectivity factor, proper bookkeeping of $|K|^2$ is critical to relate it to the radar cross-section. This point is illustrated in the following equations and discussion. Considering the backscattering cross-section

$$\sigma_b(D) = \frac{\pi^5}{\lambda^4}|K|^2 D^6,$$
(4.26)

the intrinsic reflectivity may be expressed as

$$Z = \frac{\lambda^4}{\pi^5|K|^2}\int N(D)\sigma_b(D)dD.$$
(4.27)

From these equations, it is clear that the scattering cross-section depends on the dielectric factor, whereas the intrinsic reflectivity does not, by definition. Thus, when radars observe precipitation in the ice phase, it measures the equivalent reflectivity factor, which is indeed a scaled version of the intrinsic reflectivity:

$$Z_e = \frac{|K_i|^2}{|K_w|^2} \int D^6 N(D) dD = \frac{|K_i|^2}{|K_w|^2} Z. \tag{4.28}$$

Figure 4.3a shows the reflectivity field as it appears in a typical range-height indicator (RHI) scan in stratiform precipitation. The measured reflectivity (Z_e) incorporates the range correction of the received power and a constant dielectric factor $|K_w|^2$ for all observations (eq. [2.39c]). The reflectivity factor in Figure 4.3b is instead derived by applying the dielectric factor of water $|K_w|^2$ below the melting layer and applying the dielectric factor of ice $|K_i|^2$ above (wtih a simple linear interpolation in between). In this case, Z represents the "true" sixth moment of the DSD corresponding to the

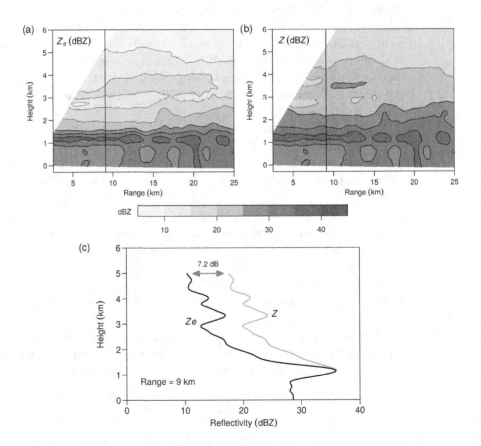

Figure 4.3 An RHI elevation scan of stratiform precipitation. The equivalent reflectivity (a) and reflectivity factor (b) are calculated by applying the proper dielectric constant to the rain and ice regions. The profiles of Z_e and Z at the range of 9 km are plotted in panel (c), illustrating the 7.2-dB difference above the freezing level.

reflectivity factor. Figure 4.3b illustrates how the reflectivity factor would look if the proper dielectric factor was applied to the measured power, in relation to the actual phase of the precipitation.

4.3   Scattering by Dielectric Spheres

Scattering theory as applied to objects is a fully developed field spanning many disciplines, from astronomy to radar meteorology. This section provides a brief introduction to scattering concepts; for further reading, refer to Bringi and Chandrasekar [21]. To fully characterize the scattering cross-sections for arbitrary scatterers, the far-field amplitude vector for the scatterer, $s(\hat{s}, \hat{i})$, must be known for all combinations of scattering ($\hat{s}$) and incident ($\hat{i}$) directions.

The general bistatic radar cross-section is proportional to the ratio of incident to scattered power for arbitrary scattering and incidence angles:

$$\sigma_{bi}(\hat{s}, \hat{i}) = 4\pi |s(\hat{s}, \hat{i})|^2. \tag{4.29}$$

The radar cross-section (also known as the *backscatter cross-section*) is a special case of the bistatic RCS where the scattering direction is back toward the incident direction (i.e., $\hat{s} = -\hat{i}$), giving

$$\sigma_b = 4\pi |s(-\hat{i}, \hat{i})|^2. \tag{4.30}$$

The total scattering cross-section is the integral of all scattered power:

$$\sigma_s = \int_{4\pi} |s(\hat{s}(\Omega), \hat{i})|^2 \, d\Omega. \tag{4.31}$$

The incident angle is fixed, and the scattering direction is varied as a function of the solid angle, Ω, over the entire sphere (with an area of 4π steradians).

The extinction cross-section for the scatterer is

$$\sigma_e = -\frac{4\pi}{k_0} \text{Im}(s(\hat{i}, \hat{i})). \tag{4.32}$$

The extinction cross-section relates to the forward-propagating wave's power that is lost to absorption and scattering (note that the incident and scattering directions are the same).

For all but the simplest cases (Rayleigh or optical scattering from spheres), the relevant cross-sections (e.g., the backscatter, total scattering, absorption) do not have simple closed-form solutions. Even for Mie scattering, which describes scattering from dielectric spheres when the size of the spheres is not small compared with the wavelength, the scattering cross-section needs to be calculated as a power series expansion [21, 134], and numerical techniques are required. More complicated geometries require full numerical electromagnetic solvers and are a field of study on their own. Fortunately, closed-form solutions for the cross-sections of spheroidal scatterers are available in the Rayleigh scattering regime.

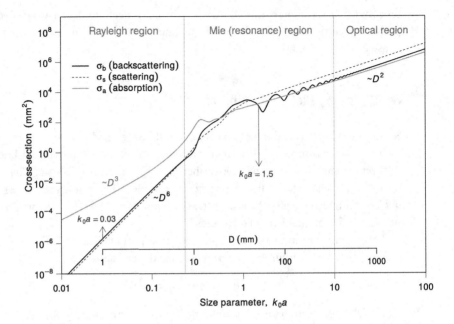

Figure 4.4 Scattering cross-sections of spherical water particles at the S band ($T = 0°C$) as a function of the size parameter. Note the different size relations in the Rayleigh and optical regions.

The scattering from spheres (and spheroids) is broadly classified into three scattering regimes: Rayleigh, Mie, and optical. Although the RCS has units of meters squared, the geometric area is not necessarily equivalent to the RCS of the object. (A metal sphere that is large with respect to the wavelength being a notable exception, where the RCS is the cross-sectional area of the sphere with radius a as $\sigma_b = \pi a^2$.)

Figure 4.4 illustrates how the radar, absorption, and scattering cross-sections vary with the diameter of a sphere of water. The size of the scatterer determines its scattering regime. Figure 4.5 shows the resulting scattering pattern for two different-size spheres: one in the Rayleigh scattering regime and one in the Mie regime. In the Rayleigh regime, the spheroid's scattering pattern matches an infinitesimal dipole [18]. The scattering pattern in the Mie regime varies with the size of the particle. The variability can be inferred from the oscillation of the cross-sections in the Mie regime in Figure 4.4. In the Mie scattering regime, the magnitude and direction of scattered power depend on the diameter of the scatterer and the radar's wavelength. The backscattering cross-section, which includes the scattering amplitude and phase from all parts of the sphere, exhibits constructive or destructive interference, resulting in variations in the cross-section as the size of the sphere varies relative to the wavelength.

For weather radar, Rayleigh scattering from spheres provides a sound basis for approximating the scattering characteristics of hydrometeors. With this understanding, numerical techniques for estimating the scattering characteristics can then be used to provide more exact solutions when Mie scattering from spheres is applicable.

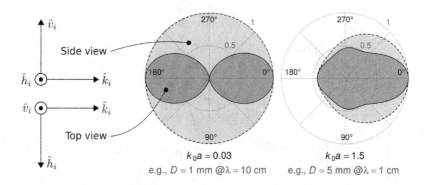

Figure 4.5 Polar graphs of scattering patterns $|s|^2$ of spherical water particles with the size parameters indicated by the arrows in Figure 4.4. The incident wave is polarized on the horizontal plane and propagates along $\hat{k}_i$ (i.e., backscattering at 180°, forward scattering at 0°). For ease of visualization, the patterns are normalized to the maximum value. In the polarization reference frames on the left, the circled dots represent vectors pointing out of the page (see also Fig. 2.2). The dark- and light-gray patterns, respectively, represent the radiation in the parallel (top view) and perpendicular (side view) planes. For a vertically polarized incident wave, the patterns would be reversed; that is, the light-gray (dark-gray) area would represent the top (side) view.

Assuming Rayleigh scattering, the scattering amplitude of a dielectric sphere is given by

$$s = s(-\hat{i}, \hat{i}) = \frac{\pi^2}{2\lambda^2} \frac{\epsilon_r - 1}{\epsilon_r + 2} D^3, \tag{4.33}$$

and the radar (or backscattering) cross-section is

$$\sigma_b = 4\pi |s|^2 = \frac{\pi^5}{\lambda^4} |K|^2 D^6, \tag{4.34}$$

where D is the sphere's diameter and $|K|^2$ is the dielectric factor of the sphere (from eq. [4.25]).

The scattering cross-section is usually displayed as a function of the size parameter $k_0 a = 2\pi a/\lambda$, where k_0 is the wavenumber of free space, and $a = D/2$ is the radius of the sphere. To facilitate the interpretation of the scattering behavior for common operational radar frequencies, in Figures 4.6 and 4.8, the corresponding scales for the particle diameter are added for all weather radar bands, from the S band to the W band. It is important to note that the dielectric constant in eq. (4.34) is a function of both frequency and temperature, as discussed in the previous section. In Figures 4.6 and 4.8, the frequency dependence related to the varying permittivity is shown using gray shading. For water (Fig. 4.6), the backscattering cross-section is insensitive to the permittivity changes with frequency in the Rayleigh region, that is, when the particle diameter is smaller than approximately 10% of the wavelength ($D/\lambda < 0.1$ or size parameter $k_0 a < 0.3$). For larger particles, interference and resonance effects perturb

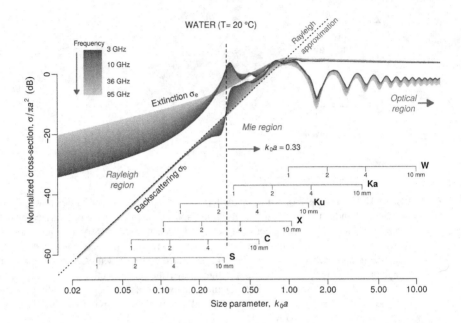

Figure 4.6 Normalized backscattering and extinction cross-sections for water spheres at 20°C temperature, plotted versus the size parameter $k_0a = 2\pi a/\lambda$. The actual diameter of the spherical particles is shown in the overplotted axes for the S to W frequency bands. The gray shading shows the impact of varying water permittivity at different frequencies in the range of 3–95 GHz. Both the frequency gray shading and the plot axes are scaled logarithmically. The Rayleigh approximation (eq. [4.34]) is also shown as a dotted line. The vertical dashed line corresponds to $k_0a = 0.33$, a critical value of the size parameter where resonance effects can be especially evident (see also Fig. 4.11 in Section 4.4 for the impact of scattering from nonspherical particles).

the scattering response of spherical particles, which becomes a complex function of the radar frequency and can be described by the Mie theory.

The extinction cross-section represents the total power loss suffered by an incident wave due to both scattering and absorption by the particle:

$$\sigma_e = \sigma_a + \sigma_s, \tag{4.35}$$

where σ_a is the absorption cross-section, and σ_s is the total scattering cross-section. The total scattering cross-section, which is the sum of all power scattered from the dielectric sphere with volume $V = \pi/6D^3$, is given as follows [21]:

$$\sigma_s = \frac{3}{2\pi} k_0^4 V^2 |K|^2. \tag{4.36}$$

For the same dielectric sphere, the Rayleigh absorption cross-section is [21]

$$\sigma_a = 9k_0 V \frac{\epsilon_r''}{|\epsilon_r + 2|^2}. \tag{4.37}$$

The absorption is proportional to the volume of the particle or D^3. From eq. (4.37), it is clear that the attenuation depends on the radar frequency and the phase of the particle through its complex relative permittivity. In particular, the extinction cross-section of a spherical water particle is shown in Figure 4.6 to be an increasing function of frequency in the Rayleigh region for a given size parameter.

For Rayleigh scatterers, the total scattering cross-section does not significantly contribute to the extinction cross-section. To demonstrate,

$$\sigma_e^{(\text{Rayleigh})} = \sigma_s + \sigma_a \tag{4.38a}$$

$$= \frac{3}{2\pi} k_0^4 V^2 \left| \frac{\epsilon_r - 1}{\epsilon_r + 2} \right|^2 + 9k_0 V \frac{\epsilon_r''}{|\epsilon_r + 2|^2} \tag{4.38b}$$

$$= \frac{3k_0 V}{|\epsilon_r + 2|^2} \left[\frac{k_0^3 V}{2\pi} |\epsilon_r - 1|^2 + 3\epsilon_r'' \right]. \tag{4.38c}$$

From the extinction cross-section, the left term (representing σ_s) has $k_0^3 V \propto D^3/\lambda^3$. In the Rayleigh regime, the drop diameter is small with respect to the wavelength. As such, $D^3/\lambda^3 \ll 1$, which becomes vanishingly small with decreasing D. Therefore, for Rayleigh scattering,

$$\sigma_e^{(\text{Rayleigh})} \approx \sigma_a. \tag{4.39}$$

As the particle size increases into the Mie and optical scattering regimes, the total scattering cross-section significantly contributes to the extinction cross-section.

The attenuation (via the extinction cross-section) depends on the radar frequency and the phase of the particle through its complex relative permittivity. The total power removed from an incident wave is due to both scattering and absorption by the particles along the propagation path. For typical rain DSDs, absorption is the dominant contributor to the overall extinction cross-section for frequencies up to 10 GHz (X band), whereas scattering may become significant at the Ku band and is dominant at the W band for drops larger than $\sim$1 mm. This is illustrated in Figure 4.7, where the ratio σ_a/σ_e is plotted as a function of the drop diameter for water.

When ice particles are considered (Fig. 4.8), the backscattering cross-section in the Rayleigh regime is lowered by ≈ 7.2 dB with respect to the corresponding cross-section of water particles as a result of the different dielectric constant (see previous discussion about Fig. 4.3). The most notable difference, however, is the dramatic decrease of the extinction coefficient in the Rayleigh regime, resulting in negligible attenuation for radar signals propagating through ice precipitation. Another relevant aspect of scattering by ice spherical particles is that Mie resonance effects do not show up until particles reach a size parameter of about 1, that is, when the particle diameter is approximately 30% of the wavelength. This means that large frozen precipitation

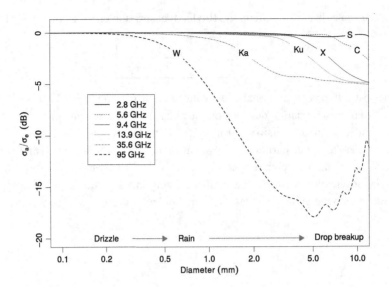

Figure 4.7 The relative contribution of absorption (σ_a) to the extinction cross-section (σ_e). At the Ka band, the scattering losses become relevant for drops larger than ~2 mm. At the W band, absorption is the main process responsible for the signal extinction only in drizzle, whereas for drops larger than 0.5–1 mm, scattering becomes the dominant process.

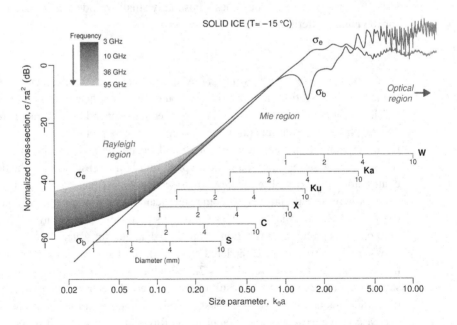

Figure 4.8 Similar to Figure 4.6 but for spheres of solid ice at $-15°C$. The value of the complex relative permittivity for ice is computed according to Maetzler [138] and ranges between $3.175 - j2.62 \times 10^{-4}$ (3 GHz) and $3.175 - j6.55 \times 10^{-3}$ (95 GHz). The resonance peaks of σ_b for size parameter $k_0a \geq 4$ result from the fine-scale structure of Mie scattering from a single spherical particle with a small imaginary part of the relative permittivity [140]. These fine-scale ripples are smoothed out for distributions of particles with varying size.

particles generally have a smaller impact in terms of non-Rayleigh scattering, in comparison with water particles of the same size. For water particles, consider the dashed vertical line in Figure 4.6 (size parameter $k_0 a = 0.33$, corresponding to a diameter of ~ 10, ~ 6, ~ 3, ~ 2 mm, respectively, for S, C, X, and Ku bands). For particles with a size close to this value, the extinction cross-section is near its maximum value, whereas the backscattering cross-sections have an enhanced oscillatory behavior, especially at low frequencies. Deviations from Rayleigh scattering may have a particular impact on the radar's polarimetric variables.

4.4   Scattering by Spheroids

The raindrop's equivolume diameter is the diameter of a sphere with the equivalent amount of water as the raindrop. This is important because as a raindrop's volume increases, the ratio of the horizontal to vertical dimensions increases as a result of aerodynamic drag. The hydrostatic forces of surface tension that try to maintain the spherical shape must balance against the aerodynamic forces. This results in a spheroidal shape where the raindrop's major axis is perpendicular to the direction of motion (Section 3.3). A spheroid is defined by the surface obtained by rotating an ellipse about one of its two main axes. If the ellipse is rotated about the minor axis, an oblate spheroid is obtained (Fig. 4.9, panel [a]), whereas if the ellipse is rotated about the major axis, the result is a prolate ellipsoid (Fig. 4.9, panel [b]). As illustrated in the figure, b is used to represent the dimension along the symmetry axis, and a is used for the orthogonal equatorial radius. The ratio between the two main axes defines the shape of a spheroid. This book follows the convention of defining the axis ratio as the

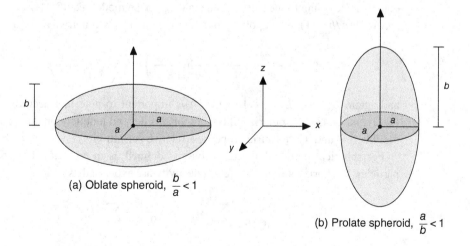

(a) Oblate spheroid, $\dfrac{b}{a} < 1$

(b) Prolate spheroid, $\dfrac{a}{b} < 1$

Figure 4.9 Oblate (a) and prolate (b) spheroids with symmetry axis along the z-axis. A spherical particle has an axis ratio of 1.

ratio of the minor to major axis (b/a for oblate spheroid, a/b for prolate spheroids), such that the axis ratio is always ≤ 1. Similar to raindrops, planar ice crystals such as plates and dendrites can be modeled to a first approximation as oblate spheroids, whereas columns and needles can be represented by prolate spheroids. By extension, the aspect ratio of generic precipitation particles with arbitrary shapes is defined as the ratio of the minor to major dimension. Note that other definitions are available in the literature, especially when ice crystals are considered, so care must be taken when comparing results from different studies.

Under the Rayleigh–Gans approximation, the scattering amplitudes for an oblate spheroid with symmetry axis along z are given by [21]

$$s_{hh} = \frac{\pi^2}{6\lambda^2} D^3 \frac{(\epsilon_r - 1)}{1 + L(\epsilon_r - 1)} \tag{4.40}$$

$$s_{vv} = \frac{\pi^2}{6\lambda^2} D^3 \frac{(\epsilon_r - 1)}{1 + L_z(\epsilon_r - 1)}, \tag{4.41}$$

where D is the equivolume diameter (the diameter of a sphere with the same volume as the spheroid), and $L_{x,y,z}$ are the three depolarization factors with $L = L_x = L_y$ and $L_x + L_y + L_z = 1$. Note that the scattering amplitudes s_{hh} and s_{vv} are the copolar elements of the dual polarization scattering matrix (see Section 4.5).

The depolarization factor along the symmetry axis z for ellipsoids can be expressed in a closed form [141]. In particular, for oblate spheroids ($0 < b/a \leq 1$):

$$L_z = \frac{1 + f^2}{f^2} \left(1 - \frac{1}{f}\tan^{-1}f\right); \quad f = \sqrt{\left(\frac{a}{b}\right)^2 - 1}, \tag{4.42}$$

where $L_{x,y} = L = \frac{1}{2}(1 - L_z)$. Figure 4.10a shows the depolarization factor for raindrops as a function of the axis ratio, with the top x-axis representing the raindrop's diameter according to the Beard–Chuang drop-shape model (eq. [3.31]). For a prolate spheroid ($a/b < 1$), the depolarization factor along the symmetry axis is given by

$$L_z = \frac{1 - e^2}{e^2} \left(-1 + \frac{1}{2e}\ln\frac{1 + e}{1 - e}\right); \quad e = \sqrt{1 - \left(\frac{a}{b}\right)^2}, \tag{4.43}$$

and again, $L_{x,y} = L = \frac{1}{2}(1 - L_z)$. It is important to note that according to eqs. (4.42) and (4.43), the polarizability components depend uniquely on the axis ratio of the spheroid and its dielectric properties (relative permittivity).

For spherical particles, the reflectivity is defined as in eq. (2.6). For spheroidal particles, the horizontal and vertical reflectivity are expressed as

$$Z_h = \frac{4\lambda^4}{\pi^4 |K|^2} \sum_i |s_{hh_i}|^2 = \frac{1}{9|K|^2} \sum_i |p_{hi}|^2 D_i^6 \tag{4.44}$$

$$Z_v = \frac{4\lambda^4}{\pi^4 |K|^2} \sum_i |s_{vv_i}|^2 = \frac{1}{9|K|^2} \sum_i |p_{vi}|^2 D_i^6, \tag{4.45}$$

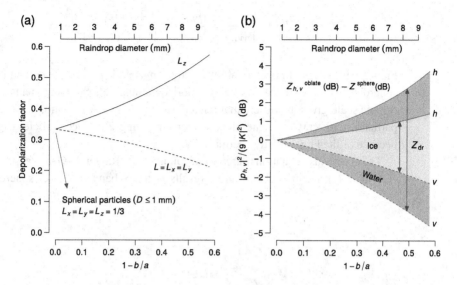

Figure 4.10 (a) Depolarization factor for oblate spheroids as a function of $1 - b/a$. The corresponding equivolume diameter of raindrops is shown on the top x-axis, assuming the axis ratio relation from the Beard–Chuang drop-shape model (eq. [3.31]). (b) Polarizability components in logarithmic scale for a water spheroid (upper and lower bounds of dark-gray area) at horizontal (solid lines) and vertical (dashed lines) polarization. These are the terms multiplying $D_i{}^6$ in the summations in eqs. (4.44) and (4.45), valid under the Rayleigh–Gans approximation. For spherical particles, these terms reduce to unity (0 dB), and the reflectivity is simply given by the sum of the sixth power of drop diameters in the unit volume. The same axis ratio applied to ice particles (upper and lower bounds of light-gray area) emphasizes the role of the dielectric factor in the differential reflectivity (vertical extension of the gray-shaded areas). The solid and dashed lines can also be seen as representing the deviation of the spheroid's reflectivity with respect to a spherical particle with the same volume.

where $|K|^2 = \left|\frac{\epsilon_r - 1}{\epsilon_r + 2}\right|^2$, and $p_{h,v}$ are polarization factors that depend on the geometric and dielectric properties of the particle:

$$p_h = \frac{(\epsilon_r - 1)}{1 + L\,(\epsilon_r - 1)} \tag{4.46}$$

$$p_v = \frac{\epsilon_r - 1}{1 + L_z\,(\epsilon_r - 1)}. \tag{4.47}$$

For spherical particles, $L_x = L_y = L_z = 1/3$, and $|p_{h,v}|^2/(9|K|^2) = 1$; then, $Z_h = Z_v = Z$, and $Z_{dr} = 0$ dB.

4.4.1 Single Particle's Z_{dr}

For a single oblate particle, the differential reflectivity Z_{dr} can be readily calculated using eqs. (4.40)–(4.42) as

$$Z_{dr} = \frac{|s_{hh}|^2}{|s_{vv}|^2} = \frac{|p_h|^2}{|p_v|^2} = \frac{|1 + L_z(\epsilon_r - 1)|^2}{|1 + L(\epsilon_r - 1)|^2}. \qquad (4.48)$$

Figure 4.10b shows the polarizability factors $|p_h|^2$ and $|p_v|^2$ for water and ice oblate spheroids. The difference between the horizontal and vertical components in a logarithmic scale gives the differential reflectivity. For a given axis ratio, the Z_{dr} of water particles (dark gray) is higher than the corresponding Z_{dr} of ice particles (light gray) because of the larger relative permittivity.

The previous equations are valid under the Rayleigh–Gans approximation ($\lambda \gg \pi D$) and produce accurate results at the S band for water spheroids with

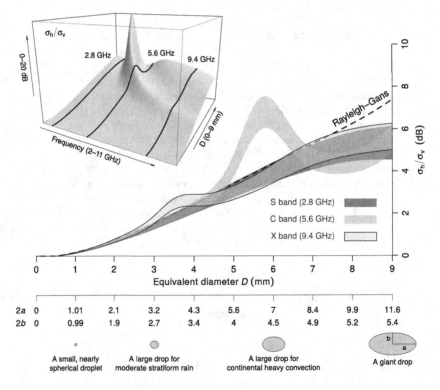

Figure 4.11 The main plot region shows the differential reflectivity of a single raindrop (oblate spheroidal particle with $Z_{dr} = \sigma_h/\sigma_v = |s_{hh}|^2/|s_{vv}|^2$) for different frequency bands at 20°C temperature. The upper limit of each gray band corresponds to scattering by a particle with zero canting angle, whereas the lower limit is obtained by considering canting angles following a Gaussian distribution with zero mean and 15° standard deviation. The axis ratio of the spheroid is modeled according to the Beard–Chuang model (eq. [3.31], Fig. 3.7). Note that the upper limit of the S-band simulation (darker gray) is very close to the Rayleigh–Gans approximation (dashed line) for particle sizes up to ∼8 mm. The gray ellipses on the bottom represent example raindrops used in the simulation ($D = 1, 3, 6, 9$ mm). The 3D surface in the top-left inset emphasizes the variation of Z_{dr} as a function of the frequency and equivalent diameter of the spheroid.

equivolume diameters up to ~8 mm. For higher frequencies (or larger particles), it is important to consider the non-Rayleigh scattering behavior. Figure 4.11 shows the single particle Z_{dr} of water spheroids obtained from electromagnetic scattering simulations at common weather radar frequencies. The three-dimensional (3D) surface representing the ratio of the horizontal to vertical radar cross-sections ($\sigma_h/\sigma_v = Z_{dr}$) shows a characteristic ridge with increasing amplitude at lower frequencies and larger diameters. At the S band (2.75 GHz), the peak is encountered at large particle sizes ($D > 9$ mm), making it a negligible possibility for scattering by raindrops. At the X band, the peak shows up at smaller particles size (around 3–4 mm), but its amplitude enhancement is quite small. This makes the theoretical behavior of S-band and X-band differential reflectivity quite similar. In the intermediate C-band frequency range, the scattering resonance peak has a large impact on rain observations, showing up with significant intensity in a portion of the particle-size spectrum that is often populated with raindrops originating from melting graupel or hail in convective precipitation.

4.5   Polarimetric Scattering Matrix

The radar's measurements are directly related to the complex amplitudes of the scattering matrix. The relationship between the scattering matrix and the dual polarization radar variables is considered here. The scattering amplitudes are a function of the characteristics and geometries of all of the scatterers within the radar volume. This implies that over time, the scattering amplitudes can vary as the distribution and properties of the scatterers change. The mean value of this distribution is the ensemble average. The ensemble average for a variable x is denoted as $\langle x \rangle$. In statistics, an ensemble is defined as a set of all possible outcomes of a stochastic process. The ensemble average is the expected value of the stochastic process. The ensemble average can be thought of as the arithmetic mean for an "infinite" number of random (and independent) reorganizations of the scatterers within the radar volume.

The incident electric field interacts with the scatterer, and in doing so, it effectively makes the scatterer a new source (the scatterer becomes its own transmitter). The electric field, at a range r from its source (e.g., a radar's transmitting antenna or a raindrop scattering an incident TEM wave), is generalized as $E(r) = E_0 s e^{-jkr}/r$, where s is the complex scattering amplitude (for the interested reader, this is equivalent to f in eqs. 1.24 and 1.35 of Bringi and Chandrasekar [21]). In the case of a scatterer, E_0 is the incident electric field's complex amplitude. The field emitted by the source is a spherically propagating wave. The amplitude and phase shift resulting from the spherical propagation are determinded by e^{-jkr}/r. The source's contribution to changes in the amplitude and phase are included in s.

The scattering properties of precipitation are a direct result of the physical properties of the collection of hydrometeors within the radar volume. The radar's observations are a direct result of the scattering properties of the hydrometeors within the radar's observation volume. The dual polarization radar observations and the dual polarization scattering properties are therefore directly related to one another and also directly related to the physical properties of the hydrometeors. In Chapter 6, the signal processing techniques for using the radar observations to estimate the dual polarization radar variables are discussed. Here, the radar variables are defined in terms of the scattering matrix elements.

4.5.1   Dual Polarization Scattering Matrix

The polarization of the transverse electromagnetic wave is defined by the electric field. For a propagating wave, the plane of polarization is perpendicular to the direction in which the wave propagates. The polarization plane is defined by two independent states. In dual polarization radar systems, the polarization states can be characterized by two different sets of bases: linear polarization or circular polarization. The two states for linear polarization are horizontal ($\hat{h}$) and vertical ($\hat{v}$). Similarly, the two circular polarization states are right-hand circular ($\widehat{RHC}$) and left-hand circular ($\widehat{LHC}$). The linear and circular polarization states are related as follows:

$$\begin{bmatrix} \widehat{RHC} \\ \widehat{LHC} \end{bmatrix} = \frac{1}{\sqrt{2}} \begin{bmatrix} 1 & j \\ 1 & -j \end{bmatrix} \begin{bmatrix} \hat{h} \\ \hat{v} \end{bmatrix}. \qquad (4.49)$$

(Note, j is the imaginary number.)

Fully polarimetric radars can control the transmitted signal polarizations and measure the two polarizations of the received signals. Because the two dimensions are independent, if only the horizontal polarization is perturbed, it has no effect on the vertical polarization, and vice versa.

For a circular polarization basis, the differential phase shift due to the aspect ratio (see Section 3.3) of rain on propagation through precipitation can cause the circular polarization to change from one state to the other (i.e., from RHC to LHC, or vice versa). When using the linear polarization states, the polarization state does not change, although the effects of differential attenuation and differential phase shift are measurable. As a result of the effects of the oblate raindrops on the circular polarization basis, linear polarization has been adopted as the standard for operational dual polarization weather radar. Linear dual polarization is considered exclusively in this book. A detailed discussion of circular polarization is found in Bringi and Chandrasekar [21].

For dual polarization weather radar, the horizontal (h) and vertical (v) states' scattering characteristics for the radar volume are represented by the scattering matrix:

$$\mathbf{S} = \begin{bmatrix} s_{hh} & s_{hv} \\ s_{vh} & s_{vv} \end{bmatrix}. \qquad (4.50)$$

The elements of the matrix $\mathbf{S}$ are referred to as the *complex-valued scattering amplitudes*. For each complex amplitude, the second subscript represents the incident polarization, and the first subscript is the scattered polarization. The copolar scattering characteristics are represented on the matrix diagonal (i.e., s_{hh} and s_{vv}), and the cross-polar scattering characteristics are represented on the matrix off-diagonal. The linear dual polarization scattering matrix defines the relationship between the incident ($e^{(i)}$) and scattered ($e^{(s)}$) electric-field components:

$$\begin{bmatrix} e_h^{(s)} \\ e_v^{(s)} \end{bmatrix} = \begin{bmatrix} s_{hh} & s_{hv} \\ s_{vh} & s_{vv} \end{bmatrix} \begin{bmatrix} e_h^{(i)} \\ e_v^{(i)} \end{bmatrix}. \tag{4.51}$$

Note that alternate notation for the electric field components is used: $E_{h,v} = e_{h,v}$ (where e is proportional to the signal's complex envelope, as discussed in Chapters 5 and 6).

The values of the scattering matrix depend on the physical characteristics of the scatterers. The scattering characteristics (and therefore the scattering matrix) are a function of the incident and scattering angles. More generally, the polarimetric scattering amplitudes are defined as $s_{xy}(\hat{s}, \hat{i})$ for a specific pair of scattering and incidence angles, $\hat{s}$ and $\hat{i}$, respectively. For weather radar applications, the most common scenario is the monostatic configuration where the incident direction is along unit vector $\hat{i}$, and the scattering direction of interest is back toward the radar $\hat{s} = -\hat{i}$. For considerations of forward scattering, such as when estimating the specific differential phase shift, K_{dp}, the incident angle would be $\hat{i}$, and the scattering angle would also be $\hat{i}$. For bistatic radar systems or scenarios involving multiple scattering, any combination of incidence and scattering angles could be considered.

4.5.2 Scattering Matrix Alignment Conventions

Figure 4.12 illustrates the two scattering coordinate reference frames: the forward-scattering alignment (FSA) and the backscatter alignment (BSA). The propagation direction of the incident wave is from the monostatic radar, and the echo from a raindrop is the backscattered wave. In the FSA frame, from the point of view of the propagating wave (either the incident or scattered wave), the propagation direction is in the direction the wave is traveling, the vertical axis points up, and the horizontal axis is the cross product of the two others ($\hat{k} = \hat{v} \times \hat{h}$). As a result of the change in propagation direction, when viewed from the radar, the sign of the propagation axis changes. Following the BSA convention, the vertical and horizontal axes remain the same for both the incident and scattered waves, as follows:

$$\mathbf{S}_{\text{BSA}} = \begin{bmatrix} 1 & 0 \\ 0 & -1 \end{bmatrix} \mathbf{S}_{\text{FSA}}. \tag{4.52}$$

The "-1" can be implemented as a $180°$ phase shift (where the amplitude is non-negative). Another side effect of this $180°$ phase shift is that the off-diagonals are equal (i.e., $\mathbf{S}_{\text{BSA}} = \mathbf{S}_{\text{BSA}}^T$, and $s_{hv} = s_{vh}$). When using the FSA convention, the off-diagonals are complex conjugates of each other (i.e., $\mathbf{S}_{\text{FSA}} = \mathbf{S}_{\text{FSA}}^H$, and $s_{hv} = s_{vh}^*$). In the FSA reference frame, the axes are constant with respect to the monostatic wave's propagation direction. In the BSA reference frame, the axes are constant with respect to the radar's antenna. For dual polarization weather radar applications (and radar applications more broadly), the BSA convention is typically used to describe the scatterer when the incident and backscattered wave use the same reference frame.

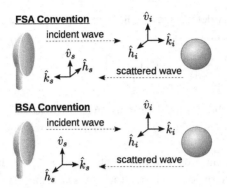

Figure 4.12 A diagram of the radar's incident wave (subscript i) on a raindrop and the back-scattered echo's wave (subscript s) from the raindrop. Two coordinate frames are commonly used for describing electromagnetic scattering. The backscatter alignment (BSA) uses a constant reference frame with respect to the antenna. The forward scatter alignment (FSA) uses a constant reference frame with respect to the wave's propagation. For radar applications, the BSA convention is typically used so that the fields of the incident and scattered waves have the same coordinate reference frame.

In some applications (e.g., numerical scattering models), the FSA convention may be preferred.

4.6　Polarimetric Radar Variables

This section defines the polarimetric radar variables Z_h, Z_{dr}, ρ_{hv}, and the linear depolarization ratio (LDR) in terms of the dual polarization scattering amplitudes. The differential phase shift Φ_{dp} and the specific differential phase shift K_{dp} are defined later in Section 4.7.

4.6.1　Reflectivity

The equivalent reflectivity factor at horizontal polarization for a collection of particles can be written in terms of the ensemble average of the squared magnitude of the copolar scattering amplitudes s_{hh} as

$$Z_h = \frac{4\lambda^4}{\pi^4 |K_w|^2} \langle |s_{hh}|^2 \rangle, \tag{4.53}$$

where the ensemble average is performed over a unit volume of 1 m^3. Analogously, the vertical reflectivity Z_v is obtained from eq. (4.53), substituting s_{hh} with s_{vv}. The realization of the ensemble average representing the unit volume is sometimes represented as a summation of the hydrometeors' scattering properties within the a unit volume (see, e.g., eq. [4.44], which could be used to calculate reflectivity from observed DSDs over discrete classes of diameter) or as an integral (e.g., eq. [4.27], where the DSD $N(D)$ represents a continuous function of the drop's size). The two notations are often used interchangeably to denote the expected value of the radar reflectivity factor. The integral form can be more useful to explicitly show the role of the DSD, especially when this is represented by an analytical function such as the exponential distribution.

4.6.2   Differential Reflectivity

In dual polarization systems, the addition of vertical reflectivity enables measurements of the differential reflectivity, defined as the ratio between the ensemble averages of the squared magnitudes of the scattering amplitudes:

$$Z_{dr} = \frac{Z_h}{Z_v} = \frac{\langle |s_{hh}|^2 \rangle}{\langle |s_{vv}|^2 \rangle}. \tag{4.54}$$

The differential reflectivity is the ratio of the scattered power of the horizontal and vertical polarizations and therefore is unitless. As with reflectivity, Z_{dr} is also typically given using a logarithmic scale (dB). For Rayleigh scattering, recall that the reflectivity is proportional to the hydrometeor's diameter (from eq. [2.6]). This means that the differential reflectivity provides a measure of the shape of the scatterers within the volume. Scatterers that are larger in the horizontal dimension than the vertical dimension (which includes oblate spheroids and large raindrops) have $Z_{dr} > 1$. For scatterers that are spherical, the horizontal and vertical sizes are the same: $Z_{dr} = 1$. For scatterers that are larger in the vertical dimension, such as prolate spheroids, $Z_{dr} < 1$.

4.6.3   Linear Depolarization Ratio

When transmitting a single linear polarization and receiving power for both polarizations, the ratio of the cross-polar to copolar received power is the linear depolarization ratio:

$$LDR_{vh} = \frac{\langle |s_{vh}|^2 \rangle}{\langle |s_{hh}|^2 \rangle} \tag{4.55}$$

$$LDR_{hv} = \frac{\langle |s_{hv}|^2 \rangle}{\langle |s_{vv}|^2 \rangle}. \tag{4.56}$$

By virtue of the scattering matrix reciprocity ($s_{hv} = s_{vh}$), the LDR measurements from the two different polarizations are related as:

$$LDR_{hv} = LDR_{vh} Z_{dr}. \tag{4.57}$$

The LDR provides a sense of the orientation of the scatterers relative to the polarization axes, which is explored further in Section 4.9. In some instances, the subscripts may not be given for LDR, and the horizontal linear depolarization ratio LDR_{vh} is implied. Analogous to reflectivity and differential reflectivity, the LDR is also often expressed using a logarithmic scale (dB).

4.6.4   Copolar Correlation Coefficient

The complex-valued correlation between the complex-valued scattering amplitudes of the horizontal and vertical polarizations (which is normalized by their power) is

$$\rho_{co} = \frac{\langle s_{hh}^* s_{vv} \rangle}{\sqrt{\langle |s_{hh}|^2 \rangle \langle |s_{vv}|^2 \rangle}}. \tag{4.58}$$

The argument (i.e., phase) of the numerator in eq. (4.58) is the backscatter differential phase shift, defined as

$$\boxed{\delta_{co} = \arg\langle s_{hh}^* s_{vv} \rangle.} \tag{4.59}$$

The magnitude of δ_{co} significantly affects the value of the correlation coefficient, especially in the melting layer or for rain/hail mixtures.

In practice, it is customary to refer to the magnitude of the copolar correlation, $|\rho_{co}|$, as the copolar correlation coefficient, which is simply denoted by the symbol ρ_{hv} for linear polarization. From eq. (4.58):

$$\boxed{\rho_{hv} = |\rho_{co}| = \frac{|\langle s_{hh}^* s_{vv} \rangle|}{\sqrt{\langle |s_{hh}|^2 \rangle \langle |s_{vv}|^2 \rangle}}.} \tag{4.60}$$

The complex-valued correlation in eq. (4.58) includes the scattering differential phase and the copolar correlation coefficient as

$$\rho_{co} = \rho_{hv} e^{j\delta_{co}}. \tag{4.61}$$

For randomly oriented scatterers in the polarization plane (i.e., $Z_{dr} = 0$ dB), ρ_{hv} can be shown to be directly related to the linear depolarization ratio (LDR $=$ LDR$_{vh}$ $=$ LDR$_{hv}$ because of symmetry in the polarization plane when $Z_{dr} = 1$ in linear scale). In particular, for simultaneous transmission [21, 142]:

$$\rho_{hv} = \frac{1 - \text{LDR}}{1 + \text{LDR}}, \tag{4.62}$$

whereas for alternate transmission [21, 143]:

$$\rho_{hv} = 1 - 2\,\text{LDR} \tag{4.63}$$

Note that LDR in these expressions use a linear scale (as opposed to dB).

This inverse relation (ρ_{hv} decreases as LDR increases) is valid for scatterers with rotational symmetry in the polarization plane, such as the case of columnar ice crystals observed at vertical incidence. In fact, small ice columns and needles can be represented by prolate particles. As the particles fall, the columns orient in a horizontal plane that is parallel to the ground. When viewed from below (a radar pointed in the vertical direction), the ice columns do not fall with a preferential orientation angle within the horizontal plane and therefore are uniformly distributed between 0 and π (see, e.g., Fig. 4.21). This results in $Z_{dr} = 0$ dB and redundancy of ρ_{hv} and LDR estimates (from eq. [4.62] and [4.63]). In this case, the observation of LDR (which requires a different dual polarization radar configuration with respect to the widely used simultaneous transmit and simultaneous receive [STSR] mode) would not bring any additional information beyond that already provided by ρ_{hv}.

At the other extreme, consider a distribution of identical oblate spheroids with a constant canting angle of $\beta = 45°$ (as defined in Fig. 2.2; not to be confused with the phase constant). In this case, $\rho_{hv} = 1$ (perfect correlation when all scatterers have the same shape and orientation), but the LDR would be close to its maximum value when considering all possible orientations for the same particle distribution (see later Fig. 4.24). This intuitively shows that the LDR in general provides complementary information with respect to ρ_{hv} when the mean canting angle of the scatterers deviates from zero, as shown later in Figure 4.23e.

The correlation coefficient is very useful for the characterization of meteorological and nonmeteorological echoes. Z_{dr} has been discussed previously to describe single-particle behavior, in particular for the case of oblate spheroids, which closely represent actual raindrops. The correlation coefficient, on the other hand, is not meaningful for a single particle because by removing the ensemble average in eq. (4.58), the result is always $\rho_{hv} = 1$. The correlation coefficient brings information about the collective behavior of an ensemble of scatterers by measuring how the scattering amplitudes of the two orthogonal channels change with respect to one another over time.

Recall the meaning of *ensemble average* discussed at the beginning of Section 4.5. To get an intuitive interpretation of the correlation coefficient, consider the simple idealized examples depicted in Figure 4.13, where the weather targets are represented

(a) $Z_{dr} = 1.6$ dB, $\rho_{hv} = 1$ (b) $Z_{dr} = 3.1$ dB, $\rho_{hv} = 0.998$ (c) $Z_{dr} = 0.6$ dB, $\rho_{hv} = 0.896$

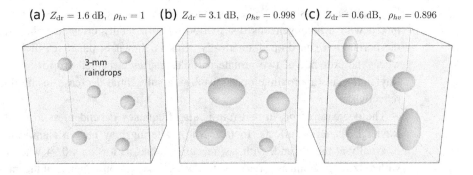

Figure 4.13 Differential reflectivity and correlation coefficient in different idealized situations under Rayleigh–Gans approximation. (a) A distribution of identical 3-mm drops approximated by oblate ellipsoids show perfect correlation ($\rho_{hv} = 1$). (b) For a distribution of raindrops with different sizes and axis ratios (larger drops become more oblate according to eq. [3.31]), ρ_{hv} is still very close to unity. In this case, the Z_{dr} value (3.1 dB) is mainly contributed by the largest 5-mm raindrop (single particle $Z_{dr} = 3.4$ dB). (c) If two oblate particles are replaced by large prolate spheroids with the same permittivity (water), the correlation drops below 0.9.

by six particles with varying sizes and shapes. For simplicity, all spheroids are oriented with their symmetry axis in the vertical direction. The radar variables Z_{dr} and ρ_{hv} for the distribution of six particles are calculated as:

$$Z_{dr} = \frac{\langle |s_{hh}|^2 \rangle}{\langle |s_{vv}|^2 \rangle} = \frac{\sum_{i=1}^{6} |s_{hh}(i)|^2}{\sum_{i=1}^{6} |s_{vv}(i)|^2} \qquad (4.64)$$

$$\rho_{hv} = \frac{|\langle s_{hh}^* s_{vv} \rangle|}{\sqrt{\langle |s_{hh}|^2 \rangle \langle |s_{vv}|^2 \rangle}} = \frac{|\sum_{i=1}^{6} s_{hh}^*(i) s_{vv}(i)|}{\sqrt{\sum_{i=1}^{6} |s_{hh}(i)|^2 \sum_{i=1}^{6} |s_{vv}(i)|^2}}, \qquad (4.65)$$

where $s_{hh}(i)$ and $s_{vv}(i)$ should be replaced by the expressions for spheroidal particles given in eqs. (4.40) and (4.41). The resulting Z_{dr} and ρ_{hv} values, reported in Figure 4.13, illustrate the peculiar capability of ρ_{hv} to recognize the degree of mixing of hydrometeors with different shapes within the radar resolution volume. In pure rain, the raindrops follow a quite well-prescribed behavior in terms of shape–size relation and fall behavior (Section 3.3). An ordered distribution within the radar resolution volume results in very high correlation (panel [b]). A near perfect correlation would result from a population of identical particles (panel [a]). For real rain, the correlation coefficient is also affected by the orientation variability of the scatterers (canting angle) and raindrop oscillations. These factors slightly reduce the observed correlation, which in general attains values above 0.98 at the S band (note that this value, which is often taken as a threshold for the identification of rain in classification algorithms, may significantly depend on the system characteristics, in particular the antenna performance). As illustrated in this simplified example, the distribution of raindrops with a given shape–size relation causes ρ_{hv} to deviate from unity. In particular, ρ_{hv} can be shown to weakly decrease (down to $\sim$0.98) with increasing Z_{dr} for Rayleigh scatterers [21]. When particles of different shape or orientation become mixed with a homogeneous distribution of horizontally oriented oblate spheroids, the correlation can drop dramatically. This is shown in panel (c) of Figure 4.13, where the presence of two prolate particles causes a marked drop in ρ_{hv} under the Rayleigh–Gans approximation, as well as a substantial decrease in the differential reflectivity.

The decrease of ρ_{hv} in real rain is also frequency dependent as a result of non-Rayleigh scattering effects. In fact, Mie scattering may have a significant impact, especially at the C band, with ρ_{hv} reaching values as low as $\sim$0.94 when raindrops with a size of 5–7 mm are present in the DSD. This is a consequence of the interference processes occurring when the ratio of the drop diameter to wavelength increases above $\sim$0.1 and the resulting differences in scattering phases for the orthogonal channels affects the correlation between the complex scattering amplitudes (eq. [4.58]). Figure 4.14 shows a typical DSD in rain. When the maximum drop diameter is smaller than $\sim$5 mm (DSD1, represented by light-gray bars), all radar variables have similar values at the S and C bands, including the correlation coefficient ($\rho_{hv} \sim$0.99; see inset table in

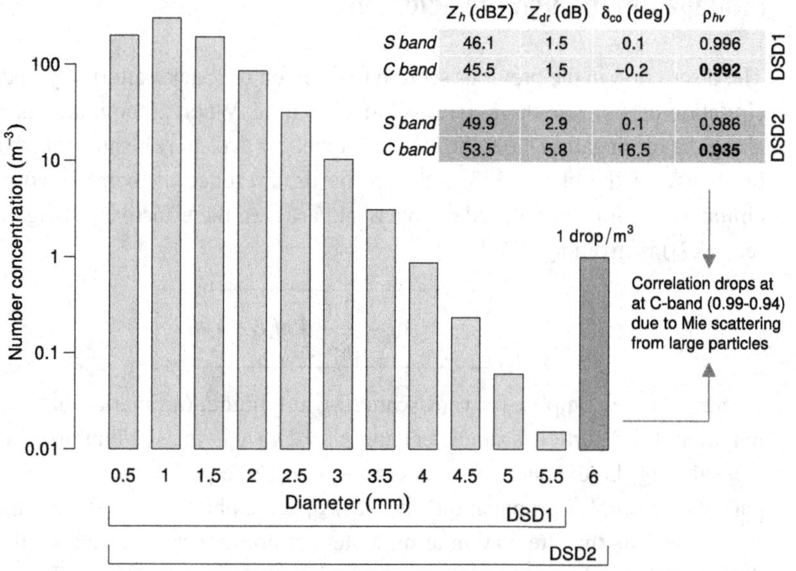

Figure 4.14 The bar plot shows a typical raindrop size spectrum approximated by a gamma distribution (light-gray bars, DSD1) and a second distribution including a single additional 6-mm drop per cubic meter (DSD2). The radar variables of both DSDs are synthesized in the upper-right table, highlighting the decrease of the correlation coefficient at the C band due to Mie scattering from large particles. Note that bimodal DSDs like the one represented here (DSD2) are often observed in convective precipitation as a result of melting ice particles.

Fig. 4.14). If larger raindrops appear in the radar resolution volume (DSD2 includes one additional 6-mm drop per cubic meter, represented by the darker-gray bar), the polarimetric radar variables show distinct values at the S and C bands. At the C band, the increase in the reflectivity and differential reflectivity is larger than that for the S band because of the 6-mm drops lying in the Mie region (sometimes referred to in the literature as the *resonance region*), as illustrated in Figures 4.6 and 4.11. Mie scattering effects also cause the correlation coefficient to drop from ~0.99 to ~0.94 (showing the lower limit of the correlation coefficient in rain at the C band), whereas ρ_{hv} remains close to ~0.99 at the S band. It is also important to consider these frequency-related differences when designing qualitative applications such as hydrometeor classification algorithms (Section 8.3).

In addition to varying shape and orientation, particles in the same radar resolution volume can also have different composition, (water/ice). In the melting layer, partially melted snowflakes have shapes, orientations, and compositions that may substantially differ from those of raindrops (completely melted particles), leading to especially low correlation coefficient values. A similar behavior can also be observed in rain/hail mixtures (Fig. 4.22).

4.7   Propagation through Precipitation

The discussion in the previous section focused on the backscattering properties of precipitation particles in the radar resolution volume. When precipitation is encountered along the propagation path, the electromagnetic waves may be attenuated, and the rate of change of the phase shift is altered by the presence of hydrometeors. For a precipitation medium composed of spherical particles, the effective propagation constant (eq. [4.7]) is given by [144]

$$k_{\text{eff}} = k_0 + \frac{2\pi}{k_0} \int f N(D) dD, \qquad (4.66)$$

where f is the complex forward-scattering amplitude, (a function of the radar wavelength and the particle's diameter) and $k_0 = 2\pi/\lambda$ is the wavenumber of free space. Considering the exponential term $\exp(-jk_{\text{eff}}r)$ in eq. (4.18), it follows that the real part of the complex propagation constant k_{eff} is the phase term, whereas the imaginary part represents the attenuation term. Note that nonspherical particles will introduce a different phase shift depending on the incident polarization plane. This is illustrated in Figure 4.15, which considers the effect of spherical and oblate particle distributions along the propagation path on the phase of the radar signals (an analogous illustration of the effect on the signal's amplitude is shown later in Fig. 4.19).

4.7.1   Differential Phase Shift (Φ_{dp} and K_{dp})

From eq. (4.66), the specific differential phase shift between the horizontal and vertical polarizations can be defined as

$$K_{\text{dp}} = \text{Re}(k_h - k_v) = \lambda \int \text{Re}(f_h - f_v) N(D) dD, \qquad (4.67)$$

where k_h and k_v are the effective propagation constants at horizontal and vertical polarizations, $f_{h,v}$ are the forward-scattering amplitudes, and K_{dp} is measured in units of radians per unit length. In the Rayleigh–Gans approximation, K_{dp} for a population of identically oriented spheroids can be written in terms of the real part of the scatterers' polarization factors p_h and p_v from (4.46) and (4.47), respectively:

$$K_{\text{dp}} = \frac{\pi^2}{6\lambda} \int \text{Re}(p_h - p_v) D^3 N(D) dD. \qquad (4.68)$$

Figure 4.16 shows the term $\text{Re}(p_h - p_v)$ as a function of the raindrop's axis ratio b/a for water and ice oblate spheroids. It is clear that $\text{Re}(p_h - p_v)$ is roughly linearly related with $1 - b/a$. Because the axis ratio for raindrops can also be approximated by a linear function of the equivalent diameter (for $D > 1$ mm; Fig. 3.7), it follows that $K_{\text{dp}} \propto D^4$ in rain. This is a remarkable result for Rayleigh scatterers, with important implications for rainfall estimation, given the similar proportionality between the rain rate and the raindrops' diameter ($R \propto D^{3.7}$; eq. [3.35]). Conversely, for a population

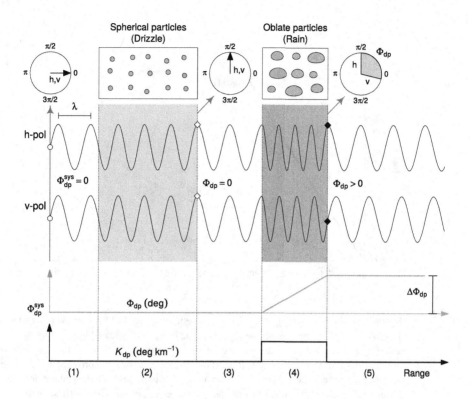

Figure 4.15 Conceptual model of differential phase shift in drizzle and rain. For ease of visualization, the initial phases of the horizontal and vertical polarizations are assumed to be the same ($\Phi_{dp}^{sys} = 0°$). Note that when looking at a fixed range location, the phase rotates counterclockwise as the wave propagates to the right in the image. In range segment (2), spherical particles introduce the same phase shift for the horizontal and vertical polarization planes, resulting in $\Delta\Phi_{dp} = 0°$ (the phase of both signals is equal to $\pi/2$ at the leading boundary of precipitation, as illustrated in the phasor diagram). When oblate raindrops are present in range segment (4), the phase of the wave at the horizontal polarization is delayed more relative to the vertical (see the phasor representation on the top right, the h-pol phase lags the v-pol phase), and $\Delta\Phi_{dp} > 0°$. The differential phase shift reflects properties of the propagation path and is not related to the backscattering properties of the precipitation in the radar resolution volume.

of high-density pristine ice crystals, the term $\text{Re}(p_h - p_v)$ can be treated as a constant in eq. (4.68), and $K_{dp} \propto D^3$; that is, K_{dp} can be directly used for the estimation of the ice water content (IWC), where the proportionality constant is controlled by the microphysical properties of ice particles (see discussion in Section 8.5.2).

In practice, K_{dp} is estimated by taking the half-range derivative of the measured differential phase shift (Section 6.3). Φ_{dp} can then be defined as

$$\Phi_{dp}(r) = \Phi_{dp}^{sys} + 2\int_0^r K_{dp}(r)dr, \qquad (4.69)$$

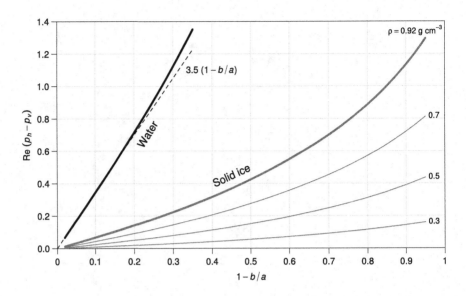

Figure 4.16 The term $\mathrm{Re}(p_h - p_v)$ in the K_{dp} equation (eq. [4.68]) is shown as a function of $1 - b/a$ for oblate spheroids. The characteristic behavior indicated by the thick solid (water) and gray (solid ice) lines takes into account both geometric factors and the dielectric properties of the precipitation particles. The dashed line indicates a useful approximation for rain used in eq. (9.20), whereas the thinner gray lines are for ice particles with decreasing density (indicated beside each curve in $\mathrm{g\,cm}^{-3}$). The effective permittivity for these particles is calculated using the Bruggeman mixture model in eq. (4.75). For water, the indicated relation applies to raindrops with equivalent diameter up to approximately 6 mm ($b/a \approx 0.65$; see Fig. 3.7). Plate-like ice crystals may have an extremely flattened shape, with the ratio of major to minor dimension reaching up to 20 (i.e., $b/a = 0.05$) when represented by oblate spheroids. This shows how a dense population of highly oblate ice crystals may result in a significant K_{dp} signal, despite the lower dielectric factor with respect to water.

where $\Phi_{\mathrm{dp}}^{\mathrm{sys}}$ is the system differential phase. In other words, Φ_{dp} measures the cumulative propagation effect induced by the presence of nonspherical particles with a preferential orientation along the path. The factor of 2 in eq. (4.69) takes into account the two-way propagation of the radar signals (radar to target and target to radar). Note that the convention used here implies that Φ_{dp} includes the system phase, whereas other definitions may only include the integral of K_{dp}. In eq. (4.69), $\Phi_{\mathrm{dp}}^{\mathrm{sys}}$ serves as an explicit reminder that the system phase is included in the radar's observations and can vary as a function of the azimuth and elevation. In either case, the derivative of Φ_{dp} in range is K_{dp} because the system phase is only a constant starting phase.

    It is important to note that even if the dual polarization powers and phases are perfectly matched to have $45°/135°$ polarization at the radar's transmitter (i.e., the two signal components are either in phase or out of phase: $\Phi_{\mathrm{dp}}^{\mathrm{sys}} = 0°$ or $\Phi_{\mathrm{dp}}^{\mathrm{sys}} = 180°$), the attenuation and differential phase shift due to propagation through precipitation will result in continuous Φ_{dp} increase as a function of range, as illustrated in Figure 4.17.

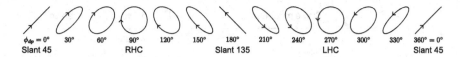

Figure 4.17 For horizontal and vertical polarizations with equal magnitude, we plot the resulting polarization shape as a function of the differential phase Φ_{dp} between the polarizations. If the system phase is not zero, the starting Φ_{dp} is altered accordingly.

The polarization of the propagating signal is generally elliptical and varies with the range from the radar.

4.7.2 Backscatter Differential Phase

The total phase difference between the two polarizations' signals measured by the radar include a combination of the integrated propagation phase shift (Φ_{dp}) and the backscatter differential phase (δ_{co}):

$$\Psi_{dp} = \Phi_{dp} + \delta_{co}.$$ (4.70)

Similar to the copolar correlation coefficient, the subscript "co" is often dropped in the literature, and the backscatter differential phase may be referred to simply as δ. In Rayleigh scattering conditions, the contribution of δ_{co} can be neglected and assume $\Psi_{dp} \approx \Phi_{dp}$. When the radar wavelength approaches the size of the hydrometeors, Mie scattering can modify the phase of the backscattered wave with respect to the phase of the incident wave, resulting in non-zero δ_{co}. Panel (a) of Figure 4.18 shows the value of δ_{co} for three radar frequencies as a function of the equivalent drop's diameter. The magnitude of δ_{co} is especially relevant at the C band because of resonance scattering by raindrops with a size of approximately 6 mm.

The backscattering phase shift is defined in terms of the scattering amplitudes in eq. (4.59). Because the radar measures the total phase shift, δ_{co} cannot be directly estimated. However, several studies suggest that δ_{co} can be related to the differential reflectivity, in particular at the X band [145–147]. Panel (b) of Figure 4.18 shows the variation of δ_{co} as a function of Z_{dr} for the C and X bands, calculated from real DSDs collected during the Iowa Flood Studies (IFloodS) campaign in Iowa. The gray-shaded areas represent the most populated regions of the (Z_{dr}, δ_{co}) polarimetric space for this specific data set, with each area encompassing approximately 90 percent of the total $\sim 5 \times 10^5$ observations. An empirical fit for the X band is also displayed [147] (dashed line):

$$\delta_{co} = Z_{dr}^{1.8},$$ (4.71)

where Z_{dr} is expressed in decibels and δ_{co} in degrees. This relation provides a simple approximation for the estimation of δ_{co} from the observed differential reflectivity. Note, however, that Z_{dr} needs to be corrected for differential attenuation in order to apply eq. (4.71). A procedure to incorporate the estimation of δ_{co} into an attenuation and differential attenuation correction is illustrated later in Section 9.4.3.

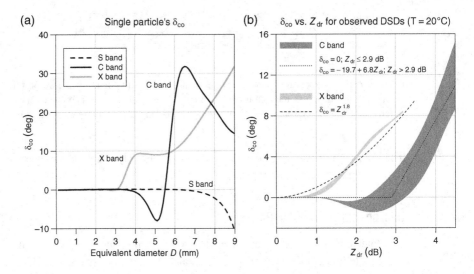

Figure 4.18 (a) Backscatter differential phase δ_{co} as a function of the raindrop's diameter for different radar wavelengths at $20°$C. (b) Backscatter differential phase δ_{co} as a function of Z_{dr}, calculated from observed DSDs during the IFloodS campaign at $20°$C. The gray shaded areas encompass approximately 90 percent of the observations. The dashed line represents the best-fit relationship between δ_{co} (in degrees) and Z_{dr} (in dB) given by Otto and Russchenberg [147] for the X band, whereas the dotted line is a piecewise linear fit for the C band.

At the S band, δ_{co} is in general negligible in rain, whereas at the C band, there is more disperse relationship between δ_{co} and Z_{dr} in comparison with the X band because of the more widely variable polarimetric response, including negative δ_{co} for large drops with diameter in the resonance region (5–7 mm). For practical purposes, the δ_{co}-versus-Z_{dr} relation can be approximated as a piecewise linear equation (dotted line in panel [b] of Fig. 4.18).

4.7.3   Attenuation

The radar signal experiences attenuation while propagating through precipitation, which is illustrated in Figure 4.19. While only briefly introduced here, an in-depth discussion on specific attenuation A and specific differential attenuation A_{dp} are presented in Section 9.3. For rain, the observed differential phase shift may be used to estimate the attenuation and differential attenuation (Section 9.4). In fact, the effective propagation constant k_{eff} is complex, so the radar signals experience both phase delay and attenuation, and the two effects are approximately linearly related for frequencies up to Ku band (refer to Fig. 9.19). Even in drizzle or light rainfall, although the DSD is typically dominated by spherical droplets, there are always few larger oblate raindrops that may induce a sufficient differential phase shift to estimate attenuation.

For the ice phase, the imaginary part of the propagation constant is relatively small, so it is possible to observe a large differential phase shift but with negligible attenuation effects (see examples in Section 8.5.1), even at typical "attenuating frequencies"

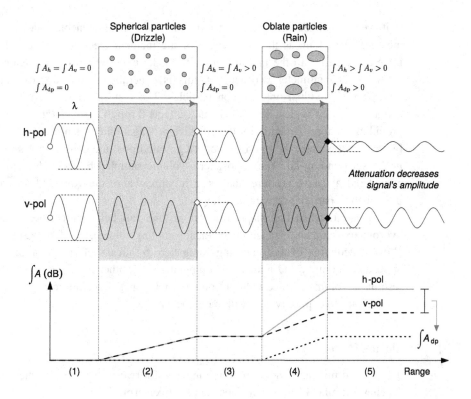

Figure 4.19 Conceptual model of attenuation in drizzle and rain. Propagation through a lossy medium such as water (characterized by a significant imaginary component of the complex permittivity) introduces attenuation, in addition to phase shift (Fig. 4.15). When precipitation is mainly composed of spherical particles, the specific attenuation is about the same on both channels (A_h and A_v), resulting in nearly zero differential attenuation ($A_{dp} = 0$). Conversely, the larger amount of water typically encountered by the horizontal propagating signal in rain or melting snow implies a larger attenuation compared with the vertical polarization ($A_{dp} > 0$).

such as the C band and X band. (Note that the term attenuating frequency refers to the effects from propagation in rain.)

4.8 Mixtures

The permittivity values illustrated in Figures 4.1 and 4.2 are used for homogeneous precipitation particles such as raindrops (pure water) or plate-like crystals (solid ice). However, many hydrometeor types are heterogeneous mixtures of two or three phases. For example, dry snow is composed of ice and air (two-phase mixture), whereas wet snow is made of ice, air, and water (three-phase mixture). In these cases, it is useful to define an effective permittivity (ϵ_{eff}) of the mixture to describe the response of the hydrometeor to an incident electromagnetic field and treat it as if it were homogeneous. This approximation of a heterogeneous material is only meaningful as long as the wavelength of the incident electromagnetic radiation is large compared with the internal structure of the material [148]. For weather radar operating at centimeter wavelengths, this is generally a good approximation for multiphase individual precipitation particles. The concept of effective permittivity can also be used to describe propagation effects through a precipitation medium by providing an approximation for the effective propagation constant k_{eff} (defined in eq. [4.66]). In this case, the hydrometeors (e.g., raindrops) are considered as inclusions in the surrounding air. The atmospheric volume sampled by the radar can then be considered a sparse mixture, that is, a mixture where the volume fraction of the precipitation particles is extremely low.

In general, when the shape, orientation, and fall behavior of the mixed-phase hydrometeors are uniform within the radar sampling volume, the effective permittivity allows calculation of the scattering parameters by means of analytical solutions as opposed to complicated numerical computations. In other instances, when the radar resolution volume is filled with different particle types (e.g., hail mixed with rain), it is necessary to explicitly deal with their coexistence.

4.8.1   Mixture Models

The most simple and widely used mixing rule for two-phase media is the **Maxwell–Garnet** formula [149], which gives the effective permittivity as

$$\epsilon_{eff} = \epsilon_e + 3 f_i \epsilon_e \frac{\epsilon_i - \epsilon_e}{\epsilon_i + 2\epsilon_e - f_i(\epsilon_i - \epsilon_e)}, \tag{4.72}$$

where ϵ_e and ϵ_i are the permittivities of the environment (background) and the inclusions, respectively, whereas f_i is the volume fraction of the inclusions in the mixture. Equation (4.72) is derived under the hypothesis of spherical inclusions in a homogeneous background medium, as illustrated in Figure 4.20. Note that eq. (4.72) reduces to $\epsilon_{eff} = \epsilon_e$ for $f_i = 0$ and to $\epsilon_{eff} = \epsilon_i$ for $f_i = 1$. For a very dilute concentration ($f_i \ll 1$), eq. (4.72) can be approximated as

$$\boxed{\epsilon_{eff} = \epsilon_e + 3 f_i \epsilon_e \frac{\epsilon_i - \epsilon_e}{\epsilon_i + 2\epsilon_e}} \tag{4.73}$$

The Maxwell–Garnet mixing formula is not symmetric; that is, it is not invariant under the permutation $\epsilon_e \longleftrightarrow \epsilon_i$ and $f_i \longleftrightarrow (1 - f_i)$. The distinction between the environment and the inclusion is essentially given by the large difference in their respective volume fractions, implying that the validity of the rule holds in general for sparse mixtures.

Another widely used mixing rule in remote-sensing applications is the **Bruggeman** formula, which takes the following form for two-phase mixtures:

$$(1 - f_i) \frac{\epsilon_e - \epsilon_{eff}}{\epsilon_e + 2\epsilon_{eff}} + f_i \frac{\epsilon_i - \epsilon_{eff}}{\epsilon_i + 2\epsilon_{eff}} = 0. \tag{4.74}$$

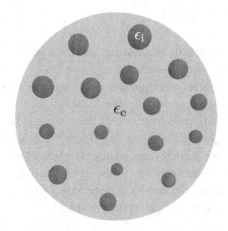

Figure 4.20 Spherical inclusions with permittivity ϵ_i (dark gray shading) embedded in a background material with permittivity ϵ_e (light gray shading). The spherical inclusions do not need to be of the same size as long as they are all smaller than the radar wavelength [148].

The explicit solution for eq. (4.74) is given by (see, e.g., [150])

$$\epsilon_{\text{eff}} = \frac{b + \sqrt{8\epsilon_e\epsilon_i + b^2}}{4}; \quad b = (2f_e - f_i)\epsilon_e + (2f_i - f_e)\epsilon_i, \qquad (4.75)$$

where $f_e = 1 - f_i$. As opposed to the Maxwell–Garnet rule (eq. [4.72]), the Bruggeman formula is symmetric; that is, it is invariant under the permutation $\epsilon_e \longleftrightarrow \epsilon_i$ and $f_i \longleftrightarrow (1 - f_i)$. Because it treats the environment and the inclusions in the same way, eq. (4.74) can be applied to mixtures with arbitrary volume fractions. Equation (4.74) can also be generalized to the case of a mixture of N materials as follows:

$$\sum_{n=1}^{N} f_n \frac{\epsilon_n - \epsilon_{\text{eff}}}{\epsilon_n + 2\epsilon_{\text{eff}}} = 0; \quad \sum_{n=1}^{N} f_n = 1. \qquad (4.76)$$

However, analytical solutions for $N > 2$ are not readily available, and eq. (4.76) may need to be solved numerically.

4.8.2 Pristine Ice Crystals and Dry Snow

Pristine ice crystals and dry snow are two-phase mixtures of ice and air. Ice is a dielectric material characterized by very small losses, and its relative permittivity can be approximated by the real part ($\epsilon_r \approx 3.17$; Section 4.2). In addition, the permittivity contrast of ice–air mixtures ($\epsilon_{\text{ice}}/\epsilon_{\text{air}}$) is low. The effective permittivity of ice particles with varying density can then be easily estimated with the Maxwell–Garnet or the Bruggeman formula, with small differences depending on whether the ice material is treated as inclusion in an air background, or vice versa. For snowflakes, the volume fraction of ice is directly related to the particle density ρ_s as

$$f = \frac{\rho_s}{\rho_{\text{ice}}}, \tag{4.77}$$

where $\rho_{\text{ice}} = 0.917 \text{ g cm}^{-3}$ is the density of solid ice, and ρ_s is generally approximated by an inverse function of the particle size (see, e.g., eq. [3.74]).

Ice particles such as pristine crystals and aggregates can generally be represented by either oblate (plates, dendrites, aggregates) or prolate (columns, needles) spheroids. Figure 4.21 shows the expected differential reflectivity under the Rayleigh approximation for oblate and prolate spheroids with varying density as a function of the axis ratio. The Z_{dr} values for oblate particles are obtained from eq. (4.48), replacing ϵ_r with the effective permittivity corresponding to the given fractional volume of ice. For columns and needles (approximated by prolate particles), the calculation of Z_{dr} is more complicated because these ice crystals tend to fall with their major axis in the horizontal plane (panel [a] in Fig. 4.21). The vertical reflectivity can be obtained from eq. (4.47), considering the depolarization factor along the minor a axis ($L_a = L_x = L_y = \frac{1}{2}(1 - L_z)$). For their horizontal reflectivity, consider a distribution of prolate particles with the same axis ratio and random orientation of the symmetry axis b in the horizontal plane. After averaging the horizontal backscatter cross-section, Z_{dr} can be shown to be expressed as follows [151]:

$$Z_{\text{dr}} = \frac{3}{8}X^2 + \frac{1}{4}X + \frac{3}{8}; \quad X = \frac{1 + |\epsilon_r - 1|(1 - L_z)/2}{1 + |\epsilon_r - 1|L_z}. \tag{4.78}$$

Equations (4.48) and (4.78) allow the theoretical calculation of Z_{dr} for arbitrary spheroidal particles with a given axis ratio and relative permittivity. These equations can also be used to provide an upper limit for the differential reflectivity of extremely oblate (e.g., plate crystals) or extremely prolate (e.g., needles) ice particles. In fact, bearing in mind that for solid ice, $|\epsilon_r| \approx 3.17$ (Fig. 4.2), in the limit of $L \longrightarrow 1$, the differential reflectivity of oblate ice particles is $Z_{\text{dr}} = 20\log_{10}(|\epsilon_r|) \approx 10$ dB, whereas for prolate ice particles (oriented with symmetry axis in the horizontal direction), $Z_{\text{dr}} = 10 \log_{10}\left[(3|\epsilon_r|^2 + 10|\epsilon_r| + 19)/32\right] \approx 4$ dB (solid black lines in Fig. 4.21). For this reason, as schematically represented in the ice-crystal habit diagram in Figure 3.18, extremely high values of Z_{dr} are more frequently observed in the atmospheric layer denoted by a prevalence of plate-like crystals (up to the $-25°C$ level), whereas at higher altitudes (where the columnar regime prevails), Z_{dr} does not typically exceed 4 dB.

4.8.3   Wet Snow

Modeling of wet snow is especially important for the study and interpretation of radar observations in the melting layer. When a snow aggregate is modeled as a three-component mixture, either the Maxwell–Garnet or the Bruggeman model can be applied in cascade. For example, a mixture of ice and air can first be derived and later used as an inclusion in a water background. The notation $MG_{1,2}$ indicates a mixture obtained with the Maxwell–Garnet model where material 2 is treated as inclusion in the background material 1 [153]. Because the Maxwell–Garnet model is

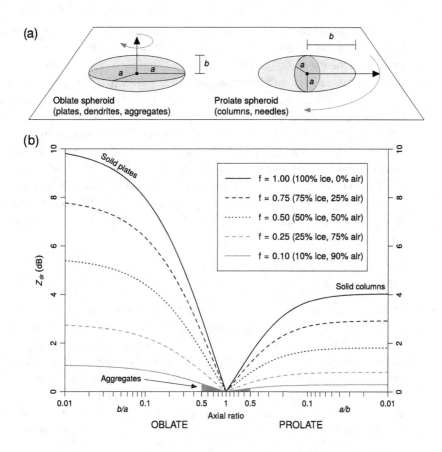

Figure 4.21 Differential reflectivity of ice particles with a varying fractional volume of ice as a function of the axis ratio (b), calculated using the effective dielectric constant from the Bruggeman formula (eq. [4.75]). Ice particles are approximated by oblate (plates, dendrites, and aggregates) and prolate (needles and columns) spheroids, as shown in in panel (a), with the axis ratio defined as the ratio between the minor and major axes of the spheroid (b/a for oblate and a/b for prolate spheroids). Note the smaller Z_{dr} values of prolate particles, which is a result of the typical fall behavior of columnar ice crystals, with the major axis randomly oriented in the horizontal plane [151]. The gray-shaded region refers to snowflake aggregates, which are characterized by small Z_{dr} values because of the very low density (typically $\rho_s < 0.1\,\mathrm{g\,cm^{-3}}$ for unrimed aggregates; see, e.g., [152]), even when the axis ratio significantly deviates from unity. This diagram is not valid for larger scatterers like graupel and hail, whose fall behavior differs from that of the smaller ice particles represented here.

noncommutative ($MG_{1,2} \neq MG_{2,1}$) and nonassociative (e.g. $MG_{12,3} \neq MG_{1,23}$), there are many possible ways to calculate the effective permittivity for a three-phase mixture. However, as noted earlier, the low dielectric contrast of ice–air mixtures makes the reversal of the two components insignificant, whereas the most impactful choice is whether water (or the water–ice mixture) is used as the inclusion or as the environment [153].

Meneghini and Liao [153] focused on the scattering behavior of mixed-phase spherical particles relying on the Mie theory. They found that any of the schemes using water as inclusion ($\text{MG}_{\text{a, iw}}$, $\text{MG}_{\text{ai, w}}$, $\text{MG}_{\text{ia, w}}$) are able to reproduce the co-polarized scattering parameters, in agreement with direct numerical computations. Liao and Meneghini [154] later extended these results to nonspherical particles represented by oblate and prolate spheroids, composed of randomly mixed ice and water. The effective permittivity of the mixed-phase spheroids was accurate enough for the computation of the copolar and cross-polar scattering parameters. These results imply that complex scattering computations can be replaced by faster analytical solutions where the scatterers are treated as homogeneous particles. The main difficulty in modeling the melting layer lies in the accurate representation of how the melted water spreads across the complex ice particles as they fall below the freezing level, modifying their shape and orientation.

4.8.4   Graupel and Hail

The effective permittivity of graupel and hail can be computed similarly to that of dry and wet snow, taking care to adapt the fractional volume of ice and water (for wet graupel) based on the proper particle densities (see, e.g., [155]). Despite the ϵ_r of hail and graupel being similar to that of some low-inertia pristine ice crystals or wet snow aggregates, their orientation and fall behavior are generally radically different, leading to specific dual polarization signatures. These typically include a negative Z_{dr} for conical graupel, with the reflectivity varying depending on the particle's size and degree of wetting, and nearly zero Z_{dr} for randomly oriented hailstones.

4.8.5   Hail Mixed with Rain

When hail particles coexist with raindrops within the radar resolution volume, the diversity of the particles' orientation and fall behavior complicates the description of the dual polarization scattering properties for the mixture. Under simplified hypotheses, this problem may be dealt with explicitly by calculating the covariance matrix of the two-component mixture, which is modeled through the superposition of the two individual covariance matrices (rain and hail). The model described by Bringi and Chandrasekar [21] (their Section 7.2.2) treats raindrops as oblate particles satisfying mirror symmetry (about a zero mean canting angle), whereas hailstones are represented by randomly tumbling particles satisfying plane isotropy (the canting angle is uniformly distributed over 0 to $360°$). In this simplified scenario, Z_{dr} (hail) = 0 dB, and ρ_{hv}(hail) = $1 - 2\,\text{LDR(hail)}$ (alternate transmission; eq. [4.63]). The radar variables of the rain–hail mixture under the Rayleigh approximation can then be expressed as a function of the reflectivity-weighted ice fraction $f_z = Z_h^{\text{ice}}/Z_h$, with $Z_h = Z_h^{\text{rain}} + Z_h^{\text{ice}}$, (wrapping in the size and number concentration of hail particles) and the radar variables of the individual components as (refer to [21] for the details):

$$\text{LDR}^{\text{m}} = f_z\,\text{LDR}^{\text{h}} + (1 - f_z)\text{LDR}^{\text{r}} \tag{4.79}$$

$$Z_{dr}^m = \frac{Z_{dr}^r}{\left[1 + f_z \left(Z_{dr}^r - 1\right)\right]} \tag{4.80}$$

$$\rho_{hv}^m = \frac{f_z \rho_{hv}^h + (1 - f_z)\rho_{hv}^r (Z_{dr}^r)^{-1/2}}{(Z_{dr}^m)^{-1/2}}, \tag{4.81}$$

where superscripts r, h, and m, respectively, stand for rain, hail, and mix. LDR^m is a simple linear combination of the two precipitation components, Z_{dr}^m is a decreasing function of the ice fraction f_z, and ρ_{hv}^m is a more complex function of the expected LDR of hail and both the Z_{dr} and ρ_{hv} of rain.

Raindrops originating from melting hail can reach large dimensions, leading to Mie scattering effects (Fig. 4.14). In such cases, the scattering phase shift δ_{co} is no longer negligible, and the correlation coefficient of the mixture becomes, after modest algebra,

$$\rho_{hv}^m = \frac{\left| f_z \rho_{hv}^h + (1 - f_z)\rho_{hv}^r e^{-j\,\delta_{co}}(Z_{dr}^r)^{-1/2} \right|}{(Z_{dr}^m)^{-1/2}}$$

$$= \frac{\left[f_z^2 \rho_{hv}^{h\,2} + 2f_z(1 - f_z)\rho_{hv}^h \rho_{hv}^r (Z_{dr}^r)^{-1/2}\cos\delta_{co} + (1 - f_z)^2 \rho_{hv}^{r\,2}(Z_{dr}^r)^{-1} \right]^{1/2}}{(Z_{dr}^m)^{-1/2}}. \tag{4.82}$$

Figure 4.22 shows the expected Z_{dr} and correlation coefficient of a mixture with a varying ice fraction at the X band. Assuming LDR $= -20$ dB for hail, $\rho_{hv}^h = 0.98$. Two different DSDs are considered in the figure, with a median volume diameter of $D_0 = 1.8$ mm and $D_0 = 3.5$ mm. The radar variables for both DSDs are calculated using scattering simulations, and the resulting Z_{dr}, ρ_{hv}, and δ_{co} of rain are used in eq. (4.82) to calculate the expected values for the mixture. For the two cases, Z_{dr} monotonically decreases as the mixture includes increasing contributions by randomly oriented hail particles. For the smaller D_0 distribution, δ_{co} is negligible, and the correlation coefficient of the mixture is approximately constrained within the interval bounded by the pure rain and pure hail ρ_{hv} values. Conversely, if there is a significant scattering phase shift ($\delta_{co} \sim 10°$ for the DSD with $D_0 = 3.5$ mm), the correlation coefficient of the mixture can attain lower values than the ρ_{hv} of the individual components. This typically results in a well-defined polarimetric signature that can be exploited in particle-classification algorithms. Mie scattering from large melting particles in the melting layer may lead to a similar reduction of the correlation coefficient, in particular at the C band.

4.9   Canting

When a raindrop is canted (i.e., the projection of its symmetry axis on the polarization plane is not aligned with the vertical electric field; see Fig. 2.2), the incident horizontally or vertically polarized electromagnetic wave will excite dipole moments

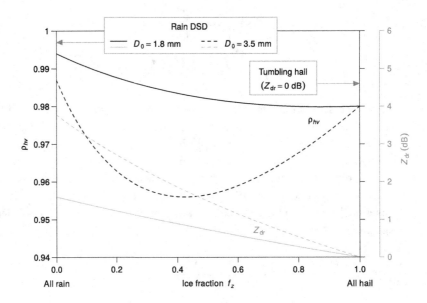

Figure 4.22 Simulated correlation coefficient and differential reflectivity of rain–hail mixtures at the X band for two DSDs denoted by moderate and large median volume diameters (D_0 = 1.8 mm and D_0 = 3.5 mm, solid and dashed lines, respectively). The differential reflectivity (gray lines) and correlation coefficient (black lines) of the mixture are calculated according to eqs. (4.80) and (4.82). In this simplified model, hail is approximated by randomly tumbling particles with plane isotropy; that is, Z_{dr} (hail) = 0 dB. Note how the correlation coefficient of the mixture can drop below 0.96 (lower than ρ_{hv} of the individual components) in the case of large D_0 (implying Z_{dr} of rain close to 4 dB and δ_{co} ~10°). Refer to Figure 4.18 for the δ_{co}-versus-Z_{dr} relation and Figure 9.11 for D_0 versus Z_{dr}.

on the major and minor axes of the particle. The resulting backscattered radiation will then have both copolar and cross-polar components, implying nonzero off-diagonal terms in the scattering matrix (eq. [4.50]) and LDR > 0 (on a linear scale). If we consider a population of scatterers, the distribution of the canting angle β can generally be described by the mean $\bar{\beta}$ and standard deviation $\sigma(\beta)$. Consider the configurations depicted in Figure 4.23. When all hydrometeors (either raindrops or ice particles) are vertically oriented (panels [a] and [c]), both the mean and standard deviation of the canting angles are zero, with no cross-polar return (LDR = 0). The most common situation encountered in real precipitation is represented in panels (b) and (d), with particles having orientations following a distribution with a zero mean canting angle and a nonzero standard deviation. In this case, the cross-polar return is generally small, and the LDR typically shows values below −25 dB. When large melting snowflakes are present in the melting layer or in rain–hail mixtures, the LDR can exceed −15 dB. These large LDR values are accompanied by a markedly low correlation coefficient. Conversely, when an ensemble of scatterers has a preferential orientation departing from the horizontal (e.g., ice crystals in panel [e] of Fig. 4.23),

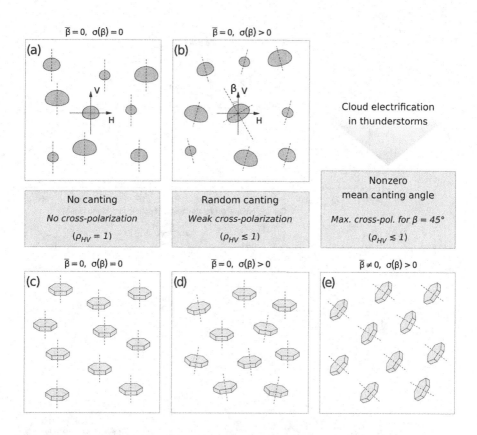

Figure 4.23 Ideal raindrops (a) and ice particles (c) with perfect horizontal orientation (canting angle $\beta = 0$ for all drops). (b) Raindrops with random orientation (zero mean canting angle). $\sigma(\beta)$ is typically on the order of $10°$. A similar fall behavior characterizes ice particles (d). (e) Under the effect of an electric field, ice crystals may assume a preferential orientation in thunderstorms, leading to a strong cross-polarization signal (high LDR). As opposed to ice crystals, raindrops have too much inertia to be reoriented by the electric field (see Section 8.5.3).

the correlation coefficient is high but the cross-polarization is strong, with the LDR reaching values above -20 dB (refer to the electrification example in Section 8.5.3).

In the Rayleigh–Gans approximation, the differential reflectivity and the linear depolarization ratio for canted spheroidal particles observed at horizontal incidence can be computed from the following equations:

$$Z_{\mathrm{dr}}(\beta) = \frac{\left|1 + (\epsilon_r - 1)(L_z \cos^2\beta + L \sin^2\beta)\right|^2}{\left|1 + (\epsilon_r - 1)(L_z \sin^2\beta + L \cos^2\beta)\right|^2} \tag{4.83}$$

$$\mathrm{LDR}_{vh}(\beta) = \frac{\left|(L - L_z)(\epsilon_r - 1)\sin\beta \cos\beta\right|^2}{\left|[1 + L_z(\epsilon_r - 1)]\cos^2\beta + [1 + L(\epsilon_r - 1)]\sin^2\beta\right|^2}. \tag{4.84}$$

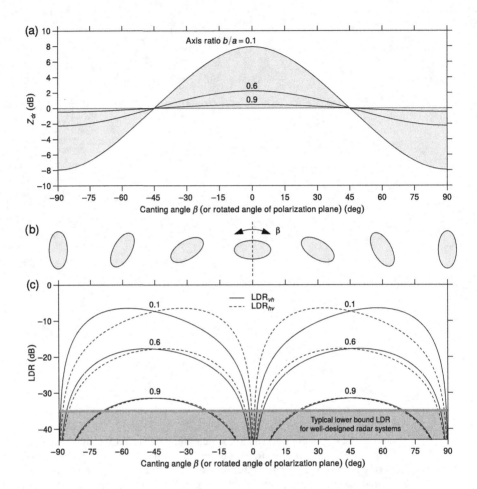

Figure 4.24 Z_{dr} (a) and LDR (c) from a single ice particle represented by an oblate spheroid ($\epsilon_r = 3.17$) (b), according to the Rayleigh–Gans approximations in eqs. (4.83) and (4.84). Note that changing the canting angle of the particle is equivalent to rotating the polarization plane around the propagation direction (Fig. 2.2). The LDR is zero ($-\infty$ in dB) if the symmetry axis of the spheroid is aligned with either the horizontal or the vertical polarization when Z_{dr} correspondingly reaches its maximum and minimum value. The different curves in panels (a) and (c) correspond to particle shapes ranging between weakly oblate spheroids (axis ratio $b/a = 0.9$) and heavily flattened particles ($b/a = 0.1$). The solid and dashed lines in panel (c), respectively, represent values for the two possible LDR configurations, that is, LDR_{vh} and LDR_{hv} (eqs. [4.55] and [4.56]), whose difference in logarithmic scale is given by Z_{dr} (eq. [4.57]). Note that the cross-polar power (proportional to the numerator in eq. [4.84]) is symmetric about $0°$ and peaks at $\beta \pm 45°$.

Equation (4.83) is illustrated in panel (a) of Figure 4.24, from which it can be clearly seen that Z_{dr} has perfectly symmetric behavior, with a maximum positive value when $\beta = 0°$, a minimum negative value for $\beta = 90°$; that is, $Z_{dr}(90°) = -Z_{dr}(0°)$ and $Z_{dr}(\pm 45°) = 0$ dB. In particular, for $\beta = 0°$, eq. (4.83) reduces to eq. (4.48). The cross-polar power (proportional to the numerator in eq. [4.84]) has an opposite behavior, with maximum values at $\beta = \pm 45$ (not shown). Note that the $\sin\beta \cos\beta$ term implies that there is no cross-polar return when the symmetry axis of the spheroid is perfectly aligned with either the horizontal or the vertical polarization. The solid and dashed lines in panel (c) of Figure 4.24 represent the two possible LDR configurations (LDR$_{vh}$: transmit h, receive v, or LDR$_{hv}$: transmit v, receive h). The difference between LDR$_{vh}$ and LDR$_{hv}$, more marked for strongly oblate spheroids, is due to the varying copolar scattering amplitudes when transmitting either horizontal or vertical polarization signals (i.e., the denominator in eq. [4.84]), and is directly related to Z_{dr} (eq. [4.57]). Practically, the lowest measurable LDR for well-designed radar systems is typically between -30 and -35 dB, posing a limit on the detection capability of weakly canted particles. (The lower limit of LDR is determined by the antenna and discussed in Section 7.4.1.)

4.10 Elevation Dependence of Radar Variables

Precipitation particles are generally nonspherical and tend to fall with a preferential orientation of their symmetry axis with respect to the zenith angle (e.g., $0°$ for raindrops and planar ice crystals, $90°$ for columnar ice crystals). This implies that the apparent shape of the particles seen by the radar (i.e., their projection on the plane perpendicular to the propagation direction) will change with the elevation angle θ, as illustrated in Figure 4.25.

For most operational applications, radars operate in volume scan mode, with a volume coverage pattern mainly consisting of mostly low-elevation angles. In this case, the effects of the elevation view angle on the polarimetric radar variables may generally be neglected. Conversely, for volume scans containing high elevations ($\theta > 20°$) or for vertical scans (RHI mode), the elevation dependence needs to be considered and can actually provide important information for quality-control procedures (Z_{dr} calibration) and microphysical retrievals (ice particles).

Figure 4.26 shows the theoretical curves of $Z_{dr}(\theta)$ obtained from electromagnetic scattering simulations for oblate spheroids with $Z_{dr}(0°)$ between 0 and 4 dB. A good analytical approximation can be derived from the general scattering amplitude expressions for a spheroid with an arbitrary orientation and direction of the incident wave, as follows [21, 156]:

$$Z_{dr}(\theta) \approx \frac{Z_{dr}(0°)}{\left[Z_{dr}^{1/2}(0°)\sin^2\theta + \cos^2\theta\right]^2}, \quad (4.85)$$

where $Z_{dr}(\theta)$ and $Z_{dr}(0°)$ are in linear scale. It is clearly seen in Figure 4.26 that all curves converge to $Z_{dr} = 0$ dB for $\theta = 90°$ because of the circular symmetry of the vertically oriented oblate spheroid. In fact, when the radar antenna points upward, the shape of the raindrop appears circular. This provides an effective way to calibrate the differential reflectivity [157], as discussed in Section 7.4. Looking at real measurements in rain (lines with open circles in Fig. 4.26), it should also be clear that a

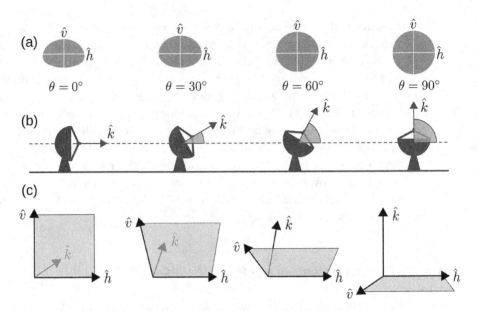

Figure 4.25 Illustration of how a raindrop appears to the radar with increasing elevation angle. (a) The raindrop's cross-section area, projected onto the horizontal ($\hat{h}$) and vertical ($\hat{v}$) polarization plane. (b) A side-view of the antenna's elevation angles with the propagation direction ($\hat{k}$). (c) A front-view illustrating how the polarization reference frame changes with elevation angle (the plane with the elevation angle is into the page).

sequence of Z_{dr} observations collected at increasing elevation angles should match the corresponding theoretical curve starting at $0°$ elevation, whenever the intrinsic Z_{dr} is uniform across the rain layer (negligible impact of precipitation processes). In this case, the departure of the observations from the corresponding theoretical curve may provide an alternative method to check the Z_{dr} calibration [158], as well as for radars not able to scan at vertical incidence.

Under the Rayleigh approximation, the scattering amplitude has a symmetric pattern along the backward and forward directions of the beam incidence, as shown in Figure 4.5. As with the elevation angle dependence of differential reflectivity, assume a vertical orientation of the spheroid's symmetry axis (zero canting angle) to get an expression for the specific differential phase shift:

$$K_{dp}(\theta) = K_{dp}(0°) \cos^2\theta. \tag{4.86}$$

This relation provides an excellent approximation for correcting the effects of the drop shape's projection as a function of elevation angle. This is useful for rainfall applications (Section 9.5) if elevations higher than $\sim 20°$ are used.

For rain, prior knowledge of the characteristic shape and fall behavior of raindrops allows us to use $Z_{dr}(\theta)$ observations for data quality control, such as Z_{dr} calibration. For pristine ice particles, however, the dominant crystal habit and fall behavior are generally unknown. In this case, the collection of radar observations at different elevation angles may prove useful to get more insights about the microphysical composition of the ice crystals within the radar resolution volume.

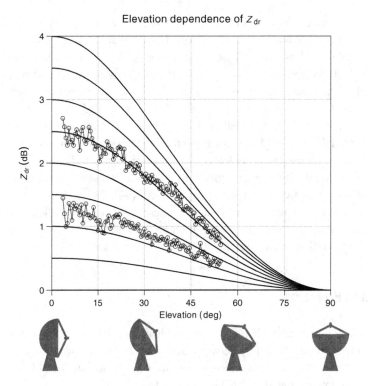

Figure 4.26 Theoretical elevation dependence of Z_{dr} for oblate spheroids. Each solid line represents the Z_{dr} change with elevation for a given initial value $Z_{dr}(0°)$ between 0.5 and 4 dB. The lines with open circles represent X-band RHI measurements in rain at two distinct times in stratiform precipitation (Italy). The observations are averaged over the slant range 600–1800 m and cut off at 55° elevation to avoid contamination by the melting layer.

In particular, the ratio of the cross-polar to copolar return is sensitive to the shape, orientation, and density of the scatterers and varies for different polarimetric configurations (e.g., H or V linear, circular, slanted linear). Polarimetric cloud radars have been used for the shape classification of ice crystals, using the depolarization ratio dependence on the elevation angle that can be exploited to distinguish between layers of oblate (plates, dendrites) and prolate (columns, needles) particles [159]. For oblate particles with preferential horizontal orientation (symmetry axis along the vertical) and a given distribution of canting angles, the LDR decreases with elevation until vertical incidence (reaching values close to the system cross-coupling). Prolate particles with a major symmetry axis randomly oriented in the horizontal plane show an opposite behavior (LDR increases with elevation), allowing a distinction between the two classes of ice crystals. For this purpose, the use of a polarization basis rotated at 45° with respect to the horizontal (slanted linear polarization) may be advantageous for cloud radars because the cross-polar return in the standard horizontal-vertical linear basis may be too weak to detect above the noise level at low-elevation angles (because of the preferred horizontal orientation of the scatterers) when the canting angle is small [160, 161].

The copolar correlation coefficient is sometimes used as a proxy for LDR, and an elevation-angle dependence can be exploited to estimate the shape and orientation of ice crystals for radars using STSR mode, based on the combined analysis of differential reflectivity and correlation coefficient (which gives similar information as the LDR when the mean canting angle is zero) [162]. Measurements collected with a 35 GHz cloud radar scanning in an elevation between $40°$ and $150°$ ($-60°$ to $60°$ with respect to the zenith) in the Netherlands allowed the identification of two separate layers (at height of 3 and 5 km, corresponding to temperatures of $-6°C$ and $-14°C$) of, respectively, prolate and oblate ice particles with nearly horizontal orientation. For both layers, the differential reflectivity is around 0 dB at vertical incidence. The layer of oblate particles is denoted by high Z_{dr} (4–5 dB) at low elevation angles and increasing ρ_{hv} with higher elevation, whereas for prolate particles, Z_{dr} reaches moderate values (2–3 dB) at low elevation angles and ρ_{hv} decreases as the antenna moves up to vertical incidence.

4.11   Selected Problems

1. Consider a $f = 2.8$ GHz plane wave that is propagating along the z-direction and interacts with the boundary between free-space and the plane of water at $z = 0$. The negative side of the z-axis is free-space (this is medium 0), and the positive side is pure water at $10°C$ (this is medium 1).
   a) Calculate the characteristic impedance of free space.
   b) Calculate the characteristic impedance of $10°C$ pure water (assume a dielectric constant $\epsilon_r = 80 - j16$).
   c) What are the magnitude and angle of the reflection coefficient Γ at the interface between the two media for a wave propagating from free-space into a plane of water at perpendicular incidence?
   d) What fraction of the power is back-scattered from the plane of water and how much is transmitted into medium 1?

2. Airborne in-situ sample collection is used extensively in cloud microphysical studies to measured the particle size distribution along a path. This particle size distribution is often used to compute radar observations such as reflectivity for cross validation with ground radar. Assume we are conducting an in-situ validation experiment with a radar and an aircraft equipped with a particle size distribution measurement probe. One common sampling method for studying microphysics is spiral descent from the cloud's top to below cloud's base. For simplicity let us assume we detect only very small particles above and below the melting layer at 4 km, and the reflectivity is estimated as $Z = \sum D^6$. Considering that above 4 km is ice and below 4 km is water, what adjustments are needed to compare the reflectivity computed from the particle size distribution and the radar's observation? Assume the radar is operating at X band and the ice particles are small aggregates with a density of 0.1 g cm^{-3}.

3. Consider the exponential DSD in panel (d) of fig. 3.3, with $D_{max} = 6$ mm, and a radar free space wavelength $\lambda_0 = 10.9$ cm (S-band). The presence of precipitation along the path will affect the propagation constant k_{eff}.
   a) Calculate k_{eff}.
   b) How much is the change with respect to the free space propagation constant?

4. Calculate the horizontal reflectivity and differential reflectivity for an exponential DSD, with $N_0 = 2000$ m^{-3}mm^{-1}, $D_0=1.5$ mm, and $D_{max} = 8$ mm, under the Rayleigh-Gans approximation and assuming the Beard-Chuang drop's axis ratio relation (eq. [3.31]). Use the relative permittivity value for 2.8 GHz at 20°C in table 4.2.

5. Consider an exponential distribution of dry graupel particles ($\Lambda = -0.55$ mm^{-1}, $N_0 = 20$ m^{-3} mm^{-1}, $D_{max} = 15$ mm) with a density of 0.5 g cm^{-3}.
   a) Calculate the equivalent reflectivity Z_e using the Maxwell-Garnett mixture model for the effective permittivity.
   b) When graupel melts, water forms and starts to percolate in the cavities. Calculate the reflectivity assuming 30% of the ice has melted.

6. Using the simple models given in eqs. 4.48 and 4.78, calculate the differential reflectivity for two pristine ice crystals populations: (Hint: approximate the solid plates by oblate spheroids and the columns by prolate spheroids.)
   a) a monodisperse population of solid plates, with major to minor axis ratio $b/a = 0.05$.
   b) a monodisperse population of columns, with axis ratio $a/b = 0.02$.

7. The linear depolarization ratio (LDR) can be measured transmitting either a horizontal or a vertical polarization, and receiving both the copolar and cross-polar components. Eg. (4.84) gives an expression for LDR_{vh} (transmit horizontal) as a function of the particle's canting angle β.
   a) How would eq. (4.84) for $LDR_{vh}(\beta)$ be modified to calculate $LDR_{hv}(\beta)$? (Hint: see fig. 4.24.)
   b) Considering eqs. (4.83) and (4.84), prove that $LDR_{vh}(\beta) = LDR_{hv}(\beta)Z_{dr}(\beta)$.

5   Introduction to Weather Radar Signals

The radar system is an important tool used to measure the scattering properties of clouds and precipitation. Radar observations are used to infer their microphysical and dynamic properties. To understand the measurements made by the radar system, and how the observations relate to the properties of the hydrometeors being observed, signal-analysis techniques are necessary to interpret the radar's signal.

This chapter discusses the link between the radar signal, the radar system and the scattering volume's characteristics. In Chapter 2, the radar equation was introduced, and the components of the radar systems were described. In Chapter 4, the scattering characteristics of precipitation were presented. The radar signal originates from the transmitter, which propagates away from the radar's antenna. The transmitted wave scatters from rain, snow, and buildings. A portion of the scattered wave (the echo) propagates back to the radar's receiving antenna, where the echoes are received (and processed) by the radar's receiver. The received wave contains information about precipitation and is typically referred to as the *signal*. The basis of the weather radar signal's properties and procedures used for modeling and analyzing them are discussed. This chapter provides the foundation for the signal-processing techniques used to estimate the dual polarization radar variables which are discussed in Chapter 6.

With single-polarization radar, only one channel is used to transmit and receive. In dual polarization radar, there are up to two transmit channels and two receive channels, resulting in a vector of measurements sampled for every range resolution volume. Through analysis and processing of these dual polarization signals, the properties of the precipitation can be obtained. In addition to the reflectivity and Doppler velocity, these dual polarization signals are used to infer the differential reflectivity, differential phase, and copolar correlation. Measurements of the cross-polar correlation and depolarization ratio are also possible when a single polarization is transmitted, and two channels are used to receive both polarizations (the copolar and cross-polar echoes). Each of these signals provides additional insight into the physical properties of the precipitation volume.

To aid in this discussion of radar signals, specific mathematical tools are required. Section 5.1 is a brief refresher of the Fourier transform and its applications to radar signal analysis. Section 5.2 briefly reviews statistical models commonly used to describe radar observations of precipitation, which are fundamental to weather radar signal processing. The statistical models provide a means to represent the characteristics of the weather radar echoes from a large radar volume filled with many hydrometeors.

Although the specific properties of each scatterer within the volume are not known, the statistical representations are well defined for an ensemble of scatterers.

In Section 5.6, a model for the radar signal is presented that takes into account the effects of the radar system, propagation, and the location and properties of the scatterers (which can vary with time). This model is used to describe, at a signal level, how a radar's transmitted pulse propagates and interacts with the scatterers and is then measured to estimate the radar variables.

5.1   A Brief Review of Signals and Systems

In this section, a review of some of the fundamental concepts frequently used to describe and manipulate radar signals is presented. These principles and their relevant details will be used throughout the discussion of radar signals and processing techniques. Although these topics are hopefully a review for many readers, these mathematical tools form an anchor for the theory of how weather radar signals are measured and processed. Mathematical expressions are truly the most concise descriptions of the radar signals, but it can sometimes be challenging to infer a deeper understanding of the application from a few equations. Practical elements are interjected in the discussions to aid readers in more quickly developing an understanding of the key elements of weather radar signals.

5.1.1   The Complex (IQ) Signal

Complex numbers are commonly used to represent weather radar signals. The weather radar's signal can be thought of as a slowly varying envelope that encloses the faster-oscillating radar carrier frequency. The envelope may have a temporal resolution on the order of microseconds, whereas the carrier signal oscillations have time periods on the order of nanoseconds; thousands of the radar carrier's cycles are within the envelope. The radar signal characterizes the magnitude and phase of the radar carrier that is within the radar's sampled volume. (The complex envelope is elaborated on in Section 5.1.5.) The complex-valued signal is composed of a real in-phase (I) and an imaginary quadrature (Q) component:

$$x = x_I + jx_Q. \tag{5.1}$$

The complex-valued signal can also be represented as a "phasor" with amplitude, a, and phase, φ:

$$x = ae^{j\varphi}. \tag{5.2}$$

The amplitude of the signal is

$$\boxed{a = |x| = \sqrt{x_I^2 + x_Q^2},} \tag{5.3a}$$

and the phase of the signal is the angle between the signal, represented as a vector in the IQ plane, and the positive real axis (i.e., the in-phase axis):

$$\varphi = \angle x = \arg(x),$$

(5.3b)

where $\arg(x)$ is typically calculated using the function $\mathrm{atan2}(x_Q, x_I)$. The power of the signal x is then

$$p = xx^* = a^2 = x_I^2 + x_Q^2.$$

(5.4)

Appendix A provides a brief refresher on the atan2 function, complex numbers, and their operations.

The complex plane can only unambiguously represent phases over a span of 2π radians. If the phase changes by $\geq 2\pi$, the measured phase will wrap and "alias" to a value within the range $[0, 2\pi)$. In the absence of any other prior knowledge about the radar's signals, it is typical to assume the phase may be positive or negative, and therefore the span of phases is $(-\pi, \pi]$ (as opposed to $[0, 2\pi)$). The aliasing effect of the phase implies that there are multiple phase shifts (i.e., $m2\pi + \varphi$, where m is an integer multiplier) that result in the same measured phase. For example, when estimating the Doppler velocity from the difference in phase between successive radar measurements of the same radar volume, the true Doppler velocity can exceed the span of the unambiguous velocities the radar can measure. With additional context (e.g., spatial continuity of the Doppler velocity field across many radar volumes), the multiplier m can be estimated to de-alias, or "unwrap," the phase and extend the resulting Doppler velocity estimate.

Consider the radar receiver's IQ samples in the complex plane and the resulting amplitude, power, and phase estimates for each sample in Figure 5.1. Manipulation and calculations using IQ samples like these, and the resulting estimates of the phase and power of the received signals, are the fundamental basis for weather radar signal processing operations. Figure 5.1 shows 32 consecutive IQ samples received by the horizontal polarization of the Colorado State University–Chicago Illinois (CSU-CHILL) S-band radar from precipitation. At a range of approximately 75 km from the radar, the receiver's IQ signal is sampled every 962 microseconds. This sampling period is the pulse repetition interval of the radar. The amplitude, power, and phase of the precipitation signal are varying slowly during the observation period. The amplitude and power are directly related (see eq. [5.4]), and the change in the signal's phase from pulse to pulse increases at a nearly constant rate. The change in phase of the signal between pulses is a measure of the rate of change in the distance between the radar and the scattering phase center. The nearly constant change in the signal's phase implies that the precipitation is moving at a nearly constant radial velocity for the time period it is observed (see Section 5.3.2). From panel (d) of Figure 5.1, the total phase shift is 1450° for the observation period. The total phase shift is related to the change in the total path length the radar wave

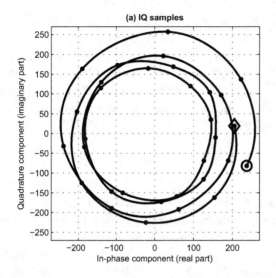

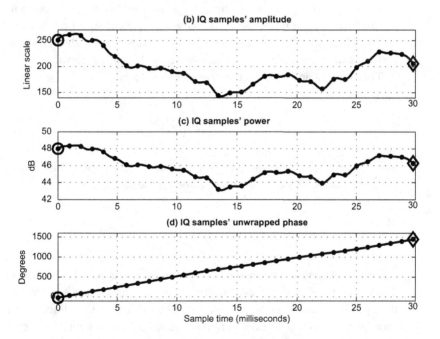

Figure 5.1 Thirty-two IQ samples from the CSU-CHILL S-band radar's horizontal polarization's observations of precipitation. The signal's amplitude, power (in decibels), and unwrapped phase are calculated from the time series of IQ samples. The start of the observed signal is denoted by the circle, and the end is marked by the diamond to show how the signal varies in the IQ space in panel (a). The radar's discrete samples are marked as dots. The solid line approximates how the continuous signal varies with time by interpolating between samples.

travels. Additionally, the phase is increasing, indicating that the distance to the precipitation is increasing and therefore is moving away from the radar system. With the CSU-CHILL's S-band wavelength of 11.01 cm, the precipitation particles' range changes by approximately $1450°/2 \cdot 0.1101 \, \text{m}/360° = 0.222 \, \text{m}$ (again, the phase is divided by 2 because the distance the wave travels is $2r$). The range changes 0.222 m over a time period of 29.8 milliseconds, implying the precipitation volume's mean radial velocity is $0.222 \, \text{m}/0.0298 \, \text{s} = 7.45 \, \text{m s}^{-1}$.

5.1.2   The Fourier Transform

The Fourier transform and its variants (e.g., the discrete Fourier transform) play a fundamental role in signal analysis and modern digital signal processing. For weather radar signals, it is sometimes more convenient to conceptualize or manipulate the observations after they have been decomposed into their Doppler frequency (or Doppler velocity) components. The Doppler velocity space may be a more natural means of working with signals. Spectral processing (so named because it processes the frequency spectrum of the signal) is widely used in signal processing. Here, the Fourier transform, along with some of its properties, are introduced. These are relied on for the discussion of weather radar signals and signal processing.

The Fourier transform of a time-varying signal x is

$$X(\omega) = \int_{-\infty}^{\infty} x(t)e^{-j\omega t} \, dt, \tag{5.5a}$$

where $\omega = 2\pi f$. The signal $x(t)$ may be real- or complex-valued, and $X(\omega)$ is complex-valued. Similarly, the inverse Fourier transform is given by

$$x(t) = \frac{1}{2\pi} \int_{-\infty}^{\infty} X(\omega)e^{j\omega t} \, d\omega. \tag{5.5b}$$

The only differences between the Fourier transform and its inverse are the scale factor $1/(2\pi)$ and the sign of the exponential function's argument (the other differences in the equations are only in the variable notations). One of the conditions for the Fourier transform of a signal to exist is that the signal $x(t)$ is absolutely integrable, which physically means it has a finite amount of energy:

$$\int_{-\infty}^{\infty} |x(t)| \, dt < \infty. \tag{5.5c}$$

All physically realizable signals meet this requirement and therefore have Fourier transforms.

The Fourier transform of the signal, $X(\omega)$, is also referred to as the *frequency spectrum* of the signal (the independent variable is ω). The power spectral density

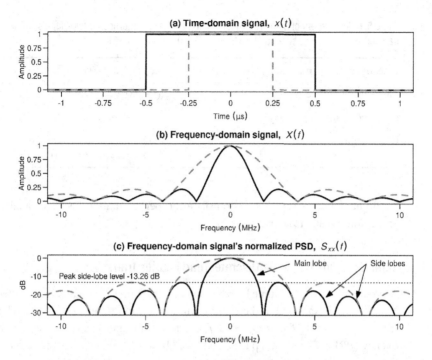

Figure 5.2 Two square pulses with different pulse lengths are shown to compare their characteristics in the time and frequency domains. The solid black line is a 1-microsecond pulse, and the dashed gray line is a 0.5-microsecond pulse. Panel (a) shows the time-domain amplitude of the rectangular pulses. The square time-domain pulses have a frequency response (scaled to a maximum amplitude of 1) characterized by the absolute value of a sinc function shown in panel (b). The power spectra for the two pulses is shown in panel (c), which is normalized to a maximum power of 0 dB. The width of the frequency response is inversely proportional to the pulse's duration—the short pulse has a larger spectrum main-lobe width. The power spectrums also highlight the peak power level of the first side lobe, which is -13.26 dB for the sinc amplitude function.

also referred to as the *power spectrum*, can be calculated from the Fourier transform of the signal as

$$S_{xx}(\omega) = \lim_{T \to \infty} \frac{1}{T} |X(\omega)|^2. \tag{5.6}$$

The power spectral density is a measurement of power per unit frequency, and therefore, is defined to be real-valued. Note that alternate sources (e.g., [163]) may generalize the power spectrum as complex-valued. In Section 6.2.3, the estimation of complex-value spectra is briefly discussed.

An example of the time-domain behavior of two rectangular pulses with different pulse lengths, the amplitude of their Fourier transforms, and their power spectra are shown in Figure 5.2. The Fourier transform of a rectangular function is a sinc function, which is the shape of the pulses' frequency spectra X (sinc$(\omega) = \sin(\omega)/\omega$).

The properties of the Fourier transform that relate the time and frequency domains ($x(t) \Longleftrightarrow X(\omega)$) are useful to describe and manipulate the radar signals. Table 5.1

Table 5.1 A summary of some relevant Fourier-transform properties for two signals and their Fourier transforms, $x_1(t) \Longleftrightarrow X_1(\omega)$ and $x_2(t) \Longleftrightarrow X_2(\omega)$.

Linearity and superposition	$c_1 x_1(t) + c_2 x_2(t) \Longleftrightarrow c_1 X_1(\omega) + c_2 X_2(\omega)$
Time-shifting property	$x(t - t_0) \Longleftrightarrow X(\omega)e^{-j\omega t_0}$
Frequency-shifting property	$x(t)e^{j\omega_0 t} \Longleftrightarrow X(\omega - \omega_0)$
Convolution	$x_1(t) * x_2(t) \Longleftrightarrow X_1(\omega)X_2(\omega)$
	$X_1(\omega) * X_2(\omega) \Longleftrightarrow x_1(t)x_2(t)$

highlights some of the Fourier-transform properties that are used for analyzing and processing weather radar signals.

5.1.3   Sampled Signals and the Discrete-Time Fourier Transform

The radar observes a volume by receiving echoes from periodically transmitted pulses. The volume is sampled at the radar's pulse repetition time (PRT) of T_s. The sample frequency, $f_s = 1/T_s$ (or $\omega_s = 2\pi f_s$), is also referred to as the pulse repetition frequency (PRF). The radar's pulsing scheme allows the radar to sample the continuous-time characteristics of the observation volume. The echo signal, $x(t)$, is sampled at $t = nT_s$ with integer pulse count n, to give the sampled signal $x[n] = x(nT_s)$. (Consider Fig. 5.1, where the solid line represents the continuous-time signals, and the radar's samples are marked as dots.)

The pulsed weather radar periodically samples the radar volume at its PRT. By sampling in time, the signal can no longer represent infinite frequencies. With a finite sample period (e.g., the radar's PRT), the maximum frequency that can be represented by the signal is $f_s/2$, according to the Nyquist theorem. The Nyquist sampling theorem states that to fully represent a signal, the signal must be sampled at a rate greater than twice the highest frequency within the signal.

To analyze the frequency spectrum of the radar's sampled signal, a variation of the (continuous-time) Fourier transform is needed. The answer is the discrete-time Fourier transform (DTFT):

$$\text{DTFT}\{x\} \Rightarrow X(\Omega) = \sum_{n=-\infty}^{\infty} x[n]e^{-j\Omega n}, \tag{5.7}$$

where $\Omega \in [0, 2\pi)$ is periodic with an unambiguous span of 2π. The corresponding frequency (radians s^{-1}) is scaled by the PRT to give $\omega = \Omega/T_s$. The inverse DTFT is

$$\text{IDTFT}\{X\} \Rightarrow x[n] = \frac{1}{2\pi} \int_0^{2\pi} X(\Omega)e^{j\omega n} \, d\Omega. \tag{5.8}$$

It is important to consider the effect of the signal's sampling rate on the range of the signal's frequencies that can be resolved. The signal's highest frequency is determined by its bandwidth. For any signal frequencies that exceed those resolved by the radar's sampling rate (i.e., $f > f_s/2$), these frequencies are aliased back into the range of the frequency spectrum that can be represented: $(-f_s/2, f_s/2]$. The aliasing effect is generally undesirable.

Although the DTFT is sampled in time, it is continuous in frequency. This requires infinite samples in time (as shown in the limits of the summation in eq. [5.7]). Time and frequency are related as $t \propto 1/\omega$, implying that as $t \to \infty$, $\omega \to 0$; for an infinitely long sample period in time, the frequency can be resolved as a continuum. Practical applications only observe signals over a finite time interval, and therefore, only a finite number of samples are available. With a finite length of time, the frequency can no longer be represented as a continuous function; instead, the frequencies are discretely sampled, each covering a "region of support" that is proportional to the total length of time.

Like the DTFT, the discrete Fourier transform (DFT) is sampled at discrete times but also is sampled in frequency. The resolution of each frequency bin, in radians per second, for a signal with sampling period T_s and N samples is

$$\Delta\omega = \frac{2\pi}{NT_s}. \tag{5.9a}$$

Similarly, the frequency resolution in hertz is

$$\Delta f = \frac{1}{NT_s}. \tag{5.9b}$$

Sampling in time limits the maximum frequency that can be represented to $1/T_s$. (Conversely, the signal's maximum frequency that must be represented can be used to determine the required sampling period.) The length of time (or number of samples) that is required is related to the desired frequency resolution (Δf).

If the DTFT is limited to a summation over a finite time, the result is the DFT, which is sampled in both frequency and time. For the radar signal x, sampled in time with a length of N samples, the DFT is calculated as

$$\mathrm{DFT}\{x\} \Rightarrow X[k] = \sum_{n=0}^{N-1} x[n]e^{-j2\pi nk/N}; \quad k \in [0, N-1], \tag{5.10}$$

and the inverse discrete Fourier transform (IDFT) is

$$\mathrm{IDFT}\{X\} \Rightarrow x[n] = \frac{1}{N}\sum_{k=0}^{N-1} X[k]e^{j2\pi kn/N}; \quad n \in [0, N-1]. \tag{5.11}$$

In the literature or descriptions of radar processing algorithms, the fast Fourier transform (FFT) is sometimes used synonymously with the DFT. The FFT and its inverse (IFFT) are computationally efficient algorithms to estimate the DFT and IDFT, which are optimal when the number of samples N is a power of 2 (i.e., $N = 64$, $N = 128$, etc.) [164].

From Parseval's theorem, the time-domain signal and frequency domain spectra are related as

$$\sum_{n=0}^{N-1}|x[n]|^2 = \frac{1}{N}\sum_{k=0}^{N-1}|X[k]|^2, \tag{5.12}$$

where the left-hand side is the total energy integrated over the N samples (via summation of the signal's power).

5.1.4   Linearity and Superposition of Signals

The linear combination of two signals in time is also a linear combination of the signals in frequency. For two Fourier-transform signal pairs $x_1(t) \Longleftrightarrow X_1(\omega)$ and $x_2(t) \Longleftrightarrow X_2(\omega)$ and two constants c_1 and c_2, then

$$c_1 x_1(t) + c_2 x_2(t) \Longleftrightarrow c_1 X_1(\omega) + c_2 X_2(\omega). \tag{5.13}$$

Consider a received signal that includes contributions from precipitation as well as clutter or noise. The individual contributions combine to give the total observed signal. In some cases, signal processing algorithms attempt to separate and filter the contributions from different sources. In particular, the contributions from noise and ground clutter are generally unwanted when trying to observe precipitation. Chapters 6 and 7 discuss how time-domain and frequency-domain signal-processing techniques use the linearity and superposition properties to distinguish and estimate the contributions of noise, clutter, and precipitation to the received signal.

5.1.5   Modulated Signals

Modulation of signals is a fundamental technique used by radar and communication systems. From Section 2.4.2, it's clear that the signal's bandwidth determines the radar's range resolution. In Chapter 4, the relation between the radar's frequency and the scattering characteristics of the precipitation was discussed. Using modulation, the signal's bandwidth can be centered around the radar's operating frequency. Through modulation, signals with bandwidths on the order of megahertz can be used with radar frequencies of gigahertz or higher. It is common to use the term *modulation* to imply increasing the signal's central frequency and *demodulation* to imply decreasing the signal's central frequency. All of the signal's useful information is contained within its bandwidth. To minimize the amount of data that must be represented (e.g., digitized and stored), the radar's received signal is demodulated from a higher frequency to a lower frequency. Modulation of the signal doesn't add new information to the signal, and therefore there is nothing lost by modulating the signal. We can use modulation to shift the center frequency between the baseband – centered at zero hertz – and the radar's operating radio frequency (RF) or potentially any other intermediate frequency (IF) used by the radar system or signal processor. It is convenient to use complex values to represent these signals.

When considering the modulation of signals, Euler's formula is useful for relating the complex exponential function to the trigonometric sine and cosine functions. A complex-valued signal at a constant frequency $\omega = 2\pi f$ (the frequency in radians per second) and time t in seconds is

$$e^{j\omega t} = \cos(\omega t) + j \sin(\omega t). \tag{5.14}$$

Using eq. (5.14), the cosine can be rewritten as the sum of two complex exponential functions, where one has a positive frequency and the other a negative frequency:

$$\cos(\omega t) = \frac{e^{j\omega t} + e^{-j\omega t}}{2}. \tag{5.15}$$

From the Fourier transform's frequency-shift property,

$$X(\omega - \omega_0) \Longleftrightarrow x(t)e^{j\omega_0 t}. \tag{5.16}$$

The frequency shift is implemented as a modulation of the time-domain signal, $x(t)$, by a frequency ω_0. Modulation by the cosine implements two frequency shifts: one positive and one negative. The total energy is divided equally between the two components. The sine is similarly split into a positive- and negative-frequency component but with a phase shift of $\pi/2$ radians (90°), which is a result of the j in the denominator:

$$\sin(\omega t) = \frac{e^{j\omega t} - e^{-j\omega t}}{2j}. \tag{5.17}$$

This mathematical representation is important to describe how RF mixer components work to modulate the RF signal (recall Section 2.6.5). The up-converter mixer acts like a multiplier for the IF and local oscillator (LO) to give the RF as its output.

Panel (a) of Figure 5.3 shows the radar transmitter's signal with a nominal bandwidth that is centered around an IF. Both the positive- and negative-frequency components are included. Similarly, panel (b) of Figure 5.3 shows the LO's positive- and negative-frequency components. Ideally, when the IF and LO signals are mixed, it results in the four components shown in panel (c) of Figure 5.3: IF+LO, IF−LO, −IF+LO, and −IF−LO. In practice, there are other harmonics of these signals that are also generated: $n \cdot$IF$+m \cdot$LO, where n and m are integer values. During the design, the radar's LO and IF frequencies are selected so that with appropriate filtering (e.g., via bandpass or lowpass filters), only the desired frequencies are selected, and all others are rejected. An example bandpass filter's pass band is shown in panel (d) of Figure 5.3. The transmitter's final signal, after mixing and filtering, is shown in panel (e) of Figure 5.3, where the filter effectively rejects the lower side-band component of the IF signal.

In the radar's receiver, the mixer is used to demodulate (down-convert) from the RF frequency to the IF frequency, nominally using the same LO at the transmitter. The modulation process is similar to that discussed for the transmitter, and the receiver's mixing and filtering steps are shown in Figure 5.4. Notice that at the end of the receiver's demodulation and filtering stages, in panel (e), the frequency spectrum is the same as the IF signal that was used at the beginning of the transmitter's up-conversion from panel (a) of Figure 5.3.

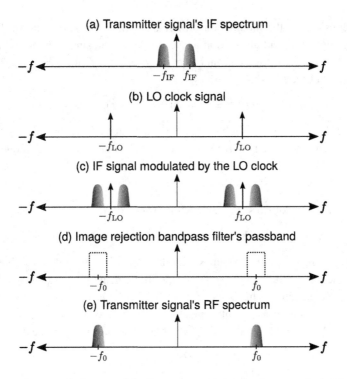

Figure 5.3 A diagram of the frequency spectrum for the transmit signal as it is up-converted and filtered before being transmitted. The transmitter signal's IF spectrum is typically generated by the transmitter's arbitrary waveform generator (AWG). The changes in the signal's amplitude through the up-conversion process are not shown.

In modern systems, the radar's IF signal (panel [e] of Fig. 5.4) is sampled by the receiver's digitizer (see Section 2.6.5). Once it is digitized, digital signal-processing techniques are used to filter and modulate the signal, generally following the steps shown in Figure 5.5. The positive- and negative-frequency components of a real-valued signal are mirrors of each other (they have the same information). For a real-valued signal with bandwidth B, its frequency spans $\left[-\frac{B}{2}, \frac{B}{2}\right] + f_{\mathrm{IF}}$, with a mirror image spanning the negative frequencies over $\left[-\frac{B}{2}, \frac{B}{2}\right] - f_{\mathrm{IF}}$. Using the complex-valued representation of the signal, only the positive-frequency components from $\left[-\frac{B}{2}, \frac{B}{2}\right] + f_{\mathrm{IF}}$ are required to capture the signal's information. Using a complex-valued bandpass filter (panel [b] of Fig. 5.5), the positive frequencies are "selected," and the negative frequencies are removed. After filtering, the complex-valued signal can be demodulated to center the signal's bandwidth around zero. The demodulation is implemented by frequency-shifting the signal by $-f_{\mathrm{IF}}$ (panel [c] of Fig. 5.5). Panel (d) of Fig. 5.5 shows the signal after demodulation, where its bandwidth is centered at zero and it spectrum spans $\left[-\frac{B}{2}, \frac{B}{2}\right]$. This is the signal's baseband representation, which contains all of the radar signal's information. This is the "signal" used for weather radar signal processing discussed in Chapter 6.

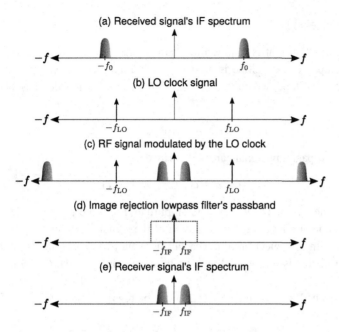

Figure 5.4 A diagram of the frequency spectrum for the received signal as its passes though the mixer and filter stages of a radar receiver's analog down-conversion process. Note that changes in the signal's amplitude are not represented.

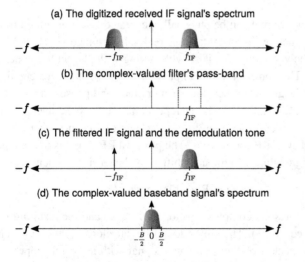

Figure 5.5 The frequency spectrum of the digitized received signal through digital filtering and demodulation. The complex-valued filter selects the positive-frequency band of the signal. It is then frequency-shifted to center the bandwidth at zero, resulting in the baseband signal that is used for all subsequent radar signal processing. The baseband signal contains all of the information that was captured by radar's received RF signal.

5.1.6   The Signal Envelope

The key information that is required to characterize the radar signal is its complex-valued envelope. The envelope is defined by its amplitude (a) and phase (φ). The radar signal's envelope, which is all of the signal's information in a baseband representation, is

$$x(t) = x_I(t) + jx_Q(t) = a(t)e^{j\varphi(t)}. \tag{5.18}$$

A radar's propagating signal can be described as

$$x(t) = a(t)\cos\left(2\pi f_0 t + \varphi(t)\right), \tag{5.19}$$

where f_0 is the carrier frequency. The amplitude a and the phase φ are both time varying and contain the unique information of the signal. The same signals a and φ can be used with any selection of the carrier frequency $\omega_0 = 2\pi f_0$. Note that the RF signal in eq. (5.19) is real-valued and the baseband signal in eq. (5.18) is complex-valued.

The real-valued RF signal in eq. (5.19) can be rewritten with the help of trigonometric formulas as

$$\begin{aligned} x(t) &= a(t)\cos\left(\varphi(t)\right)\cos\left(\omega_0 t\right) + a(t)\sin\left(\varphi(t)\right)\sin\left(\omega_0 t\right) \\ &= x_I(t)\cos\left(\omega_0 t\right) + x_Q(t)\sin\left(\omega_0 t\right), \end{aligned} \tag{5.20}$$

where x_I and x_Q are the in-phase and quadrature components of the baseband signal, respectively.

Only a real-valued signal can be physically generated and transmitted by the radar – the transmitter generates an instantaneous real-value voltage as a function of time. The envelope of the voltage source is slowly varying with respect to the period of the RF frequency f_0. The complex numbers provide a convenient representation for the envelope of the RF signal by describing its amplitude and phase, which are physical properties of the actual signal. The envelope's amplitude is the average magnitude of the RF signal. The envelope therefore represents the mean properties of the RF signal over a number of RF oscillations. The physical RF signal is the real part of the complex-valued signal that is frequency-shifted by the carrier frequency,

$$x(t) = \mathrm{Re}\left(a(t)e^{j\varphi(t)}e^{j\omega_0 t}\right). \tag{5.21}$$

Figure 5.6 illustrates the complex envelope $a(t)$, $\varphi(t)$ and the resulting modulated radar signal from eq. (5.21). The amplitude of the envelope controls the scale factor (and therefore the peaks) of the modulated signal within the envelope. The phase can be thought of as a fine-scaled shift in the timing of the modulated signal, either delaying or advancing the time of the signal's zero-crossings.

5.1.7   Propagating Wave

The weather radar is used to remotely sense precipitation. The radar system transmits a pulse of energy into the environment as a wave. The wave propagates away from

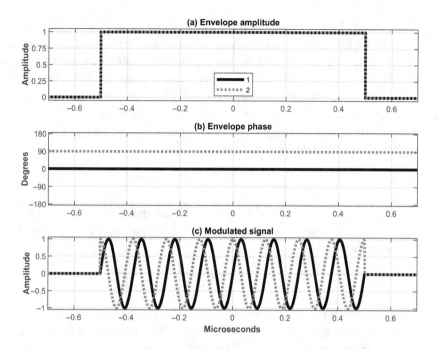

Figure 5.6 A diagram of the complex-valued envelope (defined by its amplitude and phase) and the resulting real-valued modulated signals calculated by eq. (5.21). The complex-valued envelope's amplitude (a) is shown with two different envelope phases ($0°$ and $+90°$) (b). (c) A modulation frequency of $f_0 = 8$ MHz is used in this example to show the effect of the complex-valued envelope's phase on the resulting real-valued modulated signal following eq. (5.21).

the radar until it scatters from objects such as rain. The scattered wave (the echo) then propagates away from the scatterer, where a portion of the echo is received by the radar as the received signal. To meaningfully characterize precipitation with the radar, the signal's amplitude and phase are measured as a function of time, t. The relationship between range, r, and the wave's round-trip travel time determines the distance between the radar and the echo's corresponding scatterer.

The effect of propagation is to introduce a time delay. The radar's signal travels at the speed of light, c; therefore, the one-way travel time is $\Delta t = r/c$. The scatterer interacts with a transmitted waveform that was generated by the source at a time Δt in the past (e.g., $x(t - \Delta t)$). Recalling the Fourier transform's time-shift property, a delay in time results in a proportional shift of the signal's phase:

$$x(t - \Delta t) \Longleftrightarrow X(\omega)e^{-j\omega\Delta t}. \tag{5.22}$$

Note that the magnitude of the phase shift is frequency dependent (from ω). Weather radar signals have narrow bandwidths (on the order of megahertz) with respect to the radar's carrier frequency (on the order of gigahertz). A narrow-band approximation is typically adopted, where the phase shift at the radar's carrier frequency, ω_0, is used for the signal's entire bandwidth. The range between the radar and scatterers imposes a time delay between the transmitted waveform and received echo. The round-trip time delay for a monostatic radar is $\Delta t_{2\text{way}} = 2r/c$ (the signal travels from the radar to the scatterers and then back to the radar; hence the factor of 2). A change in the range between the radar and scatterers is observed as a shift in phase. This is the basis of how Doppler velocity is measured and will be discussed in Section 5.3.2.

Consider a continuous-wave signal with frequency f_0 as a function of t. As the observation point changes its range, the phase of the signal changes as a result of the time-delay effect (neglecting the amplitude):

$$x(t,r) = e^{j2\pi f_0(t-r/c)}. \tag{5.23a}$$

After separating the time and range effects into their own exponential functions, and using the definitions of wavenumber, $k = 2\pi/\lambda$, where the wavelength is $\lambda = c/f$, the signal is

$$x(t,r) = e^{j2\pi f_0 t} e^{-jk_0 r}, \tag{5.23b}$$

where $k_0 r$ is the phase shift due to the one-way propagation at a distance r. The wavenumber k_0 corresponds to frequency f_0. The monostatic radar's signal propagates from the transmitter to the scatterer and then back to the receiver. With a two-way travel distance of $2r$, the signal's total phase shift is

$$x_{2\text{way}}(t,r) = e^{j2\pi f_0 t} e^{-jk_0 2r}. \tag{5.23c}$$

A radar transmits a finite amount of power. As described in Section 4.1.5, as the radar's wave propagates and spreads out spherically, the power density of the transmitted wave is reduced such that the total integrated power over the surface of a sphere remains constant. The antenna is used to select the direction, in terms of azimuth and elevation, along which the radar will sample in the range (see Section 2.4.3). The signal's complex amplitude at time t after propagating a distance r from the source, including the power density and the phase shift from one-way propagation, is

$$x(t,r) = \sqrt{P_t} e^{j2\pi f_0 t} \frac{e^{-jk_0 r}}{\sqrt{4\pi r}}, \tag{5.24}$$

where P_t is the transmitted waveform's power.

It is also worth mentioning that for linear dual polarization weather radar, the differential phase shift (Φ_{dp}) is measured, which is the difference in the phase between the horizontal and vertical polarizations' signals. When observing rain (especially heavy rain) at low-elevation angles, the propagating signal experiences a slightly different phase shift between the two polarizations. This phase shift accumulates as the range increases and is observed as an increase in Φ_{dp} with range. This differential phase shift is due to a slight difference in the propagation constants between the two polarizations as a result of the oblate raindrops (see Chapter 4 for a more detailed discussion of Φ_{dp}).

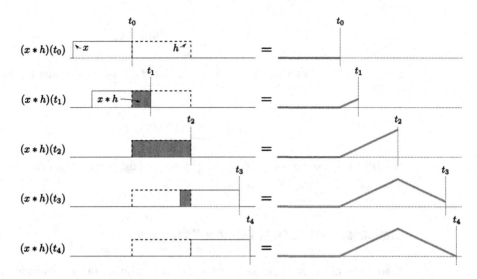

Figure 5.7 Convolution in the time-domain of a rectangular signal (x) with a rectangular filter (h). This is an idealized model of a weather radar waveform and its receiver filter. Notice that in this representation, the filter's position is fixed and the signal moves with time. The total area where x and h overlap at time t is the convolution result for that time.

5.1.8 Convolution

Filters are necessary elements of signal processing and radar systems. Filters are used to improve the detection of echoes and to remove unwanted noise and interference. The process of "filtering" a signal can be implemented as multiplication in the frequency domain or through convolution in the time domain (for completeness, this holds for linear time-invariant systems, which applies here). The convolution operation is denoted by the "$*$," which is not to be confused with the complex conjugation (which is used as a superscript). When a signal x is passed through a filter, represented by its impulse response h, the output is the convolution of x with h:

$$y(t) = x(t) * h(t) = \int_{-\infty}^{\infty} x(\tau)h(t - \tau)d\tau. \tag{5.25}$$

Convolution is commutative, so regardless of the order of x and h, the operation yields the same results:

$$y(t) = x(t) * h(t) = h(t) * x(t). \tag{5.26}$$

The convolution of a signal and filter are illustrated in Figure 5.7 for multiple instances in time.

With the availability of low-cost, high-capability digital electronics, modern radar systems use digital signal-processing techniques to sample and digitally filter the radar's signals. In discrete time, the convolution operator from eq. (5.25) is implemented as

$$y[n] = \sum_{k=-\infty}^{\infty} x[k]h[n-k].$$ (5.27)

For a filter, h, with a finite length N, the limits of the summation can be changed to $[0, N-1]$ as

$$y[n] = \sum_{k=0}^{N-1} x[k]h[n-k].$$ (5.28)

Note that it is assumed that values for $x[k]$ always exist. If no sample $x[k]$ is available, a value of zero is used to "pad" the signal.

5.2 Statistical Properties of Weather Radar Signals

If all of the physical properties of the hydrometeors in the radar volume were known (e.g., the position, orientation, velocity, thermodynamic state), Maxwell's equations would provide an exact calculation of the radar response of the volume. Using the signal model and the exact characteristics of each particle, along with the waves' path between the particle and the radar, the echo's response for each pulse could be calculated. Although such endeavors are useful when trying to understand the characteristics of hydrometeors, it is not feasible to know all of the required information for practical weather applications.

To simplify the problem, the radar's echo from precipitation can be represented using a statistical description. These models also allow well-developed signal-processing techniques to be used to estimate the properties of the precipitation from weather radar observations. By using statistical models to describe the radar's signal, uncertainty bounds for the accuracy of the estimates can also be evaluated. These statistical models are the foundation of the weather radar signals and signal processing discussed here and in Chapter 6, as well as quantifying the data quality, as described in Chapter 7.

5.2.1 The Signal Distribution

The Gaussian distribution, which is also referred to as a *normal distribution*, is frequently used to describe radar signals. The Gaussian distribution is fully characterized by two parameters: its mean, μ, and its variance, σ^2. It can be argued and experimentally observed that the complex Gaussian distribution can be used to characterize the properties of the weather radar's IQ signal [21, 165].

Radar observations of precipitation include the scattering contributions from many different particles within the radar's observation volume. Each of these particles' position, size, and motion can be modeled as statistically independent from one another. The justification for a Gaussian distribution comes from invoking the central limit theorem, which states that when independent random variables are added, the resulting distribution tends toward a Gaussian distribution [163]. In the case of weather radar, each of the scatterers represents an independent variable whose echoes are added together. The bulk motion of precipitation follows the air motion of the storms and gravity. Once these mean motion effects are removed from the precipitation, the residual motions of the particles relative to one another can be modeled as independent random variables.

The Gaussian distribution's probability density function (PDF) is

$$f_N(x|\mu,\sigma^2) = N(x|\mu,\sigma^2) = \frac{1}{\sqrt{2\pi\sigma^2}} e^{-\frac{(x-\mu)^2}{2\sigma^2}}, \tag{5.29}$$

where x is the value for which the probability density is to be calculated. A complex-valued Gaussian distribution (also referred to as a complex normal distribution) is the model used to describe weather radar signals: $CN(0,\sigma^2) = N(0,\sigma^2) + jN(0,\sigma^2)$. Note that here, the I and Q components are independent normal distributions with the same mean and variance (this is a special case of the complex normal distribution).

Consider the distribution of the IQ samples for a single range-volume cell from fixed-pointing CSU-CHILL S-band observations that is presented in Figure 5.8. The antenna position was fixed (not moving), and a constant observation volume was selected for the samples shown. For observing precipitation, statistical stationarity is broadly assumed, meaning that the underlying distributions do not change, and the distributions of raindrops and wind velocity remain constant during the observation period. For the approximately 8-s observation period shown, stationarity is a good assumption. The distribution of the IQ samples in Figure 5.8 is confirmed to match the complex Gaussian PDF models.

5.2.2   Distributions for Power, Amplitude, and Phase

The IQ observations from precipitation are well described by a complex Gaussian distribution, $x = CN(0,\sigma^2)$. In addition to representing the radar's signals as in-phase and quadrature components, the radar's signal can also be defined in terms of its amplitude and phase. The power of the signal is proportional to the square of its amplitude. Statistical distributions of the received signal's amplitude, phase, and power are frequently used to estimate and analyze weather radar signals.

The amplitude, a, of the signal $x = CN(0,\sigma^2)$ has a Rayleigh distribution following

$$f_a(a|\sigma^2) = \frac{a}{\sigma^2} e^{-\frac{a^2}{2\sigma^2}}; \quad a \geq 0. \tag{5.30}$$

The phase of the IQ signal is $\varphi = \arg(ae^{j\varphi})$, where $-\pi \leq \varphi < \pi$ or $0 \leq \varphi < 2\pi$. The zero-mean complex–Gaussian signal $x = CN(0,\sigma^2)$ has a phase that is uniformly distributed, giving a circularly symmetric distribution:

$$f_\varphi(\varphi) = \frac{1}{2\pi}; \quad 0 \leq \varphi < 2\pi. \tag{5.31}$$

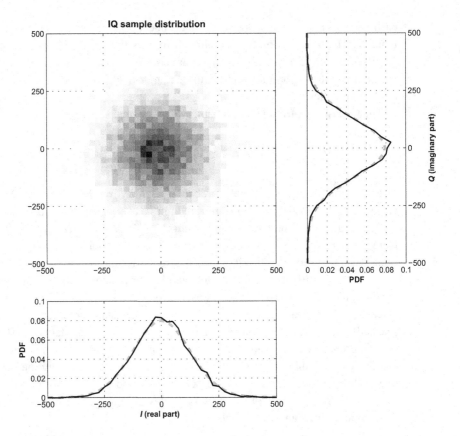

Figure 5.8 The distribution of 8192 IQ samples (which is approximately 8 seconds of observation) from a precipitating radar volume located 75 km from the CSU-CHILL radar. The joint distribution of the IQ results shows the circular symmetric characteristics expected from a complex Gaussian distribution. The conditional distribution of the in-phase (the real part of the IQ signal) and the quadrature component (the imaginary part of the IQ signal) are shown for these observations. For the I and Q distributions, the Gaussian PDF from eq. (5.29) is fit and shown as dashed gray lines.

The power of each of the signal's IQ samples is

$$p = x^* x = x_I^2 + x_Q^2. \tag{5.32}$$

The signal's power is an exponential probability distribution following

$$f_p(p|\sigma^2) = \frac{1}{2\sigma^2} e^{-\frac{p}{2\sigma^2}}; \quad p \geq 0. \tag{5.33}$$

The mean power of the exponential distribution is $2\sigma^2$ for the signal $x = CN(0, \sigma^2)$.

For a complex-valued Gaussian signal with $\sigma^2 = 1$, the probability distributions for amplitude, phase, and power are shown in Figure 5.9. The figure shows the

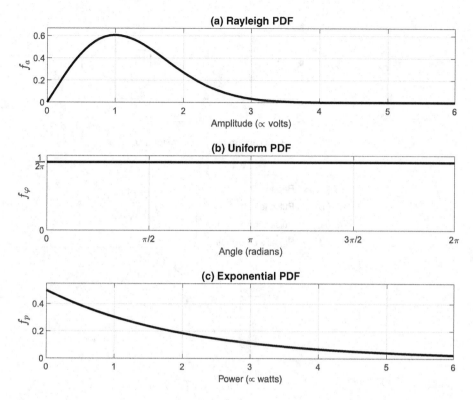

Figure 5.9 Example PDFs for the Rayleigh, uniform, and exponential statistical distributions. The Rayleigh, uniform, and exponential distributions correspond to the statistical models for the amplitude, phase, and power of a complex-valued Gaussian random variable, such as those used to model precipitation.

Rayleigh, uniform, and exponential distributions calculated from eqs. (5.30), (5.31), and (5.33), respectively.

5.3   Sampling the Radar Signal

The objective of any weather radar system is to remotely observe and quantify clouds and precipitation. For pulsed weather radar, the radar system transmits a burst of energy and then waits to receive the echoes. This cycle repeats periodically to make frequent measurements of the cloud or precipitation. The time between these pulses, T_s, is the PRT. (It is also referred to as the *pulse repetition period* or *pulse repetition interval* [PRI]). The pulse rate can also be defined by the pulse repetition frequency, or PRF, which is the number of pulses per second, $1/T_s$.

The radar transmits pulses and receives their echoes as a signal sampled in time. Panel (a) of Figure 5.10 illustrates a sequence of transmitted pulses, represented by the gray areas, each of which is followed in time by the samples of its respective received

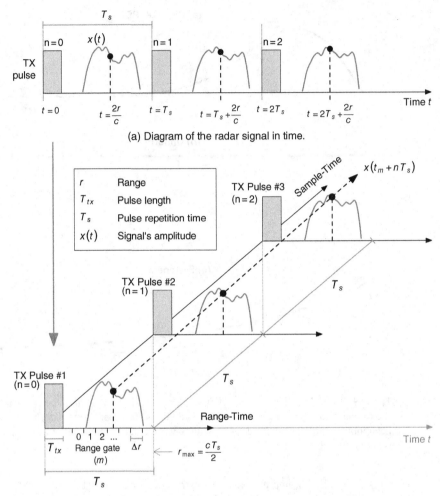

(a) Diagram of the radar signal in time.

(b) Organizing the radar signal by range-time and sample-time.

Figure 5.10 The radar signal is sampled in time, t. The observation volume's range is determined by the elapsed time from the transmitted pulse to the time the echo is received. The radar transmits a new pulse every PRT, T_s (and each pulse has an index n). For each transmitted pulse (and therefore each PRT), the received signal starts a new range line. At a fixed range gate, each pulse provides a successive sample in sample time; these are spaced apart by the radar's PRT, which enables estimates of the properties and dynamics of the hydrometeors within the selected radar volume. Note that *sample time* is also called *slow time*, and *range time* is also called *fast time*.

echoes (up to the next transmitted pulse). The received signal is sampled in time at the corresponding range resolution. For each pulse, the range-time measurement's index increases in proportion to the radar's range. (This range measurement index is also referred to as a *range gate*.) When the elapsed time equals the radar's PRT, the radar transmits another pulse. This effectively resets the sampling range to zero and then restarts the range-time measurement process for the new pulse. The radar system repeats this process continuously.

The one-dimensional received signal (one-dimensional because it's only a function of time) can be reorganized into a two-dimensional matrix of the radar's observations, where each transmitted pulse is assigned to a column, and the rows correspond to the radar range:

$$x[t] \rightarrow x[m,n]. \tag{5.34a}$$

The range-gate index, m, and pulse index, n, are related to the time index as

$$t = m\frac{2\,\Delta r}{c} + nT_s, \tag{5.34b}$$

where Δr is the range resolution (and $2\,\Delta r/c$ is the range-time sampling interval). After N pulses are sampled (N being the total number of columns in the matrix), the observation matrix is used for radar signal processing. For each radar pulse n, the time delay from the transmitted signal to the received echo determines the range between the radar and the echoes' scatterers. This sampling in range (i.e., as a function of m) is referred to as *fast-time* or *range-time* sampling. The collection of range-time samples for a given pulse is referred to as a *range line*. Note that the rate at which the signal is sampled in range should be the same as or smaller than the radar's range resolution (this implies that the range sampling rate should be the same as or greater than the bandwidth). At a fixed range sample m, the pulse-to-pulse measurements are the *sample time* measurements.

By separating each pulse in panel (a) of Figure 5.10, the signal is reorganized in two dimensions, as illustrated in panel (b), with the range time on the horizontal axis and the sample time on the diagonal axis. A constant range-gate index, m, corresponds to a fixed observation volume. An example of a constant radar range volume is shown as a black dot for each pulse in Figure 5.10. By analyzing the sample time for a fixed observation volume, the properties of the volume's precipitation, including its dynamics, can be estimated. This is fundamental to weather radar signal processing, which generally focuses on estimating signal properties using the sample-time observations. In range time, the measurements are typically assumed to be independent.

5.3.1 Unambiguous Velocity and Unambiguous Range

The echo power that the radar can detect depends on the distance between the radar and scattering volume. One could conceive that there is a range beyond which the echoes are too weak for the radar to detect. To cover those areas, changes to the radar design would be needed to increase the radar's performance. When deploying a radar system, the radar's location and the system itself are selected to cover an area of interest. The maximum range at which the radar can unambiguously observe is a system parameter that can be controlled. The unambiguous range of the radar system is determined by the PRT as:

$$r_a = \frac{cT_s}{2}.$$

(5.35)

The span of Doppler velocities that the radar can unambiguously measure using a uniform pulsing scheme is also determined by the PRT. The unambiguous velocity, which is typically defined to cover $[-v_a, v_a)$ (or $\approx \pm v_a$), is determined by the Nyquist theorem as

$$v_a = \frac{\lambda}{4T_s}.$$

(5.36)

With some rearranging of eqs. (5.35) and (5.36), it is clear that the product of the unambiguous range and unambiguous velocity is proportional to the radar's wavelength, λ:

$$r_a v_a = \frac{c\lambda}{8}.$$

(5.37)

This relationship between the unambiguous range and the unambiguous velocity illustrates the radar performance trade-off between the two parameters. For signals from ranges $r > r_a$, the radar will observe them at an aliased range of $\tilde{r} = (r - nr_a)$, where n is the lowest nonnegative integer that places $\tilde{r} \in [0, r_a)$. Similarly, for velocities v that are outside of the range $[-v_a, v_a)$, their signal is observed at an aliased velocity $\tilde{v} = (v - m2v_a)$, where m is an integer (positive, zero, or negative).

Figure 5.11 shows the unambiguous range and velocity for three frequency bands that are uniformly sampled in time. The unambiguous radar range is only a function of the PRF, whereas the unambiguous velocity depends on the PRF as well as the radar's wavelength. In Section 6.4.1, advanced signal-processing techniques are discussed to relax the range–velocity constraint. These techniques are typically used for higher-frequency radars (e.g., X band) to achieve a desired unambiguous range and velocity similar to the larger, more expensive S-band and C-band radars.

5.3.2 Doppler Velocity

From classical Newtonian physics, the total change in an object's position is the integral of the object's velocity. For a monostatic radar, the signal is measured along a single axis: range. This measurement of range extends radially from the radar. The radar's velocity estimate is therefore more precisely a measurement of radial velocity. (Chapter 10 discusses how observations from multiple radars can be combined to

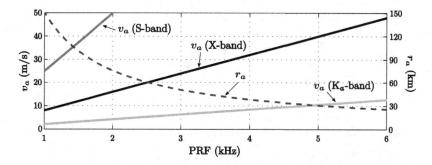

Figure 5.11 The unambiguous range and unambiguous velocity as a function of the uniform sampling PRF. The unambiguous range is only a function of the PRF, whereas the unambiguous velocity also depends on the radar's wavelength. Compare the unambiguous velocity for three radar frequencies: S band (3.0 GHz), X band (9.4 GHz), and Ka band (35.5 GHz).

estimate two- or three-dimensional velocity vectors.) The change in the scatterer's range (r) over time and its radial velocity (v) are related as

$$r(t) = r(0) + \int_0^t v(\tau)d\tau. \tag{5.38}$$

The equation follows the commonly used convention where a negative velocity indicates objects moving toward the radar (and therefore leads to a closing range as time progresses). A positive velocity indicates that the object is moving away from the radar (with an opening range over time). Note that this velocity convention is not followed universally and therefore should be confirmed for each radar.

The radar's estimate of velocity is referred to as the mean Doppler velocity. The Doppler velocity is proportional to the Doppler frequency shift that results from relative motion between the radar and scatterers. The Doppler frequency shift is due to the scatterers' motion away from, or toward, the radar. Figure 5.12 illustrates a transmitter that is generating a constant-frequency tone and a moving scatterer that Doppler-shifts the echo. The wavelength is the distance between the wave's peaks. If the transmitter and scatterer are stationary, the time it takes each amplitude peak to arrive at the receiver is determined by the signal's propagation speed, c. If the scatterer is moving toward the radar at a constant velocity (panel [a] of Fig. 5.12), the distance between the transmitter and scatterer gets smaller with time. The time interval when each amplitude peak leaves the transmitter is unchanged, and the wave's propagation speed is constant (the speed of light). The time each pulse takes to travel from the transmitter to the scatterer and back to the receiver is slightly reduced for every peak of the wave as time increases because the range is decreasing. As a result, each peak of the echo arrives at a slightly earlier time than it would if the range stayed constant. The numbers of peaks received in a period of time is the frequency of the signal. If the receiver counts more peaks in a fixed period of time, then the apparent frequency has increased. The Doppler frequency shift, f_d, is the difference between the transmitter's frequency and the apparent frequency measured by the receiver.

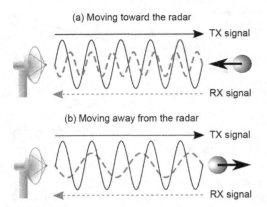

Figure 5.12 The Doppler shift increases the frequency of the echo for targets moving toward the radar and decreases the frequency of the received echo for targets moving away from the radar. The transmitted signal's frequency is identical for both cases and is only modulated by the motion of the scatterer. Note that the echo's amplitude is lower than the transmitted signal's amplitude and is a function of the scatterer's radar cross-section (RCS).

For a stationary monostatic radar with a carrier frequency of f_0 that observes a moving target at velocity v, the Doppler frequency shift is

$$f_d = f_0 \left(\frac{c-v}{c+v} - 1 \right), \tag{5.39a}$$

where c is the speed of light. For a target moving away from the radar (i.e., a positive velocity), $f_d < 0$. A target that is stationary with respect to the radar has $f_d = 0$, and a target moving toward (i.e., a negative velocity) the radar has $f_d > 0$. The relationship between Doppler frequency and Doppler velocity can be approximated as

$$\boxed{f_d \approx -f_0 \frac{2v}{c} \approx -\frac{2v}{\lambda}.} \tag{5.39b}$$

It is clear that the magnitude of the Doppler frequency shift is wavelength dependent. For S-band radars, radial velocities of $|v| = 100\,\mathrm{m\,s^{-1}}$ result in Doppler frequency shifts of $|f_d| = 2\,\mathrm{kHz}$. At the Ka band, the Doppler frequency shift is approximately 12 times greater than at the S band.

In pulsed radar systems, the mean Doppler velocity is estimated by comparing the object's position at two instances in time. The difference in the object's position is measured as a phase shift between the two observations. The radar's received echo from a moving scatterer is modeled as

$$x_{rf}(t) = s(t)e^{j2\pi f_0 t} e^{j2\pi f_d t}, \tag{5.40}$$

where $s(t)$ is the complex-scattering amplitude. The radar electronics act to demodulate the RF signal (see Section 5.1.5) to give the baseband representation, which includes the Doppler frequency shift:

$$x_{bb}(t) = x_{rf} e^{-j2\pi f_0 t} = s(t) e^{j2\pi f_d t}. \tag{5.41}$$

With a pair of samples of the radar volume separated by a period T_s, the Doppler velocity can be estimated (this is the pulse-pair Doppler velocity). Each sample measure the signal's amplitude and phase at an instance in time. The change in phase between two complex values is the angle of their dot product: $\varphi = \arg(x^* y) = -\arg(xy^*)$ (see Appendix A). As the range of the scatterer changes, so does the phase of the echo following $e^{-jk_0 2r}$ (recall eq. [5.24], with the factor of 2 accounting for the travel distance of the radar's wave to the target and back). If the scatter is moving at a radial velocity v, and it is sampled with a time interval of T_s, then the scatterer range changes $dr = r(T_s) - r(0) = vT_s$ during the interval. From the dot product of the sampled echo at two different times, the observed phase change is due to a change in the scatterer's range from $t = 0$ to $t = T_s$:

$$\arg\left(x_{bb}^*(0) x_{bb}(T_s)\right) = \arg\left(e^{-jk_0 2[r(T_s) - r(0)]}\right) = -k_0 2[r(T_s) - r(0)]. \tag{5.42}$$

Recalling $f_d \approx -\frac{2v}{\lambda}$ and $k_0 = 2\pi/\lambda$, the Doppler frequency, radial velocity, and range difference between two ranges all give the same effective result:

$$e^{-jk_0 2[r(T_s) - r(0)]} = e^{-jk_0 2vT_s} = e^{j2\pi f_d T_s}. \tag{5.43}$$

The change in the echo's phase between the two sample times is a direct measurement of the scatterer's Doppler frequency shift (with the caveat that the change in phase does not alias by exceeding $\pm\pi$ radians between measurements).

Consider the two samples of a scatterer with a constant Doppler frequency that are observed at times $t = 0$ and $t = T_s$:

$$x_{bb}[0] = s[0] e^{j2\pi f_d 0} = s[0] \tag{5.44a}$$

and

$$x_{bb}[T_s] = s[T_s] e^{j2\pi f_d T_s}. \tag{5.44b}$$

The signal's autocovariance at time lag T_s (see Section 5.4) is

$$r_{xx}[T_s] = x_{bb}^*[0] x_{bb}[T_s] = s^*[0] s[T_s] e^{j2\pi f_d T_s}. \tag{5.45}$$

When the scatterer's amplitude is constant (i.e., $s[T_s] = s[0]$), then $s[T_s] s^*[0] = |s|^2$, and eq. (5.45) reduces to

$$r_{xx}[T_s] = |s|^2 e^{j2\pi f_d T_s}. \tag{5.46}$$

To estimate the Doppler velocity, the change in the echo's phase over a time T_s is $\angle r_{xx}[T_s]$. Note that the phase is unambiguous over a span $(-\pi, \pi]$. The Doppler velocity is unambiguous over the span $[-v_a, v_a)$, where the unambiguous velocity scale factor is $v_a = \lambda/(4T_s)$ (from eq. [5.36]). To convert the phase to Doppler velocity, it is multiplied by $-v_a/\pi$. The minus sign is used to follow the convention that a positive velocity is moving away from the radar. The Doppler velocity, Doppler frequency, and their estimation using two radar pulses are related as follows:

$$v = \frac{-\lambda}{4\pi T_s} \arg\left(x^*[0]x[T_s]\right) = \frac{-\lambda 2\pi f_d T_s}{4\pi T_s} = \frac{-f_d \lambda}{2}. \tag{5.47}$$

5.4   Covariance and Power Estimation

Everything that the radar can measure or infer about the singal's propagation path and scatterers' characteristics are derived from its received signals. Within the radar electronics, the echo's signal is measured as a voltage that is proportional to the incident electric field. The envelope of the RF signal contains all of the available information. For dual polarization weather radar, the properties most frequently estimated from the radar's signals are the polarimetric Doppler radar variables: reflectivity, mean Doppler velocity, Doppler spectrum width, differential reflectivity, differential phase, and copolar correlation. The linear depolarization ratio can also be estimated for radar systems operating in alternate transmit and alternate receive (ATAR) or alternate transmit and simultaneously receive (ATSR) mode (see Section 2.3). This section provides a generalized discussion of the signal analysis tools and signal properties that are relevant to weather radar.

In this section, two important statistical tools used to process and analyze the radar signals are introduced: the covariance and correlation estimators. At each range gate, the radar variables are estimated using (N) sample-time measurements acquired at the radar's PRT. The details of how each radar variable is processed using the covariance and correlation estimates are discussed in Chapter 6. Estimates of the signal's power is the most fundamental radar measurement and, naturally, it's directly related to the weather radar's most fundamental variable: the equivalent reflectivity factor. This section also describes the estimation of the signal power and introduces the concept of additive noise in the radar observations. Additive noise is a practical part of radar signal processing and must be considered to ensure accurate estimates.

5.4.1   Covariance Estimation

Covariance estimation is fundamental to statistical signal processing. The covariance is a measure of the joint variability of random variables (or signals). The autocovariance is the covariance of a signal with itself, and the cross-covariance is the covariance between two different signals. The cross-covariance is the more general form using two signals (x and y) as a function of an independent variable and an offset. For radar processing, the independent variable is time, t, and the offset is the time delay, τ. The cross-covariance is then

$$r_{xy}(t, \tau) = \text{cov}\left(x(t), y(t + \tau)\right) = \langle x^*(t)y(t + \tau)\rangle, \tag{5.48a}$$

where $\langle\rangle$ denote an ensemble average. The ensemble average is the mean over all possible realizations of the statistical process. The autocovariance of x is then $r_{xx}(t, \tau)$, which is the cross-covariance of a signal with itself (i.e., for the signal x, $y = x$). The time offset τ is also referred to as a *time lag*, and τ can be positive or negative. The power spectral density is related to the signal's covariance function r_{xy} as

$$S_{xy}(\omega) = \left| \int_{-\infty}^{\infty} r_{xy}(\tau)e^{-j\omega\tau} \, d\tau \right|. \tag{5.48b}$$

A strict-sense stationary process is a stochastic process whose joint distribution does not change with time. For a stationary process, the covariance is only a function of the time delay, τ, simplifying the covariance function to $r_{xy}(\tau)$. A wide-sense stationary (WSS) is a weaker form of strict-sense stationary where only the first- and second-order statistical moments are required to be constant with time. To analyze a WSS process, only the mean (μ) and variance (σ^2) are required. Notice that a Gaussian distribution (see eq. [5.29]) is fully characterized by its first- and second-order statistics. The IQ signal for precipitation is a zero-mean process ($\mu = 0$), and therefore, only the variance of the signal needs to be estimated.

With stationary processes, the statistical moments can be estimated by averaging in time. A process is ergodic when samples over time can be used in place of different realizations of the same process. This is essential to the principle of radar variable estimation. During the signal's integration period, the radar signal is assumed to be WSS. For the zero-mean IQ signal, the covariance provides an estimate of the second-order moment to characterize the process as a function of τ.

For a sampled signal, the time lags, τ, are at discrete intervals in multiples of the sample period T_s following $\tau = l \cdot T_s$, where l is the time-lag index. The signal itself consists of a total of N samples in time, $t = n \cdot T_s$. The sampled cross-covariance is estimated from simultaneously sampled radar observations x and y (e.g., horizontal and vertical polarizations). For WSS signals, the unbiased cross-covariance estimator for time lags $l \cdot T_s$ is

$$\tilde{r}_{xy}[l] = \begin{cases} \frac{1}{N-|l|} \sum_{n=0}^{N-|l|-1} x^*[n+|l|]y[n]; & -(N-1) \le l < 0 \\ \\ \frac{1}{N-l} \sum_{n=0}^{N-l-1} x^*[n]y[n+l]; & 0 \le l \le N-1 \end{cases} \tag{5.49}$$

Although this estimator is unbiased, it is well known that the estimator given by eq. (5.49) is unstable and not guaranteed to satisfy the basic properties of a covariance function [166]. As an alternative, the "biased" estimator of the sampled covariance is widely used and is implemented as

$$\hat{r}_{xy}[l] = \begin{cases} \frac{1}{N} \sum_{n=0}^{N-|l|-1} x^*[n+|l|]y[n]; & -(N-1) \le l < 0 \\ \\ \frac{1}{N} \sum_{n=0}^{N-l-1} x^*[n]y[n+l]; & 0 \le l \le N-1 \end{cases} \tag{5.50}$$

The difference between the biased and unbiased covariance estimators is the biased estimator $\widehat{r}_{xy}[l] = \tilde{r}_{xy}[l]h_t[l]$, where h is a triangle window function

$$h_t[l] = \frac{N - |l|}{N}; \ -(N-1) \leq l \leq N-1. \tag{5.51}$$

The biased estimator better approximates the expected covariance characteristics as the time lag increases. In weather radar signal processing, only lags $l = 0$ and $l = 1$ are used for the pulse-pair radar variable estimators. For typical integration periods, with $N \gg 1$, the error between the two covariance estimators is negligible, given other estimator uncertainties.

5.4.2   Correlation Estimation

The correlation is a measure of the linear dependence between two signals. The copolar correlation radar variable indicates the similarity between the horizontal and vertical polarizations' signals. The correlation is a normalized version of the autocovariance and cross-covariance estimates. The correlation of two N-length sampled complex-valued signals x and y is

$$\boxed{|\rho_{xy}[l]| = 0 \leq \frac{|r_{xy}[l]|}{\sqrt{r_{xx}[0]r_{yy}[0]}} \leq 1.} \tag{5.52}$$

For a correlation of unity, the behavior of the two signals is the same. For example, if signals x and y have a correlation of $|\rho| = 1$, then $ax = y$ for a constant complex-valued scale factor, a. A correlation $|\rho| = 0$ indicates that there is no meaningful relationship between the means and variances of the signals. To say the signals are "independent" in a statistical sense requires evaluation of higher-order statistical moments as well. Gaussian-distributed signals are a notable exception, and their statistical distributions are completely characterized by their means and variances. Uncorrelated Gaussian signals are also independent.

5.4.3   Coherent and Incoherent Power Estimation

In SI units, power is measured in joules per second or watts. The signal's power estimate, along with the radar constant and observation volume's range, are all that is required to estimate the radar's equivalent reflectivity factor using eq. (2.39c). The radar's received signal is a direct measurement of the echo from the scatterers within the observation volume. The signal is sensitive to the configuration and the state of the scatterers at that instant in time.

For real-valued signals, the signal power is

$$p = x^2. \tag{5.53}$$

Typically, the received signal is filtered and converted to a complex-valued representation (see Section 5.1.5 and Fig. 5.5). The power estimated from a signal's complex-valued voltage envelope is

$$p = x^*x = xx^* = |x|^2. \tag{5.54}$$

The scatterers' configuration within the observation volume are treated as a stochastic process. The signal is the radar's measurements that is sampling this process. As the scatterers move and reshuffle, they create different independent configurations, which are realizations of the stochastic process. The average power of the complex-valued signal that samples this process is

$$\langle p \rangle = \langle x^*x \rangle. \tag{5.55}$$

The ensemble averaged power is approximated as the mean of the measured power within a fixed range volume. For N observations of the signal, its mean power estimate is

$$\tilde{p} = \frac{1}{N} \sum_{n=0}^{N-1} x^*[n]x[n] = r_{xx}[0]. \tag{5.56}$$

Note that $\tilde{p}$ is the estimate of $\langle p \rangle$. As $N \to \infty$, $\tilde{p} = p$. In many instances, you will find that p, $\langle p \rangle$, and $\tilde{p}$ are used loosely to refer to the mean power. Strictly speaking, the estimates p and $\tilde{p}$ have an associated uncertainty or error with respect to $\langle p \rangle$.

The power calculated from specific samples is exact and represents one realization of the random process. To fully characterize the underlying process that is generating the realizations, many samples are required. The ensemble average of the power is considered the true power of the random signal. Each individual sample's power is not necessarily the same as the ensemble average power, and therefore there is an associated error: $\tilde{p} - <p>$. The average error, which itself is averaged using many realizations of $\tilde{p}$ with N samples each, is the estimate of the power measurement's uncertainty (see Chapter 7 for discussions on estimator uncertainty).

The incoherently averaged power for a signal vector x with N samples is estimated as

$$p_{\text{incoh}}(x) = \frac{1}{N} \sum_{n=0}^{N-1} x_n^* x_n = \frac{1}{N} \sum_{n}^{N} |x_n|^2. \tag{5.57}$$

(Note that this is equivalent to eq. [5.56]). The incoherent power is the pulse-pair estimator for the signal power. The incoherent power estimation uses only the magnitude of the individual samples. In contrast, a coherent power estimation implies that both the amplitude and phase are considered in the average. The coherently average power for the same sampled signal vector x is estimated as

$$p_{\text{coh}}(x) = \left| \left(\frac{1}{N} \sum_{n=0}^{N-1} x_n \right)^* \left(\frac{1}{N} \sum_{m=0}^{N-1} x_m \right) \right|$$

$$= \frac{1}{N^2} \left| \sum_{n=0}^{N-1} \sum_{m=0}^{N-1} \left(N - |m - n| \right) r_{xx}[m - n] \right|, \tag{5.58}$$

where $r_{xx}[l]$ is the autocovariance function of the signal x. (Note that here, m is a second sample index, not to be confused with a range index.)

Consider two special cases of signals for the incoherent and coherent power estimates that are frequently encountered in weather radar signal processing. For the first special case, the signal is a constant value, $x = C$. (This case closely approximates the signal from a stationary scatterer, such as ground clutter.) The coherently integrated average power for the constant signal is estimated as

$$p_{\text{coh}}(C) = \left| \frac{NC}{N} \right|^2 = |C|^2. \tag{5.59a}$$

Similarly, the incoherently integrated average power for the constant signal is

$$p_{\text{incoh}}(C) = \frac{1}{N} N |C|^2 = |C|^2. \tag{5.59b}$$

When the signal is a constant value, both the coherent and incoherent power estimators are equivalent. When the results of both estimators are equal, this indicates that the signal is constant and fully coherent throughout the integration length, N.

For the second special case, the signal is a white-noise random variable, w, with all samples independent from each other (e.g., thermal receiver noise). The signal is then $x = w$ with variance σ_w^2. For $N \to \infty$, the covariance between the independent samples is zero, and therefore

$$
\begin{aligned}
r_{ww}[0] &= \sigma_w^2 \\
r_{ww}[l \neq 0] &= 0
\end{aligned} \tag{5.60}
$$

For the random white-noise signal and a large number of samples, the coherently integrated power is

$$p_{\text{coh}}(w) = \frac{1}{N^2} N \sigma_w^2 = \frac{\sigma_w^2}{N}. \tag{5.61a}$$

The incoherent estimate of power for the same white-noise signal is

$$p_{\text{incoh}}(w) = \frac{1}{N} N \sigma_w^2 = \sigma_w^2. \tag{5.61b}$$

For a signal whose samples are all statistically independent, the results of the coherent and incoherent power estimators are very different. The coherent power estimate scales the variance of the white noise by the total number of samples, N. In contrast, the incoherent power estimate is constant, with a power estimate equal to the variance of the white noise. The random white noise used as the signal is an example of a fully incoherent signal, given the $1/N$ scaling in eq. (5.61a).

The fully coherent constant signal and the fully incoherent white noise represent the two endpoints of the continuum for signal coherency. For weather radar, this continuum is characterized by the spectrum width of the signal, but it's also referred to as the coherence time or decorrelation time. The signals from clouds and precipitation are within this continuum.

5.4.4   Echo Power from a Volume of Precipitation

The echo from each hydrometeor in a radar volume has its own magnitude and phase. The echoes from each scatterer are added together and contribute to the radar's received signal. Neglecting the Doppler frequency shift for the moment, the monostatic radar's received signal is proportional to the summation of the echoes from all N scatterers:

$$s = \sum_{n}^{N} \sqrt{\sigma_n} e^{-jk2r_n}, \tag{5.62}$$

where σ is the RCS, k is propagation constant, and r is the range. Figure 5.13 illustrates the summation of the signals from multiple scatterers, where each scatterer's magnitude and phase are represented as a vector.

The power of the received signal from a volume of precipitation is

$$p = s^* s = \left(\sum_{n}^{N} \sqrt{\sigma_n} e^{-jk2r_n} \right)^* \left(\sum_{n}^{N} \sqrt{\sigma_n} e^{-jk2r_n} \right). \tag{5.63a}$$

For a collection of N scatterers in the radar observation volume, the relative motions and positions of the scatterers are independent from one another. As the scatterers move within the volume, the relative positions of the scatterers are reshuffled, giving a new realization of radar range (and therefore signal phase). For sufficiently long integration times, and as a result of the apparently random phase of each scatterer relative to the others due to their motion, the coherent integration of the scatterers within the volume resembles the formulation of the coherent integration of white noise in eq. (5.61a). The mean echo power for the volume of precipitation is simply the summation of the RCSs of the scatterers within the volume:

$$p = \langle s^* s \rangle = \sum_{n}^{N} \sigma_n. \tag{5.63b}$$

5.4.5   Noise in Observations

The radar's received "signal" is frequently used as a synonym for the radar observation. In practice, the radar observation, x, contains not only the echoes' signal, s, but also a contribution from noise, w. The radar's measured complex-valued voltage envelope can be expressed as

$$x = s + w. \tag{5.64}$$

Noise is always present in radar observation. When the signal's power is large enough with respect to the noise power, the noise can be safely neglected, with little error or bias in the estimate. For weaker echo signals, the noise power can have similar magnitude to the power of the echo. The signal-to-noise ratio (SNR) is

$$\mathrm{SNR} = \frac{\langle |s|^2 \rangle}{\langle |w|^2 \rangle}. \tag{5.65}$$

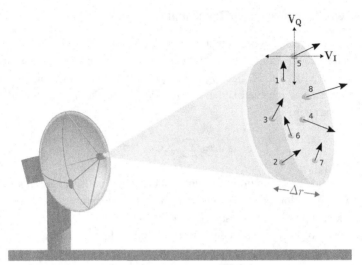

(a) Individual scattering amplitude and phase (shown in the IQ plane)

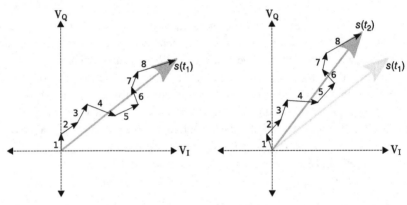

(b) Coherently integrated amplitude and phase from all scatterers at time t_1 (shown in the IQ plane)

(c) Coherently integrated scatterers at time $t_2 = t_1 + T_s$

Figure 5.13 To illustrate the volume integration of scatterers, consider a volume with eight raindrops. Each individual scatterer has its own amplitude and phase, shown as IQ voltages in panel (a). The summation of each scatterer's amplitude and phase results in the total received signal, shown as a gray vector in panels (b) and (c), for times t_1 and t_2, respectively. The positions of the scatterers change between t_1 and t_2, which changes the phase of each scatterer. The contribution of each scatterer (including the phase change due to their motion) is captured in the summation. Compare the summation vectors at t_1 and t_2 to the radar IQ time-series observations from the CSU-CHILL in Figure 5.1.

As the ratio of the signal power to the noise power decreases, the noise power becomes a significant portion of the received power ($p = |x|^2$). When the noise power becomes "significant," the receiver's observed power is no longer a good representation of the signal power because it contains a measurable bias due to the noise. The threshold that determines what is a "significant" level of noise power is set by the amount of measurement error that can be tolerated. For most weather radar observations, an SNR greater than 10–20 dB is typically considered a moderate to high SNR (the signal power is 10–100 times greater than the noise power).

The weather radar's signal s is the echo from hydrometeors within the radar volume. The noise in radar systems is typically thermal radiometric power from the radar's own electronics and noise received by the antenna from the environment. The power of the noise is sometimes quantified as a noise temperature (see Section 2.7.1). The noise temperature is the equivalent radiated power from a black body at the same temperature. (The black body implies that it is not reflecting power from other sources; the only power from the object is due to its own radiation.)

White noise is independent for all samples, and the noise is independent between the polarizations. The noise is also independent of the signal. For statistically independent signals, their powers are separable. The measurement's power (the lag-zero covariance estimate, $r[0]$) is simply the sum of the signal's and noise's power:

$$\langle |x|^2 \rangle = \langle |s|^2 \rangle + \langle |w|^2 \rangle. \tag{5.66}$$

By subtracting an estimate of the noise power from the measured power, the echo's contribution to the measured power can be estimated:

$$|s|^2 = |x|^2 - |w|^2. \tag{5.67}$$

Noise correction (or noise subtraction) is particularly helpful to improve the accuracy of the estimated echo power for low-SNR measurements. If noise subtraction is not used, the measurement's minimum power is $|w|^2$ (in the absence of an echo, $|x|^2 = |w|^2$). With a perfect estimate of the noise power, the minimum echo power that can be measured would be zero. Techniques to estimate the noise power are discussed in Section 7.3.4.

Using noise subtraction, the accuracy of the measured equivalent reflectivity factor is improved, which results in an apparent increase in the radar's sensitivity. The accuracy of all radar variables that rely on estimates of the signal's power can be improved using noise subtraction. Biases due to additive noise power only affect power-dependent measurements (i.e., Φ_{dp} and $\bar{v}$ are not biased). Although the biases for the radar variable estimators are reduced, all estimators incur an increasing uncertainty as the SNR decreases. (The uncertainty of the estimators is discussed in Chapter 7.)

The noise and echo signal are uncorrelated, therefore the observation's autocovariance (r_{xx}) is the sum of the contributions from both the echo (r_{ss}) and noise (r_{ww}) at each sample lag l:

$$r_{xx}[l] = r_{ss}[l] + r_{ww}[l]. \tag{5.68}$$

The observation's autocorrelation (from eq. 5.52) is then

$$|\rho_{xx}[l]| = \frac{|r_{ss}[l] + r_{ww}[l]|}{r_{ss}[0] + r_{ww}[0]}. \tag{5.69}$$

The noise (w) is uncorrelated, and therefore, $r_{ww}[l] \to 0$ for $l \neq 0$ as the number of measurements (N) used to estimate r_{xx} increases. In effect, the contribution of noise

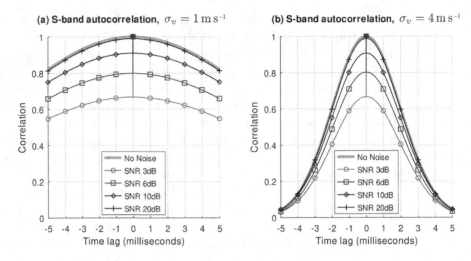

Figure 5.14 The measurement autocorrelation estimates for two different spectrum widths and varying SNR, assuming a larger number of samples (i.e., $N \to \infty$). The correlation and time lags shown are calculated assuming a 3-GHz ($\lambda = 10$ cm) S-band radar for signal spectrum widths of 1 and 4 m s^{-1}. At a time lag of zero, the correlation is 1 for all curves; for all time lags not equal to zero, the correlation is a function of spectrum width and SNR. The markers indicate the time lags associated with a PRT $T_s = 1$ ms.

is only observed at $r_{xx}[0]$ and is negligible for other time lags. As the SNR decreases, the zero-lag sample is increasingly biased by the additive noise power.

Figure 5.14 shows the effect of additive noise on the autocovariance estimates. Simulations for two different echo spectrum widths and multiple SNRs are presented. The figure shows the measurements' autocorrelation (see eq. [5.52]). The markers indicate the correlation that would be estimated at multiples of a the pulse repetition period for $T_s = 1$ ms. The spectrum-width estimates are commonly calculated using the ratio of $r_{xx}[1]$ and $r_{xx}[0]$ (see Section 6.1.1). It's clear that the spectrum-width estimate will be biased at low to moderate SNRs without accurate noise correction to approximate the signal's covariance $r_{ss}[0]$ (and using $r_{ss}[1] \approx r_{xx}[1]$).

If the SNR is high, then $r_{ss}[0] + r_{ww}[0] \approx r_{ss}[0]$, and the noise's power has little bias on the observed power and its associated radar variables. From Figure 5.14, for an SNR of 20 dB, there is little bias in the magnitude of the correlation (or covariance) estimates. The exact SNR threshold of when noise can be "neglected" depends on the tolerance to estimator's bias, but typically 15–20 dB is sufficient. Chapter 7 quantifies the effects of noise on the bias and variance of the radar variable estimates.

5.5 Sampling the Polarimetric Signature

Radar polarimetry was introduced in Section 2.1, and the dual polarization scattering matrix was elaborated in Section 4.5.1. In this section, weather radar polarimetry is presented in the context of the radar's signals.

A linear dual polarization system has two orthogonal signal bases: the horizontal and vertical polarizations. To implement full polarimetric sampling, the radar system must be able to transmit and receive signals for both polarizations. All electromagnetic waves, and therefore all radar signals, have a polarization state. This is true even if the radar is not designated as a polarimetric radar. This book focuses on linear polarization, which is typically used for weather radar. Dual polarization radar is used to estimate the parameters of the linear dual polarization scattering matrix (from eq. [4.50]):

$$\mathbf{S} = [S] = \begin{bmatrix} s_{hh} & s_{hv} \\ s_{vh} & s_{vv} \end{bmatrix}. \tag{5.70}$$

To investigate methods for sampling the scattering matrix, first, the radar's transmitted dual polarization waveforms are defined as $[M] = \begin{bmatrix} m_h & m_v \end{bmatrix}^T$. In a dual polarization system, the transmitter subsystems' phase and power for each polarization ultimately determine the transmitted signal's polarization state. The radar system itself may have a difference between the amplitudes and phases of the two orthogonal polarizations. The phase ϕ_{sys} is used is used to represent the difference in the system's phase between the horizontal and vertical polarizations. With this, the actual transmitted signals are modified by the transmitter's gain (a_h and a_v) and system phase shift to give the complex-valued waveform of the transmitted pulse:

$$m_h = a_h \tag{5.71a}$$

$$m_v = a_v e^{j\phi_{sys}}. \tag{5.71b}$$

The ideal linear dual polarization waveforms, when simultaneously transmitted, have $a_h = a_v = 1$ and $\phi_{sys} = 0°$ or $\phi_{sys} = 180°$. Slight differences in the amplitudes, phases, or both between the horizontal and vertical polarizations will result in an elliptical polarization. Changes in the amplitude will also vary the angle and eccentricity of the polarization. Ultimately, this is not a concern. As long as both polarizations are excited by the transmitter and the radar receives echoes for both polarizations, the scatterers' dual polarization response can be characterized.

For the moment, assume that the transmitters and receivers of each polarization are perfectly isolated from each other. The horizontal and vertical received signals, e_h and e_v, for a system with perfect polarimetric isolation and no losses is represented as

$$\begin{bmatrix} e_h \\ e_v \end{bmatrix} = \begin{bmatrix} 1 & 0 \\ 0 & 1 \end{bmatrix} \begin{bmatrix} m_h \\ m_v \end{bmatrix}. \tag{5.72}$$

The copolar signals are defined as the same polarization for both the transmitted and received signals. The cross-polarization signals have different polarizations for the transmit and receive signals. For example, receiving the horizontally polarized transmitted signal when the horizontal state is transmitted is a copolar signal. Receiving the horizontal transmitted signal when the vertical state is transmitted is a cross-polarization signal. Now consider varying the amplitude and phase of the diagonal (i.e., copolar) elements of the matrix, which are shown as unity in eq. (5.72). For this system, where there is no cross-polarization, only two measurements are needed to characterize the matrix which represents the signal's paths between the polarimetric transmitter and receiver. This is true because the cross-polar elements of the matrix are zero in eq. (5.72). This is the ideal model for the simultaneous transmit and simultaneous receive (STSR) dual polarization radar configuration. For the STSR sampling mode, both polarizations are transmitted at the same time for every PRT.

The radar system can be calibrated using a target or calibration path with a known scattering matrix. With the calibration complete, in STSR mode, the radar is used to estimate the copolar properties of the scatterers, s_{hh} and s_{vv}, as follows (here, it's assumed the cross-polar scattering is negligible, and therefore $s_{hv} = s_{vh} = 0$):

$$\begin{bmatrix} e_h \\ e_v \end{bmatrix} = \begin{bmatrix} s_{hh} & 0 \\ 0 & s_{vv} \end{bmatrix} \begin{bmatrix} m_h \\ m_v \end{bmatrix} = \begin{bmatrix} s_{hh}m_h \\ s_{vv}m_v \end{bmatrix}. \tag{5.73}$$

If the cross-polarization terms are nonzero, there are four terms of the scattering matrix that must be measured to fully characterize the relationship between the dual polarization transmitter and receiver. Only two measurements of the scattering matrix can be made at one time. To measure all four parameters of the full scattering matrix, a way to isolate the four terms is required. The simplest technique is to transmit a single polarization at a time and receive both polarizations. This is accomplished through time multiplexing: by sending one polarization for one PRT followed by the alternate polarization on the next PRT. This sequence then repeats, but ideally, only one polarization is transmitted at a time so that the copolar and cross-polar signals can be easily distinguished for each transmitted signal. This is the alternate transmit and simultaneous receive (ATSR) dual polarization configuration.

In the ATSR sampling mode, the horizontal and vertical polarizations are transmitted on alternate pulses. The received signal vector in a horizontal-only transmission mode, which transmits a horizontally polarized signal and receives its copolar (e_{hh}) and cross-polar (e_{vh}) signal, is represented as

$$\begin{bmatrix} e_{hh} \\ e_{vh} \end{bmatrix} = \begin{bmatrix} s_{hh} & s_{hv} \\ s_{vh} & s_{vv} \end{bmatrix} \begin{bmatrix} m_h \\ 0 \end{bmatrix} = \begin{bmatrix} s_{hh}m_h \\ s_{vh}m_h \end{bmatrix}. \tag{5.74a}$$

Similarly, for the ATSR vertically polarized transmission mode, its vertical copolar (e_{vv}) and cross-polar (e_{hv}) signals are

$$\begin{bmatrix} e_{hv} \\ e_{vv} \end{bmatrix} = \begin{bmatrix} s_{hh} & s_{hv} \\ s_{vh} & s_{vv} \end{bmatrix} \begin{bmatrix} 0 \\ m_v \end{bmatrix} = \begin{bmatrix} s_{hv}m_v \\ s_{vv}m_v \end{bmatrix}. \tag{5.74b}$$

The notation used for the STSR and ATSR received signals are slightly different. The received signals' subscripts follow the convention of the scattering matrix described in Section 4.5.1, where the left subscript is the receiver's polarization, and the right subscript indicates the transmitter's polarization. For the STSR, the received copolar signals are denoted as e_h and e_v, whereas for ATSR, the copolar signals are e_{hh} and e_{vv}, respectively. The different notations are used to identify the operating mode and the received signals' content. This notation is also adopted in Chapter 6 to differentiate the signal-processing methods for the two modes.

The STSR and ATSR results are the same when the cross-polar terms are negligible (i.e., s_{vh} and s_{vh} are ≈ 0). If there are significant contributions from the cross-polarization terms, this may not be the case. For the STSR mode, with the full scattering matrix, the cross-polar scattering terms also contribute to the received signals:

$$\begin{bmatrix} e_h \\ e_v \end{bmatrix} = \begin{bmatrix} s_{hh} & s_{hv} \\ s_{vh} & s_{vv} \end{bmatrix} \begin{bmatrix} m_h \\ m_v \end{bmatrix} = \begin{bmatrix} s_{hh}m_h + s_{hv}m_v \\ s_{vh}m_h + s_{vv}m_v \end{bmatrix}. \tag{5.75}$$

The cross-polarization effect is represented here as only being from the scattering matrix itself, but in practice, the radar system, and in particular the antenna, can be a source of cross-polarization. For accurate estimation of the dual polarization properties of precipitation when operating in STSR mode, the antenna cross-polarization (and the rest of the system) must be well controlled. For STSR radar systems, ideally, at least 30 dB of isolation should exist between polarizations for the radar system to reduce the bias in dual polarization observations [20].

If only one transmitter and one receiver are available to sample the polarimetric signature, as in the alternate transmit and alternate receive (ATAR) sampling mode, four sequences are required to sample the full polarimetric response. In practice, only three samples are needed because the off-diagonals of the scattering matrix are complex conjugates of one another (i.e., $s_{hv} = s_{vh}^*$).

5.6 Modeling the Life of the Dual Polarization Waveforms to the Received Signals

The weather radar is used as a tool to measure the properties of the precipitation scattering matrix, $[S]$. The properties of the scattering matrix are presented as the radar variables, such as reflectivity and differential reflectivity. In the process of measuring the scattering matrix, numerous elements of the radar system and environment interact with the radar's signal. Each of these elements affects the measured dual polarization radar signals and therefore influences the radar's estimates of the precipitation's properties. This section provides a detailed look at these interactions and how they manifest in polarimetric radar measurements.

The polarimetric radar's received signal is modeled as the product of the dual polarization representations of the radar components, the propagation channel, and the scatterer. The signal flow is shown graphically in Figure 5.15 and corresponds to the mathematical model for the received signal:

$$[E] = [R][F_{rx}][T_{rx}][S][T_{tx}][F_{tx}][M]. \tag{5.76}$$

The model's terms are ordered from right to left to correspond to the actual sequence the signal follows: the waveform $[M]$ is transmitted from its antenna $[F_{tx}]$, which propagating through the environment $[T_{tx}]$ to the precipitation scattering volume $[S]$, and then the scattered echo propagates toward the receiver through the environment $[T_{rx}]$ to the receiving antenna $[F_{rx}]$, where the echo is amplified and filter by the

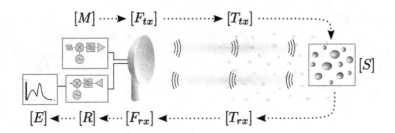

Figure 5.15 Diagram of the radar signal's path, starting with its generation in the radar's transmitter, $[M]$, and ending with the digitized IQ signal, $[E]$, where it is ready for radar signal processing. The diagram shows a monostatic radar configuration where the radar uses a common antenna for transmission and reception (i.e., $[F_{tx}] = [F_{rx}]$). The path and channel effects for the propagating signal are also along a common path for the monostatic configuration (i.e., $[T_{tx}] = [T_{rx}]$). At the scatterers, $[S]$, the signal's path changes from "transmitting" to "receiving" in this conceptual diagram.

radar's receiver electronics $[R]$ giving the measured echo signal $[E]$. The matrix representation is used as a compact way to represent the dual polarization interactions of each modeled component. This section describes each of the model's elements in detail.

The radar measurements of each polarization are represented as the complex-valued signal vector $[E]$:

$$[E] = \begin{bmatrix} e_h \\ e_v \end{bmatrix}. \tag{5.77}$$

The complex-valued samples include the effects that all of the modeled elements have on the dual polarization signal: from the radar system's transmitter, through propagation and scattering, and the radar's receiver. $[E]$ represents the dual polarization signal that is measured by the weather radar and it is the signal used in subsequent processing to extract the properties of the precipitation.

The goal of presenting the dual polarization signal model in detail is to understand the dual polarization interactions experienced by the signal, from the pulse generated in the radar's transmitter until the echo is measured in the receiver. Understanding these interactions is useful for radar system designers, radar operators, and analysts of the radar's data. This model will help elucidate how the quality of the dual polarization data can be influenced by the system, scatterers, and the radar's operating mode – whether it is STSR or ATSR. (Note that data quality is the subject of Chapter 7.) The model also provides a general and concise framework that can be used to simulate weather radar signals and systems. For this discussion, realistic simplifications are quickly adopted to discuss typical polarimetric weather radar systems and observations. With proper selection of model terms, the effect of the antenna's integrated cross-polar ratio (ICPR$_2$; eq. [2.48]) or the cross-polar scatterer terms s_{hv} and s_{vh} from eq. (5.70) can be evaluated.

The elements of eq. (5.76) can vary in range time to model the radar signal as a function of radar range (this would typically only change $[T_{tx}]$, $[T_{rx}]$, and potentially

[S] if the precipitation is spatially varying). Each radar pulse can be modeled for a time-varying scatterer [S] (to generate results similar to Fig. 5.1). It's also assumed that the effects of the transmitter, antennas, and receiver are constant values. Treating the radar system's elements and the propagation channel as constant but potentially unknown quantities is a common assumption for typical weather radar integration times. This is the focus of the discussion throughout the remainder of this section.

5.6.1   The Transmitter and the Antenna

As shown in Figure 5.15 and eq. (5.76), the radar's signal starts in the radar's transmitter subsystem, represented by [M], and is launched into the environment through the transmitting antenna, represented by $[F_{tx}]$.

The ideal weather radar's transmitted waveform is a pulse with rectangular amplitude. The pulse's peak amplitude is the square root of the transmitter's power P_t (recall that a signal's power is proportional to the square of the signal's voltage). The model for an ideal transmitted pulse, centered around $t = 0$ with a pulse duration of T_{tx} (not to be confused with the transmit channel matrix $[T_{tx}]$), is

$$m(t) = \sqrt{P_t}\,\mathrm{rect}\left(\frac{t}{T_{tx}}\right),\tag{5.78}$$

where $\mathrm{rect}(x) = 1$ for $|x| \leq 0.5$ and 0 otherwise. The transmitted pulse is centered around $t = 0$ to simplify the calculation of the range to the center of the observation volume. The ideal rectangular pulse is not truly realized because of the finite bandwidth of the radar system components. If the primary goal of the signal's model is to evaluate performance not directly related to the detailed characteristics of the pulse's shape, the rectangular pulse is a very good approximation.

The transmit waveforms for both polarizations may be the same or different, depending on the radar architecture. The pulses can be represented by an amplitude and phase for each polarization:

$$[M] = \begin{bmatrix} m_h \\ m_v \end{bmatrix} = \begin{bmatrix} \sqrt{P_{th}}\,e^{j\varphi_{txh}} \\ \sqrt{P_{tv}}\,e^{j\varphi_{txv}} \end{bmatrix}.\tag{5.79}$$

The amplitude of each is determined by their respective polarization's transmitter power P_{th} and P_{tv}. The phase for the individual pulses, φ_{txh} and φ_{txv}, can also be modeled, which is useful when evaluating effects such as the system differential phase or the pulse-to-pulse phase differences (e.g., the magnetron's random start phase), which are discussed in Chapter 6.

For STSR-mode radars, the ideal transmitter can be modeled as

$$[M_{\mathrm{STSR}}] = \sqrt{P_t}\begin{bmatrix} 1 \\ e^{j\varphi_{tx}} \end{bmatrix},\tag{5.80}$$

where the relative phase shift between the two polarizations is $\varphi_{tx} = \varphi_{txv} - \varphi_{txh}$ (note this is a contributing factor to ϕ_{sys}). the ATSR mode, the transmitted pulses are modeled as

$$[M_{\text{ATSR}}] = \sqrt{P_t} \begin{bmatrix} 1 & 0 \\ 0 & e^{j\varphi_{tx}} \end{bmatrix}. \tag{5.81}$$

Here, for the ATSR mode, each column is used to represent a different pulse in time. The left column, $[1 \ 0]^T$, transmits the horizontal pulse, and the right column, $[0 \ e^{j\varphi_{tx}}]^T$, transmits the vertical pulse of equal power and a phase shift relative to the horizontal pulse. Additional information on the STSR and ATSR modes can be found in Section 2.2.

The transmitter's waveform becomes a propagating signal via the transmitting antenna, $[F_{tx}]$. For an ideal dual polarization antenna, the horizontal and vertical copolar gains are equal in magnitude and phase, and the cross-polar gain, f_{tcx}, is zero. In practice, the cross-polarization coupling f_{tcx} is nonzero, but it is typically multiple orders of magnitude smaller than the copolar gains (see Section 2.6.3 for discussion of the antenna subsystem). A differential phase shift between the horizontal and vertical copolar signals (and the copolar and cross-polar signals) is typically assumed. These differences in phases between the horizontal and vertical polarizations of the antenna (and transmitter) contribute to the starting differential phase shift, Φ_{dp}, at $r = 0$.

A generalized dual polarization model for the transmitting antenna is

$$[F_{tx}] = \begin{bmatrix} f_{thh} & f_{tcx} \\ f_{tcx}^* & f_{tvv} \end{bmatrix}, \tag{5.82}$$

where each element is complex-valued. The idealized transmitting antenna can be modeled as

$$[F_{tx}] = \sqrt{G_t} \begin{bmatrix} 1 & 0 \\ 0 & e^{j\varphi_{ftdp}} \end{bmatrix}. \tag{5.83}$$

The transmitting antenna's boresight gain, G_t, is also found in the weather radar equation, which assumes a homogeneous volume of scatterers. A differential phase shift, φ_{ftdp}, is included with the vertical copolar signal. Like the transmit waveforms, the horizontal polarization is used as the reference point for the antenna's differential phase.

The antenna's gain and phase are dependent on the azimuth and elevation angles with respect to the antenna's boresight (the boresight generally being the peak gain). When modeling point targets or reflectivity gradients within the antenna's observation volume, the angle-dependent gain and phase to each point or subvolume can be used.

5.6.2   Propagation and Scattering

After the transmitted waveform leaves the antenna, it propagates through the channel (e.g., the clear air, clouds or precipitation) from the radar to the scatterer, the incident propagating waveform scatters, and then the scattered echo propagates back to the radar system. The propagation channel and scattering components of the signal model are not properties of the radar system itself. This fact should be emphasized: the purpose of weather radar is to remotely measure or infer the properties of the channel and scatterers, which includes $[T_{tx}]$, $[S]$, and $[T_{rx}]$ from eq. (5.76). The scattering volume (represented by $[S]$) is the point where the signal's path switches from being part of the "transmitted waveform" to being part of the "received echo."

The scattering characteristics of hydrometeors are discussed in detail in Chapter 4. Duplicated from eqs. (4.50) and (5.70), the scatterer is modeled as

$$[S] = \begin{bmatrix} s_{hh} & s_{hv} \\ s_{vh} & s_{vv} \end{bmatrix}. \tag{5.84}$$

For all precipitation scatterers (and in fact, all scatterers typically encountered) the cross-polarization terms are related as $s_{vh} = s_{hv}$ in the backscatter alignment convention (see Section 4.5.2).

As the radar's pulse propagates in the "channel" from the radar to the scatterer, the environment can act on the signal. The spherical power loss, attenuation from rain, and differential phase shift are some of the factors that can affect the signal during propagation. The general form of the propagation matrix is

$$[T] = \begin{bmatrix} T_{hh} & T_{cx} \\ T_{cx}^* & T_{vv} \end{bmatrix}. \tag{5.85}$$

After extensive research in the 1980s, the forward-propagating cross-polarization term T_{cx} has since been assumed to be negligible for horizontal and vertical polarizations. This is certainly true for propagating in clear air, and it is generally a good assumption for Rayleigh scattering. There are exceptions to this situation, such as propagation through oriented ice particles.

The copolar terms of the propagation matrix (T_{hh} and T_{vv}) include the phase shift due to propagation and the spherical spreading power losses, (as discussed in Section 5.1.7), which leads to an idealized model for the wave's free-space propagation:

$$[T] = \frac{e^{-jk_0 r}}{\sqrt{4\pi r}} \begin{bmatrix} 1 & 0 \\ 0 & 1 \end{bmatrix}, \tag{5.86}$$

which assumes the wavenumber, k, is constant over all ranges. The wavenumber changes slightly with variations of the composition of the atmosphere (e.g., inclusion of ice or rain). The variation of the wavenumber k from the free-space wavenumber, k_0, is small; however, dual polarization radar receivers are able to measure the small difference in the phase shift between the polarizations. The phase shift of the channel $[T]$ is the integrated effect over a one-way path length. (The two-way path is accounted for using both $[T_{rx}]$ and $[T_{tx}]$.) The relationship between the horizontal and vertical polarizations' wavenumbers is included in the weather radar's differential phase shift, Φ_{dp}.

For monostatic radar systems, the effect of propagation on the transmitted signal is the same as the effect of propagation on the received signal (i.e., $[T] = [T_{tx}] = [T_{rx}]$). The common propagation channel, $[T]$, includes the path-integrated effects from the transmitting antenna to the scatterer (or the scatterer to receiving the antenna). The one-way propagation channel is modeled as

$$[T] = \frac{e^{-jk_0 r}}{\sqrt{4\pi}r} \begin{bmatrix} 1 & 0 \\ 0 & e^{j\Phi_{dp}/2} \end{bmatrix}, \tag{5.87}$$

where Φ_{dp} is in radians. Attenuation effects are not included here and are covered in Chapter 9. Attenuation would be implemented as additional scale factors for each polarization.

5.6.3   The Receiver and Antenna

The receiving antenna, $[F_{rx}]$, and the radar receiver, $[R]$, are the remaining terms to complete the description of the dual polarization radar signal model of eq. (5.76). Like the transmitting antenna, the receiving antenna selects the azimuth and elevation angles to "listen" for the pulse's echoes from the scatterer. The echoes' power is collected by the receiving antenna and routed to the radar's receiver, where it can be amplified, filtered, modulated, and ultimately, sampled.

The receiving antenna is modeled like the transmitting antenna, but it includes additional terms for the antenna's power-collection area (see Section 2.5.1). The dual polarization receiving antenna model ($[F_{rx}]$) can be generalized as

$$[F_{rx}] = \frac{\lambda}{\sqrt{4\pi}} \begin{bmatrix} f_{rhh} & f_{rcx} \\ f_{rcx}^* & f_{rvv} \end{bmatrix}. \tag{5.88}$$

The ideal receiving antenna, which has negligible cross-polar coupling and equivalent copolar gain for both polarizations, is

$$[F_{rx}] = \frac{\lambda}{\sqrt{4\pi}} \begin{bmatrix} \sqrt{G_r} & 0 \\ 0 & \sqrt{G_r}e^{j\varphi_{frdp}} \end{bmatrix}. \tag{5.89}$$

The same discussions and treatments of the transmitting antenna apply to the receiving antenna. The ideal receiving antenna is characterized by its boresight power gain, G_r, and its copolar differential phase, φ_{frdp}. Most weather radar systems are monostatic, and the transmitting and receiving antennas are the same; therefore, $\varphi_{fdp} = \varphi_{ftdp} = \varphi_{frdp}$. Following the derivation for the receiving antenna's contribution to the radar equation in Section 2.5.1, the representations of the monostatic radar system's antennas only differ by a scale factor:

$$[F_{rx}] = [F_{tx}]\frac{\lambda}{\sqrt{4\pi}}. \tag{5.90}$$

With this convention for $[F_{rx}]$, eq. (5.76) is directly related to the weather radar equation (2.39d) to calculate the received power.

The radar's receiver can be generalized in a similar manner to the other elements of the dual polarization system. For well-designed radar systems, the receiver has negligible cross-polar terms. A typical dual polarization radar receiver can be modeled as

$$[R] = \begin{bmatrix} \sqrt{G_{rx}} & 0 \\ 0 & \sqrt{G_{rx}}e^{j\varphi_{rdp}} \end{bmatrix}. \tag{5.91}$$

The gain and phase shift of the receiver's filters can be included in $[R]$. Ideally, the gain is the same for both of the copolar signals. A differential phase shift between the two copolar signals may exist in the receiver's hardware (but nominally, this phase shift is zero).

5.6.4 An Ensemble Model for an Ideal Dual Polarization Radar

An ensemble model considers the average response of the dual polarization radar signals using the statistical properties of the radar's hardware, the propagation channel, and the scatterers. For the ensemble model, the weather radar's measurements are assumed to be ergodic, and each of the model's elements is assumed to be statistically stationary. This is an effective method of evaluating the radar's performance and the effects of the model's different elements on the dual polarization measurements of precipitation. Each parameter in the radar signal's ensemble model is either a constant or a wide-sense stationary random variable.

Starting with the signal model of eq. (5.76), the received polarimetric signal vector, $[E]$, is evaluated for a monostatic radar with a common antenna for both transmit and receiver, therefore $G = G_r = G_t$. In this formulation, the only element that includes cross-polarization terms is the scattering matrix, $[S]$. The received signals are calculated with the listed simplifications as follows:

$$\begin{bmatrix} e_h \\ e_v \end{bmatrix} = \frac{\lambda e^{-jk_0 2r}}{\left(\sqrt{4\pi}\right)^3 r^2} \begin{bmatrix} \sqrt{G_{rx}} & 0 \\ 0 & \sqrt{G_{rx}}e^{j\varphi_{rdp}} \end{bmatrix} \begin{bmatrix} \sqrt{G} & 0 \\ 0 & \sqrt{G}e^{j\varphi_{fdp}} \end{bmatrix}$$

$$\begin{bmatrix} 1 & 0 \\ 0 & e^{j\Phi_{dp}/2} \end{bmatrix} \begin{bmatrix} s_{hh} & s_{hv} \\ s_{vh} & s_{vv} \end{bmatrix} \tag{5.92}$$

$$\begin{bmatrix} 1 & 0 \\ 0 & e^{j\Phi_{dp}/2} \end{bmatrix} \begin{bmatrix} \sqrt{G} & 0 \\ 0 & \sqrt{G}e^{j\varphi_{fdp}} \end{bmatrix} \begin{bmatrix} m_h \\ m_v \end{bmatrix}.$$

Consider a monostatic STSR radar's observation volume that is uniformly filled with light rain. Equation (5.92) is used with the following additional simplifications:

- The cross-polarization scattering is neglected (i.e., $s_{hv} = s_{vh} = 0$).
- Both polarizations simultaneously transmit the same power and phase:
  $\sqrt{P_t} = m_h = m_v$.
- The antenna's and receiver's differential phase shifts are zero, $\varphi_{rdp} = \varphi_{fdp} = 0$ (with negligible cross-polar terms, these are easily calibrated out without affecting the measurement).
- The receiver gain is $G_{rx} = 1$.

The simplified signal model for an ideal monostatic STSR radar that is observing light rain is:

$$\begin{bmatrix} e_h \\ e_v \end{bmatrix} = \frac{G\sqrt{P_t}\lambda e^{-jk_0 2r}}{\left(\sqrt{4\pi}\right)^3 r^2} \begin{bmatrix} 1 & 0 \\ 0 & e^{j\Phi_{dp}/2} \end{bmatrix} \begin{bmatrix} s_{hh} & 0 \\ 0 & s_{vv} \end{bmatrix} \begin{bmatrix} 1 & 0 \\ 0 & e^{j\Phi_{dp}/2} \end{bmatrix}. \tag{5.93}$$

After the matrix multiplications, the horizontal polarization's signal is

$$e_h = \frac{G\sqrt{P_t}\lambda e^{-jk_0 2r}}{\left(\sqrt{4\pi}\right)^3 r^2} s_{hh}, \tag{5.94a}$$

and the signal for the vertical polarization is

$$e_v = \frac{G\sqrt{P_t}\lambda e^{-jk_0 2r}}{\left(\sqrt{4\pi}\right)^3 r^2} s_{vv} e^{j\Phi_{dp}}. \tag{5.94b}$$

The ensemble of the signal's power and the correlation of the copolar signals can then be evaluated. (These estimation techniques are covered in detail in Chapter 6.) The power of the horizontal polarization's signal is

$$p_{hh} = \langle e_h^* e_h \rangle = \frac{\lambda^2}{(4\pi)^3 r^4} G^2 \langle |s_{hh}|^2 \rangle P_t = \frac{\lambda^2 G^2 P_t}{(4\pi)^3 r^4} \langle |s_{hh}|^2 \rangle. \tag{5.95a}$$

Similarly, the power for the vertical polarization's signal is

$$p_{vv} = \langle e_v^* e_v \rangle = \frac{\lambda^2}{(4\pi)^3 r^4} G^2 \langle |s_{vv}|^2 \rangle P_t = \frac{\lambda^2 G^2 P_t}{(4\pi)^3 r^4} \langle |s_{vv}|^2 \rangle. \tag{5.95b}$$

The differential reflectivity is the ratio of the horizontal to vertical power:

$$Z_{dr} = \frac{p_{hh}}{p_{vv}} = \frac{\langle |s_{hh}|^2 \rangle}{\langle |s_{vv}|^2 \rangle}. \tag{5.95c}$$

The covariance between the horizontal and vertical polarizations' copolar signals is

$$r_{hv} = \langle e_h^* e_v \rangle = \frac{\lambda^2}{(4\pi)^3 r^4} G^2 \langle s_{hh}^* s_{vv} \rangle e^{j\Phi_{dp}} P_t. \tag{5.96}$$

The power of the covariance is

$$|r_{hv}| = \frac{\lambda^2}{(4\pi)^3 r^4} G^2 |\langle s_{hh}^* s_{vv} \rangle| P_t = \frac{\lambda^2 G^2 P_t}{(4\pi)^3 r^4} |\langle s_{hh}^* s_{vv} \rangle|. \tag{5.97}$$

The observation's differential phase (Ψ_{dp}) is the phase of the cross-covariance between two polarizations:

$$\Psi_{dp} = \angle r_{hv} = \Phi_{dp} + \angle\langle s_{hh}^* s_{vv} \rangle = \Phi_{dp} + \delta_{co}, \tag{5.98}$$

where δ_{co} is the copolar backscatter differential phase shift from eq. (4.59). The copolar correlation between the polarizations' signals is calculated as

$$\rho_{hv} = \frac{|r_{hv}|}{\sqrt{P_{hh}P_{vv}}} = \frac{|\langle s_{hh}^* s_{vv}\rangle|}{\sqrt{\langle |s_{hh}|^2\rangle\langle |s_{vv}|^2\rangle}}. \tag{5.99}$$

This ideal model shows how the measured radar echoes are used to estimate the dual polarization radar variables. The horizontal echo power is proportional to the radar reflectivity via the weather radar constant (compare to eq. [4.53]). The differential reflectivity Z_{dr} and copolar correlation ρ_{hv} are directly calculated from the echo without the weather radar constant (see eqs. [4.54] and [4.60], respectively). To simplify this model, the individual contributions to the system differential phase were set to zero, but would normally include the cascaded effects of the transmitter, antennas, and the receiver. This ϕ_{sys} adds to the observed differential phase ψ_{dp} of eq. (5.98).

The modeled power (eq. [5.95]) shows a form similar to the point-target radar equation if $|s_{hh}|^2$ and $|s_{vv}|^2$ are the total RCS, σ. The weather radar equation is simply an extension of the point-target radar equation (see Section 2.5.2), and they only differ in how the RCS is represented. The model provides a framework to evaluate the effects of the radar system, channel, and scatterers. In particular, the signal power estimated for copolar (and cross-polar) signals will vary when nonzero cross-polar terms are included in the model elements. The effects of the cross-polar terms differentiate the measurements in the ATSR or STSR modes, as indicated in Section 5.5.

5.7 Signal Correlation

The correlation coefficient is a measure of the similarity between two signals or the same signal separated by a time delay. For weather radar, the signal can represent different dimensions: time, space, polarization, or frequency. The size, count, and positions of the hydrometeors within a volume are described by their statistical distributions. Two different observation volumes, or even the same volume at two different times, can be described by the same statistical distribution but with physically different hydrometeors or the same hydrometeors but with relative positions. This section considers how changes in these dimensions affect the correlation of the signals.

The degree to which the signals are correlated is important for understanding how information about the signal can be used to quantify properties of clouds and precipitation. The magnitude of the correlation (e.g., $|\rho|$) can span from zero for completely uncorrelated signals to a magnitude of 1 for linearly dependent signals. A standard threshold of $|\rho| \leq e^{-1} = 0.368$ is used to indicate the separation of uncorrelated samples. Using only second-order statistical characterizations, which is the case for weather radar signals, uncorrelated samples are independent.

The temporal decorrelation considers the relative motions of the particles within the radar volume. The motion of the hydrometeors causes their positions to reconfigure. For longer time periods, particles move into and out of the radar's observation volume, but this is considered an effect of spatial correlation. The temporal decorrelation time is primarily related to the diversity of the hydrometeors' radial velocities with respect to the radar.

The radar observation volume is sampled in space: range, azimuth, and elevation. The azimuth and elevation extents are determined by the antenna. Mechanically scanning radar systems continuously move along one axis during the integration period. As a result of this scanning, the observation volume changes slightly during the observation period. The observation range remains fixed during the integration time. In some instances, the radar may "oversample" in the range, where the radar range is sampled more densely than the range resolution, leading to overlap between successive range gates.

Dual polarization radars sample the polarimetric scattering characteristics of clouds and precipitation. The correlation of the polarimetric signals can provide information about the scatterer's properties. This information is used to infer the size, shape, and state (e.g., ice, water, mixed) of the hydrometeors.

Finally, the radar system transmits and receives a narrow band of RF or microwave frequencies to measure the properties of clouds and precipitation. Often, this is represented as a pulsed sinusoid. The radar's specific frequency is assigned to ensure that no two radars within a given geographic area occupy the same spectrum (to ensure they do not interfere with one another). This is particularly useful for dense radar networks. The observations from nonoverlapping frequency bands are statistically independent [167]. Some radar designs can implement "frequency-diverse" observation schemes to increase the number of independent samples available to improve reflectivity estimates. The echoes from a radar volume filled with hydrometeors are uncorrelated when measured with nonoverlapping frequency bands.

5.7.1   Temporal Correlation

From Section 5.3.2, the mean Doppler velocity of the radar observation volume is estimated using the autocovariance at a time lag of one PRT. If the scattering volume does not change between the two samples use to estimate the velocity, the scatterering volume is perfectly correlated with $|\rho| = 1$, and the measured change in the phase is only due to the scatterer's velocity. In reality, the precipitation within the observation volume are moving relative to each other which results in a reconfiguration of the hydrometeors' relative positions. This means the scattering volume's signature, $[S]$, is changing with time leading to temporal decorrelation of the signal. This also means that the Doppler velocity measurements include both the change in the scatterers' phase and contributions from their motion. The decorrelation time is the time when the correlation for a given time delay falls below $e^{-1} = 0.368$.

Within an observation volume, all of the raindrops are falling toward the ground and are being pushed by the air's motion (i.e., the wind). These raindrops will have minor variations in their motion (due to differences in size, shape, and aerodynamic forces). The variations of the particles' motions lead to changes in their relative positions over time. This reshuffling ultimately results in hydrometeor configurations whose radar echoes are statistically independent given a sufficient long time delay. The temporal correlation, as a function of time lag τ for a Gaussian Doppler velocity spectrum, is

$$\rho[\tau] = \exp\left(\frac{-8\pi^2\sigma_v^2\tau^2}{\lambda^2}\right), \tag{5.100}$$

where σ_v is the Doppler spectrum width. The decorrelation time is

$$\boxed{T_D = \frac{\lambda}{2\pi\sigma_v\sqrt{2}}.} \tag{5.101}$$

The number of independent samples is then approximated by the ratio of the integration time $(N \cdot T_s)$ and the decorrelation time as

$$\boxed{N_I = \frac{NT_s}{T_D},} \tag{5.102}$$

subject to the requirement that $1 \leq N_I \leq N$. (There cannot be more independent samples than actual samples.) To acquire more independent samples, longer dwell times are typically required. Faster sampling rates do not typically provide more independent signal samples when observing precipitation (they can, however, provide more independent samples of noise). From eq. (5.100), smaller wavelengths are more susceptible to variations in position and therefore decorrelate faster. It's also clear that a shorter-wavelength radar will have more independent samples than a longer-wavelength radar for the same integration time.

5.7.2 Spatial Correlation

The spatial correlation between the signals from two volumes, V_1 and V_2, is a function of their overlap. As the second observation volume (V_2) moves with respect the first (V_1), the fraction of volume that are common between the two varies from 1 to 0. The spatial correlation is $|\rho_s| = 1$ when the volumes are the same (i.e., $V_1 = V_2$). When there is no overlap between the volumes, $|\rho_s| = 0$. For a radar that is not scanning, the spatial correlation between sample-time observations at a constant range is $|\rho_s| = 1$. If the radar's range-time samples are equal to the range resolution, then the spatial correlation is $|\rho_s| \approx 0$ for $r(V_1) \neq r(V_2)$. In practice, the pulse's amplitude can extend beyond the half-power range resolution (albeit with small magnitudes), which can result in a degree of correlation between neighboring range volumes.

Consider the spatial decorrelation as a function of separation in range, assuming a rectangular-shaped range resolution, Δr. The concentric azimuth and elevation angular coverage are the same for the two volumes. The beam width is a function of range, where the antenna's cross-sectional area is approximated as $A = b_\theta b_\phi = \pi r^2 \theta_1 \phi_1/4$ (from Section 2.4.4). The spatial correlation between the two observation volumes at ranges r_1 and r_2, with $r_1 < r_2 < r_1 + \Delta r$, is

$$\rho_s(r_1, r_2) = \frac{V_1 \cap V_2}{\sqrt{V_1 V_2}} \approx \frac{r_1^3 + 3r_1^2 \Delta r + 3r_1 \Delta r^2 + \Delta r^3 - r_2^3}{\Delta r \sqrt{(3r_1^2 + 3r_1 \Delta r + \Delta r^2)(3r_2^2 + 3r_2 \Delta r + \Delta r^2)}}. \quad (5.103)$$

For $r_1 > \Delta r$, the spatial correlation is well approximated:

$$\rho_s(|r_2 - r_1|) \approx \begin{cases} 1 - \frac{|r_2 - r_1|}{\Delta r} & |r_2 - r_1| \leq \Delta r \\ 0 & |r_2 - r_1| > \Delta r. \end{cases} \quad (5.104)$$

The spatial correlation between two observation volumes also varies with changes in the azimuth and elevation angles. For scanning radar systems, the rate of change in the azimuth is generally constant, and therefore the sampled volume for each successive pulse is changing. The spatial correlation in azimuth and elevation limits the coherent integration period for scanning radars. The antenna's scanning speed $\dot{\psi}_{\mathrm{ant}}$ determines the antenna's pointing angle change, ψ_{ant}, during the integration time ($N \cdot T_s$) as

$$\psi_{\mathrm{ant}} = \dot{\psi}_{\mathrm{ant}} N \cdot T_s. \quad (5.105)$$

Scanning introduces spatial decorrelation (which is combined with the temporal decorrelation effects) and increases the estimated Doppler spectrum width.

The antenna's half-power beamwidth (θ_{HPBW}) defines the azimuth and elevation extent of the V_6 radar volume. Consider changes in the azimuth and elevation for a circular symmetric antenna pattern about the boresight, which is a typical antenna configuration for weather radar. The antenna's half-power beamwidth in the azimuth and elevation are the same: $\theta_{\mathrm{HPBW}} = \phi_1 = \theta_1$. (See Section 2.4.4.) With a circular antenna beamwidth, the spatial decorrelation due to scanning only depends on the separation angle between the two volumes:

$$\psi_{\mathrm{ant}} = \sqrt{(\theta(2) - \theta(1))^2 + (\phi(2) - \phi(1))^2}, \quad (5.106\mathrm{a})$$

where θ and ϕ are the elevation and azimuth observation angles, respectively. For a monostatic radar, the one-way antenna pattern's gain is used to evaluate the correlation of observations from two different pointing angles. The idealized Gaussian antenna pattern's normalized one-way gain is:

$$G_n(\theta) = \exp\left(-4 \ln 2 \frac{\theta^2}{\theta_{\mathrm{HPBW}}^2}\right), \quad (5.106\mathrm{b})$$

where θ is the angle relative to the antenna's boresight direction (because the Gaussian antenna pattern is circularly symmetric, only the off-boresight angle is required to define the pattern's shape).

The spatial correlation between signals from the circular antenna's observation volumes at two different pointing angles (separated by an angle ψ_{ant}) is

$$\rho_s(\psi_{\mathrm{ant}}) \approx \exp\left(\frac{-1.38 \psi_{\mathrm{ant}}^2}{\theta_{\mathrm{HPBW}}^2}\right). \quad (5.106\mathrm{c})$$

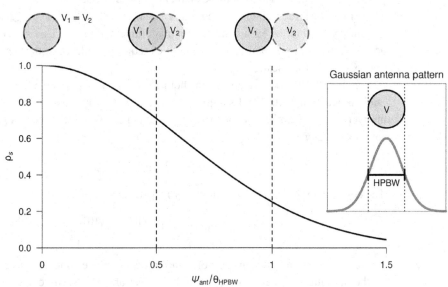

Figure 5.16 Illustration of the spatial decorrelation between observation volumes at different angles from eq. (5.106c). The separation angle between the two observation volumes is ψ_{ant}.

The correlation of signals from two observation volumes at varying angular separation (normalized by the antenna's half-power beamwidth) is illustrated in Figure 5.16. Note that as a result of the overlapping tails of the Gaussian antenna pattern, which extend beyond the HPBW, $\rho_s > 0$ when $\psi_{\text{ant}} = \theta_{\text{HPBW}}$. The spatial correlation outside of the HPBW will vary as a result of the side lobes for realistic antenna patterns. For scanning radar systems, the integration time should be configured such that the observation volume does not spatially decorrelate from the start to the end of the measurement's integration period.

5.7.3 Polarimetric Correlation

With dual polarization radar, the properties of the scattering matrix can be interrogated to infer details of the hydrometeors within the observation volume. The scattering matrix, $[S]$, has four complex-valued terms, of which three of them are unique: s_{hh}, s_{vv}, and $s_{hv} = s_{vh}$. The covariance of the unique scattering matrix elements gives the following:

$$R_{SS} = \begin{bmatrix} s_{hh} \\ s_{vv} \\ s_{hv} \end{bmatrix}^* \begin{bmatrix} s_{hh} & s_{vv} & s_{hv} \end{bmatrix} = \begin{bmatrix} |s_{hh}|^2 & s_{hh}^* s_{vv} & s_{hh}^* s_{hv} \\ s_{vv}^* s_{hh} & |s_{vv}|^2 & s_{vv}^* s_{hv} \\ s_{hv}^* s_{hh} & s_{hv}^* s_{vv} & |s_{hv}|^2 \end{bmatrix}. \tag{5.107}$$

The matrix R_{SS} is complex-conjugate symmetric $\left(\text{i.e., } R_{SS} = R_{SS}^H\right)$, with six unique products to consider. Three of these elements are real-valued: $|s_{hh}|^2$, $|s_{vv}|^2$, and $|s_{hv}|^2$. The remaining three elements are complex-valued: $s_{vv}^* s_{hh}$, $s_{hv}^* s_{hh}$, and $s_{hv}^* s_{vv}$. The three complex-valued terms each contribute a magnitude and phase. This means that there are up to nine unique (real-valued) quantities required to describe the full covariance matrix (R_{SS}) of the scatterer.

For the STSR mode of operation, both the horizontal (h) and vertical (v) polarizations are transmitted and sampled simultaneously. This results in the horizontal and vertical polarizations' copolar signals: e_h and e_v. The STSR mode's polarization sampling vector is

$$\mathbf{P} = \begin{bmatrix} e_h[t] & e_v[t] \end{bmatrix}^{\mathrm{T}}. \tag{5.108}$$

The covariance of the polarization vector is then calculated as

$$R_{\mathrm{STSR}} = \mathbf{P}^* \mathbf{P}^{\mathrm{T}} = \begin{bmatrix} r_{hh}[0] & r_{hv}[0] \\ r_{vh}[0] & r_{vv}[0] \end{bmatrix}, \tag{5.109}$$

where the time lag between the polarization covariance estimates are zero (a result of the simultaneous sampling mode). The covariance matrix is complex-conjugate symmetric, and the cross-covariance terms are related as $r_{vh} = r_{hv}^*$. Three of the R_{STSR} elements are unique (two are real-valued, and one is complex-valued for four quantities). These three covariance estimates enable estimation of four of the six standard dual polarization radar variables: reflectivity, differential reflectivity, differential phase shift, and copolar correlation. (Doppler velocity and Doppler spectrum width are dynamics measurements that characterize changes in the signal using copolar samples from two different times.)

For a dual polarization radar system operating in the ATSR mode, where all of the signals from the two transmitting and two receiving polarizations are observed, there are four signals that must be sampled: e_{hh}, e_{vh}, e_{vv}, and e_{hv}. The left subscript indicates the received polarization, and the right subscript indicates the transmitted polarization. For the ATSR mode, to sample the four polarization components, the radar makes measurements at two instances in time. The ATSR mode's polarization vector is

$$\mathbf{P} = \begin{bmatrix} e_{hh}[t] & e_{vh}[t] & e_{vv}[t + T_s] & e_{hv}[t + T_s] \end{bmatrix}^{\mathrm{T}}, \tag{5.110}$$

where T_s is the radar's PRT. The covariance of the ATSR polarization vector is as follows:

$$R_{\mathrm{ATSR}} = \mathbf{P}^* \mathbf{P}^{\mathrm{T}} = \begin{bmatrix} r_{hh,hh}[0] & r_{hh,vh}[0] & r_{hh,vv}[-T_s] & r_{hh,hv}[-T_s] \\ r_{vh,hh}[0] & r_{vh,vh}[0] & r_{vh,vv}[-T_s] & r_{vh,hv}[-T_s] \\ r_{vv,hh}[T_s] & r_{vv,vh}[T_s] & r_{vv,vv}[0] & r_{vv,hv}[0] \\ r_{hv,hh}[T_s] & r_{hv,vh}[T_s] & r_{hv,vv}[0] & r_{hv,hv}[0] \end{bmatrix}. \tag{5.111}$$

The covariance matrix for the ATSR mode, R_{ATSR}, is also complex-conjugate symmetric. From the reciprocal nature of the scattering matrix, e_{vh} and e_{hv} are also interchangeable. Note that the covariance estimate for the ATSR polarimetric covariance matrix includes a temporal element (i.e., time lags 0 and T_s). For the purpose of determining uniqueness, covariance estimates are only counted once if they exist for both time lags, leaving a total of six unique elements: $r_{hh,hh}$, $r_{vv,vv}$, $r_{vh,vh}$, $r_{vv,hh}$, $r_{vh,hh}$, and $r_{hv,vv}$. Within these, there are a total of nine unique values (magnitudes or phases), which is consistent with the full characterization of the scattering matrix from eq. (5.107).

5.7.4 Frequency Correlation

The coherent integration of a volume is the summation of the echoes of all scatterers within the volume. Assuming the amplitudes are unity for N scatterers, the received signal is

$$e(f) = \sum_{n=0}^{N-1} e^{j4\pi r_n f/c}, \tag{5.112}$$

using the relationship $k = 2\pi f/c$. Following eq. (5.63b), the ensemble averaged covariance of the scattering volume, observed at frequency f, is

$$\langle e^*(f)e(f) \rangle = \left\langle \sum_{n=0}^{N-1} e^{-j4\pi r_n f/c} \sum_{n=0}^{N-1} e^{j4\pi r_n f/c} \right\rangle = N. \tag{5.113}$$

If the radar's frequency is changed, the wavelength changes, and therefore the phase shift between the radar and each scatterer within the volume also changes. The coherent integration of the volume at two different frequencies, f and $f + df$, leads to differences in the signals' phase. Ultimately, the covariance between the two signals is also reduced following:

$$\langle e^*(f)e(f+df) \rangle = \left\langle \sum_{n=0}^{N-1} e^{-j4\pi r_n f/c} \sum_{n=0}^{N-1} e^{j4\pi r_n f/c} e^{j4\pi r_n \, df/c} \right\rangle = N|\rho_f|, \tag{5.114}$$

where ρ_f is the frequency correlation. The term $e^{j4\pi r \, df/c}$ acts as a random phase in the ensemble average. The magnitude of the random phase depends on the size of the frequency shift relative to the bandwidth of the signal (e.g., if $df = 0$, then $e^{j4\pi r \, df/c} = 1$, and there is no impact on the correlation). The frequency correlation of two signals with uniform power spectral density over a bandwidth B, and separated in frequency by df, is approximated as:

$$|\rho_f| \approx \begin{cases} 1 - \frac{|df|}{B} & |df| \le B \\ 0 & |df| > B \end{cases}. \tag{5.115}$$

When the frequency separation is greater than the bandwidth, the two signals do not share any common frequency spectrum and therefore, the two signals are uncorrelated.

5.8   Selected Problems

1. An S-band radar with with a center frequency of $f_c = 2.7$ GHz makes two consecutive measurements of a radar volume: $x(t = 0) = 0.2 + j0.7$ and $x(t = 0.001) = -0.6 + j0.4$, where t are in seconds.
   a) what are the amplitudes and phases of the two measurements?
   b) what is the power of each measurement?
   c) what is the mean velocity measured using the two pulses?

2. An X-band radar ($f_c = 9.4$ GHz) is deployed to complement a network of S-band radars and fill an observation gap in a mountainous region. The S-band ($f_c = 2.7$ GHz) radars operate with a constant pulse-repetition frequency that achieves a unambiguous velocity of $v_a = 25\,\text{m s}^{-1}$.
   a) What is the S-band radar's pulse repetition period?
   b) What is the S-band radar's unambiguous range?
   c) To achieve the same unambiguous velocity, what pulse repetition frequency must the X-band radar use?
   d) What is the X-band radar's unambiguous range?

3. What is the maximum radar center frequency that can provide an unambiguous range of 100 km and an unambiguous Doppler velocity of $\pm 17.5\,\text{m s}^{-1}$? What is the associated uniform pulse repetition period?

4. Precipitation in a storm is moving away from the radar with a velocity of $20\,\text{m s}^{-1}$. What Doppler frequency shifts are observed at S-band (2.7 GHz) and Ka-band (35.6 GHz)?

5. A signal model of an idealized monostatic radar operating using STSR mode and observing a volume with an antenna that has no cross-polar coupling is given in eq. (5.93). Now, consider replacing the idealized antenna with an antenna that has a cross-polar coupling of -20 dB compared to the copolar gain (with zero phase shift).
   a) Except for the antenna's cross-polar gain, use the same simplifications as eq. (5.93) to calculate e_h and e_v.
   b) Calculate the horizontal and vertical copolar powers, p_{hh} and p_{vv}, and Z_{dr}.
   c) What is the ratio of the estimated Z_{dr} with the non-ideal antenna to the scatterer's intrinsic Z_{dr} (eq. [5.95c]) in decibels?
   d) If ATSR mode is used instead of the STSR mode, using $[M]$ from eq. (5.81) with $\varphi_{tx} = 0$, calculate $p_{hhhh} = e_{hh}^* e_{hh}$, the horizontal copolar power for a horizontal transmitted pulse and received echo, as well as the vertical polarization's copolar power $p_{vvvv} = e_{vv}^* e_{vv}$.
   e) Using the ATSR mode, estimate Z_{dr} and compare it to the intrinsic Z_{dr}.

6. An S-band radar ($f_c = 2.7$ GHz) observes a homogeneous precipitation cell with a spectrum width of $\sigma_v = 2\,\text{m s}^{-1}$ using an integration time of 30 ms with a PRF of 1 kHz.
   a) What is the unambiguous velocity for the radar?
   b) What is the decorrelation time between consecutive samples?
   c) How many independent samples are observed during the integration period?

c) If the radar was instead X-band ($f_c = 9.4$ GHz) what is the decorrelation time between consecutive samples?

7. A Ku-band radar ($f_c = 13.6$ GHz) with a half-power beamwidth of 1 degree is scanning at 18 revolutions per minute (RPM). For a measured spectrum width of $\sigma_v = 2$ m s^{-1} with a PRF of 1 kHz,

a) what is the pulse-to-pulse correlation due to scanning at 18 RPM?

b) What is the intrinsic spectrum width of the precipitation volume assuming the contribution from scanning and the scatterers are independent?

6 Weather Radar Signal Processing

The goal of a meteorological radar is to measure the properties of precipitation (ideally at high-range resolution and over a significant distance). Originally, these measurements focused on measuring the echoes' power, which is proportional to the size and number of particles in the resolution volume. Coherent processing techniques were then added to radar systems to measure the mean Doppler velocity. With the introduction of digital signal processing, advanced methods for processing the full Doppler spectrum have become possible. Signal-processing techniques have maintained a very basic duality: any computation done in the time domain can also be done in the frequency domain. The computation method that is ultimately selected is based on the ease of implementation and the desired sophistication. Signal processing is a vast field, and the discussion in this chapter is limited to weather radar signal processing as applied to dual polarization radars. Even this subfield has seen great advances in the last two decades, especially since the introduction of solid-state transmitters. This chapter is devoted to discussing the fundamental principles of dual polarized radar signal processing, with a transition to advanced techniques.

Digital signal processing has revolutionized the modern radar and communication fields, and weather radar is no exception. Weather radars have always placed high demand on, and benefited from, the rapid development of digital signal-processing technology. What is practiced in weather radar signal processing has also changed over time. For example, in the early 1980s, weather radars used wire-wrapped boards for signal processing, whereas very quickly, modern digital processing entered the field in the early 1990s. The use of bulky signal processors was the main reason the developers of the 1980s pursued single-receiver, alternate-sampling-mode architectures for dual polarization measurements. However, when lower-cost digital receivers and signal processors became available, it was cheaper to pursue simultaneous-sampling architectures, which are used in most operational dual polarization implementations today. The signal-processing techniques available for implementation with weather radar continue to advance with improving technology. Although the sophistication of the signal-processing algorithms may change, the fundamental signal-processing principles remain the same.

In Chapter 5, a dual-polarization model for the radar's measure signals was developed, which included the radar system, propagation through the environment, and scattering from the observation volume. This process is repeated at the radar's pulse-repetition frequency (PRF) to generate a time series of samples of the echo's signal. The radar system collects coordinated time series of measurements from the horizontal and vertical polarizations, and signal-processing techniques are used to estimate the statistical properties of these received signals. The measurements are directly related to the scattering properties of the precipitation. Dual polarization radar systems typically provide six "standard" radar variables that are derived from linear dual polarization signals: reflectivity, differential reflectivity, differential phase, copolar correlation, velocity and spectrum width. Only three of these standard measurements rely on dual polarization: differential reflectivity, differential phase, and copolar correlation. Multiple variations of estimators for these radar variables exist, but all of these fall into two categories: covariance (time-domain) estimators and spectral (frequency-domain) estimators. The most common technique to estimate these variables is the pulse-pair processor (a covariance estimator), which relies on the zero- and one-lag covariance estimates of the received signals. With a pair of pulses, both the zero- and one-lag covariances can be estimated.

To improve the sensitivity, dynamic range, and/or accuracy of the radar's measurements, more advanced sampling techniques that are commonly found in modern weather radar systems are also introduced. These materials are well developed in the field of radar signal processing and are provided here as a self-contained discussion for weather radar remote sensing. Multiple PRFs are used to extend the unambiguous velocity of the radar system while maintaining a desired unambiguous range. Phase codes are discussed as a technique that may be implemented for short-range radars to detect "second-trip" echoes (echoes from beyond the radar's unambiguous range). With the increasing availability of solid-state transmitters, pulse-compression techniques are introduced as a method to achieve sensitivity levels comparable with those of other high-voltage transmitters.

6.1   Covariance Estimators

The covariance can be estimated between all available polarimetric signals' samples. The convention followed here is as follows: if two subscripts are used, STSR-mode sampling is assumed. Recall that the left subscript is the received polarization, and the right subscript is the transmitted polarization. If two comma-separated pairs are used for the subscript, this denotes the covariance of different polarimetric scattering matrix elements, and the ATSR-mode measurements follow this convention.

In practice, the covariance estimates are made with a finite number of samples. These are only estimates of the actual intrinsic properties of the scatterers. In Chapter 7, the effects of finite numbers of samples on the accuracy of these estimators will be considered in more detail.

Two polarization states are sampled over time as sequences $x(t)$ and $y(t)$. At a fixed radar range r, the signals are sampled every pulse-repetition period, T_s. Recall from Section 5.7 that the covariance estimates assume that the process is statistically wide-sense stationary (WSS). For signals that are WSS, the covariance only depends on the time lag, $\tau = l \cdot T_s$. With this and eq. (5.50), the covariance of the two signals is calculated as

$$\widehat{r}_{xy}[l] = \begin{cases} \dfrac{1}{N} \displaystyle\sum_{n=0}^{N-|l|-1} x^*[n+|l|]y[n]; & -(N-1) \leq l < 0 \\[4mm] \dfrac{1}{N} \displaystyle\sum_{n=0}^{N-l-1} x^*[n]y[n+l]; & 0 \leq l \leq N-1 \end{cases} \quad . \tag{6.1}$$

The covariance of WSS signals as $N \to \infty$ are conjugate symmetric, resulting in $r_{xy}[l] = r_{xy}^*[-l]$.

Figure 6.1 illustrates the polarization and temporal sampling schemes for weather radars using simultaneous transmit and simultaneous receive (STSR), alternate transmit and simultaneously receive (ATSR), and a copolar-only alternate transmit and alternate receive (ATAR) mode. The diagram illustrates the transmitted and received signals as a function of the sample-time index (the sample time increases from the bottom left to top right along the diagonal time axis). These sampling strategies (both in polarization and time) are followed to measure the signals that are used to estimate the radar variables.

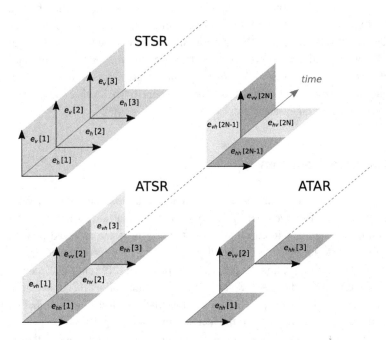

Figure 6.1 Schematic illustration of polarization modes STSR, ATSR, and one possible implementation of ATAR that only observes copolar signals. The black vertical and horizontal arrows indicate the time of the transmitted polarization. The shaded areas indicate the windows over which polarizations are received (dark gray indicates that the copolar signal is received, whereas light gray indicates the received cross-polar signals.) In STSR mode, the received signals are dominated by the copolar echoes.

In STSR mode, two measurements for the copolar signals are available: e_h and e_v. From Section 5.7.3, these two STSR measurements provide three unique covariances estimates: r_{hh}, r_{vv}, and $r_{hv} = r_{vh}^*$. At a time lag of zero, the phases of r_{hh} and r_{vv} equal zero (by definition of $r_{xx}[0]$), which yields a maximum of four unique measurement quantities (three powers and a phase). These four quantities enable us to infer the reflectivity, differential reflectivity, copolar correlation, and differential phase. These four quantities from STSR do not provide any information about the motion (i.e., mean velocity and spectrum width) of the precipitation. By using the autocovariance of one of the polarizations at one time lag (typically $r_{hh}[T_s]$ is used for the best accuracy, although additional time lags can be used), two additional quantities (a power and phase) are available to estimate the mean Doppler velocity and spectrum width.

For the ATSR mode, four measurements of the polarimetric signals are available: e_{hh}, e_{vh}, e_{vv}, and e_{hv}. These four measurements yield six unique covariance estimates. The copolar signal properties are the covariance estimates: $r_{hh,hh}$, $r_{vv,vv}$, and $r_{hh,vv}$ (for comparison, their STSR-mode counterparts are r_{hh}, r_{vv}, and r_{hv}, respectively). ATSR mode yields three additional covariance estimates using the cross-polar measurements: $r_{vh,vh}$, $r_{hv,hv}$, and $r_{vh,hv} = r_{hv,vh}^*$.

All covariance estimators typically include provisions for noise correction where relevant because this is a useful step in radar signal processing (see Section 5.4.5 for a discussion of noise in radar observations). Although subtracting noise power from the observation generally yields more accurate results, additional checks of the calculated values are required to ensure stable and accurate estimates of the radar variables. An example of such checks is that if the powers are negative (which is not physically realizable), then the radar variables are not calculated. Another example is estimates that have noise-corrected powers of zero in the denominator, which also present problems.

6.1.1 STSR Pulse-Pair Estimators

Weather radar signal processing uses the radar observations as estimates of the scattering-matrix elements, **S**. In STSR operation, the two polarization signals, e_h and e_v, are sampled simultaneously for each pulse (see Section 5.5). The sampled signals are measurements of the scattering-matrix elements s_{hh} and s_{vv}, respectively. The $r_{hh}[0]$, $r_{hh}[T_s]$, $r_{vv}[0]$, and $r_{hv}[0]$ covariance estimates are all that is required to calculate the six STSR dual polarization radar variables.

The scatterer's and noise's contribution to the idealized STSR weather radar's horizontal copolar signal is

$$e_h \propto s_{hh} + w_h, \qquad (6.2a)$$

and the idealized STSR vertical copolar signal is

$$e_v \propto s_{vv} + w_v. \qquad (6.2b)$$

Additive noise for the horizontal and vertical polarizations are included as w_h and w_v, respectively. A model of the radar's signal was developed and discussed in Section 5.6, including the radar hardware and propagation effects. Here, the radar and propagation elements are omitted to focus on the scatterer's properties (and the impact of additive noise).

With these ideal STSR signals, the three STSR covariance estimates at a time lag of zero are as follows:

$$r_{hh}[0] = \langle e_h^*[t]e_h[t] \rangle \propto |s_{hh}|^2 + |w_h|^2 \tag{6.3a}$$

$$r_{vv}[0] = \langle e_v^*[t]e_v[t] \rangle \propto |s_{vv}|^2 + |w_v|^2 \tag{6.3b}$$

$$r_{hv}[0] = \langle e_h^*[t]e_v[t] \rangle \propto s_{hh}^*s_{vv}. \tag{6.3c}$$

The effects of the transmitter, receiver, antenna, and propagation path are commonly compensated for through calibration, and they have been omitted here to highlight the relationship between the observation and the scattering volume's properties. In addition to the three lag-zero covariance estimates, the covariance is estimated between pairs of pulses that are sampled at a time of T_s apart. The "pulse-pair" estimators measure the dynamics of the observation volume to give the mean Doppler velocity and Doppler spectrum width. The STSR-mode lag-one covariance estimate is:

$$r_{hh}[T_s] = \langle e_h^*[t]e_h[t + T_s] \rangle \propto \rho[T_s]|s_{hh}|^2, \tag{6.4}$$

where $\rho[T_s]$ is the temporal decorrelation. Note that like $r_{hv}[0]$, the additive noise is uncorrelated between samples in $r_{hh}[T_s]$, and therefore the additive noise contribution tends toward zero.

Finally, for a sequence of N STSR samples, e_h and e_v, the covariance estimates that are required to calculate the radar variables are as follows:

$$r_{hh}[0] \approx \widehat{r}_{hh}[0] = \frac{1}{N} \sum_{n=0}^{N-1} e_h^*[n]e_h[n] \tag{6.5a}$$

$$r_{vv}[0] \approx \widehat{r}_{vv}[0] = \frac{1}{N} \sum_{n=0}^{N-1} e_v^*[n]e_v[n] \tag{6.5b}$$

$$r_{hv}[0] \approx \widehat{r}_{hv}[0] = \frac{1}{N} \sum_{n=0}^{N-1} e_h^*[n]e_v[n] \tag{6.5c}$$

$$r_{hh}[\tau = T_s] \approx \widehat{r}_{hh}[l = 1] = \frac{1}{N} \sum_{n=0}^{N-2} e_h^*[n]e_h[n + 1]. \tag{6.5d}$$

Power, SNR, and Reflectivity

Signal power and noise power are important for accurately estimating the magnitude-based radar moments. The power of the received signal, with noise correction, is

$$p_x = r_{xx}[0] - n_x, \tag{6.6}$$

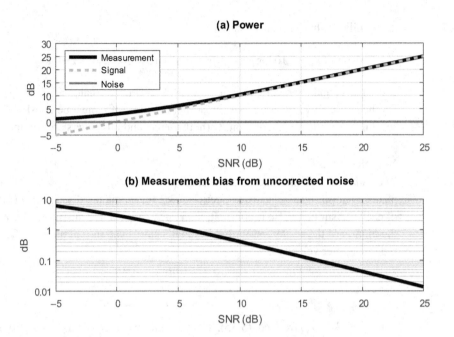

Figure 6.2 The measurement, signal, and noise power for a range of SNR are shown in panel (a). In panel (b), the bias of the measurement compared with the signal power without noise correction (the error is the ratio of measured power to signal power). As the SNR increases, the measurement and signal power are approximately the same. For regions of "high SNR," the biasing effects of additive noise can typically be neglected, and the measurement is the signal. For "low SNR," the measurement and signal power can differ significantly, and noise correction (i.e., noise subtraction) is necessary.

where the noise power is $n_x = |w_x|^2$, and in these equations, x may be either h or v for the appropriate polarization. The signal-to-noise ratio (SNR) is simply

$$\text{SNR}_x = \frac{p_x}{n_x}. \tag{6.7}$$

Figure 6.2 illustrates the measured power's bias due to uncorrected additive noise. As the SNR increases, the additive noise has less contribution, and the measured power becomes asymptotic to the signal's power. With sufficiently high SNR, the additive noise can be neglected altogether. As shown in Figure 6.2, when the SNR is greater than 15 dB (without noise correction), the measured power's bias due to additive noise is approximately 0.1 dB. The threshold for defining a "sufficiently high SNR" depends on the measurement accuracy requirements. Chapter 7 discusses estimating the noise power and the effect of additive noise on the radar variables.

For a dual polarization radar, the horizontal reflectivity in dBZ is

$$10 \log_{10}(Z_h) = 10 \log_{10}(r_{hh}[0] - n_h) - 10 \log_{10}(C_h) + 20 \log_{10}(r), \tag{6.8a}$$

and similarly, the vertical reflectivity is

$$10\log_{10}(Z_v) = 10\log_{10}(r_{vv}[0] - n_v) - 10\log_{10}(C_v) + 20\log_{10}(r), \qquad (6.8b)$$

for range, r, and the radar constants, C_h and C_v, which are calculated from eq. (2.39d) for their respective polarizations. Note that the units of r must be consistent with the formulation of the radar constant, C (both meters and kilometers are commonly used).

Velocity

The radar system measures the mean Doppler velocity as a change in the signal's phase from pulse to pulse. (See Section 5.3.2 for a detailed discussion.) If the phase of each transmitted pulse is constant, then any change in the measured phase is due to the motion of the scatterers. With the signal's autocovariance at time lag T_s, the mean radial Doppler velocity is

$$\bar{v} = \frac{-v_a}{\pi} \angle (r_{hh}[T_s]), \qquad (6.9)$$

with $v_a = \lambda/(4T_s)$ from eq. (5.36). The convention is that a positive velocity indicates that the scatterer is moving away from the radar. Although the majority of weather radars adopt this convention, it is not a universal standard (e.g., some vertical profiling radars use positive velocity to indicate scatterers falling toward the radar).

The mean Doppler velocity estimate is reported within the radar's unambiguous velocity range: $-v_a \leq \bar{v} < v_a$. For scatterers whose true radial Doppler velocities are outside of this range, the measured mean velocity will alias into this range. The true and aliased velocities are related by adding or subtracting integer multiples of $2v_a$ to $\bar{v}$. If more information about the target's true velocity is known, the unambiguous velocity spectrum's range can be "reassigned" to be, for instance, $[0, 2v_a)$ or $(-2v_a, 0]$, but only ever spanning $2v_a$. Section 6.4.1 discusses techniques to extend the unambiguous range for estimating the mean velocity.

Spectrum Width

The STSR Doppler spectrum width estimate assumes a Gaussian-shaped Doppler spectrum (and therefore a Gaussian-shaped autocorrelation function). The Doppler spectrum width, σ_v, is also denoted using w as the variable name. With the Gaussian model, the radar signal's autocorrelation is related to the signal's spectrum width (eq. [5.100]) as

$$\rho[\tau] = \exp\left(\frac{-8\pi^2\sigma_v^2\tau^2}{\lambda^2}\right). \qquad (6.10)$$

The radar measurements are used to estimate the autocovariance at time lags $\tau = T_s$ and $\tau = 0$, which then calculate the signal's autocorrelation (from eq. [5.52]):

$$\rho_{xx}[\tau] = r_{xx}[\tau]/r_{xx}[0]. \qquad (6.11)$$

The STSR spectrum width estimator is

$$\sigma_v = \frac{v_a\sqrt{2}}{\pi}\left(\ln\left|\frac{r_{hh}[0]-n_h}{r_{hh}[T_s]}\right|\right)^{\frac{1}{2}}. \tag{6.12}$$

For the spectrum width estimate in (6.12), the lag-zero copolar covariance estimate $r_{hh}[0]$ requires noise correction for low and moderate SNRs, or else the spectrum-width estimate is biased (overestimated). The cross-covariance between the horizontal and vertical polarization signals at a zero time lag (i.e., $r_{hv}[0]$) is inherently noise-free (the noise in the horizontal and vertical polarizations are uncorrelated). For precipitation, where the copolar correlation is high, an alternative formulation of the spectrum-width estimator may be used:

$$\sigma_v = \frac{v_a\sqrt{2}}{\pi}\left(\ln\left|\frac{r_{hv}[0]}{r_{hv}[T_s]}\right|\right)^{\frac{1}{2}}. \tag{6.13}$$

Differential Reflectivity

The differential reflectivity is simply the ratio of the two polarization signals' power:

$$Z_{dr} = \frac{Z_h}{Z_v}. \tag{6.14}$$

The radar constants for the horizontal and vertical polarizations are typically similar for weather radar systems but not identical. Given the range of differential reflectivities, and its sensitivity to changes in the hydrometeors' properties, relative calibration of the radar constant between the horizontal and vertical polarization is critical to providing accurate results.

The differential reflectivity is estimated from the measurements of both polarizations as

$$Z_{dr} = \frac{(r_{hh}[0]-n_h)}{(r_{vv}[0]-n_v)}\frac{C_v}{C_h}. \tag{6.15}$$

The ratio of the radar constants is typically fixed, and therefore C_v/C_h is treated as a single calibration constant that must be estimated and corrected to achieve the desired differential reflectivity accuracy. Calibration of Z_{dr} is discussed in more detail in Chapter 7.

Differential Phase

The propagating differential phase shift, Φ_{dp}, is the path-integrated phase difference between the horizontal and vertical polarizations due to the radar waves' propagation between the radar and scattering volume. For observations in precipitation, slight differences between the horizontal and vertical polarizations' propagation constants are measured as a difference in phase between two channels. As the path length increases, the accumulated phase difference increases, and therefore, Φ_{dp} is a range-integrated quantity.

The measured differential phase shift between the polarizations is estimated from the cross-covariance between the signals from the two channels as

$$\Psi_{dp} = \frac{180}{\pi} \left[\angle (r_{hv}[0]) - \phi_{sys} \right] \tag{6.16}$$

where the radar system's differential phase, ϕ_{sys}, is ψ_{dp} at $r = 0$.

The differential backscattering phase shift, δ_{co} is included in the observation of $r_{hv}[0]$. (See Section 4.7.2 for discussion of δ_{co}.) For precipitation in the Rayleigh scattering regime, δ_{co} can be neglected giving $\Phi_{dp} \approx \Psi_{dp}$. For observations in rain, a positive increase in the differential phase with range is observed. The rate of change of Φ_{dp} with range (i.e., the estimate of K_{dp}) is used to quantify precipitation; therefore, accurate estimation and subtraction of ϕ_{sys} are typically for display purposes (this constant disappears in the K_{dp} estimation range derivative).

Copolar Correlation Coefficient
The copolar correlation coefficient, ρ_{hv}, is the magnitude of the correlation between the signals from the two polarization, with $0 \leq \rho_{hv} \leq 1$. It is the cross-correlation of the horizontal and vertical copolar observations and is sometimes denoted as $|\rho_{co}|$. The copolar correlation is calculated as

$$\rho_{hv} = \frac{|r_{hv}[0]|}{\sqrt{(r_{hh}[0] - n_h)(r_{vv}[0] - n_v)}}. \tag{6.17}$$

Noise correction must be applied with care because this may result in the estimate of ρ_{hv} being greater than unity as a result of errors in the estimator. Theoretically, a correlation coefficient has a magnitude from 0 to 1. When using noise correction, estimates of $\rho_{hv} > 1$ occur for low-SNR measurements with errors in the estimated noise power. In practice, these erroneous estimates are rounded to $\rho_{hv} = 1$ (or a higher value, such as 1.01, to indicate the estimators' uncertainty to the user).

6.1.2   ATSR Pulse-Pair Estimators

In ATSR operation, the polarimetric radar measures four signals (compared with only two in STSR mode). A uniform sampling interval T_s is assumed for the discussion of the ATSR radar variable estimates and signal processing. Using the ATSR sampling mode, the dual polarization variable estimators must be adjusted to account for the time delay between the copolar samples of the two polarizations. For ATSR, the cross-polarization terms from the scattering matrix **S** are included following the discussion in Section 5.5 (in particular, see eq. [5.74]).

The horizontal and vertical polarizations are alternately transmitted, as shown in Figure 6.1. Each pulse is separated by the pulse repetition period, T_s. The successive horizontal pulses are therefore separated by $2T_s$ (the same is also true for the vertical polarization's pulses).

Similar to the STSR signals discussed in Section 6.1.1, the effects of the radar hardware and propagation path are omitted here to focus on the scatterer properties and the effect of additive noise. Refer to Section 5.6 to consider the effects of the hardware and propagation. The ideal ATSR horizontal polarization's copolar and cross-polar measurements are as follows:

$$e_{hh}[t] \propto s_{hh} + w_h \tag{6.18a}$$

$$e_{vh}[t] \propto s_{vh} + w_v. \tag{6.18b}$$

The vertical polarization's copolar and cross-polar ATSR signals are as follows:

$$e_{vv}[t + T_s] \propto s_{vv} + w_v \tag{6.18c}$$

$$e_{hv}[t + T_s] \propto s_{hv} + w_h. \tag{6.18d}$$

The following ATSR covariance estimates are used to estimate the six standard dual polarization variables: $r_{hh,hh}[0]$, $r_{vv,vv}[0]$, $r_{hh,hh}[2T_s]$, $r_{vv,vv}[2T_s]$, $r_{hh,vv}[T_s]$, and $r_{vv,hh}[T_s]$. The cross-polarization radar variables are estimated using the additional covariance estimates: $r_{hv,hv}[0]$, $r_{vh,vh}[0]$, and $r_{vh,hh}[0]$. The ATSR-mode covariance estimates used to estimate the dual polarization radar variables are:

$$r_{hh,hh}[0] = \langle e_{hh}^*[t]e_{hh}[t] \rangle \propto |s_{hh}|^2 + |w_h|^2, \tag{6.19a}$$

$$r_{vv,vv}[0] = \langle e_{vv}^*[t]e_{vv}[t] \rangle \propto |s_{vv}|^2 + |w_v|^2, \tag{6.19b}$$

$$r_{vh,vh}[0] = \langle e_{vh}^*[t]e_{vh}[t] \rangle \propto |s_{vh}|^2 + |w_v|^2, \tag{6.19c}$$

$$r_{hv,hv}[0] = \langle e_{hv}^*[t]e_{hv}[t] \rangle \propto |s_{hv}|^2 + |w_h|^2, \tag{6.19d}$$

$$r_{hh,vv}[T_s] = \langle e_{hh}^*[t]e_{vv}[t + T_s] \rangle \propto \rho[T_s]s_{hh}^*s_{vv}, \tag{6.19e}$$

$$r_{vv,hh}[T_s] = \langle e_{vv}^*[t]e_{hh}[t + T_s] \rangle = r_{hh,vv}^*[T_s], \tag{6.19f}$$

$$r_{vh,hh}[0] = \langle e_{vh}^*[t]e_{hh}[t] \rangle \propto s_{vh}^*s_{hh}, \tag{6.19g}$$

$$r_{hh,hh}[2T_s] = \langle e_{hh}^*[t]e_{hh}[t + 2T_s] \rangle \propto \rho[2T_s]|s_{hh}|^2, \tag{6.19h}$$

$$r_{vv,vv}[2T_s] = \langle e_{vv}^*[t]e_{vv}[t + 2T_s] \rangle \propto \rho[2T_s]|s_{vv}|^2. \tag{6.19i}$$

The ATSR-mode horizontal and vertical channel noise powers are $n_{hh} = |w_h|^2$ and $n_{vv} = |w_v|^2$, respectively.

Reflectivity and Differential Reflectivity

The estimates of power, reflectivity, and SNR follow the same pattern as the STSR estimators described in Section 6.1.1. For the ATSR sampling mode, the signal subscripts are modified to indicate the received and transmitted polarizations, as the first and second subscripts, respectively. The covariance and power estimators for the STSR and ATSR modes also use different subscripts to differentiate their associated operating modes. The horizontal copolar signal's power is

$$p_{hh} = r_{hh,hh}[0] - n_{hh}, \tag{6.20a}$$

and the horizontal cross-polar signal's power (transmitting the horizontal polarization and receiving the vertical polarization) is

$$p_{vh} = r_{vh,vh}[0] - n_{vv}. \tag{6.20b}$$

Note that the vertical channel's noise power is subtracted in p_{vh}. This assumes that the radar's configuration for measuring the vertical cross-polar and copolar signals are identical (if not, a representative noise estimate should be used for noise correction). The vertical copolar signal's power is

$$p_{vv} = r_{vv,vv}[0] - n_{vv}, \tag{6.20c}$$

and the vertical cross-polar signal's power is

$$p_{hv} = r_{hv,hv}[0] - n_{hh}. \tag{6.20d}$$

The horizontal copolar reflectivity factor, in dBZ, is then estimated following

$$10\log_{10}(Z_{hh}) = 10\log_{10}\left(r_{hh,hh}[0] - n_{hh}\right) - 10\log_{10}(C_{hh}) + 20\log_{10}(r), \tag{6.21a}$$

where the noise power, range correction, and radar constant for the ATSR mode are applied. The vertical reflectivity factor follows as

$$10\log_{10}(Z_{vv}) = 10\log_{10}\left(r_{vv,vv}[0] - n_{vv}\right) - 10\log_{10}(C_{vv}) + 20\log_{10}(r). \tag{6.21b}$$

Note that the radar constants for the STSR and ATSR modes (e.g., for the horizontal polarization, these are C_h and C_{hh}, respectively) may be the same or may differ slightly, depending on how the radar's hardware implements the operating mode (consider Section 2.2).

The estimation of differential reflectivity is the same as for STSR mode; it is the ratio of the horizontal and vertical reflectivity:

$$Z_{dr} = \frac{Z_{hh}}{Z_{vv}} = \frac{\left(r_{hh,hh}[0] - n_{hh}\right)}{\left(r_{vv,vv}[0] - n_{vv}\right)}\frac{C_{vv}}{C_{hh}}. \tag{6.22}$$

Again, the ratio of the radar constants is typically replaced by a calibration constant specifically for differential reflectivity (see Section 7.4.4).

Spectrum Width

Like the STSR mode, the spectrum width is calculated from covariance estimates at two time lags. For the ATSR mode, the autocovariance at time lags 0 and $2T_s$ are commonly used:

$$\sigma_v = \frac{\lambda}{2\pi 2T_s\sqrt{2}}\left(\ln\left|\frac{r_{hh,hh}[0] - n_{hh}}{r_{hh,hh}[2T_s]}\right|\right)^{\frac{1}{2}}. \tag{6.23}$$

The time between the copolar samples of the same polarization is $2T_s$, and therefore, the unambiguous velocity range, v_a, in the ATSR mode is effectively reduced by a factor of 2 compared to the same pulse repetition time in STSR mode.

Copolar Correlation Coefficient

For the STSR mode, estimating the copolar correlation coefficient is a simple and direct calculation because the two copolar signals are sampled simultaneously (eq. [6.17]). For the ATSR mode, the horizontal and vertical copolar signals (i.e., e_{hh} and e_{vv}) are sampled at different times, separated by T_s. The correlation of these ATSR copolar signals therefore includes the effects of both the copolar correlation coefficient (ρ_{hv}) and temporal correlation:

$$|\rho_{co}[T_s]| = \frac{|r_{hh,vv}[T_s]|}{\sqrt{r_{hh,hh}[0]r_{vv,vv}[0]}} = \rho_{hv}|\rho[T_s]|. \tag{6.24}$$

The temporal and copolar correlation effects are independent of one another and therefore are separable. Recall that the temporal correlation, $|\rho[T_s]|$, is directly related to the observation volume's spectrum width (Section 5.7.1).

In the ATSR mode, the temporal correlation effect must be estimated, and then corrected for, to isolate the copolar correlation coefficient's contribution:

$$\rho_{hv} = \frac{|\rho_{co}[T_s]|}{|\rho[T_s]|}. \tag{6.25}$$

The numerator, $|\rho_{co}[T_s]|$, is calculated from the ATSR copolar covariance estimates in eq. (6.24), which includes the copolar correlation as well as temporal correlation. The temporal correlation ($|\rho[T_s]|$) must be estimated seperately to cancel its effect and recover only the copolar correlation's contribution from the estimator. Temporal correlation is estimated using the autocovariance estimates at time lags 0 and $2T_s$ because the autocovariance at time lag T_s is not directly estimated in ATSR mode and therefore must be calculated indirectly.

Using the Gaussian temporal correlation model from eq. (5.100), the correlations at two different time lags (T_1 and T_2) are related following

$$|\rho[T_2]| = \left(|\rho[T_1]|\right)^{(T_2/T_1)^2}. \tag{6.26}$$

The available ATSR-mode autocovariance estimates at time lags 0 and $2T_s$ are used to estimate the time lag T_s correlation (shown here for the horizontal polarization measurements):

$$|\rho[T_s]| \approx \left(|\rho[2T_s]|\right)^{(T_s/(2T_s))^2} = \left(\frac{|r_{hh,hh}[2T_s]|}{r_{hh,hh}[0]}\right)^{1/4}. \tag{6.27}$$

Note that the horizontal and vertical polarizations' temporal correlation effects are similar as a result of the same scatterers being observed by both polarizations. Combining eqs. (6.24, 6.25, 6.27), the copolar correlation estimate in ATSR mode, with noise correction omitted for clarity, is calculated as

$$\rho_{hv} = \frac{|r_{hh,vv}[T_s]|}{\sqrt{r_{hh,hh}[0]r_{vv,vv}[0]}} \left(\frac{|r_{hh,hh}[2T_s]|}{|r_{hh,hh}[0]|}\right)^{-1/4}. \tag{6.28}$$

Velocity and Differential Phase

The differential phase is the phase of the horizontal and vertical copolar signals' cross-covariance at a zero time lag (i.e., $r_{hh,vv}[0]$). In ATSR mode, the direct calculation of $r_{hh,vv}[0]$ is not available because the two copolar signals are sampled at different times. The phase of the lag-zero covariance must therefore be estimated using the other available measurements, similar to the ATSR-mode copolar correlation coefficient.

As with the STSR mode, two successive measurements from the same polarization can be used to estimate the mean Doppler velocity. The ATSR sampling mode alternates the transmitted polarization with each PRT, and therefore the unambiguous Doppler velocity range is decreased by a factor of 2 (the sampling rate between copolar observations is $2T_s$ in ATSR mode). The horizontal polarization's mean Doppler velocity is estimated as

$$\bar{v}_{hh} = \frac{-\lambda}{4\pi(2T_s)} \angle r_{hh,hh}[2T_s]. \tag{6.29}$$

It is desirable to be able to measure the mean Doppler velocity within the full unambiguous velocity span for the radar's PRT T_s (the same span available in STSR mode). For observations of precipitation, the scatterers observed by the horizontal and vertical polarizations are highly correlated. As a result, the horizontally and vertically polarized copolar signals can be combined to estimate the mean Doppler velocity over the full unambiguous velocity (rather than half the span when only a single polar-ization's measurements are used). However, using both copolar signals to estimate velocity requires a correction for the phase difference between the horizontal and vertical polarizations. This difference may be due to the system differential phase (ϕ_{sys}) or the signal's Φ_{dp}, which varies with range and weather. Figure 6.3 shows the phase measurements using the cross-covariance between the ATSR mode's copolar signals, highlighting the factors included in the measurements.

Within a given range volume, the propagation and scattering differential phase shift, Ψ_{dp}, and the system differential phase, ϕ_{sys}, can be assumed to be constant. The mean Doppler velocity results in a time-varying phase from pulse to pulse. For a constant velocity, the velocity-induced phase shift between each PRT, ϕ_{vel}, is also constant. As shown in Figure 6.3, the covariance estimates for both the horizontal-to-vertical sequence ($r_{hh,vv}[T_s]$) and the vertical-to-horizontal sequence ($r_{vv,hh}[T_s]$) allows the polarizations' differential phase shift to be separated from the phase shift due to Doppler velocity. The phase of the cross-covariance between the horizontal-to-vertical samples is

$$\Psi_1 = \angle r_{hh,vv}[T_s] = \phi_{vel} + \phi_{sys} + \Psi_{dp}. \tag{6.30}$$

Similarly, the phase of the cross-covariance between the vertical-to-horizontal signals (still with positive time lags) is

$$\Psi_2 = \angle r_{vv,hh}[T_s] = \phi_{vel} - \phi_{sys} - \Psi_{dp}. \tag{6.31}$$

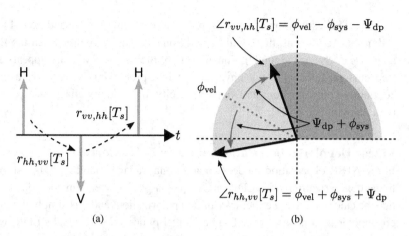

Figure 6.3 (a) In ATSR mode, the copolar measurements are sampled by alternately transmitted polarizations, with each pulse separated in time by the PRT, T_s. The cross-covariance of the copolar measurements are used to estimate the differential phase shift and the mean Doppler velocity. (b) The ATSR mode covariances' phase includes contributions from Ψ_{dp}, ϕ_{sys}, and the mean Doppler velocity [168]. By adding and subtracting the two covariance estimates, $r_{vv,hh}[T_s]$ and $r_{hh,vv}[T_s]$, the contributions of the differential phase shift, $\Psi_{dp} + \phi_{sys}$, can be separated from the velocity-induced phase shift, ϕ_{vel}.

To estimate the total differential phase shift and system phase, the difference of Ψ_1 and Ψ_2 removes the velocity's effect:

$$\Psi_1 - \Psi_2 = 2(\Psi_{dp} + \phi_{sys}). \tag{6.32}$$

In ATSR mode, the propagation differential phase, which is unambiguous over 180° (instead of 360° for STSR mode) is

$$\Phi_{dp} \approx \Psi_{dp} = \frac{180}{\pi}\left(\frac{\Psi_1 - \Psi_2}{2} - \phi_{sys}\right). \tag{6.33}$$

The sum of Ψ_1 and Ψ_2 removes the polarizations' differential phase, and the phase shift due to the Doppler velocity is estimated as

$$\phi_{vel} = \frac{\Psi_1 + \Psi_2}{2}. \tag{6.34a}$$

With the cross-covariance between the successive horizontal and vertical copolar signals in ATSR mode, the mean Doppler velocity can be estimated without sacrificing the unambiguous velocity span, to give the ATSR mean Doppler velocity estimator:

$$\bar{v} = \frac{-\lambda}{4\pi T_s}\phi_{vel}. \tag{6.35a}$$

The ATSR estimation of mean Doppler velocity involves the addition of two phase estimates. As a result, there will be additional phase ambiguity when the Φ_{dp} phase aliases (i.e., "wraps around"). The aliasing of the Doppler velocity appears as a large step change, but this is not physical – it is a processing artifact. By using spatial continuity constraints on the Doppler velocity, the velocity aliasing can be corrected (i.e., "unwrapped") [21].

Linear Depolarization Ratio and Cross-Polar Correlation

In the ATSR observation mode, two additional radar variables can be estimated: the linear depolarization ratio (LDR) and the cross-polar correlation coefficient. The LDR may be calculated from either transmitted polarization, and it is simply the ratio of the cross-polar to copolar power. The horizontal polarization channel's LDR, with noise correction, is estimated as

$$\mathrm{LDR}_{vh} = \frac{(r_{vh,vh}[0] - n_{vv})}{(r_{hh,hh}[0] - n_{hh})}\frac{C_{hh}}{C_{vh}}, \tag{6.36a}$$

and similarly, the vertical polarization channel's LDR is

$$\mathrm{LDR}_{hv} = \frac{(r_{hv,hv}[0] - n_{hh})}{(r_{vv,vv}[0] - n_{vv})}\frac{C_{vv}}{C_{hv}}. \tag{6.36b}$$

Nominally, the cross-polar signals' powers are the same for both LDR measurements, and only the copolar powers differ. The horizontal and vertical LDR measurements are related through the differential reflectivity (Z_{dr}) as

$$Z_{dr} = \frac{\mathrm{LDR}_{hv}}{\mathrm{LDR}_{vh}}. \tag{6.36c}$$

Similar to Z_{dr}, the formulation of LDR includes the ratio of the radar constants. Like Z_{dr}, the ratio is typically treated as a single correction factor that must be estimated and applied. The observational Z_{dr} accuracy for weather radar is on the order of tenths of a decibel, whereas the accuracy of reflectivity is on the order of a decibel. This necessarily implies that independent estimation and correction of the radar constants for reflectivity are not sufficient for differential reflectivity. The differential reflectivity is typically calibrated using dedicated methods.

In ATSR mode, the cross-polar correlation coefficient is a direct estimate because the copolar and cross-polar estimates are measured simultaneously in time. Using the transmitted horizontal polarization, the cross-polar correlation coefficient is

$$\rho_{cx} = \frac{|r_{vh,hh}[0]|}{\sqrt{(r_{vh,vh}[0] - n_{vv})(r_{hh,hh}[0] - n_{hh})}}. \tag{6.37}$$

6.2   Spectral Processing

With the availability of low-cost, high-performance computing resources, spectral-processing techniques are increasingly used in weather radar signal processing. In particular for ground-clutter filters or interference rejection, spectral processing can be more natural and computationally efficient to implement.

From the perspective of observing weather phenomena, spectral observations provides a more detailed characterization of the clouds and precipitation, particularly when observing at near vertical; the Doppler spectrum's velocities correspond to different particle sizes by virtue of their terminal velocities. This can be helpful in characterizing microphysical processes within the observation volume (see Section 3.5).

In covariance processing, three parameters are used to characterize the observation volume's Doppler spectrum: power, mean Doppler velocity, and Doppler spectrum width. This second-order characterization implicitly assumes a Gaussian-shaped Doppler spectrum. Although spectral-processing techniques can also be used to estimate mean power, mean Doppler velocity, and Doppler spectrum width, the measured Doppler spectrum provides a higher-order characterization of the observation volume.

If the measured Doppler spectrum is not symmetric, for example, having a combination of two Gaussian spectra, then the three-parameter covariance processing estimates do not adequately represent these features. For example, if stationary ground clutter and moving rain are within a common observation volume, the pulse-pair's three-parameter model of the measurement is not necessarily the best representation of the two distinct processes within the observation volume, and more terms are required. The Doppler spectrum estimates can include up to N terms (where N is the number of Doppler bin samples).

6.2.1   The Doppler Spectrum

In Section 5.1.2, the Fourier transform was introduced to estimate the power spectrum of a signal. The discrete Fourier transform of the radar's echo subdivides the signal's power into "bins" according to its Doppler velocity. For N samples in the coherent integration period, the velocity resolution of the Doppler spectrum is

$$\Delta v = \frac{2v_a}{N} = \frac{\lambda}{2T_s N}. \tag{6.38}$$

Consider a signal $x(t)$, from a fixed observation volume that is uniformly sampled in time. The spectrum of the signal, X, is estimated using the discrete Fourier transform (DFT) (and an appropriate window function, h) as

$$\widehat{X}[k] = \sum_{n=0}^{N-1} x[n]h[n]e^{-j2\pi nk/N}; \quad k \in [0, N-1]. \tag{6.39}$$

The velocity at the center of each bin in the Doppler spectrum is

$$v[k] = \begin{cases} 2v_a \frac{k}{N} & 0 \le k < N/2 \\ 2v_a \left(\frac{k}{N} - 1\right) & N/2 \le k \le N - 1 \end{cases}. \tag{6.40}$$

The mean power of the time-domain signal is defined to be the zeroth moment, or sum, of the power spectrum (see Section 6.2.2). From Parseval's theorem in eq. (5.12), the sum of the time-domain signal's power is equated to the mean of the signal's power spectrum. To relate this to the signal's mean power estimator (eq. [5.56]), both sides of eq. (5.12) must be divided by N giving:

$$\frac{1}{N} \sum_{n=0}^{N-1} |x[n]|^2 = \frac{1}{N^2} \sum_{k=0}^{N-1} |X[k]|^2. \tag{6.41a}$$

The left-hand side now has the form of the lag-0 autocovariance, $r_{xx}[0]$, which is the mean power estimate of the signal x. The right-hand side is the summation of the signal's power spectrum. Therefore, the power spectrum estimate is defined as

$$\boxed{S_{xx}[k] = \frac{1}{N^2} |X[k]|^2,} \tag{6.41b}$$

and the covariance and spectral estimators for the signal's mean power are equated as:

$$r_{xx}[0] = \sum_{k=0}^{N-1} S_{xx}[k]. \tag{6.41c}$$

It is also worth noting that the Fourier transform of a Gaussian-shaped function also has a Gaussian shape. Consider the Gaussian time-domain signal centered around $t = 0$:

$$g(t) = \frac{1}{\sqrt{2\pi\sigma^2}} \exp\left(-\frac{t^2}{2\sigma^2}\right). \tag{6.42}$$

The Fourier transform of the Gaussian signal is

$$G(\omega) = \exp\left(-\frac{\omega^2\sigma^2}{2}\right) = \exp\left(-\frac{\omega^2}{2\sigma_f^2}\right), \tag{6.43}$$

where $\sigma_f = 1/\sigma$. A narrow shape in the time domain is a wide shape in the frequency domain, and vice versa. This relationship between the time and frequency domains is important to the Doppler spectrum-width estimator, which assumes a Gaussian time-domain correlation and Gaussian spectrum shape.

When estimating the signal's spectrum (i.e., eq. [6.39]), a window function is typically used to smooth the estimated spectrum's frequency response and, more importantly, reduce the side-lobe levels that result from the DFT of the signal from a finite observation period with N samples. Examples of four different window functions commonly found in weather radar spectral processing are shown in Figure 6.4. The rectangular (or Boxcar) window is

$$h[n] = 1; \quad 0 \le n \le N - 1. \tag{6.44a}$$

The Hamming window is

$$h[n] = 0.54 - 0.46 \cos \left(\frac{2\pi n}{N-1} \right); \quad 0 \le n \le N - 1. \tag{6.44b}$$

The Hann window, which is sometimes referred to as a *Hanning window* or *raised cosine window*, is

$$h[n] = 0.5 - 0.5 \cos \left(\frac{2\pi n}{N-1} \right); \quad 0 \le n \le N - 1. \tag{6.44c}$$

The Blackman–Harris window is

$$h[n] = 0.35875 - 0.48829 \cos \left(\frac{2\pi n}{N-1} \right) + 0.14128 \cos \left(\frac{4\pi n}{N-1} \right)$$

$$- 0.01168 \cos \left(\frac{6\pi n}{N-1} \right); \quad 0 \le n \le N - 1. \tag{6.44d}$$

Various other window functions are also available [169, 170], each optimized to provide various spectral characteristics for particular applications. In addition to these windows, a triangle window function is given in eq. [5.51] and a Tukey window is defined later in eq. [6.95]. For weather radar spectral estimation and signal processing, the window functions presented here are well suited for most applications. When estimating the signal's reflectivity from the power spectrum, the gain of the window function should be considered. For spectral estimation, it may be desirable to normalize the vector's coefficients so that the resulting window function has a unity gain (i.e., $\sum^N |\widehat{h}[n]|^2 = 1$):

$$\widehat{\mathbf{h}} = \frac{\mathbf{h}}{\sqrt{\mathbf{h}^H \mathbf{h}}}, \tag{6.45}$$

where the Hermitian transpose, denoted by the superscript H, indicates that the vector is both complex conjugated and transposed.

In Figure 6.4, the amplitude weighting function and resulting power spectrum of the four different window functions are shown for $N = 32$. The boxcar window shows the narrowest main lobe of the four window functions but also has the highest power in its side lobes. At the other end, the Blackman–Harris window has the lowest side-lobe levels of the window functions shown, but it has the widest spectral main lobe. Note that the Blackman–Harris window weights the samples near the beginning and end with very small values; this almost removes their contribution to the overall spectrum, with the samples in the middle dominating the calculated spectrum. Although this provides good control of the spectrum's side-lobe levels, it effectively reduces the number of samples used in the estimate. With the rectangular window, the number

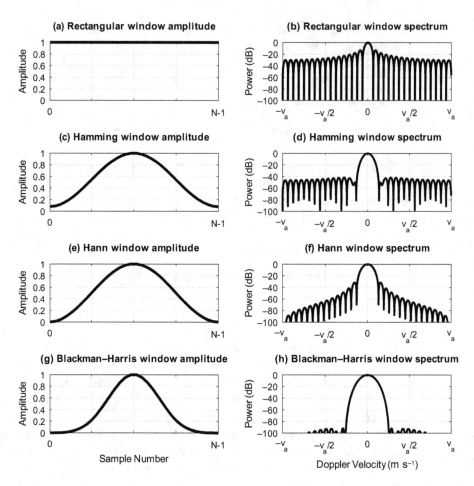

Figure 6.4 The amplitude and power spectrum (for $N = 32$) for four common window functions. The amplitude and power spectrum are normalized to a maximum of 1 for illustration and comparison. As N increases, the width of the window spectrum's main lobe decreases (following eq. [6.38]). The rectangular window has the greatest number of samples with an amplitude above 0.5, which also results in the narrowest main lobe, shown in panels (a) and (b), respectively. Compare this to the Blackman–Harris window, which has the narrowest amplitude width (number of samples with amplitude > 0.5), which results in the broadest width in the Doppler spectrum, shown in panels (g) and (h), respectively. The Hamming and Hann windows have a similar number of samples, > 0.5, but the shape of the windows near the edges is different. The different shape, in particular, the "tails" of the window function, result in much different spectral side lobe performance between the two windows. The Hamming window's side-lobes are nearly flat in panel (d). The Hann window's spectrum in panel (f) has slightly higher side-lobe levels near the main lobe compared with the Hamming window, which it trades for significantly lower far side lobes.

of independent Doppler velocity bins is equal to N. When the window modifies the amplitude of the input signal, it effectively reduces the number of samples used in the Fourier transform. This introduces a degree of correlation between neighboring spectral bins, which manifests as a broadening effect observed in the spectrum of Figure 6.4, which "smooths" the estimated spectrum.

The selection of the window function is typically most important when precipitation and ground clutter exist in the same observation volume (see Section 7.3.6 for discussion of ground-clutter filters). Ground clutter is centered around a $0 - \mathrm{m\,s}^{-1}$ Doppler velocity. If the ground clutter is static, its true spectrum width is zero. The estimated spectrum width then depends on the number of samples used in the integration period and the window function, as shown in the left-hand side of Figure 6.4. Consider Figure 6.5, an example of simultaneously observed ground clutter and precipitation (the rain's mean Doppler velocity is $-v_a/2$, and it has a negligible spectrum width). For this illustrative example, the precipitation's echo power is 30 dB less than the ground clutter's echo power. With the rectangular window, the rain's echo spectrum is masked by the ground clutter's spectrum and therefore not detectable. The Hamming and Hann windows have similar performance as one another, but the actual results depend on the mean Doppler velocity of the precipitation (when $\bar{v}$ is near zero, the Hamming window performs better; when $\bar{v}$ is closer to $\pm v_a$, the Hann window performs better). The Blackman–Harris window provides the best suppression of the clutter's spectral side lobes at the expense of a broad main lobe which can mask weak precipitation echoes with near-zero velocities.

Without any prior knowledge of the precipitation's spectrum, the Hamming window provides a good compromise between the first side-lobe power level and the integrated side-lobe levels. For strong narrow-band targets (e.g., ground clutter), the resulting side-lobe power from the ground clutter may be higher than the precipitation power that we are interested in. For cases where ground clutter is a particular concern, the Blackman–Harris window can improve the detection and estimation of precipitation but should be combined with longer integration periods to mitigate the window's effect on the spectrum width (this will also improve the detection and estimation of precipitation when its mean Doppler velocity is near zero).

The Doppler power spectra are coherently integrated using N samples. A new spectrum is estimated at a rate of $N \cdot T_s$. With a sequence of power spectra, a time series is available for each Doppler bin: $\widehat{S}_{xx}[k,t]$. Like the time-domain covariance power estimates, M power spectra can be incoherently averaged at each Doppler bin:

$$\bar{S}_{xx}[k] = \frac{1}{M} \sum_{m=0}^{M-1} \widehat{S}_{xx}[k, t + mNT_s]. \tag{6.46}$$

A total of $N \cdot M$ samples are used to estimate $\bar{S}_{xx}$.

Within the radar's Doppler spectrum, the power of white noise is constant. The noise power of a Doppler bin is thus the density multiplied by the size of the bin

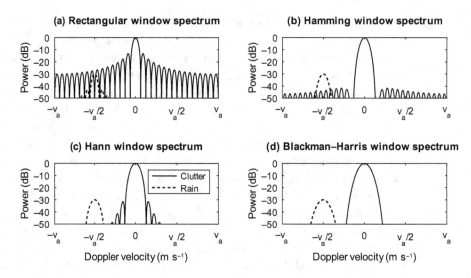

Figure 6.5 Example power spectra of ground clutter and precipitation (for $N = 32$), assuming each has small intrinsic spectrum widths. The same precipitation and clutter signals are used to consider the effects of different window functions. The precipitation's power is 30 dB below the clutter's echo power. For this simple example, the rectangular window does not sufficiently suppress the clutter spectrum's side-lobe levels, which have higher power than the rain (and therefore the rain's spectrum cannot be detected). For the three other windows, the clutter's side lobes are sufficiently low enough so that the peak of the precipitation's spectrum can be estimated in the presence of the clutter.

(i.e., $\Delta v \propto \Delta f$). The total noise power in the Doppler spectrum's bin is reduced by increasing N to integrate more samples and decrease Δv. In practice, the signal's power also has a spectrum width (and therefore a spectral density). There is a limit to the SNR improvement that can be achieved within a Doppler spectral bin. At some point, reducing Δv will decrease the total noise power and total signal power equally within the bin, and there is no additional improvement in the SNR.

To detect precipitation in the estimated spectrum for low SNRs, the optimal selection of N depends on the rain's intrinsic spectrum width [171]:

$$N_{opt} \approx \frac{K_{opt}\lambda}{2\sigma_v T_s}, \tag{6.47}$$

where K_{opt} is a constant that depends on the window used [171] ($K_{opt} = 0.7$ for a rectangular window, $K_{opt} = 0.97$ for the Hamming window, $K_{opt} = 1.06$ for a Hann window, and $K_{opt} = 1.41$ for the Blackman–Harris window). If $N > N_{opt}$, the resolution of the Doppler spectral bins is fine enough so the signal's power is spread among multiple velocity bins, and there is no further SNR improvement available within the signal's spectral bin (at this point, for the purposes of signal detection, it is better to incoherently average multiple spectra by increasing M). If $N < N_{opt}$, the Doppler spectral bin is wider than the signal spectrum, and therefore the SNR within the signal's bin is sub-optimal. While the number of samples N used within the integration time are fixed to a constant value, the precipitation processes typically observed by the radar should guide its selection.

Figure 6.6 shows estimated power spectrums (spectrograms) of precipitation observed by the Colorado State University–Chicago Illinois (CSU-CHILL) S-band radar. For a statistically stationary process, the choice of N is a trade-off between Doppler resolution and coherent observation time. The spectrum width of the expected processes also provides guidance for the choice of N to ensure that the spectral shape of the process is sufficiently resolved in the Doppler spectrum domain. At some point, more samples will not provide any new information about the spectral shape of the process.

The echo's power and the dynamics of how the power changes with time are useful for understanding the behavior of the scatterers within the radar's observation volume. The reorganization of hydrometeors is only possible if there is relative motion between them in addition to the volume's mean velocity. The distribution of these relative velocities results in a spread in the Doppler spectrum, which is measured as the Doppler spectrum width, σ_v.

6.2.2 Spectral Estimators

As with the covariance estimators, the echo's power spectrum can be used to estimate the radar variables. The power (and therefore reflectivity), mean Doppler velocity, and Doppler spectrum width of the signal are the spectral moments of the power spectrum [163]. The mth spectral moment of a function $f(x)$ is calculated as

$$\mu_m = \int_{-\infty}^{\infty} (x - c)^m f(x)\, dx. \tag{6.48}$$

Spectral moments about the mean (i.e., $c = \mu_1$) are referred to as "central moments." Spectral moments about zero ($c = 0$) are "raw moments." (The weather radar variables are sometimes referred to as the radar moment estimates.) The power is the zeroth raw moment, mean Doppler velocity is the first raw moment, and the spectrum width is the second central moment.

Power
The equivalent radar reflectivity factor is proportional to the total power of the Doppler spectrum (or the zeroth moment of $|S|$):

$$p = \sum_{k}^{K} |S[k]|, \tag{6.49}$$

where K is the length of S. The power estimator is equivalent to the covariance power estimates discussed in Section 6.1.

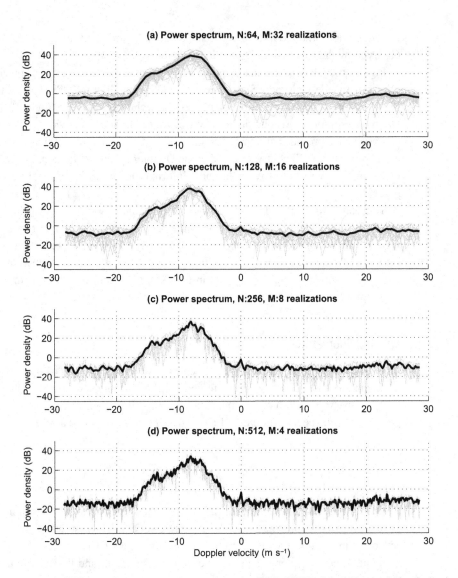

Figure 6.6 The Doppler spectra estimates from the CSU-CHILL S-band horizontal polarization. These observations of precipitation are made in fixed-pointing mode. In total, 2048 samples (approximately 1 s of data) were acquired. The total number of samples is divided by the window length (N) to generate M realizations of the sampled process. A Hann window was used when estimating the power spectrum.

Mean Velocity

The mean Doppler velocity (the first moment of $|S|$, normalized by the power) is

$$\bar{v} = \frac{1}{p} \sum_{k}^{K} |S[k]| v[k], \tag{6.50}$$

where v is the velocity at the center of each of the spectra's bins (recall eq. [6.40]). For the same signal, the spectral mean Doppler velocity estimate can deviate from the covariance pulse-pair mean Doppler velocity estimate if the Doppler spectrum is not symmetric.

The velocity and spectrum width estimators are sensitive to bias from Doppler velocity aliasing. If the Doppler spectrum of the signal is partially aliased (if part of the signal is near $-v_a$ and the other part is near v_a), the estimated mean Doppler velocity from eq. (6.50) will be biased.

The prospect of estimator bias due to aliasing may be alleviated by representing the velocity spectra on the complex plane. The velocity is continuous on the complex-valued unit circle instead of discontinuous on the real-valued plane. (Recall that Doppler velocity is estimated from the signal's phase shift between pulses.) Using the complex-valued representation of the velocity, the spectral estimator's bias can be largely mitigated for observations where the signal's spectrum is aliased. The modified spectral moment estimator for velocity is

$$\bar{v} = \frac{v_a}{\pi} \arg\left[\sum_{k}^{K} |S[k]| \exp\left(j\pi v[k]/v_a\right) \right]. \tag{6.51}$$

Spectrum Width
The spectrum width, the square root of the second central moment of $|S|$ normalized by the power, is calculated as

$$\sigma_v = \sqrt{\frac{1}{p} \sum_{k}^{K} |S[k]| \left(v[k] - \bar{v}\right)^2}, \tag{6.52}$$

which has the same units as the mean Doppler velocity.

Like the power estimate, the spectrum-width estimator is subject to bias from additive noise power. Noise subtraction may be implemented prior to estimating σ_v. Like the mean Doppler velocity, the spectral spectrum-width estimator will be biased when the signal's spectrum aliases. To minimize the Doppler aliasing bias effect, the spectrum width can also be estimated using the complex plane and recentering the spectrum (with mean velocity $\bar{v}$) before estimating the spectrum width. As long as the signal's spectrum does not exceed the ambiguous sampling limits (i.e., $\pm v_a$), the bias due to phase wrapping is mitigated. The modified spectral estimator for spectrum width is

$$\sigma_v = \sqrt{\frac{v_a^2}{p\pi^2} \sum_{k}^{K} |S[k]| \left[\arg\left(\exp[j\pi(v[k] - \bar{v})/v_a] \right) \right]^2}. \tag{6.53}$$

6.2.3   STSR Dual Polarization Spectral Estimators

Up to this point, each radar variable is expressed as an integrated estimate, using all available samples in the coherent processing interval. The radar variable estimators can be adapted to consider each velocity bin of the Doppler spectrum. The spectral estimates are similar in form to the covariance radar variable estimates. Note, the different sampling times of each polarization in ATSR mode can result in a phase shift due to the precipitation dynamics, which must be accounted for to estimate the copolar correlation and differential phase. Here, only the STSR spectral dual polarization radar variables are considered.

The complex-valued spectrum of a WSS signal is the Fourier transform of the signal's autocovariance r_{xx} [163]. Similarly, the complex-valued cross-spectrum is the Fourier transform of the covariance of two signals: r_{xy} [172]. To estimate the dual polarization spectra for the observation volume, complex-valued spectra $(\widetilde{S})$ are estimated using the biased covariance estimates from the dual polarization measurements (see eq. [5.50]). Following the definition in Section 5.1.2, the magnitudes of the complex-valued spectra are their PSDs (i.e., $S = |\widetilde{S}|$). The phase of each bin of the Doppler spectra is required to estimate the copolar correlation coefficient and differential phase. The covariance vectors are used to estimate the complex spectra which have time lags spanning $[-(N-1)T_s, (N-1)T_s]$, for a total of $2N-1$ time lags. With the covariance estimator from eq. (6.1), the horizontal polarization's copolar spectrum estimate [173] is

$$\widetilde{S}_{hh}[k] = \frac{1}{2N-1} \sum_{n=-(N-1)}^{N-1} r_{hh}[n]h[n]e^{-j2\pi nk/(2N-1)} \qquad (6.54)$$

where $k \in [-(N-1), N-1]$ and h is a desired unitary window-weighting function. Similarly, the vertical copolar spectra are

$$\widetilde{S}_{vv}[k] = \frac{1}{2N-1} \sum_{n=-(N-1)}^{N-1} r_{vv}[n]h[n]e^{-j2\pi nk/(2N-1)}, \qquad (6.55)$$

and the horizontal and vertical copolar covariance spectra are

$$\widetilde{S}_{hv}[k] = \frac{1}{2N-1} \sum_{n=-(N-1)}^{N-1} r_{hv}[n]h[n]e^{-j2\pi nk/(2N-1)}. \qquad (6.56)$$

To estimate the dual-polarization spectral moments, the copolar covariance, $\widetilde{S}_{hv}$, includes critical information. Using the spectral estimates of the signal (e_h or e_v) instead of the covariance for each Doppler bin would result in a single independent value, which would necessarily have a correlation of 1. (This would be the same as calculating the correlation for $N = 1$.) An average of multiple samples is necessary for a meaningful correlation estimate. The signals' covariance r_{hv} is an average over the CPI, and this is used to estimate the complex-valued spectra. The other spectra are estimated following the same process for the calculation of the dual polarization spectra. The same spectral-processing techniques discussed previously for estimating the moments can be used to estimate power, mean velocity and spectrum width. Note that the length of the complex spectra is $2N - 1$ rather than N; therefore, the size and spacing of the Doppler bins must correctly account for the number of samples used.

Three dual polarization spectra can be estimated from the STSR observations. The estimation at each Doppler bin has a similar form to the STSR covariance estimates in Section 6.1.1. Noise correction is not shown in the following equations but can easily be included in the estimators. The spectral differential reflectivity is estimated as

$$S_{Z_{dr}}[k] = \frac{|\widetilde{S}_{hh}[k]|}{|\widetilde{S}_{vv}[k]|}. \tag{6.57}$$

Corrections may be applied if the radar constants for the horizontal and vertical polarizations are not equal. The spectral differential propagation phase is

$$S_{\Phi_{dp}}[k] = \frac{180}{\pi} \left[\arg\left(\frac{\widetilde{S}_{hv}[k]}{\sqrt{\widetilde{S}_{hh}[k]\widetilde{S}_{vv}[k]}} \right) + \phi_{sys} \right], \tag{6.58}$$

and the spectral copolar correlation is

$$S_{\rho_{hv}}[k] = \frac{|\widetilde{S}_{hv}[k]|}{\sqrt{|\widetilde{S}_{hh}[k]||\widetilde{S}_{vv}[k]|}}. \tag{6.59}$$

6.3 Specific Differential Phase (K_{dp}) Estimation

The specific differential phase shift, K_{dp}, is the rate of change in the phase difference between the two polarizations as the wave propagates. The typical units of K_{dp} are degrees per kilometer ($°km^{-1}$). The scattering characteristics that relate to K_{dp} are explored in Section 4.7, which shows it is the real part of the difference between the horizontal and vertical polarizations' forward-scattering amplitudes.

An ideal radar observing Rayleigh scatterers without measurement errors would measure the two-way range integral of K_{dp}, that is, the differential phase shift due to the propagation through a precipitation medium between the radar and the target:

$$\Phi_{dp}(r) = \phi_{sys} + 2 \int_0^r K_{dp}(r)\,dr, \tag{6.60}$$

where ϕ_{sys} is the system differential phase. Besides this propagation differential phase shift, non-Rayleigh scattering by large particles contributes an additional differential phase shift from backscattering (δ_{co}):

$$\Psi_{dp}(r) = \Phi_{dp}(r) + \delta_{co}(r). \tag{6.61}$$

Finally, radar observations are also affected by measurement errors (ϵ). The actual differential phase shift estimated by signal processing (eqs. [6.16, 6.33] for the covariance estimators in STSR and ATSR mode) is then contributed by three terms:

$$\widehat{\Psi}_{dp}(r) = \Phi_{dp}(r) + \delta_{co}(r) + \epsilon; \ \ 0° \leq \widehat{\Psi}_{dp} < 360° \ (\text{STSR})$$
$$0° \leq \widehat{\Psi}_{dp} < 180° \ (\text{ATSR}). \quad (6.62)$$

The estimation of K_{dp} from the observed differential phase shift may require different approaches, depending on the specific radar system and application. In particular, several techniques have been introduced for the retrieval of K_{dp} in rain. As with any slope estimation, K_{dp} comes with the challenge of finding a compromise between accuracy and resolution. The easiest way to estimate slope is by the finite difference method. Under the hypothesis that both ϵ and δ_{co} are negligible, K_{dp} could be estimated following eq. (6.60):

$$\widehat{K}_{dp} = \frac{\Phi_{dp}(r_2) - \Phi_{dp}(r_1)}{2(r_2 - r_1)} \approx \frac{\widehat{\Psi}_{dp}(r_2) - \widehat{\Psi}_{dp}(r_1)}{2(r_2 - r_1)}, \quad (6.63)$$

where r_1 and r_2 are the ranges of two observation volumes along the range profile. The finite difference method becomes more accurate if the path length ($r_2 - r_1$) becomes large. However K_{dp}, which is proportional to rainfall rate, loses its resolution if estimated over a longer path length. In general light rain, that has small K_{dp}, is more uniform compared to heavy rain. Thus there is an interest to estimate higher K_{dp} accurately but also with good resolution. Another issue that impacts a slope estimation is the jump in differential phase as it crosses cycles of 360 degrees (STSR mode) or 180 degrees (ATSR mode) inherent to any phase measurement. Several techniques have been developed to mitigate the above general problems in K_{dp} estimation with the purpose of automatically determining the accuracy and spatial resolution of K_{dp} estimation.

A simple approach to determine K_{dp} is to estimate the slope of the range profile of $\widehat{\Psi}_{dp}$ applying a piecewise linear regression to a given number N of consecutive range bins. Similar results can also be efficiently achieved using finite impulse response (FIR) filters to smooth out the measurement noise [174]. Methods that smooth the K_{dp} profile excessively result in an underestimation of the precipitation's peak K_{dp} and potentially introduce an estimation bias. Adaptive K_{dp} estimation techniques [175] are able to strike a balance between smoothing the profile to reduce the measurement noise and achieving small-scale slope estimation of K_{dp} in heavy precipitation.

The unambiguous range for the differential phase shift is either 360° (STSR mode) or 180° (ATSR mode). Although the 360° range is generally sufficient to avoid phase folding, in ATSR mode, phase folding is more likely to occur during intense precipitation. In addition, the system differential phase, that is, ϕ_{sys} in eq. (6.60), is typically adjusted near the lower limit in order to use the full unambiguous range and also avoid folding issues in simultaneous-operation mode. The phase-wrapping discontinuity (e.g., the phase crossing the 360°–0° boundary) appears only when the phase is projected on the real domain. Considering Φ_{dp} as an angular variable in the complex plane instead (Appendix A), the phase is naturally continuous across this boundary, and K_{dp} can be estimated using the angular $\widehat{\Psi}_{dp}$, defined as

$$\theta(r) = e^{jm\widehat{\Psi}_{dp}(r)}, \tag{6.64}$$

where $\widehat{\Psi}_{dp}$ is expressed in radians, $m = 1$ for the STSR mode, and $m = 2$ for the ATSR mode in order to transform $\widehat{\Psi}_{dp}$ to a circular variable over the full $0° - 360°$ range. Taking the derivative of eq. (6.64) with respect to r:

$$\theta'(r) = jme^{jm\widehat{\Psi}_{dp}(r)}\frac{d\widehat{\Psi}_{dp}(r)}{dr}, \tag{6.65}$$

the range derivative of $\widehat{\Psi}_{dp}$ can be written as $d\widehat{\Psi}_{dp}(r)/dr = -j\,\theta'(r)/[m\,\theta(r)]$. Note that we're interested here in the angles and their variation (the imaginary part of the complex numbers) and the ratio $\theta'(r)/\theta(r)$ is a purely imaginary number for a continuous angular $\widehat{\Psi}_{dp}$ profile. Therefore, neglecting numerical errors in the derivatives, an estimate of K_{dp} is obtained as

$$\widehat{K}_{dp} = \frac{1}{2m}\,\text{Im}\left[\frac{\theta'(r)}{\theta(r)}\right]. \tag{6.66}$$

In order to deal with noisy observations, smooth splines provide a way to obtain stable function estimates over N range bins. For generic observations y_i, a spline solution $s(r)$ is sought by minimizing the following error function:

$$\sum_{i=1}^{N}\left[y_i - s(r_i)\right]^2 + \lambda\int\left[s''(r)\right]^2 dr, \tag{6.67}$$

where i is the index of the range gate, the first term ensures the goodness of fit (distance from the observations), and the second term (the second-order derivative of the solution s) with the smoothing factor λ to penalize excessive fluctuations. Using the equivalent trigonometric form of the angular $\widehat{\Psi}_{dp}$,

$$\theta(r) = \cos\left[m\widehat{\Psi}_{dp}(r)\right] + j\sin\left[m\widehat{\Psi}_{dp}(r)\right], \tag{6.68}$$

smooth splines can be independently obtained for the real and imaginary parts of $\theta(r)$, that is, $\cos\left[m\widehat{\Psi}_{dp}(r)\right]$ and $\sin\left[m\widehat{\Psi}_{dp}(r)\right]$. Figure 6.7 shows a range profile of $\widehat{\Psi}_{dp}$ observed by a C-band radar with alternate transmission mode ($m = 2$) during heavy rainfall (panel [a]). By either setting $y_i = \cos\left[2\widehat{\Psi}_{dp}(r_i)\right]$ or $y_i = \sin\left[2\widehat{\Psi}_{dp}(r_i)\right]$ in (6.67), the real and imaginary components of the angular phase can be readily fitted by the smooth splines (s_r and s_i respectively, represented by solid lines in panel [b]), despite the phase wrapping in the observations. Computation of the range derivative

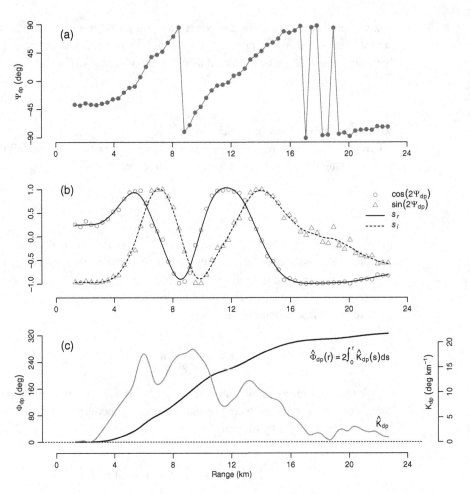

Figure 6.7 Estimation of K_{dp} using smoothed splines on the angular differential phase. (a) Observed differential phase shift from the C-band Settepani radar in Italy during torrential rain. The alternate-transmission operation mode and heavy rainfall led to multiple phase wrapping. (b) The real and imaginary part of the angular Ψ_{dp} with the corresponding spline fits. (c) The retrieved K_{dp} and unwrapped Φ_{dp}, showing a total phase shift in excess of $320°$ and K_{dp} up to $\sim 20\,°\text{km}^{-1}$.

of the fit curves (s_r' and s_i') finally allows us to retrieve the estimate of K_{dp} according to eq. (6.66):

$$
\begin{aligned}
\widehat{K}_{dp} &= \frac{1}{2m}\text{Im}\left\{\frac{s_r' + js_i'}{(s_r + js_i)}\right\} \\
&= \frac{1}{2m}\text{Im}\left\{\frac{1}{(s_r^2 + s_i^2)}\left[j(s_r s_i' - s_i s_r') + (s_r s_r' + s_i s_i')\right]\right\} \\
&= \frac{1}{2m}\frac{s_r s_i' - s_i s_r'}{(s_r^2 + s_i^2)}
\end{aligned}
\tag{6.69}
$$

The resulting K_{dp} estimate is shown in Figure 6.7 (panel [c]), together with the corresponding unfolded Φ_{dp} retrieved by range integration of K_{dp}. Note that the real part of the term in curly brackets in eq. (6.69) is zero for a noiseless linearly increasing differential phase (e.g., for $\widehat{\Psi}_{dp}(r) = kr$, then $s_r s_r' + s_i s_i' = -k\cos(kr)\sin(kr) + k\sin(kr)\cos(kr) = 0$). It is non-zero in practice for noisy discrete observations with associated numerical errors arising from computation of the derivatives. In this sense, it is considered an extra term associated with the estimation error.

The global smoothing realized through the minimization of eq. (6.67) may lead to underestimation of the K_{dp} peaks and overestimation of K_{dp} in the surrounding regions where the precipitation intensity is weaker. It is possible to introduce variable weights in eq. (6.67) in order to adaptively change the degree of smoothing. Wang and Chandrasekar [175] used two range-varying weights in the minimization of eq. (6.67) to cope with statistical fluctuations and avoid excessive smoothing in regions of the profiles with large K_{dp}. The error function then becomes

$$\sum_{i=1}^{N} w^2(r_i)\left[y_i - s(r_i)\right]^2 + \lambda \int q^2(r)[s''(r)]^2 dr, \tag{6.70}$$

where w is chosen as the inverse of the standard deviation of the angular Φ_{dp} and can be approximated by $w^{-2} \approx 1 - |\rho_{hv}|$ (refer to eqs. 29–32 in [175] for the details). To decrease the smoothing in regions of large K_{dp}, the q term is chosen to be inversely proportional to K_{dp}, i.e. $q = 1/(2K_{dp})$, where the factor 2 is in place to compensate the two-way phase shift profile. An initial estimate of K_{dp} to calculate q can be obtained either from the observed reflectivity through a power-law relation ($K_{dp} = aZ_h^b$) or by a first minimization of the nonadaptive error function (eq. [6.67]). The former approach (reflectivity-based estimate) may be suitable for S-band systems, whereas the latter is preferable for higher-frequency radar, where the attenuation cannot be neglected. The general smoothing factor λ can be chosen based on empirical evaluation and depending on the specific application. In practice, λ may range between $1/\Delta r$ and $10/\Delta r$, where Δr is the range resolution. The contamination of the observed Ψ_{dp} by δ_{co} is in general smoothed out with this method, owing to the w weighting in eq. (6.70) and the expected lower correlation in regions with a significant backscattering differential phase.

Recent research has also been focusing on a new class of K_{dp} estimators with physical constraints. The rationale behind these methods is to take advantage of the theoretical behavior of Φ_{dp} in rain to improve the retrieval of K_{dp} from noisy observations. By definition, K_{dp} is positive in rain (eq. [4.68] and Fig. 4.16), and Φ_{dp} is a monotonically increasing function of range (eq. [6.60]).

A variational technique can be used with a monotonicity constraint to calculate K_{dp} in rain [176]. Similarly, a linear programming technique can be used for the retrieval of Φ_{dp} in the presence of measurement fluctuations [177]. This method relies on the minimization of an objective function, that is the L_1 norm:

$$\sum_{i=1}^{N} \left| f_i - \widehat{\Psi}_{dp}(r_i) \right|, \tag{6.71}$$

where f_i represents the desired fit to the differential phase observations $\widehat{\Psi}_{dp}(r_i)$, subject to a monotonicity constraint. The monotonicity constraint is enforced by requiring a nonnegative range derivative, estimated using a five-point Savitzky–Golay second-order polynomial derivative filter. The minimization of the L_1 norm with range derivative constraints leads to the retrieval of the Φ_{dp} profile from the radar's Ψ_{dp} observations. An estimate of δ_{co} is available as a residual of the Φ_{dp} estimate from the Ψ_{dp} input.

Although the mathematical methods are different, there are similarities between the smoothed spline approach (eq. [6.67]) and the linear programming approach (eq. [6.71]). In the first method, the goodness of fit is ensured by the quadratic term (L_2 norm), whereas the linear programming approach uses the L_1 norm. In addition, both methods implement a constraint: the smoothed spline uses a smoothness constraint (second derivative in eq. [6.67]), whereas the linear programming uses a monotonicity constraint.

These types of retrieval approaches allow the introduction of additional physical constraints. For example, a polarimetric self-consistency constraint can be incorporated through parameterizations of the form $K_{dp} = a Z_h^b$ to better match the theoretical expectation of higher K_{dp} in correspondence with regions of high reflectivity [177]. The implementation of this kind of additional physical constraint should be carefully evaluated for radars operating at attenuating frequencies [178, 179]. Synergistic approaches pursue the simultaneous estimation of Φ_{dp}, K_{dp}, and δ_{co}, as well as attenuation-corrected reflectivity and differential reflectivity. The use of the extended Kalman filter [178] is aimed at an optimal retrieval of the radar variables sequentially along the range dimension, exploiting the redundancy among the radar variables (i.e., a self-consistency constraint), and a specific relationship between δ_{co} and Z_{dr}.

Although physically constrained methods have the potential to improve the K_{dp} behavior, these methods are generally designed (i.e., physically constrained) for rain. In ice or for nonmeteorological scatterers, the constraints may not be valid. Therefore, care must be taken to properly select the portions of the range profile in which to apply the retrieval, which can be done via hydrometeor classification (see Chapter 8). The standard finite-differences, linear regression, and smoothed-spline approaches are not subject to physical constraints and can be applied irrespective of the microphysical composition of the precipitation.

6.4   Advanced Sampling Methods

Alternatives to the uniform sampling scheme can be used to overcome the technical limitations of a radar system. One of the most common reasons to seek nonuniform sampling schemes is to extend the unambiguous velocity of the radar without reducing the radar's unambiguous range. This is particularly useful for weather radars using higher operating frequencies (such as X band), but these techniques also benefit S- and C-band radars.

Phase-coding techniques can be used to detect and disambiguate echoes from weather beyond the radar's unambiguous range. For magnetron-based transmitters, the starting phase of each pulse is random because of how the magnetron operates. To estimate Doppler velocity, the starting phase of the transmitted pulse must be measured and subsequently corrected for in the received signal.

6.4.1   Extending the Unambiguous Velocity

The simplest weather radar sample-time schemes use a constant pulse repetition period. With a constant PRT, the unambiguous Doppler velocity and unambiguous range of the radar system are tightly coupled. From Section 5.3.1, recall that:

$$r_a v_a = \frac{c\lambda}{8},$$ (6.72)

and

$$v_a = \frac{\lambda}{4T_s}.$$ (6.73)

This means that the radar uniquely resolves ranges from zero up to a maximum range r_a and resolves velocities spanning $\pm v_a$. Significant weather radar signal-processing research and effort have been put into relieving the restrictions of the unambiguous range–Doppler velocity dilemma. Although the range–velocity constraint cannot technically be violated, there are techniques available to bend the rules and provide relief.

In weather applications, the mean velocity of storms and clouds in the atmosphere can be significant (wind speeds of >50 m s^{-1} are frequently observed in severe weather or in high-level clouds in the free troposphere). Although the mean velocity of the storms may be large, the spectrum widths within the observation volumes are typically up to a few meters per second (these may be larger for sheer layers or severe convective systems).

With a uniform sampling rate, the entire Doppler spectrum spanning $[-v_a, v_a)$ is unambiguously resolved. The precipitation's Doppler velocity spectrum may exceed this range, resulting in aliasing of the Doppler spectrum components, which can lead to inaccurate estimates of mean Doppler velocity. While the precipitation's mean Doppler velocity may be large, its spectrum widths are well bounded. Using multiple PRTs, the velocities can be de-aliased to significantly enhance the radar's unambiguous mean Doppler velocity span. Note that the Doppler spectrum width estimators are still restricted by the individual PRTs used.

For eq. (6.72), T_s is the only value available to control the radar's sampling. To unambiguously resolve ranges or Doppler velocities in excess of the limits r_a and v_a, multiple values for T_s must be used. For each PRT, which have different r_a and v_a, the radar estimates the Doppler velocity and echo power over its unambiguous range. As a result of the different PRTs, each pulse has different limits where aliasing occurs, and therefore the two measurements can be compared to determine the correct de-aliasing factor to apply so that the range and velocities from the different PRTs match. Typically, the maximum range of the PRTs are selected so that range de-aliasing is not required and the focus is on extending the unambiguous velocity. With two or more PRTs, the radar's ability to unambiguously resolve targets in range and velocity can be extended beyond the capabilities with a single PRT. Note, the maximum spectrum width that can be resolved, and the estimator's performance, are still restricted by the v_a of each PRT. By combining the estimates from the two PRTs, the unambiguous extent of the mean Doppler velocity estimates can be significantly extended.

Using PRTs T_1 and T_2, the mean Doppler velocity can be "unfolded" to unambiguously estimate the mean velocity over a span $\pm v_u$ with

$$v_u = \frac{\lambda}{4(T_2 - T_1)}. \tag{6.74}$$

Note that the variable subscript is u to differentiate the unfolded unambiguous velocity, which relies on multiple PRTs, from the unambiguous velocity of a single PRT, v_a. The unambiguous velocity of the individual PRTs are v_{a1} and v_{a2} for T_1 and T_2, respectively. The unfolded unambiguous velocity, v_u, is the same as if the radar's PRT is $T_u = T_2 - T_1$. The PRT ratio is typically selected using an integer velocity extension factor, K. The velocity extension ratio is

$$\kappa = \frac{T_1}{T_2} = \frac{K}{K+1}; \quad 0 < \kappa < 1. \tag{6.75}$$

Two common sampling techniques used by weather radar systems to increase the unambiguous velocity are the dual PRF and staggered-PRT sampling methods. (Although the discussion is limited to two PRTs/PRFs in this chapter, it may be extended to more [180].) The uniform sampling timing is shown in panel (a) of Figure 6.8. The dual PRF technique shown in panel (b) uses a sequence of samples at one PRF ($f_1 = 1/T_1$), followed by a second sequence at a different PRF ($f_2 = 1/T_2$). For the dual PRF sampling scheme, a time delay may be used between the two sequences without any impact on the theory of operation. The staggered-PRT technique in panel (c) of Figure 6.8 alternates PRTs T_1 and T_2 for each successive pulse. For both techniques, the convention used is $T_2 > T_1$.

The radar's maximum range requirement, where the velocity estimate must include the span $\pm v_u$, determines the minimum PRT. The maximum range therefore determines T_1 (recalling $T_1 < T_2$). With T_1 selected, the velocity extension factor can be calculated, and T_2 can subsequently be defined. The unfolded velocity is related to each PRT's unambiguous velocity as

$$\begin{aligned} v_u &= (K+1)\frac{\lambda}{4T_2} = (K+1)v_{a2} \\ &= K\frac{\lambda}{4T_1} = Kv_{a1}. \end{aligned} \tag{6.76}$$

As an example, consider an X-band (9.4-GHz) radar with a maximum range of $r_a \geq 60$ km and $v_u \geq 30$ m s^{-1} (neglecting any additional time required by the radar for transmitting the pulse, internal calibration, etc.) The unambiguous range is used

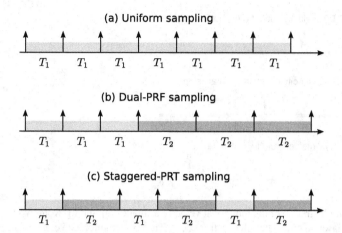

Figure 6.8 Examples radar transmit timing for uniform sampling (a), $\kappa = 2/3$ dual PRF sampling (b), and $\kappa = 2/3$ staggered-PRT sampling (c). For the dual PRF and staggered-PRT sampling schemes, $T_1 = \kappa T_2$.

to estimate T_1 following $r_a = cT_1/2$, with $T_1 \geq 400\,\mu$s. This PRT results in $v_{a1} = 19.95\,\mathrm{m\,s^{-1}}$. To achieve the unfolded velocity of $\pm 30\,\mathrm{m\,s^{-1}}$, $v_u \leq Kv_{a1}$ results in $K \geq 1.5$. Recalling that K must be an integer, selecting $K = 2$ gives $v_u = 39.9\,\mathrm{m\,s^{-1}}$, and $T_2 = 600\,\mu$s to meet both the r_a and v_u requirements for the example X-band radar.

To estimate the unfolded velocity, the radar's velocity estimates are required for each PRT. The mean Doppler velocities can be estimated for each PRT using the pulse-pair covariance techniques from Section 6.1 (pulses separated by T_1 and T_2 give velocities $\bar{v}_1$ and $\bar{v}_2$, respectively). For the dual PRF technique, the spectral-processing techniques from Section 6.2 are also commonly used separately for each sequence of pulses associated with the two PRTs. The pulse-pair mean velocity is estimated for both PRTs as

$$\bar{v}_1 = \frac{-\lambda}{4\pi T_1} \angle r_{hh}[T_1], \tag{6.77a}$$

and

$$\bar{v}_2 = \frac{-\lambda}{4\pi T_2} \angle r_{hh}[T_2]. \tag{6.77b}$$

With velocity estimates for each PRT, the unfolded velocity can be calculated as

$$\tilde{v} = -\frac{v_u}{\pi} \angle \left(e^{-j\pi \bar{v}_2/v_{a2}} e^{j\pi \bar{v}_1/v_{a1}} \right). \tag{6.78a}$$

With the lag T_1 and lag T_2 covariance estimates (shown here using the horizontal polarization), the unfolded mean Doppler velocity can be directly estimated as

$$\bar{v} = \frac{-\lambda}{4\pi (T_2 - T_1)} \arg \left(r_{hh}[T_2] r_{hh}^*[T_1] \right). \tag{6.78b}$$

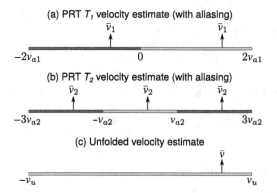

Figure 6.9 An illustration of the individual velocity estimates, $\bar{v}_1$ and $\bar{v}_2$, from two PRTs, which are aliased to cover the full spectrum $\pm v_u$. The correct unfolded velocity $\bar{v}$ in panel (c) is easily identified as the velocity that is common between the two PRTs' individual velocity estimates, which are shown replicated according to each PRTs' unambiguous velocities in panels (a) and (b).

The estimate of $\bar{v}$ can be refined by using the constraint that the mean Doppler velocity from each PRT is from the same observation volume, and therefore their de-aliased velocities should agree. The integer de-aliasing factor must be determined for each PRT, n and m, are estimated as

$$n = \text{round}\left(\frac{\bar{v} - \bar{v}_1}{v_{a1}}\right) \tag{6.79a}$$

and

$$m = \text{round}\left(\frac{\bar{v} - \bar{v}_2}{v_{a2}}\right) \tag{6.79b}$$

Using the initial mean velocity estimate from eqs. (6.77a) or (6.77b), and using the de-aliasing factor constraints from eqs. (6.79a) and (6.79b), the unfolded mean velocity estimate is refined as

$$\bar{v} = \frac{n v_{a1} + \bar{v}_1 + m v_{a2} + \bar{v}_2}{2}, \tag{6.79c}$$

with appropriate checks to ensure that the solution for $\bar{v}$ is within the span $[-v_u, v_u)$. Figure 6.9 illustrates the velocity-unfolding process using a 2/3 staggered-PRT (or dual PRF), where $K = 2$.

After de-aliasing, the final span of the unfolded mean Doppler velocity estimate is $[-v_u, v_u)$. Note that for a constant T_1, by increasing K, the maximum unambiguous unfolded velocity, v_u, also increases. As K becomes larger, the susceptibility to unfolding errors also increases and therefore, the velocity estimate uncertainty for both PRTs must also improve to maintain the same level of accuracy in the unfolding process. The uncertainty of the velocity estimates is directly related to the Doppler spectrum width. (Velocity estimator accuracy is covered in Section 7.1.) The staggered-PRT measurements are interleaved, and therefore, there is a high degree of correlation between the two velocity estimates $\bar{v}_1$ and $\bar{v}_2$. The staggered PRT will perform as well as or better than the dual PRF technique when unfolding the mean Doppler velocity.

When the velocity estimates $\bar{v}_1$ and $\bar{v}_2$ are uncorrelated, which is typical for the dual PRF scheme, the velocity unfolding estimate in eq. (6.79c) has improved accuracy compared with eq. (6.78b) for estimates that are de-aliased to the proper sector of the spectrum through correct selection of n and m. The improved accuracy comes from the averaging of the two unfolded, uncorrelated velocities in eq. (6.79c). The improvement is less significant for the staggered PRT because the velocity estimates are highly correlated (the two staggered-PRT velocity estimates use the same samples for both estimates, whereas the dual PRF uses different sets of samples). Both velocity estimates $\bar{v}_1$ and $\bar{v}_2$, including their estimator error, must be within a span of approximately $2v_u/(2K+1)$ of each other to dealias the velocity to the correct segment in the unfolded velocity span $[-v_u, v_u)$. (Consider Fig. 6.9.) For accurate estimation of $\bar{v}_1$ and $\bar{v}_2$, the precipitation's Doppler spectrum must be well sampled for all PRTs' unambiguous velocity spans, or $\sigma_v/v_a < 0.6$. See Section 7.1 for discussion of estimate error.

The dual PRF technique is more common in practice than the staggered-PRT technique. Ground-clutter filtering and other signal-processing techniques using the dual PRF method are straightforward because the radar is effectively using two different uniform sampling schemes that can be processed as independent sequences, and then the results can be averaged and used to unfold the velocity. The dual PRF uses one integration time at one PRF and then switches to another integration time at the second PRF (the time to switch between the two PRFs is not critical for performance as long as the scanning does not result in significant spatial decorrelation).

With the staggered-PRT technique, it is not straightforward to implement spectral processing as a result of the nonuniform sampling. This leads to different system and processing accommodations for the staggered-PRT operation. The staggered PRT can introduce additional technical challenges because the PRT must alternate between T_1 and T_2 from pulse to pulse, which is not conducive to some transmitter systems (e.g., magnetron).

6.4.2 Sample-Time Phase Coding

The phase of the radar's transmitted pulse (i.e., the "range-time" signal) can be a systematically varied, which is a means of implementing pulse compression. The range-time signal resolves in range, whereas the sample-time signal resolves each range cell in time (or Doppler velocity). Here, phase codes are defined as known phase shifts that are applied on a "pulse-by-pulse" basis (i.e., varying in sample time).

Magnetrons, as a by-product of how they operate, typically have a random starting phase, and as a result, they implement a random phase code for the transmitted signal. Coherent radar systems that use magnetron transmitters must implement a method of measuring the transmitted signal's phase for each pulse. The technique of measuring and correcting for the transmitter's phase using the radar's receiver is called *coherent on receive*. For radars using power amplifiers, the phase of the input signal to the amplifier is directly controlled and is termed *coherent on transmit*. Whether the transmitter's phase is measured or directly controlled, it is then used to correct and "decode" all received samples associated with that transmitted signal.

For radar systems that use magnetrons as their transmitter (and require Doppler capability), the transmitted signal's phase must be estimated, and the received signal must be decoded by the transmitted phase. For other radar systems with fully coherent transmitters (e.g., solid-state power amplifiers), phase codes can be used to improve the radar's performance for certain applications. Phase coding can help reduce the bias of radar moments from "second-trip" signals (these are echoes from pulses transmitted before the current PRT's transmitted pulse).

To implement phase coding, the radar's transmitted pulse m is phase-shifted on a pulse-by-pulse basis. The phase shift $\phi_{pc}[n]$ may be done intentionally by programming the transmitter's phase, or it can be a by-product of the radar system's design. Like the transmitted pulse, the ideal signal $x[t]$ is also phase-shifted by the code's phase for each particular pulse number, n. The received phase-coded signal from the associated pulse is

$$x_{\text{coded}}[n,t] = x[t]e^{j\phi_{pc}[n]} \propto m \cdot e^{j\phi_{pc}[n]}. \tag{6.80}$$

When ϕ_{pc} is a constant value for all pulses, the transmitter is coherent from pulse to pulse, and no phase correction is required in the receiver for coherent Doppler velocity estimation. (This assumes the receiver is also phase-locked with the transmitter.) The phase-coded signal can be decoded by multiplying it by the complex conjugate of its associated transmit phase-shift code. The decoded signal is then

$$x_{\text{decoded}}[n,t] = x_{\text{coded}}[t]e^{-j\phi_{pc}[n]} = x[t]e^{j\phi_{pc}[n]}e^{-j\phi_{pc}[n]} = x[t], \tag{6.81}$$

which implies that with a known pulse phase code, the ideal signal can be recovered. For echoes received and decoded by the radar using an incorrect phase code (e.g., second-trip signals from earlier pulses' echoes that are from beyond the radar's ambiguous range), the receiver's "decoding" step effectively applies a new phase code to the unknown received echo.

When two or more signals are contained in the measurement, but each is phase-coded differently, an estimation of the two signals can be performed. The estimators' effectiveness is limited by the signal-to-interference-plus-noise ratio. While an individual signal is being estimated, the other signal acts as a source of interference. Consider a random phase-coded transmitter and the phase correction implemented by the receiver using the current pulse's phase (the current pulse is n). The second-trip echo is coded by pulse $n-1$, but the received signal is decoded using n:

$$x_{\text{decoded}}[n,t] = x[t]e^{j\phi_{pc}[n-1]}e^{-j\phi_{pc}[n]} = x[t]e^{j(\phi_{pc}[n-1]-\phi_{pc}[n])}. \tag{6.82}$$

The phase of each sample of ϕ_{pc} is uniform random; therefore, $\phi_{pc}[n-1] - \phi_{pc}[n]$ is also uniform random on the unit circle. This implies that the Doppler spectrum of the second trip has the properties of white noise. Note that the second-trip power level is unchanged. By using a random phase code, the second trip behaves as range-dependent white noise. The noise power can be estimated and subtracted on a range-by-range basis. The second trip reduces the detection signal-to-noise-plus-interference ratio (where the interference is the second-trip echo). If the second-trip echo is to be recovered, the correct phase code must be used for decoding the received signal (i.e., $\phi_{pc}[n-1]$):

$$x_{\text{decoded}}[n-1,t] = x[t]e^{j\phi_{pc}[n-1]}e^{-j\phi_{pc}[n-1]} = x[t]. \tag{6.83}$$

As the radar's unambiguous spectrum narrows (i.e., the span of $\pm v_a$), the likelihood that the two signals' spectra will overlap one another increases. In such cases, it is generally best to transform the unwanted signal into white noise. This is optimal in the sense that the unwanted signal will, on average, have the lowest possible peak power in the decoded spectrum. Signal processing and data quality control techniques for estimating signals in the presence of additive white noise are well developed and can be readily applied here.

When using a random phase code, the radar system's coherent leakage signals and echoes from prior transmitted pulses are effectively turned into white noise. Random phase codes turn signals that are not coherent with the transmitted pulse (but coherent with the sampling clock) into white noise. Systematic phase codes typically move signals that are not the transmitted pulse (but coherent with the sampling pulse) to other spectral bins following a pattern. Radar systems with large unambiguous velocity spectrum spans (e.g., S-band systems may have $50\,\text{m s}^{-1}$ of unambiguous spectrum), systematic codes can enable multiple coded signals to be separated and recovered with minimal interference between them. This has been demonstrated to recover and separate signals within and outside of the radar's unambiguous range using systematic phase codes [181, 182]. Systematic phase codes are preferable when they can completely separate the signals of interest in the Doppler spectral domain. When the code can separate the multiple signals, it provides the highest possible SNR for the recovered signals, as long as their spectra do not overlap.

6.5 Pulse Compression

Weather radars have traditionally used a short-duration transmit pulse from a high-peak-power transmitter. For the short pulse, the radar's frequency is constant, whereas the pulse's amplitude is varied to form the pulse's envelope. The short pulse is a simple and effective method for detecting weather phenomena. To achieve the typical weather radar sensitivities and range resolutions, high-peak-power transmitter subsystems are required. The bandwidth of this amplitude-modulated pulse is related to the duration of the pulse as $B = 1/T_{tx}$. The time-bandwidth product (TBP) of the short pulse is $T_{tx} \cdot B = 1$. The unity TBP pulse (i.e., $T_{tx} \cdot B = 1$) is the most common waveform for pulsed weather radar systems, and therefore the pulse's duration T_{tx} and its bandwidth B are directly related to the radar's range resolution.

The radar's sensitivity is ultimately determined by its SNR. To increase the radar's sensitivity, either the signal power must be increased or the noise power must be decreased. For unity TBP pulsed radars, the only means of increasing sensitivity while maintaining the same range resolution (and without affecting other radar performance parameters) is to increase the transmitted power, P_t. Implementing a higher-power transmitter may have its own set of technical (and economic) challenges.

Pulse-compression techniques decouple the pulse bandwidth (i.e., range resolution) from the pulse's duration (i.e., transmitted energy). Pulse compression can be used to increase the signal's energy which improves the radar's sensitivity without the sacrifice in range resolution that would be required for a unity TBP pulse [183]. Pulse-compression techniques are used to replace higher peak-transmit power with lower peak-power transmitters that transmit the same total energy per pulse. Pulse compression results in time–bandwidth products greater than 1, where the TBP is the pulse-compression gain. By transmitting longer radar pulses, the total amount of transmitted energy is increased, but this energy is spread out over time (which equates to being spread out in range). To take advantage of the increased pulse energy, the pulse compression's long-duration waveform must be decoded to make it act like a short-duration pulse. By taking the total energy that is transmitted and shifting it to add up within a shorter time period, the waveform is "compressed," which increases the apparent power of the signal. (Recall that power is energy per unit time. Compressing the same energy into a shorter period of time thus has the effect of increasing the power.)

To correctly realign the waveform's signal so that all its components are aligned to the same time, the transmitted waveform must be uniquely encoded over its length. Think of this as assigning unique colors or unique tags to the waveform for all points along its duration. With each part of the waveform uniquely identified, the receive echoes can be compressed to achieve a range resolution that is much finer than the pulse's duration. The distribution of the pulse-compression waveform's and filter's energy is illustrated in Figure 6.10. The number of segments in Figure 6.10 is the TBP. The total length is T_{tx}, and the size of each segment is $1/B$.

Pulse compression is implemented by selecting a waveform–filter pair. The waveform (the transmitted signal) is uniquely "encoded" throughout the range-time pulse. Linear frequency modulation (LFM) is a commonly used waveform for pulse compression, where the radar's transmitted frequency linearly increases throughout the pulse's duration. The compression of the waveform (or its echoes) is implemented by convolving the echo from the waveform with its corresponding pulse-compression filter, such as a "matched filter." The matched filter is a time-reversed, complex-conjugate version of the waveform and is the filter that maximizes the output SNR for its specific waveform.

When using pulse-compression techniques, the combination of the transmitter's waveform and receiver's filter is critical to the overall sensitivity and performance of the radar system. The convolution of the received signal with the pulse-compression filter affect both the echo and noise. The pulse-compression waveform's echo power increases by the TBP gain factor. The noise power of the signal remains unchanged by the pulse-compression filter (the pulse-compression filter nominally only modifies the phase of the signal). From eq. (5.61a), when averaging N samples of noise, the mean power is reduced by a factor of N. The convolution is a summation of N = TBP

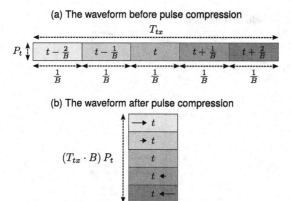

(a) The waveform before pulse compression

(b) The waveform after pulse compression

Figure 6.10 An illustration of how pulse compression increases the radar system's apparent transmitted power. (a) The actual pulse-compression waveform's energy is spread out over a pulse length of T_{tx} with a peak power P_t. For determining the ideal gain of the pulse compression, the pulse is divided into sections $1/B$ in duration, giving a total of $T_{tx} \cdot B$ sections. The different sections are indicated as different gray levels, which correspond to different ranges of frequencies of the LFM chirp. (b) When the waveform is compressed by the pulse-compression filter, the sections of the filtered waveform are effectively shifted into a single section with width $1/B$. The energy from all sections add together to achieve the performance of a unity-TBP radar transmitter with power $(T_{tx} \cdot B)P_t$ and a pulse length of $1/B$.

samples, not an average. Therefore, there is no change in the noise power as a result of the pulse-compression filter.

With pulse-compression techniques, radar systems are no longer bound to high-peak-power transmitters to achieve radar sensitivity requirements. Recalling that the total transmitted energy is $P_t \cdot T_{tx}$, with pulse compression, there is an additional trade-off available for radar system designers between the transmitted pulse's duration, T_{tx}, and the peak transmit power, P_t. Pulse compression allows radar engineers to use solid-state transmitters, which typically have lower peak powers, to achieve the desired radar sensitivity.

Pulse compression is an extensive topic with numerous possibilities for technical implementation. They all fundamentally accomplish the same goal of providing more signal energy with a fixed "code" that can be used to focus the long-duration transmit waveform through signal-processing techniques into a shorter, localized response. The following discussion is an introduction to this important concept of pulse compression for weather radar systems.

6.5.1   Blind Range

As a side effect of the relatively long pulse durations used in pulse compression, a "blind range," or minimum range at which targets can be detected, is imposed. (Note that unity time–bandwidth radar systems also have blind ranges, but these are typically

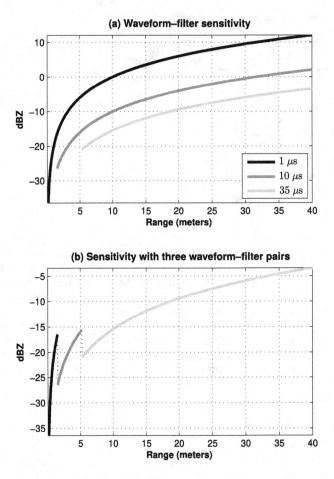

Figure 6.11 (a) Example sensitivity curves for three pulse widths (T_{tx}), all with the same bandwidth (pulse compression is used). Note that the blind range increases as the pulse width increases. (b) A combination of the three pulses can be selected to achieve the highest sensitivity over all ranges by relying on the shorter pulse widths to fill in the blind range of the long pulse widths.

the result of transmit/receive duplexer switching delays.) The additional blind range imposed by the transmitted pulse's duration may be significant (many kilometers). The relative increase in sensitivity also has an increase in blind range (see Fig. 6.11). To mitigate the radar's blind range, multiple pulses, each with a different time–bandwidth TBP, can be used to provide full range coverage. When using multiple pulses, the pulse with the maximum sensitivity should be used for all ranges where the filter's response is valid (e.g., not in the blind range).

Pulsed radars with TBP = 1 use a single waveform to detect clouds and precipitation. This implies a single sensitivity curve whose sensitivity varies with $1/r^2$, as shown in panel (a) of Figure 6.11. Pulse-compression radars can adopt multiple pulses with varying sensitivities. The traditional $1/r^2$ limitation on sensitivity is fundamentally broken, and a "sawtooth"-shaped sensitivity is the new paradigm. Additional parameters are available to radar designers and operators to tune the performance of weather radar beyond those strictly found in the radar equation.

The multiple pulses can be implemented by time-multiplexing the pulses or by using frequency multiplexing. When using frequency multiplexing, the individual waveforms are used with slightly different center frequencies (so that there is no overlap between the waveforms' spectra). The shorter pulses, which are used to fill in the blind range, are to be transmitted last. When using multiple frequency-diverse pulses, the pulses are transmitted sequentially, one directly after another, from longest to shortest T_{tx}. Note that the additional time to transmit the shorter pulse at the end of the transmit sequence slightly increases the blind range of the longer pulses. The multiple frequency-diverse pulses are then received simultaneously and must be filtered and processed individually. In practice, a short-duration (typically a unity-TBP pulse) and a long-duration pulse are typically sufficient for meeting the radar's sensitivity requirements.

6.5.2   Radar Sensitivity with Pulse Compression

Weather radar sensitivity was explored in Section 2.7. Here, radar sensitivity is revisited in the context of pulse compression to evaluate how the radar's waveform and filter affect its sensitivity. From the weather radar (eq. [2.38b]), the received echo power is

$$P_r = P_t \frac{\lambda^2 G_r G_t}{(4\pi)^3 r^4} \left(\frac{cT_{tx}}{2} \frac{\pi \theta_1 \phi_1}{8 \ln 2} r^2 \right) \bar{\eta}. \tag{6.84}$$

If the transmit waveform's duration increases while all other terms are kept constant, the received echo power also increases. The weather radar's sensitivity is proportional to the transmitter's peak power, P_t. A longer transmitted pulse results in a higher total power at the receiver. (Remember the assumption here that the observation space is uniformly filled with homogeneous targets with reflectivity $\bar{\eta}$.) After pulse compression, the ideal range resolution of the pulse is directly related to the bandwidth of the waveform following $c/(2B)$. In eq. (6.84), the range extent of the pulse is implied to be $cT_{tx}/2$, which, for pulse compression, can be multiple kilometers. From Section 2.4.2, the range resolution for TBP $= 1$ waveforms is $\Delta r = \frac{c}{2B} = \frac{cT_{tx}}{2}$. This is an important subtlety in the assumption of the range resolution when determining the radar constant with pulse compression. In general, the range resolution should always be approximated by

$$\Delta r = \frac{c}{2B}, \tag{6.85}$$

with the caveat that the actual final bandwidth of the filtered waveform should be used for B when available. The waveform's window function and filter can modify the bandwidth compared with the nominal value.

From eq. (2.39c), the reflectivity is estimated from the received power and the radar constant $Z_e = P_o r^2 / C$. The weather radar constant from eq. (2.39d) is modified to use the bandwidth (B) as a proxy for the range resolution, and an explicit accounting of the pulse compression's TBP, waveform loss (L_w), and filter loss (L_f) are added:

$$C = \frac{T_{tx} \cdot B}{L_w L_f} \frac{P_t G^2 \pi^3 |K_w|^2 \Delta r \, \theta_1 \phi_1}{10^{18} \lambda^2 512 \ln 2}. \tag{6.86}$$

Here, the TBP's bandwidth B is the ideal bandwidth, not the effective half-power bandwidth due to amplitude window function or other effects. Effects that reduce the realized bandwidth (and therefore the pulse-compression gain) are accounted for in L_w and L_f, which will be discussed in detail in Section 6.6.3. Also note that G_{rx} from (2.39d) is replaced by the total pulse-compression gain $\frac{T_{tx} \cdot B}{L_w L_f}$. With further simplification of the radar constant in eq. (6.86), the bandwidths from the TBP and the range resolution cancel, and only T_{tx} remains, which is consistent with eq. (6.84). When working with real radar systems, an additional scale factor may be required to relate the digitizer's measured power to the actual power at the receiving antenna, which can be different as a result of the amplifiers and components in the receiver. The power at the receiving antenna is what is accounted for by the radar equation derivation in Section 2.5.

The radar's sensitivity is directly related to $\text{SNR} = P_r/P_n$. The minimum detectable reflectivity is determined by the receiver's noise power (see Section 2.7.2). The noise power is proportional to the bandwidth, B, from eq. (2.54):

$$P_n = k_B T_0 F B. \tag{6.87}$$

where k_B is Boltzmann's constant, $T_0 = 290°$ Kelvin input noise reference temperature, and F is the receiver's noise factor. For a TBP $= 1$ pulse, increasing the transmitter's pulse duration reduces the receiver's bandwidth. For pulse compression, the bandwidth and pulse length are decoupled (keeping TBP > 1). Reducing the bandwidth can increase the radar's sensitivity, but it is at the expense of range resolution (following $\Delta r = \frac{c}{2B}$).

Consider the waveform and transmitter performance characteristics in the detection of a single-point target (this could be a small aircraft, a single bird, or a building). This represents a single target whose radar cross-section (RCS) is constant, and the scatterer remains completely within the range volume for the bandwidths considered. At any instant in time, the instantaneous echo power is proportional to $P_t \sigma$. When using pulse-compression, the pulse-compression gain factor is also included: $T_{tx} \cdot B$. The SNR for a point target is

$$\text{SNR}_{\text{point}} = \frac{P_r}{P_n} \propto \frac{P_t \sigma (T \cdot B)}{B} = P_t T_{tx} \sigma. \tag{6.88}$$

For the point target, the bandwidth does not alter the SNR because the single target is always completely contained within the range volume.

Weather phenomena are volume targets whose RCSs are assumed to be distributed uniformly over radar's observation volume. The distributed weather target's RCS scales with the size of the radar volume: $\sigma = \bar{\eta} V_6$. The effects of the transmitter and waveform's properties on the SNR for the volume of scatterers can be evaluated.

With eqs. (6.86) and (6.87) for the signal and noise power, respectively, the SNR for a volume of scatterers is

$$\text{SNR}_{\text{volume}} = \frac{P_r}{P_n} \propto \frac{P_t \bar{\eta}(T_{tx} \cdot B)}{B^2} = \frac{P_t T_{tx} \bar{\eta}}{B}. \tag{6.89}$$

For a volume of distributed targets, decreasing the bandwidth increases the SNR via a larger radar volume. The bandwidth determines the range resolution, and therefore $V_6 \propto \Delta r \propto 1/B$.

For both the point target and distributed-volume target, pulse compression can increase the observation's SNR. The relative performance between two idealized waveform–filter pairs can be compared for detecting weather echoes. The ratio of SNRs from two waveform–filter pairs for the same radar system, indicated by the superscripts $^{(1)}$ and $^{(2)}$, is

$$\frac{\text{SNR}^{(2)}}{\text{SNR}^{(1)}} = \frac{P_n^{(1)} P_r^{(2)}}{P_r^{(1)} P_n^{(2)}} = \frac{B^{(1)} T_{tx}^{(2)}}{T_{tx}^{(1)} B^{(2)}}. \tag{6.90}$$

This follows eq. (6.89), which assumes a uniformly distributed volume of scatterers and also assumes all other radar system parameters remain constant. The relative change in the sensitivity, referenced to a 1-MHz and 1-μs pulse (i.e., a unity time–bandwidth product), is shown in Figure 6.12 for a range of pulse durations and bandwidths. (Depending on the radar system's performance requirements, the pulse width and bandwidth can extend beyond those shown in the figure.) The circle represents the 1-MHz/1-μs pulse's reference point's gain (at 0 dB). This reference point is arbitrary; what matters is the alteration of sensitivity between two points to determine the radar's change in sensitivity in response to changes in the waveform–filter pair's bandwidth and pulse width. Figure 6.12 can be used as a first-order design evaluation for pulse-compression waveform–filter pairs.

The solid black line in panels (a) and (b) of Figure 6.12 represents the continuum of TBP $= 1$ pulses as the pulse's bandwidth (or range resolution) is changed. Because the time and bandwidth are coupled for unity-TBP pulses, a change to the pulse parameter provides a gain (or loss) to both the total signal energy via the pulse width and the noise power via the receiver's bandwidth. For time-bandwidth combinations in the shaded areas above the TBP $= 1$ line, pulse-compression techniques must be used. In practice, the SNR actual gains vary slightly as the waveform and filters are tuned to optimize performance in other areas (this is covered in detail in the following discussions). For waveform–filter pairs with $T_{tx} \cdot B > 1$, the time and bandwidth are decoupled. The same pulse compression performance information are shown in both panels of Figure 6.12. The range resolution and bandwidth have an inverse relation and the blind range and pulse width are proportional to each other. By using pulse-compression techniques (i.e., $T_{tx} \cdot B > 1$), two independent parameters (the bandwidth and pulse width) are available to determine the radar system's performance. Typically, the range resolution sets the bandwidth parameter. For $T_{tx} \cdot B = 1$, only a single parameter is available to the radar designer or operator to tune the radar's sensitivity which is typically constrained by the resulting range resolution.

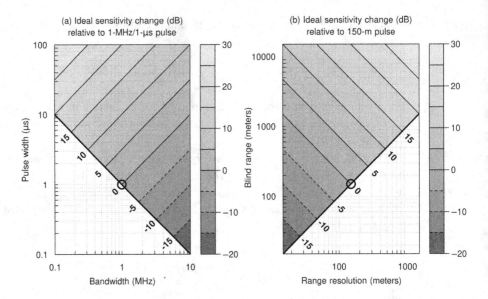

Figure 6.12 Pulse-compression design curves for idealized waveform–filter pair performance. The figures show the relative sensitivity change compared with a reference 1-MHz and 1-μs pulse. The relative change for alternative reference points can be calculated by simply taking the difference between two points on these panels. The blind range ($cT_{tx}/2$) and range resolution ($c/(2B)$), where c is the speed of light, in panel (b) are directly related to the pulse width (T_{tx}) and bandwidth (B) in panel (a). The solid black line indicates TBP = 1; the shaded gray areas are TBP > 1, which require pulse-compression techniques; and the white area is TBP < 1, which does not provide any gains.

6.6   Pulse-Compression Waveforms and Filters

6.6.1   Linear Frequency-Modulated Waveforms

By modulating the frequency of the transmitted waveform throughout the pulse, the pulse's instantaneous frequency can be used as a means to help resolve the targets in a range smaller than that allowed by the pulse width. Specifically, LFM can be used to generate the transmitted waveform, where the total bandwidth B is linearly swept over the pulse's duration T_{tx}. The frequency can linearly increase or linearly decrease. Typically, a linear increase is used. If the LFM waveform is audible, it sounds like a chirp. Hence, LFM waveforms are commonly referred to as "chirps." The bandwidth and pulse duration are decoupled so that $B > 1/T_{tx}$ (and $TBP > 1$). The range resolution is approximated by $\Delta r = c/(2B)$. Once the range resolution is determined, the radar's sensitivity can be improved by increasing T_{tx}. LFM waveform–filter pairs can provide pulse-compression gain with TBP $\gg$ 1.

For an LFM waveform centered at the radar's operating frequency f_c, the radar's instantaneous transmit frequency is

$$f(t) = f_c + \frac{B}{T_{tx}}t, \tag{6.91}$$

where the time, t, spans $[-T_{tx}/2, T_{tx}/2]$. The LFM chirp is a simple and effective pulse-compression waveform. At baseband (where $f_c = 0$), the LFM waveform's instantaneous phase is calculated by

$$\phi_{\text{LMF}}(t) = 2\pi \frac{B}{2T_{tx}} t^2. \tag{6.92}$$

A linear increase in frequency is a quadratic increase in phase. The LFM can be combined with an amplitude taper, commonly implemented as a window function $w[t]$, give a complex-valued (i.e., IQ) waveform:

$$m[t] = w[t] e^{j\phi_{\text{LFM}}[t]} = w[t] \exp\left(j2\pi \frac{B}{2T_{tx}} t^2 \right). \tag{6.93}$$

The instantaneous frequency of the waveform is simply the derivative of the phase. The baseband LFM waveform's instantaneous frequency (in Hertz) is

$$f_{\text{LFM}}(t) = \frac{1}{2\pi} \frac{d\phi_{\text{LFM}}}{dt} = \frac{B}{T_{tx}} t, \tag{6.94}$$

which matches eq. (6.91) with $f_c = 0$. The rate at which the chirp's frequency changes is B/T_{tx}.

Pulse-compression techniques enable longer pulse widths to be used without sacrificing range resolution; however, the transmitted pulse's energy still covers a long duration, T_{tx}. The pulse's "compression" is performed by convolution of the waveform with the pulse-compression filter. Ideally, this would result in all of the waveform's energy being localized into a single range gate with a width of the range resolution. Because of the finite lengths of the waveform and filters, the pulse compression is not perfect, and a small fraction of the waveform's energy manifests as range side lobes over the extent of the pulse's duration. Careful design of the pulse-compression waveform and filter can control the range side lobes to acceptable levels and is discussed in Section 6.6.3. If the range side lobes are not adequately managed, weaker echoes, such as cloud or light rain, can be obscured by the range side lobes of strong echoes, such as ground clutter.

A larger class of nonlinear frequency modulation (NLFM) waveforms may also be used for pulse compression. The NLFM waveform's frequency may vary as piecewise linear, quadratic, or even have higher-order terms [184–186]. In general, NLFM waveforms can achieve lower-range side-lobe levels compared with similar TBP LFM waveforms when using a matched filter. NLFM waveforms effectively replace the amplitude taper typically used by LFM waveforms to control side-lobe levels with a "phase" taper. Without using an amplitude taper, the pulse-compression SNR is improved slightly. By using LFM waveforms with optimized "mismatched" pulse-compression filters [187, 188], negligible differences between the side-lobe performance of LFM and NLFM waveforms can be achieved. The LFM waveforms have superior sidelobe performance in the presence of Doppler shifts [188]. For these two reasons, only LFM waveforms are covered here.

6.6.2   Pulse-Amplitude Shaping

The most basic transmitted waveform, such as the unity-TBP pulse, is a constant-frequency pulse whose performance is defined by an amplitude-weighted envelope. The length and shape of the pulse control the waveform's frequency spectrum. Defining the pulse's envelope, with a focus on the rising and falling edges, is referred to as windowing, amplitude tapering or apodization. For pulse compression, amplitude tapers are also used to control range side lobes, but the performance is also coupled to the selection of B and T_{tx}.

The unity time–bandwidth pulse provides a single degree of freedom for tuning the radar performance (for TBP $= 1$, T_{tx} and B are the inverse of one another). With the amplitude taper, an additional parameter is introduced for tuning the waveform performance, but its effect is coupled with the selection of B (or T_{tx}).

The amplitude taper is typically implemented with one of the various window functions (e.g., the window functions used for spectral processing may be used). The "frequency-domain" response of the window function is representative of its range response when used as an amplitude taper for pulse compression. (The time domain response is the range-response for unity-TBP pulses.) The Tukey window has a taper fraction (α) as its design parameter and is an ideal generalized amplitude taper. For a pulse from time $t \in [0, T_{tx}]$, the Tukey window is given by

$$w[t] = \begin{cases} 0.5 \left[1 - \cos \left(\frac{2\pi t}{\alpha T_{tx}} \right) \right] & 0 \le t < \frac{\alpha T_{tx}}{2} \\ 1 & \frac{\alpha T_{tx}}{2} \le t \le T_{tx}(1 - \frac{\alpha}{2}) . \\ 0.5 \left[1 - \cos \left(\frac{2\pi t}{\alpha T_{tx}} - \frac{2\pi}{\alpha} \right) \right] & T_{tx}(1 - \frac{\alpha}{2}) < t \le T_{tx} \end{cases} \qquad (6.95)$$

When $\alpha = 0$, the Tukey window is a rectangular window. When $\alpha = 1$, the Tukey window is a Hann window (sometimes called a *Hanning window*). Examples of the amplitude taper and the resulting effect on the spectrum of the pulse are shown in Figure 6.13.

For the unity time–bandwidth pulsed radar system, the transmitted waveform is simply the amplitude taper $m[t] = w[t]$. The amplitude taper is typically used to modify the frequency spectrum of the signal, as shown in Figure 6.13, and tune the waveform's performance for specific radar hardware or applications.

The weather radar equation assumes the total transmitted energy is $P_t \cdot T_{tx}$, which implies a rectangular pulse amplitude. The waveform loss, L_w, is the ratio of the transmitted energy compared with an ideal rectangular pulse. To calculate L_w, it's assumed that the waveform, m, has an amplitude in the range $[0, 1]$, with a peak of 1. The ideal rectangular pulse has a constant-unity amplitude. The power is proportional to the amplitude-squared, and the waveform's loss is calculated as

$$L_w = \frac{N}{\sum\limits_{n=0}^{N-1} |m[n]|^2}. \qquad (6.96)$$

The "loss" terms are always greater than or equal to 1. The waveform loss compensates for the actual amount of energy in the transmitted signal compared with the idealized

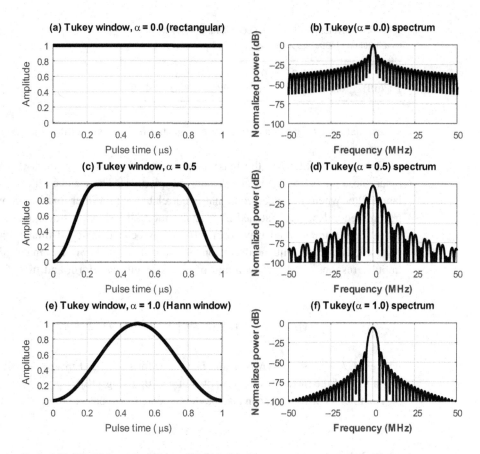

Figure 6.13 The Tukey window amplitude and its power spectrum are shown for various taper fractions, α. A 1-μs pulse is used to illustrate the relationship between the amplitude of the pulse over time on the left, and the corresponding frequency spectrum of the pulse on the right. The time and frequency may be scaled appropriately for other pulse on the right widths following the relationship $t \propto 1/f$ (e.g., doubling the duration halves the frequency spectrum).

rectangular pulse that is assumed by the weather radar equation. (This correction for the radar constant applies to both pulse-compression systems and TBP $= 1$ pulses.) The waveform losses for the Tukey window with various taper fractions are shown in Table 6.1.

6.6.3 Pulse-Compression Filters and Performance

Pulse compression is one tool available to meet performance requirements for weather radar applications. The high-level goals when designing pulse compression waveform-filter pairs are to maximize gain and minimize range side lobe levels while achieving a desired range resolution. The width of the compressed waveform's main-lobe is the range resolution. Outside the main-lobe, the side lobes are the unwanted byproducts

Table 6.1 Waveform loss, L_W, in decibels for a Tukey window as a function of taper fraction, α

α	0.0	0.2	0.4	0.6	0.8	1.0
L_w (dB)	0.00	0.62	1.29	2.08	3.05	4.30

of residual energy that was not focused into the main-lobe and that can extend over the length of the pulse.

The matched filter has the maximum chance of detecting a signal (for weather radar, these are echoes from clouds and precipitation). The matched filter is a correlation filter where the received signal is correlated with a filter that is a copy of the transmitted pulse. The matched filter is specific to the signal it's designed to detect. For a radar's transmitted waveform $m[t]$, its associated matched filter is the time-reversed, complex-conjugated version of itself: $h = m^*[-t]$. The matched filter's impulse response is the convolution of the N-length waveform and filter following eq. (5.28):

$$y[n] = \sum_{k=0}^{N-1} x[k]h[n-k] = \sum_{k=0}^{N-1} m[k]m^*[k-n], \qquad (6.97)$$

where $-(N-1) \leq n \leq N-1$ and $m[t] = 0$ for $t < 0$ and $t > N-1$. The peak magnitude of the matched filter output is at $n = 0$, giving $y[0] = \sum_{k=0}^{N-1} |m[k]|^2$.

The filter's convolution can also be implemented as a matrix multiplication:

$$\mathbf{y} = \mathbf{Xh}, \qquad (6.98)$$

where $\mathbf{X}$ has the same number of columns as the length of the filter, $\mathbf{h}$ (note that N is used here to denote the filter length and is not related to the number of sample-time measurements). The signal vector x is the range-time signal with a length of M. Each row corresponds to the signal's samples that are to be compressed to give the pulse compression response for a given range gate. The signal matrix is constructed as follows:

$$\mathbf{X} = \begin{bmatrix} x_1 & 0 & \cdots & 0 & 0 \\ x_2 & x_1 & \cdots & 0 & 0 \\ \vdots & & \cdots & & \vdots \\ x_N & x_{N-1} & \cdots & x_2 & x_1 \\ \vdots & & \cdots & & \vdots \\ x_M & x_{M-1} & \cdots & x_{M-N+2} & x_{M-N+1} \\ \vdots & & \cdots & & \vdots \\ 0 & 0 & \cdots & x_M & x_{M-1} \\ 0 & 0 & \cdots & 0 & x_M \end{bmatrix}. \qquad (6.99)$$

As constructed, the signal x is "zero padded" to implement the full convolution, and therefore the signal matrix $\mathbf{X}$ has $M + N - 1$ rows and N columns.

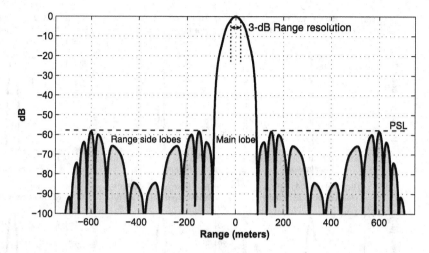

Figure 6.14 The matched-filter range response for an LFM waveform with a pulse width of $T_{tx} = 5$ μs, a bandwidth of $B = 5$ MHz, a range-time sample rate of 5 Msps, and a Hann amplitude taper (i.e., Tukey window with $\alpha = 1$). The resulting range resolution is approximately 53 m (determined by the -3-dB, or half-power, main-lobe width). The total width of the mainlobe is 187 m. For the side-lobe region (the area outside of the mainlobe), the PSL is -58.4 dB, and the ISL is -53.4 dB.

To evaluate the impulse response of the waveform ($\mathbf{m}$) and filter ($\mathbf{h}$) pair, the waveform is used as the signal vector (i.e., $\mathbf{x} = \mathbf{m}$). Here, it's assumed that the filter and waveform both have the same number of samples, and therefore, $M = N$. The waveform–filter pair's impulse response is their convolution, and the result $\mathbf{y}$ has $2N - 1$ samples. To implement the convolution using eq. (6.98), the signal matrix $\mathbf{X}$ has $2N - 1$ rows and N columns. The Nth position ($y[N] = \mathbf{x}^T\mathbf{h} = \mathbf{m}^T\mathbf{h}$) of the impulse response is typically the center of the impulse response's "main lobe" with the peak output power. (Note, this is equivalent to solving eq. (6.97) for $n = 0$.)

Figure 6.14 illustrates the match-filtered impulse response of an LFM pulse with a nominal TBP $= 25$. The main-lobe area is identified as the center area where the range response peaks. The side lobes are the area outside of the main lobe, and their extent is determined by the waveform's and filter's lengths. The range response shown, represented by the main lobe and side lobes, is the ideal response observed by the radar for echoes from a point target. Ground clutter, such as a building or water tower, can be represented as point targets that can have a large RCS, and therefore the clutter's side-lobe levels may be higher than the noise-power level and thus detectable by the radar.

The peak side-lobe level (PSL) is the power level of the highest-range side lobe measured relative to the mainlobe's peak. The integrated side-lobe level (ISL) is the integral of all power in the range side lobes normalized by the total power of the pulse-compression range response:

$$\text{ISL} = \frac{\int_{\text{sidelobes}} |y(r)|^2 \, dr}{\int_{-\infty}^{\infty} |y(r)|^2 \, dr}. \tag{6.100}$$

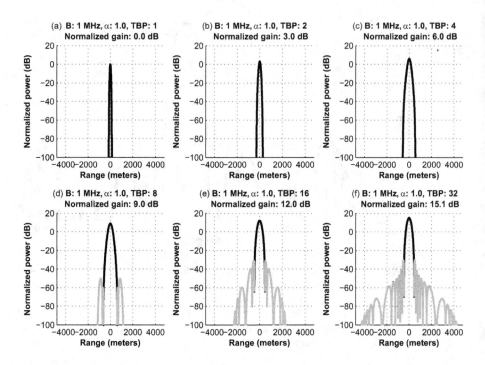

Figure 6.15 The main lobe and side lobes of the match-filtered pulse-compressed signals are highlighted (delineated by the first null in the range response) for a variety of time–bandwidth products. The sampling bandwidth for these plots was set to be 100 times the pulse's bandwidth, B. The high sampling rate mitigates any aliasing effects resulting from bandwidth modification by the taper and provides high fidelity to visualized the ideal pulse-compressed response. The peak power for each TBP indicates the relative gain compared with a rectangular pulse with TBP = 1. All pulses are implemented with an $\alpha = 1$ Tukey amplitude taper.

For some evaluations of PSL and ISL, the transition from the "main-lobe" to "side-lobe" areas may not be strictly determined by the first null in the range response. Part of the main lobe may be included as side lobes; or vice versa, part of the range side lobes may be considered part of the main lobe for the calculations. For these estimates of PSL and ISL, the main lobe is identified as the area around the center peak that extends to the first null (or change in the sign of the slope) on either side. See Figure 6.15. Note that the range-time sampling rate should be selected to be greater than or equal to the waveform's bandwidth (i.e., following the Nyquist sampling rate). Although the pulse may be designed with a bandwidth B, the effects of the taper can alter the transmitted signal's spectrum. The sampling rate is a critical parameter when designing the pulse-compression waveform and filter pairs.

As the TBP increases, the side-lobe levels decrease. When implemented in radar systems, system phase noise typically limits the realizable side-lobe levels. The matched filter ideally should account for the amplitude and phase distortions of the transmitter and receiver systems to maximize the waveform–filter pair's performance. (Sampling the transmitted signal using the radar's receiver, after the up- and down-conversion through the radar's RF subsystem, can provide this measurement.) Although it is not universal, there is typically a minimum TBP required to realize the pulse-compression gains. Consider a Tukey window with $\alpha = 1$ taper effect on a TBP $= 1$ waveform. The Tukey window effectively decreases the time duration of the pulse, which in turn increases the frequency bandwidth of the pulse. If instead, a long-duration LFM pulse is selected and then a taper is applied, the effect of the taper is to reduce the overall time and bandwidth of the LFM waveform. In between the unity-TBP and high-TBP waveforms, there is a transition where the taper's effect goes from increasing the transmitted signal's bandwidth (by reducing the pulse width) to decreasing the signal's bandwidth (by reducing the amplitude of the LFM waveform where the minimum and maximum frequencies occur). Figure 6.16 shows the effect of the TBP on the range resolution. Compare the responses shown here to the effects of α on the window function shown in Figure 6.13. These two figures help illustrate the relationship between the level of the side lobes, the main lobe width, and the TBP gain. As a rough guideline, the minimum TBP is on the order of 10 to begin to realize the benefits of pulse compression (otherwise, a unity-TBP pulse with similar T_{tx} should be considered and compared against for performance).

For strong scatterers, such as hail or ground clutter, the range side lobes can be significantly higher than the precipitation within the side lobes' ranges. Without proper design and control of the pulse-compression side lobes, they can mask the precipitation signal the radar is supposed to detect. There are alternatives to the matched filter that can offer improved pulse-compression side-lobe performance. The minimum integrated side-lobe filter [187] can provide superior side-lobe performance over the matched filter, with only a slight reduction in the pulse-compression gain. The "mismatched" filters can provide a better control than the amplitude taper alone for managing range side lobe levels and gain (which includes TBP, L_w and L_f). For the Hann window (i.e., $\alpha = 1$), the matched filter is the minimum-ISL filter (when estimated using the L2-norm). The disadvantage of the Hann window is the larger waveform loss (L_w) and reduced range resolution compared with the nominal range resolution for the waveform's total bandwidth. A waveform with a lower taper fraction and minimum-ISL filter could meet the range side-lobe performance requirement while providing better range resolution and higher processing gain. The pulse-compression gain, range resolution, and range side-lobe levels typically must be considered together when designing and optimizing the waveform filter pairs. (A more in-depth discussion of minimum ISL filters can be found in [187].)

The pulse-compression output vector is $[y[1] \cdots y[2N - 1]]^T$, which is a direct result of each row of the signal matrix in eq. (6.99). Each row corresponds to a range of the pulse-compression impulse response. If the center of the impulse response is N, and l is the sample distance relative to the peak of the main lobe, then the main lobe indices are defined as $|l| \leq l_{\text{isl}}$. The indices defining the range side lobe are $|l| > l_{\text{isl}}$.

To minimize the integrated side-lobe power, the regions defining the side lobes are selected by tuning l_{isl}. The number of samples identified as the main lobe is $m_{\text{isl}} = 2l_{\text{isl}} - 1$. For all of the rows of $\mathbf{X}$ associated with the side lobes, a matrix $\mathbf{X}_f$ is constructed. The rows of $\mathbf{X}_f$ are a subset of the full waveform convolution matrix $\mathbf{X}$, only $\mathbf{X}_f$ has the rows of the main lobe excluded. Matrix $\mathbf{X}_f$ has $2N - 1 - m_{\text{isl}}$ rows

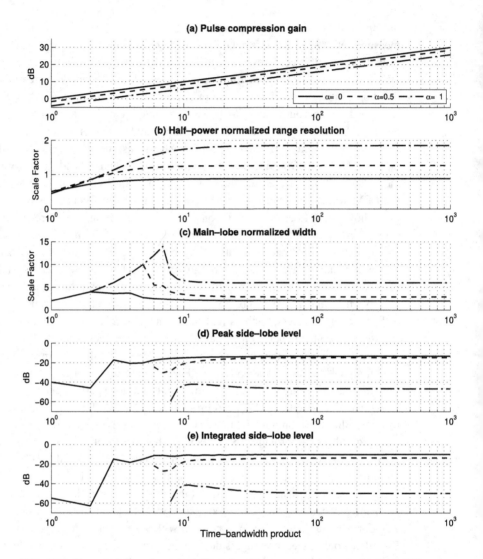

Figure 6.16 Characteristics of matched-filtered, pulse-compressed signals for three different Tukey taper coefficients are shown. For estimation of these parameters, the sampling rate is set to be 100 times the fundamental bandwidth, B. The gain in (a) is relative to to the power of a unity-TBP rectangular pulse. The half-power range resolution in (b) and the main-lobe width in (c) are normalized to the transmitted signal's nominal range resolution, neglecting the effect of the taper (e.g., $B = 1$ MHz, $\Delta r = 150$ m). The estimates of the PSL in (d) and the ISL in (e) define the side lobes as the range response outside of the main lobe (the main lobe is the area between the first nulls, as shown in Fig. 6.15).

and N columns. Although the filters are optimal in terms of integrated side-lobe levels for the inputs, the resulting filter may significantly attenuate the main lobe if l_{isl} is too small (or conversely, it can result in undesirable grating lobes within the "main-lobe" region). Selection of a truly optimal pulse-compression filter can be an iterative process. The optimal selection of l_{isl} can vary with changes in the waveform parameters, including the sample rate, α, T_{tx}, and B. For the specific waveform $\mathbf{x} = \mathbf{m}$, the filter that results in the minimum integrated side-lobe levels, in a mean-squared-error sense, is calculated as

$$\tilde{\mathbf{h}} = \mathbf{C}^{-1}\mathbf{m}. \tag{6.101}$$

The filter $\tilde{\mathbf{h}}$ is the un-normalized optimal filter, and $\mathbf{C}$ is the covariance of all time lags (i.e., ranges) over which the side-lobe power is to be minimized. The covariance matrix $\mathbf{C}$ is given as

$$\mathbf{C} = \mathbf{X}_f^H \mathbf{X}_f. \tag{6.102}$$

The optimal minimum-ISL pulse-compression filter is $\mathbf{h}$, which is $\tilde{\mathbf{h}}$ normalized to unity power:

$$\mathbf{h} = \frac{\tilde{\mathbf{h}}}{\sqrt{\tilde{\mathbf{h}}^H \tilde{\mathbf{h}}}}, \tag{6.103}$$

where ^{H} is the complex-conjugate transpose. The minimum-ISL filter in eq. (6.101), normalized to unity gain by eq. (6.103), is then convolved with the signal vector using eq. (6.98).

For the estimation of filter loss, integrated side-lobe levels, and peak side-lobe levels, it is assumed that $\mathbf{m}$ is normalized to unity power following the form in eq. (6.103). With this normalization, the performance metrics for the waveform–filter pair are easily estimated. (It's assumed that $\mathbf{C}$ and $\mathbf{X_f}$ are constructed from the power normalized waveform, $\mathbf{m}$ for these calculations.) The filter's loss is the ratio of the matched filters peak gain to the peak of the filter's gain is

$$L_f = \frac{1}{|\mathbf{m}^T\mathbf{h}|^2}. \tag{6.104}$$

The integrated side-lobe level is calculated from the waveform, filter, and matrix $\mathbf{C}$ as

$$\text{ISL} = \frac{|\mathbf{h}^H\mathbf{Ch}|}{|\mathbf{m}^T\mathbf{h}|^2}. \tag{6.105}$$

The denominator of the ISL is the peak filter gain at time lag zero. The peak side-lobe level for the waveform–filter pair is

$$\text{PSL} = \frac{\max |\mathbf{X}_f\mathbf{h}|^2}{|\mathbf{m}^T\mathbf{h}|^2}. \tag{6.106}$$

For the waveform and minimum-ISL filter pair in Figure 6.17, the main lobe is defined as $l_{\text{isl}} = 3$ (i.e., $m_{\text{isl}} = 7$ range samples). The ideal TBP is 46 (or 16.6 dB) for $T_{tx} = 20$ μs and $B = 2.3$ MHz. With a sample rate of 3 Msps, this corresponds to a range of 50 m per sample, and therefore the main lobe, for the purpose of evaluating side-lobe levels, is 350 m wide. It's important to note that because the minimum integrated side-lobe optimization acts on a sampled signal, the sampling rate is also

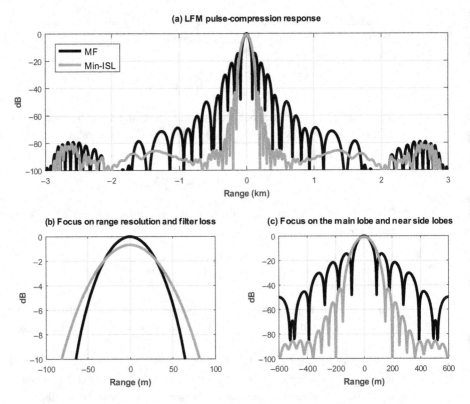

Figure 6.17 Comparison of the pulse-compression responses for an LFM waveform with $T_{tx} = 20\,\mu s$, $B = 2.3$ MHz, a range-time sample rate of 3.0 Msps, and a Tukey window with $\alpha = 0.4$. The impulse responses using a matched filter and a minimum-ISL mismatched filter (using $l_{isl} = 3$) are shown. The minimum-ISL filter has slightly less pulse-compression gain as a result of the mismatch, which is the filter loss.

a parameter to consider when tuning the filter's performance with l_{isl} (the sample rate modifies the selected main-lobe width). The actual half-power main-lobe width is illustrated in panel (b) of Figure 6.17. The waveform taper loss is $L_w = 1.29$ dB, which is independent of the filter. The filter loss for this minimum-ISL filter is $L_f = 0.68$ dB. The minimum-ISL filter's ISL is -67.1 dB, and its PSL is -79.8 dB. Compare this to the matched filter using the same ranges for the main lobe, the PSL is -17.7 dB and the ISL is -13.1 dB. The matched filter's loss is $L_f = 0$ dB by definition.

It is worth noting again that in practice, the effects of the radar system's components, such as the filter, amplifiers, mixers, and data converters, can distort the ideal waveform. These effects can modify the amplitude and phase of the received waveform. Although these effects are typically small for well-designed radar systems, they increase side-lobe levels. If the distortions can be measured, they can be incorporated into the pulse-compression filter design to maximize the waveform–filter pair's performance. The filter can be modified to account for the system's distortion so that it better matches the radar's true signal measured at the receiver. Although the

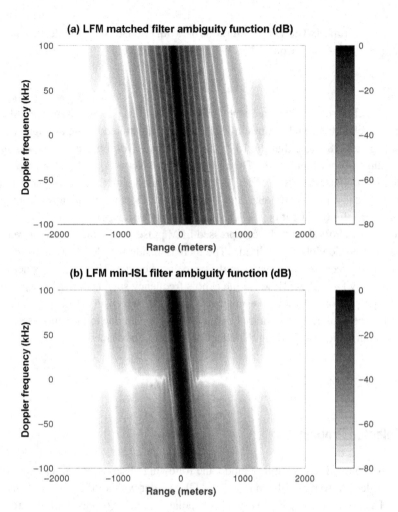

Figure 6.18 Example range Doppler ambiguity function for LFM waveform using the matched filter (a) and the minimum-ISL filter (b). The waveform and filters are the same ones shown in Figure 6.17. The shift in the apparent range, as a result of the Doppler frequency, is described by eq. (6.107). As the Doppler frequency moves away from $f_d = 0$, the range side-lobe levels also increase. For most ground weather radars, $|f_d| \ll 10$ kHz.

deterministic system distortion can be measured and corrected for, the effects of time-varying distortions or oscillator phase noise may limit the realized sidelobe levels to levels higher than those predicted by (6.105) and (6.106).

Now consider the effects of Doppler-shifted echoes on the pulse-compression filter's output. The LFM waveforms rely on a linearly varying frequency over time to encode and decode the signal. When compressing the pulse, the instantaneous frequency of the echo is used to determine the exact range of the target(s). Because of this relationship between range and frequency, any Doppler shift between the radar and the target results in an apparent change in the target's range. This is known as *range migration.* The range migration is a function of the Doppler frequency shift (recall $f_d \approx -2v/\lambda$ from eq. [5.39b]) by approximately

$$\delta r = -\frac{T_{tx}}{B} \frac{c}{2} f_d. \tag{6.107}$$

The effect of Doppler frequency shifts on the LFM pulse-compression filters (both the matched filter and minimum-ISL filter) are illustrated by the ambiguity functions in Figure 6.18. Note that the LFM pulse's impulse response is shifted (or migrated) in range proportional to the Doppler frequency. Very large Doppler frequencies are shown to illustrate their effect. For ground based radar's the Doppler frequencies are typically $|f_d| \ll 10$ kHz, and therefore the Doppler frequency shifts generally have a negligible affect on the LFM pulse-compression performance. For all Doppler frequency shifts shown, the compressed LFM pulse shows only minor increases in the near-range side lobe's amplitude. The increased side-lobe levels are a result of a mismatch between the frequencies in the echo and the filter. The filter's frequency span is nominally $[-B/2, B/2]$, while the echo's frequency span is $[-B/2 + f_d, B/2 + f_d]$. The proportion of frequency mismatch is f_d/B, which is a fraction of energy that cannot be compressed, resulting in a loss of peak gain due to uncompensated Doppler range migration (neglecting the effect of the amplitude taper):

$$L_D \approx \left(\frac{B}{B - |f_d|}\right)^2. \tag{6.108}$$

6.7   Selected Problems

1. Using the STSR sampling mode, the following observations of a range cell are collected from an S-band ($f_c = 2.7$ GHz) radar with a PRT of $T_s = 1$ ms. The horizontal (x_h) and vertical (x_v) polarizations' IQ measurements are sampled simultaneously as time-series:

$$x_h = \begin{bmatrix} -28.79 - j32.61 & 20.78 - j44.75 & 44.43 - j0.63 & 13.39 + j29.07 \end{bmatrix}$$

$$x_v = \begin{bmatrix} -7.64 - j20.35 & 20.49 - j18.08 & 24.66 + j10.38 & 0.52 + j21.25 \end{bmatrix}.$$

Using these baseband IQ measurements, consider the following:
a) What is the mean power, p (in decibels) for each polarization?
b) What is the differential reflectivity, Z_{dr} (in decibels)?
c) What is the unambiguous Doppler velocity, v_a?
d) What is the mean Doppler velocity, $\bar{v}$ (in meters per second) for the horizontal polarization?
e) What is the Doppler spectrum width, σ_v (in meters per second) for the horizontal polarization?
f) What is the estimate of the copolar correlation, ρ_{hv}?
g) What is the estimate of the differential propagation phase, Φ_{dp}?

2. For an X-band radar (9.4 GHz):
   a) What PRT is required for an unambiguous range of 50 km?
   b) For a uniform PRT with an unambiguous range of 50 km, what is the unambiguous velocity?
   c) If the radar uses a dual-PRF scheme, what are the two PRTs required to achieve an unambiguous range of 50 km and an unambiguous velocity $v_u \geq 35\,\mathrm{m\,s}^{-1}$ assuming $\kappa = 3$ (e.g., eq. [6.75])? What is the actual v_u (eq. [6.74])?
   d) How does this change if the frequency is changed to Ku-band (13.6 GHz)?
3. A C-band radar (5.6 GHz) uses a dual-PRF scheme to measure $v_1 = 12.1\,\mathrm{m\,s}^{-1}$ with PRT $T_1 = 1000\,\mu s$ and $v_2 = -6.3\,\mathrm{m\,s}^{-1}$ $T_1 = 1500\,\mu s$.
   a) what are the unambiguous velocities of each individual pulse and the combined dual-PRF observation?
   b) what are the phases of each PRTs' estimate?
   c) what velocity is estimated by taking the difference of the phases?
   d) Using the constraint that the velocity unwrapping must be an integer-multiple of the respective PRTs' unambiguous velocities, what is the estimated unwrapped velocity?
4. What is the ideal matched-filtered pulse compression gain of a LFM pulse with $B = 5$ MHz and $T_{tx} = 50\,\mu s$?
5. What is the change in the SNR (in decibels) between a pulse compression LFM waveform-filter pair with $B = 5$ MHz and $T_{tx} = 50\,\mu s$ compared to a waveform-filter pair with $B = 10$ MHz and $T_{tx} = 100\,\mu s$?
6. A 100 W X-band radar uses a 1 MHz and 1 μs waveform to achieve a minimum sensitivity of 20 dBZ at a range of 10 km.
   a) What transmit power is required for the unity-TBP pulse to achieve a minimum sensitivity of 5 dBZ at a range of 10 km?
   b) With the 100 W transmitter, what LFM pulse length is required to achieve a minimum sensitivity of 5 dBZ at a range of 10 km?
   c) What is the blind range of the LFM waveform-filter pair?
   d) Assuming negligible time to switch between the transmit and receiver state, what is the maximum uniform PRF that the radar can operate at to observe precipitation at a maximum range centered at 40 km without aliasing? (Note that the radar cannot receive while transmitting.)

7 Data Quality, Data Science, and Engineering of Weather Radars

In the early days of weather radar, most of the interpretation was in the form of black-and-white pictures and grease pencil marks on displays, with the extensive experience and background knowledge of the forecaster playing a critical role in the interpretation of the observations. The forecasters associated with each radar had developed an intricate knowledge of the storm features as seen in that radar, as well as knowledge of the radar's data-quality issues. Interpretation techniques advanced further in the 1970s and 1980s to include color displays and more elaborate observational tools.

Many of the issues with radar data quality were inherently handled by forecasters because of their knowledge and experience. Then three advancements came simultaneously: (a) the dual polarization radar, (b) affordable advanced signal processing, and (c) low-cost and high-quality displays. The demand for data quality with dual polarization was so high that it exposed other problems with existing radars. The low-cost signal processing enabled real-time cleanup of observations with filters (e.g., ground-clutter or image-processing filters). Dual polarization, with its extensive set of additional information about hydrometeors, brought the need for automated algorithms, with all these leading to addressing the topic of data quality in a systematic manner, which are discussed in this chapter. Modern automated algorithms and the widespread use of artificial intelligence make the topic of data quality more important than ever; thus, a chapter is devoted to this topic.

Data quality can refer to a diverse range of topics related to radar, but here it generally relates to the accuracy of weather observations. In this chapter, data quality considers signal processing, data analysis, operational considerations, and techniques to enhance the accuracy of, and removal of artifacts from, the radar measurements in the context of observing precipitation and atmospheric processes. The required accuracy of the radar measurements is dictated by the application, which includes hydrometeor classification (Chapter 8) and rainfall estimation (Chapter 9). Conversely, the application's performance is eventually determined by the actual level of measurement accuracy achieved. Errors in dual polarization observations have a direct effect on their interpretation.

This chapter discusses signal-processing and quantitative-analysis techniques to maximize the accuracy of weather radar observations of precipitation (which includes removing artifacts). This combines knowledge of precipitation processes, scattering theory, radar systems, and signal processing as a means of interpreting radar observations.

Table 7.1 Typical dual polarization radar measurement accuracy. Specific radar requirements may vary by application.

Z_e	1 dB
$\bar{v}$	$1 \, \mathrm{m \, s^{-1}}$
σ_v	$1 \, \mathrm{m \, s^{-1}}$
Z_{dr}	0.2 dB
Φ_{dp}	$2°$
ρ_{hv}	0.01 for $\rho_{hv} = 0.99$

To characterize the error, the first- and second-order statistics are used directly from the radar variable estimators and the intrinsic properties of the precipitation in the observation volume. The accuracy of any measurement or estimate is typically quantified by two sources of error: (a) systematic bias and (b) random errors. The bias is a constant value that can be removed (if known). Ideally, the bias component is removed by some calibration process. The standard deviation quantifies the range of "random" errors that often (but not always) span positive and negative excursions around the mean value.

Table 7.1 shows typical accuracy requirements for operational dual polarization radar that are assumed for application algorithms such as a rainfall estimation and hydrometeor classification. These requirements have emerged from historical use of radar observations and metrics related to user satisfaction.

In general, users are able to manage the known sources of measurement contamination. In fact, when the source of a data-quality issue is known, it can be dealt with by appropriate processing of the observations or by flagging the affected data. The main problem is when a data-quality issue is not known or is transient. This is one of the main objectives of this chapter: to raise awareness, identify potential data-quality issues and understand the techniques that may or may not be used to mitigate them. In some instances, radar data are made available to users with very little detail about the processing techniques used. Although these processing techniques are designed to "improve" radar data in a general sense, the data-quality algorithms may operate under specific assumptions that can have undesirable effects for certain applications (e.g., if the goal is to use ground clutter as part of a relative calibration assessment, a ground-clutter filter process could dramatically affect the results).

With single-polarization systems, many data-quality issues may simply go undetected. With dual polarization, a level of self-consistency in the radar observations is expected as a result of the scattering properties of precipitation, which brings a higher level of awareness to any potential issues. Polarimetry, by providing additional parameters sensitive to small changes, both in the environment and in the radar system, is a valuable tool for easy detection and characterization of a wide range of quality issues.

7.1   The Accuracy of Radar Variable Estimators

Dual polarization weather radars measure the amplitude and phase of the echoes as a function of time and range. All measurements have some uncertainty, and the radar's measurements of amplitude and phase are no different. Timing measurement is a very stable and mature technology, and issues with timing measurement are typically not a concern and hence are not discussed, but this should be kept in mind as an infrequent but potential source of error. The constant movement and reshuffling of the hydrometeors within the radar volume results in fluctuations of the echo. By observing the same radar volume over a longer period of time, the radar collects echoes from multiple different configurations of the precipitation (i.e., realizations of the statistical process), and thus reduces the error in the radar variable estimates for that volume. Considering only the estimators themselves (i.e., independent of attenuation, beam blockage, calibration errors, etc.), the accuracy of the estimators for the radar's measurements are evaluated here. In particular, the discussion is focused on the simultaneous transmit and simultaneous receive (STSR) pulse-pair estimators because these are currently the most commonly used estimators for the majority of operational weather radars.

The integration time is the duration over which signal samples are collected to use in the estimates of the properties of each radar range volume. The integration time is sometimes referred to as the *coherent processing interval* (CPI) when spectral-processing techniques are used. A coherent operation simply means that both the amplitude and phase are used from each measurement in the estimators (which, from eq. [6.41], is how the spectrum is calculated). The number of samples N within an integration time is typically on the order of 10s – 100s. N is commonly a power of 2 for historical reasons (e.g., 32, 64, 128) to leverage efficient fast Fourier transform (FFT) processing algorithms, although modern fast algorithms can generally accommodate all practical sample lengths. It should also be noted that unless otherwise stated, all the estimators discussed here are for uniform pulse-repetition times (PRTs).

To estimate the reflectivity factor, the radar constant is a key characteristic of the radar system that needs to be determined precisely, but this determination has an associated uncertainty. For the topic of weather radar calibration, the discussion is typically focused on the accuracy of the radar constant, and in the case of dual polar-ization radar, the difference in the radar constant between the horizontal and vertical polarizations is also of interest.

The normalized spectrum width is a parameter used as an independent variable to characterize the estimator errors and is defined as

$$\sigma_{vn} = \frac{\sigma_v}{v_a}; \quad 0 \leq \sigma_{vn}. \tag{7.1}$$

The normalized spectrum width is a useful metric for evaluating the performance of estimators. Recalling from eq. (5.102) that $N_I = N T_s / T_D$ (with $N_I \leq N$), and sub-stituting eq. (5.101) for the decorrelation time T_D, the number of independent samples is related to the Doppler spectrum width of the process and the radar's unambiguous velocity, v_a, as:

$$N_I = \frac{\sigma_v N \pi}{v_a \sqrt{2}} = \sigma_{vn} \frac{N \pi}{\sqrt{2}}; \quad 1 \leq N_I \leq N, \quad (7.2)$$

where N represents the number of pulses used (the total integration time is $N T_s$). Sampling more often (by increasing the radar's pulse repetition frequency) will increase v_a, but it will not correspondingly increase N_I. For most weather radar observations, the signal's samples are correlated, which in fact, is necessary for quality Doppler velocity estimates. The number of independent (or uncorrelated) samples is a function of the spectrum of the precipitation within the observation volume (see eq. [7.2]).

The number of independent samples, N_I, is a fundamental factor that determines the accuracy of the radar variable estimators. Although N_I is not intuitive in terms of the properties of the clouds and precipitation, it is the basis of the uncertainty estimates. (N_I must be calculated from the radar operating characteristics as well as the spectrum width of the observation.) For a given radar configuration, the uncertainty can be related to the spectrum width and integration time (as shown in Figs. 7.1–7.3).

When estimating the statistical properties of a signal, the uncertainty of the estimate is a function of the number of independent samples used. A low spectrum width (and therefore a low number of independent samples along with a high number of actual samples) provides a high degree of temporal correlation. The high temporal correlation improves the estimates of mean Doppler velocity and spectrum width, which correlate pairs of adjacent pulses. Alternatively, a higher number of independent samples (and therefore a higher spectrum width) will result in reduced uncertainty in the estimates of Z, Z_{dr}, ρ_{hv}, and Φ_{dp}.

In the case of the STSR covariance estimates, the moments (with the exception of velocity and spectrum width) are direct estimates, and no models are used. The Gaussian spectral shape model is used to estimate the Doppler spectrum width. The assumption is that the random motion of particles in the radial direction is equally distributed around the mean Doppler velocity. For observations at near vertical incidence (where the falling speed is particle-size dependent), the Gaussian distribution may not be strictly valid. It was shown in Chapter 3 that there is a size-versus-fall-velocity relationship of the hydrometeors. The vertical-pointing observation geometry is typical for vertical-profiling radar systems (as well as airborne and spaceborne precipitation radars). The overall effect is that the Doppler spectrum is asymmetric and not Gaussian shaped. For such a case, there is a difference between the results of the spectral estimates and pulse-pair estimators for the mean Doppler velocity and Doppler spectrum width.

To characterize the estimator accuracy, the bias and the standard deviation are used. The bias is the mean difference between the estimated value (denoted as the variable with a hat) and true value. The standard deviation is a measure of the average spread of the estimated values around the mean. The standard deviation is commonly used instead of the variance because the standard deviation has the same units as the estimator (i.e., $m\,s^{-1}$ instead of $m^2 s^{-2}$). The derivations of the bias and standard deviation for the radar variables are omitted here. The dual polarization weather radar estimators' performance characteristics were derived using perturbation analysis [21, 165, 189].

7.1.1 STSR Pulse-Pair Estimator Standard Deviation

The uncertainty of the estimates is quantified using the standard deviation of the estimators. Perturbation analysis linearizes the equations and uses only the lower-order terms from series expansions. Therefore, these are approximations that are best suited for a certain range of spectrum widths and signal-to-noise ratios (SNRs). As the spectrum widths approach the limits, or SNRs become too small, the assumptions used in the perturbation analysis are no longer valid, and these estimates for the standard deviations (SD) and biases are not appropriate given their assumptions. These estimates are valid for $0.04 < \sigma_{vn} < 0.60$ [189].

The standard deviations of the measured power and reflectivity are the same because they are related through the radar constant. The standard deviation of the measurement [189], which is normalized by the true power in linear units and is therefore a scale factor, is given by:

$$\frac{SD(\widehat{p})}{p} = \frac{SD(\widehat{Z})}{Z} = \sqrt{\frac{2SNR + 1}{N \cdot SNR^2} + \frac{1}{N_I}}. \tag{7.3a}$$

For power (and reflectivity) on a logarithmic scale, the standard deviation [21] is:

$$SD(\widehat{p}\,(dB)) \approx 10\log_{10}\left(1 + \frac{SD(\widehat{p})}{p}\right) \quad (dB). \tag{7.3b}$$

The number of samples, N, is used to indicate the reduction of the noise-power uncertainty within the estimate (note that the noise is independent for all N samples). The uncertainty of the precipitation's signal, in absence of noise, depends on the number of independent samples.

The standard deviation of differential reflectivity, which is represented using a logarithmic scale here, is as follows [189]:

$$SD(\widehat{Z}_{dr}\,(dB)) = \frac{10}{\ln 10}\left(\frac{2SNR_h + 1}{N \cdot SNR_h^2} + \frac{2SNR_v + 1}{N \cdot SNR_v^2} + \frac{2(1 - \rho_{hv}^2)}{N_I}\right)^{1/2} \quad (dB). \tag{7.3c}$$

The number of independent samples, N_I, is calculated from the spectrum width via eq. (7.2). The SNRs (with the h and v subscripts denoting their respective polarization) are in linear units.

The velocity standard deviation is [189]:

$$\text{SD}(\hat{v}) = \left[\frac{v_a^2}{2\pi^2 \rho^2[T_s]} \left(\frac{2\text{SNR}(1 - \rho^2[T_s]) + 1}{(N-1)\text{SNR}^2} + \frac{(1 - \rho^2[T_s])^2}{2N_I} \right) \right]^{1/2} \text{m s}^{-1},$$

(7.3d)

where $\rho[T_s]$ is the temporal correlation at a time lag of T_s (this is typically a sample lag of 1). The lag-one covariance for the velocity estimate uses $N-1$ samples, which is reflected in the noise uncertainty. The number of independent samples for the velocity estimate (with $N - 1$ samples) is also approximately N_I for typical weather radar observations.

The standard deviation for spectrum width is [165]:

$$\text{SD}(\hat{\sigma}_v) = \frac{v_a}{4\pi^{3/2} \rho[T_s] \sqrt{N_I} \sqrt{2}} \sqrt{(1 - \rho^2[T_s])^2 + \frac{2(1 - \rho^2[T_s])}{\text{SNR}} + \frac{(1 + \rho^2[T_s])}{\text{SNR}^2}} \text{ m s}^{-1}.$$

(7.3e)

The differential phase standard deviation is [189]:

$$\text{SD}(\hat{\phi}_{\text{dp}}) = \frac{1}{\rho_{hv} \sqrt{2}} \left(\frac{\text{SNR}_h + \text{SNR}_v + 1}{N \cdot \text{SNR}_h \cdot \text{SNR}_v} + \frac{1 - \rho_{hv}^2}{N_I} \right)^{1/2} \quad (\text{rad}).$$

(7.3f)

The standard deviation for copolar correlation is [189]:

$$\text{SD}(\hat{\rho}_{hv}) = \left[\frac{(1 - 2\text{SNR}_h)\rho_{hv}^2}{4N\text{SNR}_h^2} + \frac{(1 - 2\text{SNR}_v)\rho_{hv}^2}{4N\text{SNR}_v^2} \right.$$

$$\left. + \frac{\text{SNR}_h + \text{SNR}_v + 1}{2N \cdot \text{SNR}_h \cdot \text{SNR}_v} + \frac{(1 - \rho_{hv}^2)^2}{2N_I} \right]^{1/2}.$$

(7.3g)

Although K_{dp} is not a directly estimated radar variable, K_{dp} is derived from the radar measurements and therefore is subject to measurement uncertainty. The standard deviation for the K_{dp} estimate can be approximated as [21, 190]:

$$\text{SD}(\hat{K}_{\text{dp}}) = \frac{1}{\Delta r \sqrt{M(M-1)(M+1)/3}} \text{SD}(\hat{\Psi}_{\text{dp}})$$

(7.3h)

$$\approx \sqrt{\frac{3}{M}} \frac{\text{SD}(\hat{\Psi}_{\text{dp}})}{L}; \quad M \gg 1,$$

where Δr is the range spacing, M represents the number of range gates, and $L = M\Delta r$ is the path length.

Considering the number of independent samples from eq. (7.2), a higher PRF does not necessarily increase N_I, and therefore it typically has a limited effect on the standard deviation of the estimators for reflectivity, differential reflectivity, copolar correlation, and differential phase. A higher sampling rate does not necessarily improve the accuracy of the signal's power estimates. A higher PRF does provide more independent estimates of noise, which reduces the estimator uncertainty for low-SNR observations. A higher sampling rate will also reduce the uncertainty of the velocity and spectrum-width estimates by increasing the temporal correlation between samples.

7.1.2   STSR Estimator Standard Deviation with High SNR

From the estimator's standard deviation in eq. (7.3), it is clear that for low and moderate SNRs, the uncertainty in the radar variable estimates increases as a result of the additive noise's contribution. For high SNRs (typically characterized as $> 20\,\text{dB}$), the contribution of the noise can largely be ignored. For high-SNR observations, the standard deviations of the estimators are simplified as follows:

$$\text{SD}(\widehat{Z}) = \text{SD}(\widehat{p}) \approx \frac{p}{\sqrt{N_I}}, \tag{7.4a}$$

$$\text{SD}(\widehat{Z}\,(\text{dB})) = \text{SD}(\widehat{p}\,(\text{dB})) \approx 10\log_{10}\left(1 + \frac{1}{\sqrt{N_I}}\right) \quad (\text{dB}), \tag{7.4b}$$

$$\text{SD}(\widehat{v}) \approx \sqrt{\frac{v_a^2}{2\pi^2\rho^2[T_s]}\frac{\left(1 - \rho^2[T_s]\right)^2}{2N_I}} \quad \text{m\,s}^{-1}, \tag{7.4c}$$

$$\text{SD}(\widehat{\sigma}_v) \approx \frac{v_a(1 - \rho^2[T_s])}{4\pi^{3/2}\rho[T_s]\sqrt{N_I}\sqrt{2}}, \tag{7.4d}$$

$$\text{SD}(\widehat{Z}_{\text{dr}}\,(\text{dB})) \approx \frac{10}{\ln 10}\sqrt{\frac{2(1 - \rho_{hv}^2)}{N_I}} = 4.343\sqrt{\frac{2(1 - \rho_{hv}^2)}{N_I}} \quad (\text{dB}), \tag{7.4e}$$

$$\text{SD}(\widehat{\phi}_{\text{dp}}) \approx \sqrt{\frac{1 - \rho_{hv}^2}{2N_I\rho_{hv}^2}} \quad (\text{rad}), \tag{7.4f}$$

$$\text{SD}(\widehat{\rho}_{hv}) \approx \frac{(1 - \rho_{hv}^2)}{\sqrt{2N_I}}. \tag{7.4g}$$

It is important to remember that these standard deviations are approximations. These are derived using variational analysis with up to second-order terms and assume the measurements have high SNRs. Again, these approximations assume $0.04 < \sigma_{vn} < 0.60$ [189]. By definition, ρ_{hv} cannot be greater than 1, and therefore the copolar correlation standard deviation is an approximation of an asymmetric distribution.

The standard deviation of the estimators for high-SNR scenarios are shown for S-band and C-band radars in Figures 7.1 and 7.2, respectively. A range of common precipitation spectrum widths are used to illustrate the effect of the number of independent samples within the integration time. For the dual polarization radar variables (Z_{dr}, Φ_{dp}, and ρ_{hv}), the copolar correlation of 0.99 is assumed for the purposes of

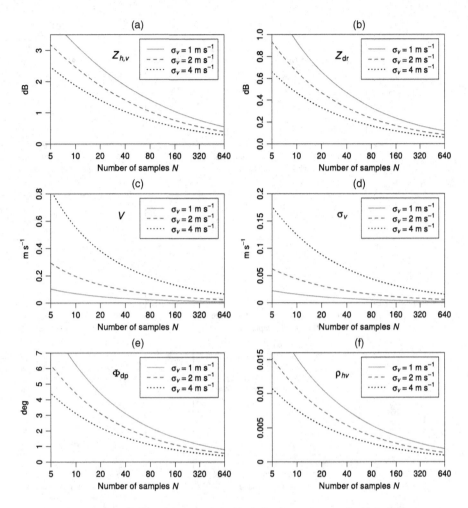

Figure 7.1 Standard deviation of the STSR dual polarization estimates for high-SNR observations at the S band (2.9 GHz) in typical rain conditions ($\rho_{hv} = 0.99$). The standard deviation of the STSR dual polarization estimates of (a) reflectivity, (b) differential reflectivity, (c) velocity, (d) spectrum width, (e) differential phase and (f) copolar correlation for high-SNR observations at the S band (2.9 GHz) in typical rain conditions ($\rho_{hv} = 0.99$). A PRT of 1 ms (PRF = 1000 Hz) is used. From eq. (7.2), the number of independent samples are $N_I \approx 0.09N$, $N_I \approx 0.17N$, and $N_I \approx 0.34N$ for $\sigma_v = 1$, $\sigma_v = 2$, and $\sigma_v = 4$ m s^{-1}, respectively. For this specific PRT, the numbers on the x-axis can also be read as the integration time in milliseconds.

estimating the error. Under a uniform pulsing scheme, note that the spectrum width and velocity uncertainty increase with more independent samples, whereas uncertainty is reduced for the other radar variables. From eq. (7.4), as the signal's intrinsic ρ_{hv} decreases, the uncertainty of the corresponding dual polarization radar variables will increase.

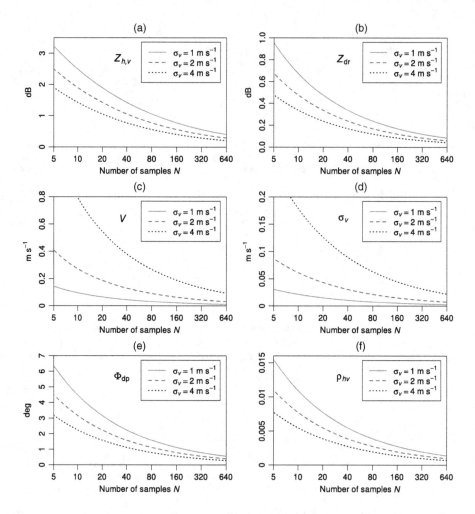

Figure 7.2 Similar to the description for Figure 7.1, but for C band (5.6 GHz). In this case, the number of independent samples is $N_I \approx 0.17N$, $N_I \approx 0.33N$, and $N_I \approx 0.66N$, respectively, for $\sigma_v = 1$, $\sigma_v = 2$, and $\sigma_v = 4\ \mathrm{m\,s^{-1}}$.

7.1.3   STSR Pulse-Pair Estimator Bias from Noise

Noise subtraction is used to remove its biasing effect, thereby increasing the overall accuracy of the estimates. In Chapter 6, the radar estimators are presented with noise subtraction. Section 7.3.4 discusses techniques for estimating the noise in the radar observations.

   This section discusses the biasing effects of noise, without noise subtraction, for the different estimators. This analysis is helpful for understanding the range of SNRs and degree of bias when noise is not corrected in the estimators. Similar relationships can be derived for the errors in the noise estimation. Typically, the errors in noise-power estimates are small, and the radar estimators' uncertainty is dominated by the standard deviation of the estimates, not residual bias. These derivations assume that a large number of (independent) samples is used to estimate the biases.

In Section 5.4.5, a description of the additive noise in radar observations was introduced, and with the inclusion of noise, the observation's SNR can be calculated. The quality of the radar measurements is affected by the prevailing SNR. From Section 7.1.1, it is clear that the uncertainty of the radar variable estimates increases as the SNR decreases. The accuracy of the estimates includes both the standard deviation (which is directly related to the variance of the estimator) and also the estimator's bias.

The mean power estimator is the most basic quantity used in a number of the radar variable estimators. The observed echo power is

$$p = |s|^2 + |w|^2 = |s|^2 \left(1 + \frac{1}{\text{SNR}}\right), \tag{7.5}$$

where $|s|^2$ and $|w|^2$ are the signal and additive noise powers, respectively. To estimate the reflectivity, the signal-processing goal is to estimate $|s|^2$ and reduce or eliminate the bias of additive noise, hence the subtraction of noise power from the measurement. On a logarithmic scale (e.g., in decibels), the bias in the estimated reflectivity (or the signal power) due to the uncorrected additive noise power is

$$\text{bias}\left(\widehat{Z}(\text{dB})\right) = 10 \log_{10} \widehat{Z} - 10 \log_{10} Z = 10 \log_{10} \left(1 + \frac{1}{\text{SNR}}\right). \tag{7.6}$$

The bias term derived for a logarithmic scale becomes a multiplier in linear scale (e.g., a gain term such as the radar constant). The reflectivity bias from additive noise must be corrected (subtracted) with the power in the linear scale.

For the differential reflectivity, the same process used for the reflectivity estimator is followed for each polarization. The differential reflectivity estimator can be written as follows:

$$\widehat{Z}_{\text{dr}} = \frac{|s_h|^2}{|s_v|^2} \frac{1 + \frac{1}{\text{SNR}_h}}{1 + \frac{1}{\text{SNR}_v}} = Z_{\text{dr}} \frac{\text{SNR}_v\,(\text{SNR}_h + 1)}{\text{SNR}_h\,(\text{SNR}_v + 1)}, \tag{7.7a}$$

where $|s_h|^2$ and $|s_v|^2$ denote the horizontal and vertical polarization signal powers. If the noise power is the same for both the horizontal and vertical channels, then $\text{SNR} = \text{SNR}_v$ and $\text{SNR}_h = \text{SNR} \cdot Z_{\text{dr}}$:

$$\widehat{Z}_{\text{dr}} = \frac{\text{SNR}\, Z_{\text{dr}} + 1}{\text{SNR} + 1} \tag{7.7b}$$

The differential reflectivity estimator's bias is then calculated as follows:

$$\text{bias}\left(\widehat{Z}_{\text{dr}}\right) = \frac{\text{SNR}\, Z_{\text{dr}} + 1}{\text{SNR} + 1} - Z_{\text{dr}} = \frac{1 - Z_{\text{dr}}}{\text{SNR} + 1}. \tag{7.8}$$

The differential reflectivity bias in decibels (which is a scale factor in linear scale) can be written as follows:

$$\text{bias}\left(\widehat{Z}_{\text{dr}}\ (\text{dB})\right) = 10\log_{10}\left(\widehat{Z}_{\text{dr}}\right) - 10\log_{10}(Z_{\text{dr}}) \tag{7.9a}$$

$$= 10\log_{10}\left(\frac{\text{SNR}\ Z_{\text{dr}} + 1}{Z_{\text{dr}}\ (\text{SNR} + 1)}\right)\ (\text{dB}). \tag{7.9b}$$

From eq. (6.12), the spectrum-width estimator can be written as follows:

$$\widehat{\sigma}_v = \frac{v_a\sqrt{2}}{\pi}\sqrt{\ln\left|\frac{|s|^2 + |w|^2}{|s|^2\ \rho[T_s]}\right|} = \frac{v_a\sqrt{2}}{\pi}\sqrt{\ln\left|\frac{\text{SNR} + 1}{\text{SNR}\ \rho[T_s]}\right|}. \tag{7.10}$$

The signal's intrinsic spectrum width (i.e., if the observation has infinite SNR) is

$$\sigma_v = \frac{v_a\sqrt{2}}{\pi}\sqrt{\ln\left|\frac{1}{\rho[T_s]}\right|}. \tag{7.11}$$

The noise-induced bias of the spectrum-width estimator is

$$\text{bias}\left(\widehat{\sigma}_v\right) = \frac{v_a\sqrt{2}}{\pi}\left[\sqrt{\ln\left|\frac{\text{SNR} + 1}{\text{SNR}\ \rho[T_s]}\right|} - \sqrt{\ln\left|\frac{1}{\rho[T_s]}\right|}\right]. \tag{7.12}$$

The STSR copolar correlation estimator (eq. [6.17]) includes noise subtraction. For a large number of samples, the ρ_{hv} estimator without noise correction is as follows:

$$\widehat{\rho}_{hv} = \frac{p_{hv}}{\sqrt{p_h p_v}} = \frac{\rho_{hv}\sqrt{|s_h|^2|s_v|^2}}{\sqrt{\left(|s_h|^2 + |w_h|^2\right)\left(|s_v|^2 + |w_v|^2\right)}}$$

$$= \frac{\rho_{hv}}{\sqrt{\left(1 + \frac{1}{\text{SNR}_h}\right)\left(1 + \frac{1}{\text{SNR}_v}\right)}}. \tag{7.13}$$

Note that SNR_h and SNR_v may differ as a result of the scatterer's Z_{dr}, due to different signal powers for each polarization or differences in the noise power of each polarization channel. The copolar correlation bias is:

$$\text{bias}\left(\widehat{\rho}_{hv}\right) = \rho_{hv}\left[\frac{1}{\sqrt{\left(1 + \frac{1}{\text{SNR}_h}\right)\left(1 + \frac{1}{\text{SNR}_v}\right)}} - 1\right] \tag{7.14}$$

$$= \rho_{hv}\left[\frac{\sqrt{\text{SNR}_h\text{SNR}_v} - \sqrt{(\text{SNR}_h + 1)(\text{SNR}_v + 1)}}{\sqrt{(\text{SNR}_h + 1)(\text{SNR}_v + 1)}}\right].$$

With $\text{SNR} = \text{SNR}_h = \text{SNR}_v$, the copolar correlation bias is simplified to:

$$\text{bias}\left(\widehat{\rho}_{hv}\right) = \frac{-\rho_{hv}}{\text{SNR} + 1}. \tag{7.15}$$

The STSR radar variable estimates that are sensitive to bias from additive noise are shown in Figure 7.3. Note that the copolar correlation (ρ_{hv}) and spectrum width (σ_v) have a clear relationship between the SNR and their biases. They can also be used as a proxy for SNR if the signal's ρ_{hv} can be assumed to be close to 1 (note,

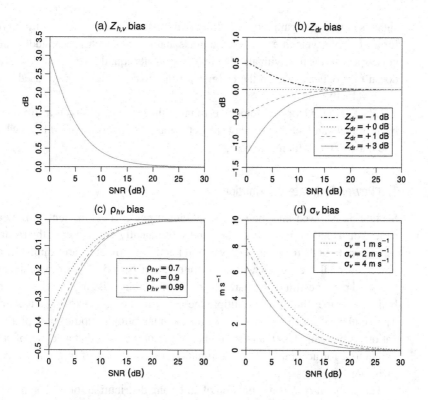

Figure 7.3 Additive noise-induced estimator biases without noise subtraction in the estimators. The spectrum-width bias is wavelength and sample-rate dependent (from v_a) and is shown for the S band (2.9 GHz) with a PRT of $T_s = 1$ ms. The biases are also a function of the precipitation's intrinsic signal properties. The ρ_{hv} bias curves assume SNR = SNR_h = SNR_v. The reflectivity and differential reflectivity biases, from eqs. (7.6) and (7.9) respectively, are shown in decibels and are added to the signal's values, also in decibels. (For a linear scale, the lines shown act as multipliers.)

the copolar correlation of stratiform rain is >0.99). The impact of additive noise on the estimators is important to understand because almost all weather radar signal processors use noise subtraction as part of their data processing and quality control. Noise subtraction uses estimates of the noise power, which are also subject to their own estimation uncertainties. Fortunately, the residual biases in the radar variable estimators from imperfect noise subtraction are small. Any reasonable estimate of the radar's noise power can substantially reduce the estimator bias and improve the accuracy of the radar variable estimates.

The velocity estimate is not biased by noise, but its standard deviation is SNR dependent. The noise-dependent bias for velocity is

$$\text{bias}(\widehat{v}) = 0. \tag{7.16}$$

Recall that the velocity is an estimate of the phase change between successive PRTs. Considering the echo's Doppler spectrum, the mean velocity is the first moment of the signal's power spectrum (eq. [6.50]). For additive white noise, its spectrum is constant for all Doppler velocities. When the noise and signal spectrum are added, the white-noise spectrum is flat, with all velocity components equally weighted, which therefore doesn't introduce a bias in the estimate of the first spectral moment used to estimate mean velocity.

Like the mean Doppler velocity estimator, the estimate of Φ_{dp} is also a phase-based measurement and therefore is not subject to the bias effects of the additive white noise. The bias of Φ_{dp} is therefore zero.

7.1.4   The Power-Estimate Distribution

Section 5.2 showed that the radar's individual power measurements follow an exponential distribution. In Section 5.4.3 and subsequently in Chapter 6, the radar's power estimators (and therefore the equivalent reflectivity estimates) are simply the averaged power over all the pulses within the integration period. The probability of radar's averaged power estimate is characterized by the Erlang distribution. The Erlang distribution describes the sum of multiple independent (i.e., N_I) exponential distributions. The number of independent samples for the integration period of precipitation observations is typically less than the number of samples (recall that a degree of correlation is required for accurate velocity estimation). The number of independent samples is therefore $1 \leq N_I \leq N$.

The probability density function of an Erlang distribution for N_I samples (which is a special case of the gamma distribution with shape parameter α) is as follows:

$$f_E(x; N_I, \alpha) = \frac{\alpha^{N_I} x^{N_I-1} e^{-\alpha x}}{(N_I - 1)!} \quad \alpha \geq 0, \tag{7.17}$$

where "!" indicates the factorial operator ($N! = \prod_{n=1}^{N} n$ and $3! = 3 \cdot 2 \cdot 1$). The mean of the Erlang distribution is N_I/α, indicating that the Erlang distribution is the sum of the N_I exponential distributions, not the mean of the N_I distributions (the power estimator implements a mean of the samples). As the radar's mean power probability distribution, the Erlang distribution must be modified to match the power estimator by by normalizing eq. (7.17) by N_I. For a signal with a true mean power of p and N_I independent samples, the probability density of the estimated power, $\hat{p}$, is

$$f_{Em}(\hat{p}; N_I, p) = N_I \frac{(\hat{p} N_I)^{(N_I-1)} e^{(-\hat{p} N_I/p)}}{p^{N_I} (N_I - 1)!}. \tag{7.18}$$

The mean of the modified Erlang distribution (eq. [7.18]) is p, and the standard deviation of eq. (7.18) is $p/\sqrt{N_I}$. This matches with the power estimator's standard deviation in eq. (7.4). For completeness, the mean power estimator's cumulative distribution function (CDF) is

$$F_{Em}(\hat{p}; N_I, p) = 1 - \sum_{n=0}^{N_I-1} \frac{1}{n!} e^{(-\hat{p} N_I/p)} \left(\hat{p} \frac{N_I}{p} \right)^n. \tag{7.19}$$

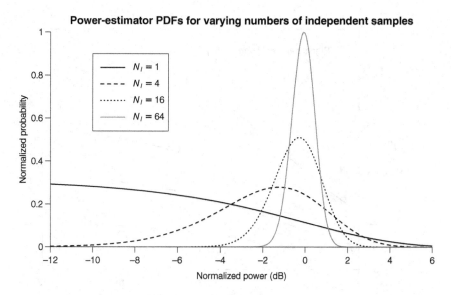

Figure 7.4 The distribution of the mean power estimators for four different numbers of independent samples, all assuming a mean power of 0 dB. The probability distribution functions (PDFs) represent the estimator uncertainty due to a finite number of independent samples. These curves are the PDFs for the modified Erlang distribution in eq. (7.18). With $N_I = 1$, the result is an exponential distribution. Using a logarithmic scale, the shape of the distribution remains the same, but the mean is shifted (added or subtracted) by the measurement's actual mean power in decibels.

Figure 7.4 illustrates the probability distribution of the estimated mean power as the number of independent samples is increased. Notice that as N_I increases, the shape of the distribution starts to look more Gaussian-like in the logarithm (decibels) scale. For a large enough number of independent samples, the mean power estimator's uncertainty is well approximated by the standard deviation from eq. (7.4b) centered around a mean power, p.

7.2   Data Quality Using Radar Variables

Although the human eye is adept at detecting features in radar observations when viewing a single variable, it is often convenient to use the information in the observations themselves to quickly detect precipitation and filter clutter and noise (i.e., remove regions that do not contain precipitation, as discussed later in Section 7.3.5). Data quality control algorithms can more quickly focus attention on regions of interest. Quantitative metrics are also important for automatic data processing tools. Numerous algorithms have been custom tailored in specific applications and radar systems to add automatic quality control of radar data to be used for subsequent processing (e.g., hydrometeor classification or rainfall rate estimation). This section introduces

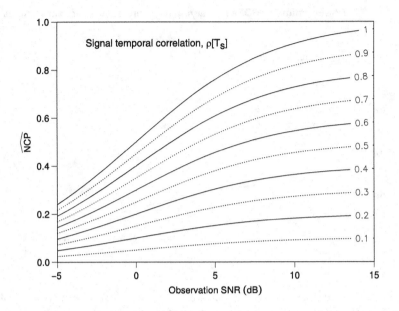

Figure 7.5 The diagram shows the relation between the estimated normalized coherent power ($\widehat{NCP}$) and the observation's SNR for measurements of a signal with additive noise. The contour lines represent different temporal correlation coefficients of the precipitation's signal (e.g., a proxy for the signal's spectrum width). The NCP estimates assume a larger number of samples (e.g., $N \to \infty$).

a few simple but effective metrics that are commonly used for automatic radar data quality control.

7.2.1   Signal Quality Index

Radar processors and visualization tools can "filter" observations using the statistics of the observations themselves. Two common terms are used to describe the same metric: signal quality index (SQI) and normalized coherent power (NCP). The NCP, shown here using the horizontal polarization, is estimated as follows:

$$\widehat{NCP} = \widehat{SQI} = \frac{|r_{hh}[T_s]|}{r_{hh}[0]}. \tag{7.20}$$

The NCP may also be calculated using the vertical polarization's observations. Notice that the NCP estimator deliberately does not include noise subtraction.

The NCP is directly related to the pulse-pair spectrum width estimate. The spectrum width, when also neglecting noise subtraction, is

$$\sigma_v = \frac{v_a\sqrt{2}}{\pi}\sqrt{\ln\left|\frac{r_{hh}[0]}{r_{hh}[T_s]}\right|} = \frac{v_a\sqrt{2}}{\pi}\sqrt{\ln\left|\frac{1}{NCP}\right|}. \tag{7.21}$$

The signal's normalized coherent power is the temporal correlation of the signal: NCP $= \rho[T_s]$. Accounting for the additive noise in the observations, the estimated NCP is

$$\widehat{\text{NCP}} = \rho[T_s]\frac{\text{SNR}}{\text{SNR} + 1}, \tag{7.22}$$

which is illustrated in Figure 7.5. The bias of NCP is due to the additive noise, and assuming a large number of samples ($N \to \infty$), the estimated NCP's bias is:

$$\text{bias}(\widehat{\text{NCP}}) = \widehat{\text{NCP}} - \text{NCP} = \frac{\rho[T_s]}{\text{SNR} + 1}. \tag{7.23}$$

From Figure 7.5 and Section 5.4.5 (see Fig. 5.14), the NCP can be used as a proxy for the measurement's SNR under assumptions about the temporal correlation. For an infinite number of samples, noise-only observations have SNR $= 0$ and NCP $= 0$ (recall that white noise has $\rho[T_s] = 0$ for $T_s \neq 0$).

With only a finite number of samples available to estimate NCP, the covariance estimates have a nonzero uncertainty (from the standard deviation estimates in Section 7.1.1). The estimates of the NCP are nonnegative in the range $[0, 1]$. If the numerator of the NCP estimate (i.e., $r_{hh}[T_s]$) is nonzero, the estimate of the NCP is also nonzero. With a finite number of samples available to estimate the covariance, the nonzero standard deviation of the power estimates indicates that $r_{hh}[T_s] > 0$ for a typical number of samples in the integration period. This is another bias effect introduced by a finite number of samples in the estimate that is not captured in eq. (7.23).

The NCP/SQI term (as the name "signal quality index" might imply) is frequently used to filter data for either display or additional processing. The number of samples in the integration period N, which is used to estimate the radar variables, is directly related to the standard deviation of the covariance estimates. To use NCP as a signal-quality filter, the estimate of $\widehat{\text{NCP}}$ is tested against a threshold value, $\text{NCP}_{\text{thres}}$. The threshold depends on the number of samples used in the radar estimates N and the desired confidence level at which observations that pass the threshold test for whether or not the observation contains a signal. Observations with $\widehat{\text{NCP}} > \text{NCP}_{\text{thres}}$ contain a signal with some desired confidence level. Similarly, $\widehat{\text{NCP}} \leq \text{NCP}_{\text{thres}}$ are noise-only observations with the corresponding degree of confidence. The NCP threshold is determined for a desired probability of filtering noise-only observations (i.e., observations with a complete absence of signal power). If the NCP threshold is set so that 90 percent of noise-only observations are suppressed, then it implies that 90 percent of the data that is not filtered contains an echo's signal. This observation may be dominated by noise (with a very low SNR). It does not mean that 90 percent of observations with a signal are detected. In fact, the higher the confidence level for the measurements that do pass the threshold test actual contain an echo (and noise-only observations are rejected), the more likely that observations with very weak signals or higher spectrum widths will not pass the threshold test. A higher NCP is required

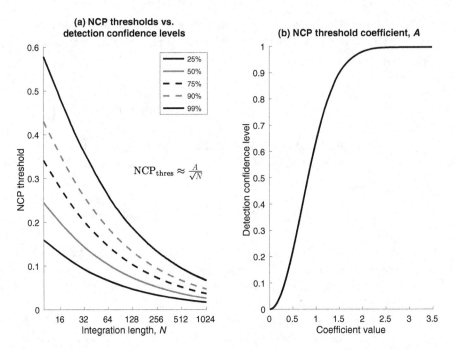

Figure 7.6 (a) The NCP threshold for various confidence levels as a function of the number of samples N in the integration period. (b) The NCP$_{thres}$ coefficient A and the associated confidence level that the observations included echos assuming the echoes have low normalized spectrum width and for $N \geq 32$. See Table 7.2 for a subset of values from the curve.

which translates to a higher SNR from Figure 7.5. The appropriate threshold depends on N as well as the user's (or application's) tolerance for missing data or residual noise-only measurements in the observations.

To identify and filter noise-only observations (or observations that have noise-like properties, such as random phase-coded second-trip echoes), the threshold for testing the NCP is

$$NCP_{thres} \approx \frac{A}{\sqrt{N}}, \tag{7.24}$$

where A is a constant that is determined based on the confidence level that the measurement contains a signal. N is the number of sample-time pulses in the integration period. Panel (a) of Figure 7.6 shows the NCP threshold as a function of N for multiple confidence levels. The curve in panel (b) gives the coefficient, A, which is used to calculate NCP$_{thres}$ for a specified confidence level. The coefficients are valid for $N \geq 32$ (and they are reasonable approximations for $N \geq 16$). For observations that contain only noise, a threshold of NCP$_{thres} = 1/\sqrt{N}$ ($A = 1$) filters approximately 63.5 percent of noise-only measurements. For the NCP threshold curves in Figure 7.6, the coefficient values and their corresponding confidence levels are listed in Table 7.2.

Table 7.2 Confidence level (CL) and its corresponding coefficient, *A*, for estimating the NCP threshold with eq. (7.24).

CL	25%	50%	75%	90%	99%
A	0.53	0.83	1.17	1.51	2.14

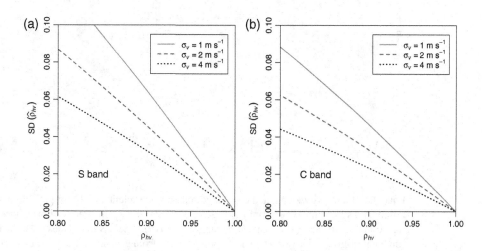

Figure 7.7 Standard deviation of the correlation coefficient, showing the increased estimator accuracy for signals with high-ρ_{hv} values. The plots, derived from eq. (7.4g), assuming $N = 50$ integrated samples and PRT = 1 ms, are valid for S-band (a) and C-band (b) radars.

7.2.2 Using Copolar Correlation to Identify Good Signal Quality

For single-polarization radars, the SQI (NCP) is a very important parameter that allows the user to filter a range of unwanted echoes, including noise, multiple-trip echoes, and radio-frequency interference (RFI; Section 7.3). SQI is the temporal correlation of the signal for a given polarization. For dual polarization systems, the estimation of the correlation between orthogonal polarizations expands the available options to control the quality of the observations. As a matter of fact, ρ_{hv} is generally the primary radar variable that is used for separating the precipitation echoes from any other nonmeteorological signal.

The copolar correlation coefficient in rain is extremely high, typically reaching above 0.98 (smaller values may occasionally be observed at the C band, as discussed in Section 4.6.4). For these high correlation values, the associated standard deviation is very small, as illustrated in Figure 7.7. Similar values of ρ_{hv} are also normally observed in dry snow, whereas mixed-phase precipitation, such as melting snow or rain mixed with hail, shows lower values, typically in the range of 0.8–0.97 (Sections 3.6.1, 3.6.3, 8.1.1).

On the other hand, most nonmeteorological targets result in low ρ_{hv}, often lower than 0.8, as a result of their size, the diversity of shapes and composition of the

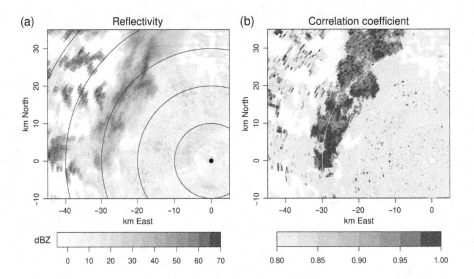

Figure 7.8 Reflectivity Z_h (a) and copolar correlation coefficient ρ_{hv} (b) for precipitation observations in complex orography (PPI scan at $2°$ elevation). The precipitation echoes around the 30-km range are mostly denoted by high ρ_{hv} (> 0.9), whereas ρ_{hv} values lower than ~ 0.8 (the lightest-gray color represents $\rho_{hv} \leq 0.825$) arise from ground clutter. Note that echoes closer than ~ 25 km are from antenna side-lobe contamination, whereas stronger reflectivity values at ranges beyond ~ 35 km (reaching 60 dBZ and more) are from ground scatterers within the main lobe (see also Fig. 7.17).

scatterers within the radar resolution volume. For example, ground scatterers (terrain, vegetation, buildings) are characterized by a large spatial variability at the scale of ~ 100 m or more, resulting in a significant decorrelation between the signals in the orthogonal horizontal and vertical channels during low-elevation radar scans. As an example, Figure 7.8 shows the observed reflectivity and copolar correlation coefficient fields during a plan position indicator (PPI) scan in complex terrain, with clearly distinguishable precipitation and ground-clutter echoes. This example shows how ρ_{hv} could be used to identify and mask suspicious data.

The contrasting behavior of ρ_{hv} for different types of targets is further discussed in Section 7.3.5 for ground clutter (Fig. 7.14) and insect/bird echoes (Fig. 7.16). The copolar correlation is widely exploited in classification algorithms either as a standalone criterion, or in combination with other parameters, such as SQI and the texture of the other dual polarization radar variables Z_{dr} and Φ_{dp}.

7.3   Observing in a Real Environment

In addition to the radar system's characteristics discussed in previous sections, when observing in a real environment, a number of factors that could potentially affect the quality of the polarimetric observations must also be considered. In particular,

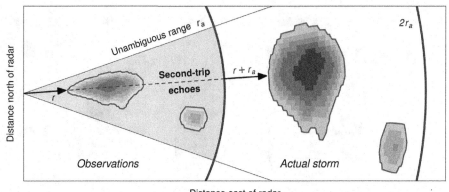

Figure 7.9 Illustration of second-trip echo from a storm beyond the radar's unambiguous range r_a. The echo appears elongated along the radial direction and shows a decreased magnitude with respect to the actual storm. A hypothetical storm beyond the $2r_a$ range would result in a third-trip echo.

the following subsections consider the effects induced by second-trip echoes, radio-frequency interference (RFI), nonuniform beam filling resulting from the presence of precipitation gradients, the possible contamination by nonmeteorological echoes, and a multiple-scattering artifact that is often observed in hailstorms (three-body scattering).

7.3.1 Range Overlay from Echoes Beyond the Unambiguous Range

The PRT determines the radar's unambiguous range r_a (eq. [5.35]), however, the transmitted pulse will keep traveling beyond r_a. The weather or clutter echos beyond r_a from pulse 1 may still be detected by the receiver after the next pulse (pulse 2) has been transmitted. Without specific processing of the pulse sequence, pulse 1's echos from beyond r_a are erroneously attributed to pulse 2. Echoes from the previous pulse (and scatterers beyond range r_a) are referred to as *second-trip echoes*, which are illustrated in Figure 7.9. Second-trip echoes appear weaker than expected (first-trip) echoes because of the wrong range correction (eq. [2.39c]) in the radar reflectivity calculation. If the echo originates from $2r_a < r < 3r_a$, it will be a third-trip echo, and so on.

Multiple-trip echoes are generally only an issue at low elevations because for high-elevation angles, the radar range volumes beyond r_a will be at a much higher altitude than plausible to encounter any weather target. Multiple-trip echoes are also more commonly observed by short-range radars, such as X-band radars and those at higher frequencies, that utilize higher PRFs.

Figure 7.10 shows an example of a multiple-trip echo in the reflectivity field (panel [a]) observed by an X-band radar with a magnetron transmitter, scanning at a 1.2° elevation angle. Magnetron transmitters have a random start phase, making

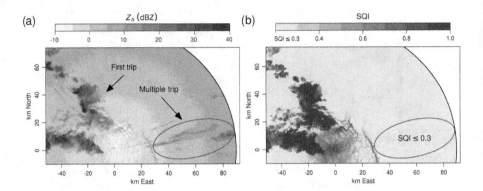

Figure 7.10 Reflectivity (a) and SQI (b) from a Colorado State University–Chicago Illinois (CSU-CHILL) radar X-band PPI scan at $1.2°$ elevation, with unambiguous range $r_a = 90$ km. The elongated pattern in the eastern sector is the result of multiple-trip echoes (second trip and third trip) from a storm located further east around the range of 160–250 km.

the second-trip echo power spectrum similar to white noise. Thermal noise also has the same properties as white noise. Unlike thermal noise, the second trip echoes are spatially varying and can have relatively high power levels. The white noise is uncorrelated and therefore has a low SQI, as shown in panel (b) of Figure 7.10. For magnetron systems, multiple-trip echoes can then be effectively masked using a simple threshold on the SQI (typically adopted SQI thresholds are in the range of 0.2–0.3).

For systems with transmitters that are coherent from pulse to pulse (which can include klystron- or solid-state-based systems), the second-trip echoes remain coherent (as if the radar were intending to measure the echo from the actual range where the second-trip echoes occur). For a fully coherent system, either random phase coding or a more sophisticated phase-coding technique can be used to suppress (or separate) echoes from beyond the unambiguous range ($r > r_a$) from the first-trip observations that are from $r \leq r_a$. Phase-coding techniques were introduced in Section 6.4.2.

7.3.2   Radio-Frequency Interference

One common challenge in weather radar data quality is radio-frequency interference (RFI). Often, weather radars operate in frequency bands close to, or shared with, other telecommunication systems. In most European countries, for example, C-band systems share the frequency band with radio local area network (RLAN) and wireless local area network (WLAN) [191, 192].

When the average interference power exceeds the receiver's noise, the radar system's sensitivity decreases. The coexistence of weather radar and telecommunication systems may then affect radar observations, which can be identified by the appearance of radial streaks during PPI scans. When the interference sources are few, the affected radials can be easily identified through image processing in the polar domain, where the RFI echo appears as a straight line. Then the weather signal along the affected radial can be approximated through interpolation from neighboring radials. At the signal-processing level, most RFI detection algorithms compare the amplitude of each pulse to that of its neighbors to remove anomalously strong signals. This processing can be especially effective for pulsed interference sources with a low duty cycle, which introduce a spike in the IQ times-series data for a given range gate. When the duty cycle of the interference source increases, multiple adjacent pulses can be affected, making detection challenging in the sample-time domain. Extending the processing to the two-dimensional range-time/sample-time domain may improve the RFI detectability. In particular, RFIs with a lower interference-to-signal ratio can be detected without increasing the false-alarm rate [193].

Because RFI are not coherent with the radar receiver, these can also be detected and masked by applying a threshold on the SQI value, similar to the multiple-trip suppression described in a previous section. It is important to note that by doing so, weak weather echoes can also be removed (especially at long ranges). Therefore, the SQI threshold needs to be carefully set to the lowest value that will allow RFI removal for a specific context.

7.3.3 Gradients in Precipitation and Nonuniform Beam Filling

Nonuniform beam filling (NUBF) can be an important issue for radar with relatively small antennas and large beamwidths, whose observation volumes cover large spatial extents. It can also affect measurements from ground-based radar when strong precipitation gradients are present. In fact, the increasing beamwidth with range may affect the quality of polarimetric observations at large distances, with a potential impact on the quantitative estimation of rainfall [194], self-consistency relations [195], and also qualitative applications such as hydrometeor classification. The effects of NUBF show up, for example, in range-height indicator (RHI) scans through the melting layer, as a result of gradients associated with the transition between rain and snow, or near the boundaries of convective structures (e.g., isolated storms, squall lines, mesoscale convective systems), in correspondence with strong cross-beam reflectivity gradients.

The presence of NUBF artifacts is normally revealed in the polarimetric observations by radial streaks with a low correlation coefficient. NUBF may also significantly affect the differential phase observations, introducing a perturbation of the radial profile [196, 197]. The affected profiles may in turn produce spurious K_{dp} estimates, with an impact on the subsequent quantitative rainfall estimation. Simple analytical formulas for the NUBF-induced errors can be derived under the hypothesis of perfectly matching two-way antenna beam patterns (idealized as axisymmetric and Gaussian) at horizontal and vertical polarizations, that is, solely considering errors arising from cross-beam gradients. In cases where the reflectivity (in logarithmic scale) and the differential phase shift Φ_{dp} vary linearly in both cross-beam directions, the biases of the polarimetric radar variables Z_h, Z_{dr}, Φ_{dp}, and ρ_{hv} are shown to be proportional to the square of the 3-dB antenna beamwidth [198]. Considering the cross-radial distances (the arc distances) $s_a = \theta r$ and $s_e = \phi r$, respectively, in azimuth and elevation for a typical antenna with beamwidth in units of length $b = \theta_{HPBW} r$, the following relations are derived:

$$\Delta Z(dB) \approx 0.01 b^2 \left[\left(\frac{dZ}{ds_e} \right)^2 + \left(\frac{dZ}{ds_a} \right)^2 \right] \tag{7.25a}$$

$$\Delta Z_{dr}(dB) \approx 0.02 b^2 \left(\frac{dZ}{ds_e} \frac{dZ_{dr}}{ds_e} + \frac{dZ}{ds_a} \frac{dZ_{dr}}{ds_a} \right) \tag{7.25b}$$

$$\Delta \Phi_{dp}(deg) \approx 0.02 b^2 \left(\frac{d\Phi_{dp}}{ds_e} \frac{dZ}{ds_e} + \frac{d\Phi_{dp}}{ds_a} \frac{dZ}{ds_a} \right) \tag{7.25c}$$

$$|\rho_{hv}^{(m)}| \approx |\rho_{hv}| \exp \left\{ -1.37 \times 10^{-5} b^2 \left[\left(\frac{d\Phi_{dp}}{ds_e} \right)^2 + \left(\frac{d\Phi_{dp}}{ds_a} \right)^2 \right] \right\}, \tag{7.25d}$$

where Z and Z_{dr} are in logarithmic scale; the differential phase shift is in degrees; and b, s_a, and s_e are in units of length. The exponential term on the right side of eq. (7.25d) is the reduction factor of the measured correlation coefficient. Equations (7.25a) through (7.25d) explicitly show that for a given precipitation gradient, the errors in the polarimetric radar variables grow with the physical size of the beam (b); that is, larger errors are expected with increasing range. Note that eqs. (7.25a) through (7.25d) can be written in polar coordinates, as in Ryzhkov et al. [156], simply by replacing b, s_a, and s_e with θ_{HPBW}, ϕ, and θ, respectively, because the range r cancels out.

For example, according to eq. (7.25a), a reflectivity gradient of $\sim$20 dB km^{-1} (e.g., near a storm's boundary) would introduce an error of $\Delta Z \sim$4 dB on the observations of reflectivity at a 60-km range with a $1°$-beamwidth antenna, compared to an error of $\Delta Z \sim$1 dB for the same antenna and gradient at a 30-km range. It is worth noting that a $2°$-beamwidth antenna at a 30-km range would behave just like the $1°$ antenna at a 60-km range (the two beams have the same physical size).

From eqs. (7.25c) and (7.25d), the error for Φ_{dp} and ρ_{hv} depends on the gradient of Φ_{dp}. At first glance, it may appear that, all else being equal, higher-frequency radars will experience a larger error on Φ_{dp} as a result of NUBF. Owing to the frequency scaling of the differential phase (which is one of the main advantages of higher-frequency radars), the same physical gradient in precipitation will result in a larger Φ_{dp} gradient when observed with a shorter-wavelength radar. However, this has little consequence for practical applications because the errors due to NUBF scale with frequency linearly, just like the parameter itself, so the relative error is unchanged. Note that unlike reflectivity, the value of Φ_{dp} does not depend on the local precipitation amount or type, but instead is the result of the precipitation profile along the whole path from the radar to the target. This implies that large gradients of Φ_{dp} may also induce NUBF errors in regions of fairly uniform precipitation.

Another radar variable affected by NUBF is the correlation coefficient, which can show a drop in the presence of strong gradients. A nonuniform Φ_{dp} affects the integration over the ensemble of scatterers contained in the radar volume, decreasing the magnitude of the covariance between the horizontal and vertical polarizations' copolar signals. For example, if the differential phase shift gradient in either azimuth or elevation is $25\,^\circ\,km^{-1}$ for S band, an observation at a 60-km range with a 1° antenna beamwidth would reduce the correlation coefficient by a factor of 0.99. For the same storm, beamwidth and range but at X band, the differential phase shift gradient would be $\sim 90\,^\circ\,km^{-1}$, leading to a drop in ρ_{hv} by a factor of 0.88.

In practice, an angular Φ_{dp} gradient of at least $\sim$40 deg deg^{-1} observed in the polar domain with a 1°-beamwidth antenna (equivalent to $36\,^\circ\,km^{-1}$ at a 60-km range or $18\,^\circ\,km^{-1}$ at a 120-km range) in either the azimuth or elevation cross-radial direction is needed to start perceiving a deterioration in the ρ_{hv} observations (a reduction factor of 0.98). Such large gradients are not frequently present in radar observations but can be detected, for example, in heavy precipitation, when the radar beam is partly in rain and partly in the ice phase (strong elevation gradient), or near the boundary of a squall line (strong azimuth gradient). This shows how, as with other unwanted echoes and artifacts, ρ_{hv} is generally the best radar variable to look at to ensure the selection of good-quality data from meteorological targets.

Note that the measured ρ_{hv} also depends on the radar system's characteristics; therefore, the actual observations may depart slightly from these simple theoretical predictions. In practice, the maximum ρ_{hv} the radar system can measure is less than 1. Although the system's copolar correlation is ideally asymptotic to 1, slight imperfections in the radar's two polarization channels and additive noise set a limit on the observable copolar correlation. For well-calibrated polarimetric weather radars observing high SNRs, the system's maximum ρ_{hv} is typically $\gg 0.995$.

The previous discussion provided general formulations valid under the hypothesis of a Gaussian and axisymmetric antenna pattern. It should be noted that the specific quality of the antenna (Section 7.4.1) may play a role in enhancing errors due to NUBF. Accurate dual polarization measurements require well-matched copolar beam patterns, a low side-lobe level, and good cross-polarization isolation. For the characterization of the CSU-CHILL antenna [194], the following antenna performance metrics were considered: the system linear depolarization ratio (LDR) limit in stratiform rain, the reduction of artifacts as a result of cross-beam gradients, and the reduction of ground-clutter echoes as a result of the improved side-lobe performance. In particular, a mismatch of the horizontal and vertical antenna patterns may cause the appearance of gradient-induced artifacts. Although the main lobes of the two orthogonal polarizations' antenna patterns in general overlap quite accurately, mismatches may exist in the side lobes. Given the small power received from directions outside the main lobe (most weather radars have a peak side-lobe level below -25 dB), the side-lobe mismatch is not an issue in most weather situations. Contamination of the radar variables may only arise in the presence of large precipitation gradients, when the main lobe is pointing in regions with no precipitation or very low reflectivity, whereas the side lobes fall into nearby regions with strong echoes. In this case, the polarimetric errors in differential variables Z_{dr} and Φ_{dp} may become nonnegligible, while the reflectivity remains generally low enough to not alter reflectivity-based quantitative applications.

Regions prone to side-lobe contamination can be located through the convolution of the nonuniform observed reflectivity field with the two-way antenna patterns. Note that the observed reflectivity field is already the result of a convolution between the intrinsic reflectivity and the antenna pattern. This double convolution, although resulting in a small underestimation of the intrinsic gradients, allows for the establishment of a procedure to mask radar observations where the use of Z_{dr} and Φ_{dp} measurements should be avoided. The tolerable magnitude of reflectivity gradients will depend on the actual antenna patterns. As an alternative to (or in conjunction with) this convolution procedure, ρ_{hv} could then be used to mask regions where side-lobe contamination may occur. In particular, data masking using a simple threshold on ρ_{hv} is extensively used for K_{dp} calculation (see, e.g., [175]).

7.3.4   Radiometric Noise in Radar Measurements

In Chapter 2, the noise in a radar system was discussed and shown to be related to both the bandwidth of the radar receiver and temperature. The noise temperature observed by the radar system can be dominated by the radar system's components themselves, but it also includes thermal emissions radiated by the environment. The earth and sun are two commonly observed sources of emissions that influence the radar's received noise power. As one might expect, this noise power can vary with elevation and azimuth but also with time (the temperature of the environment has a diurnal and seasonal cycle).

With noise correction, the accuracy of the noise estimates affect the accuracy of the moment estimators as well as the effective minimum SNR that can be used for reliable moment estimates. Noise subtraction is used to reduce the bias of power-based radar variable estimates. If the noise estimate is not accurate, a residual bias in the estimator remains. The better the noise-power estimate, the lower the bias in the reflectivity estimates at low SNRs.

The radar's observed noise-power level typically depends on what the antenna is pointing at. Both ground clutter and precipitation can vary the radar's measured noise power. The temperature and emissivity of the objects within the antenna's beam, as well as the antenna's temperature itself, are sources of thermal noise observed by the weather radar. The thermal noise power is typically lowest when the sky is clear and the antenna is pointed at a high-elevation angle. Similarly, the thermal noise power is typically highest when pointed horizontally along the ground.

The sun is an example of a thermal noise source that is routinely found in radar data, and in fact, it is also used as a source for some radar calibration activities. In radar observations that happen to scan through the sun, the increased radiometric noise can create a "sun spike" observed in the power or reflectivity (assuming data-quality filters do not remove it). The "spike" is due to an increase in the radiometric noise-power level along the entire radial when the sun is within the antenna's main lobe.

Spatial variations in noise power are observed with changes in the radar's elevation angle as a result of the length of the atmosphere along the radial. The atmosphere has a nonzero emissivity and is at a nonzero temperature; therefore, it is a source of radio-freqency (RF) noise.

As the elevation angle increases in clear-sky conditions, there is less "warm" atmosphere along the antenna's observing direction, and therefore the noise temperature contribution from the environment is generally decreased. Noise-power measurements at high-elevation angles can be used as an approximation for the thermal noise at lower elevations (high-elevation angles are desirable to ensure that anthropogenic and environmental noise are largely avoided). Although the noise estimates at high elevations may slightly underestimate the measurement noise at lower elevations, the resulting noise measurements are more stable and not widely varying as a result of the clutter environment below. To minimize the estimator biases from additive noise in the very low SNR observations, noise estimates along the particular observation direction at the time of the observation are needed.

To estimate noise power, the measurements must be free of signal (or at least have a signal power that is low enough not to corrupt the noise estimate). With SNRs > 20 dB, the effects of noise can be usually neglected, and the measurement is considered the signal (rather than signal plus noise). Conversely, for determining noise power, SNRs < −20 dB would indicate measurements where the signal can be neglected. In practice, estimates of the SNRs are not available when determining noise power (if they were, that would suggest prior knowledge about the signal, which is ultimately the radar's goal). To estimate the noise power, techniques to identify noise-only measurements are required. When these techniques are coupled with the probability that the majority of the selected measurements are unbiased by signal power, accurate estimates of the radar's noise power can be made.

The radar's noise-power level varies spatially as the radar scans, but also varies over time as the temperature of the system and environment changes (and even longer term, as components age and degrade). This implies that the noise-power estimates must periodically be updated for the system. Consider the spatial noise power estimates shown in Figure 7.11 from the D3R radar during clear air-conditions during its deployment at the National Aeronautics and Space Administration (NASA) Wallops Flight Facility in Wallops Island, Virginia (radar located at 37.9334° North, 75.4709° West). At low elevations, ground clutter is in the view of the antenna's beam and results in the noise power increasing by almost 1.2 dB compared with the higher-elevation observations of clear air. A number of buildings and a close tree line are observed from 210° to 360° azimuth, whereas a relatively clear line of sight over an airstrip is observed from 0° to 210°. As the elevation angle increases, the noise-power level decreases because the surface clutter is not in the beam, and there is less warm atmosphere between the radar and the cold of space. Similar spatial effects are observed in the thermal noise when observing precipitation, especially for convective systems, where the temperature and emissivity of the precipitation can vary the noise power by 1 dB or more.

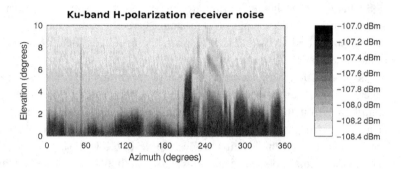

Figure 7.11 An example of spatially varying noise power. Observations of Ku-band, horizontally polarized radiometric noise during clear conditions at Wallops Flight Facility in Wallops Island, Virginia, on March 7, 2015, using the NASA D3R radar [15]. The radiometric noise measurements are shown for elevations from 0° to 10° over the full 360° of azimuth. The nearly 1.2-dB variation is primarily due to the temperatures and emissivity of the structures and vegetation within the antenna's beam.

   A number of techniques are available to estimate the radar's noise power. These range from periodic calibration activities that are controlled by radar operators, to more advanced data processing methods that can be used while the radar is operationally scanning. Some of the radar noise estimation procedures that can be found in operational and research weather radar are:

- Passively observe with the radar receiver on but the transmitter turned off. If the radar does not transmit, there is no source to generate echoes. This is a simple and very effective method of estimating the radiometric noise power. This can be included as part of a daily calibration activity controlled by the radar operator to estimate and update the radar's noise power estimate if needed.

- Use areas scanned during an observation plan that are known to be clear air. A common technique for estimating system noise is to use measurements from regions that are known to be free of scatterers (e.g., observations above a certain altitude or those obtained by pointing the radar in a direction determined by an operator to be clear air). The result is effectively the same as using passive observations.

- Algorithmically identify observation regions that are noise by identifying echo-free ranges that exhibit constant and consistent power levels [199, 200].

- Precipitation has a relatively high intrinsic copolar correlation. This fact is used as a data filter to separate precipitation from clutter and low-SNR observations. Copolar correlation can be used to identify observations that are free of echoes and therefore dominated by noise power [199].

For all the observation volumes that are identified as only containing noise, their power estimates can also be averaged incoherently to reduce uncertainty (i.e., eq. [5.57]). For "blue-sky" noise observations, the radar can observe a fixed elevation in clear conditions and incoherently integrate for as long as desired. The noise power from thermal noise sources (e.g., ground clutter or the atmosphere) can be assumed to be constant along the radial. For noise estimation, the incoherent integration can be performed not only in the sample time but also in range time.

For noise estimates during operational scanning, the noise can be estimated along the radial following a similar process as the blue-sky noise estimates, but this requires that noise-only range gates be identified so that the noise estimate is not biased by the echoes from precipitation. For moderate- to high-elevation angles, this is easily achieved by selecting ranges well above the troposphere, where precipitation does not exist. For low elevations, algorithms (or radar operators) can identify "signal-free" areas with high confidence. The noise in radar measurements is generally uncorrelated; therefore, $N_I = N$ for the purposes of estimating noise. Recalling Figure 7.4, as N_I increases, the power on a logarithmic scale tends toward a Gaussian distribution whose mean is approximated as the median power. With this, the standard deviation of the noise estimated can be calculated from eq. (7.4b).

Using the observations themselves, a histogram of their power provides a simple but effective method of estimating the noise power. The noise power is generally constant throughout the radial, whereas the echo power varies. If the radial has a fraction of noise-only measurements, this results in a high density in the histogram around the noise, whereas the precipitation power is typically spread over a large range of values. Figure 7.12 shows the estimated reflectivity and signal power as a function of range at a 4.5° elevation. Along the radial, there are regions that include ground clutter, precipitation, or only noise. The histogram shows the relative number of measured powers sampled along the radial. The radial's noise power can be directly estimated from the histogram. With an estimate of the noise-power level, the noise power can be subtracted from the observation to estimate the signal power. Note that although the noise power is constant as a function of range, the "noise" level in the radar reflectivity increases with range (this is due to the r^2 dependence between power and reflectivity from eq. [2.39c]). Ideally, after noise subtraction, the minimum detectable signal can be measured on average for SNRs down to approximately $-10 \log_{10} \sqrt{N}$ dB due to the reduced variance of the power estimate. Practical sensitivity improvements using noise subtraction are typically limited to SNRs between -10 dB to -20 dB, depending on the radar system.

Figure 7.13 shows STSR-mode observations at a low-elevation angle (0.5°). The radar observations contain a region with precipitation, but a higher contribution of ground clutter is observed at most ranges as a result of the radar's low-elevation angle and the antenna's 1.0° beamwidth. The copolar correlation (without any noise correction) is plotted with respect to the signal's power. As the received power decreases, the copolar correlation of the observation also decreases because of the low

SNR in both polarizations. The noise is uncorrelated between the two polarizations, and therefore, the estimated copolar correlation also decreases (recall eq. [7.14]). The observations with the lowest ρ_{hv} are frequently noise-only observations.

7.3.5   Nonmeteorological Echoes

Clutter is a general term for any unwanted echo, and for weather radar, nonmeteorological echoes are clutter. (For an air traffic control radar, precipitation would be one of the forms of clutter.) Sources of weather radar clutter include ground, ocean, or lake surfaces; aircraft; birds; insects; vehicles; ships; wind turbines; and more. For weather radar, the most prevalent type of clutter is from the ground (Earth's surface, including vegetation) and human-made structures, such as buildings or towers. Radars operating along shorelines are also subject to clutter from large lakes and seas, but this clutter is typically weak relative to "ground clutter." The same can be said for insects, which are frequently observed in clear-air, but have low reflectivities. Ground clutter can be removed through Doppler spectrum filtering, relying on the expected zero velocity of the stationary clutters' scatterers (Section 7.3.6).

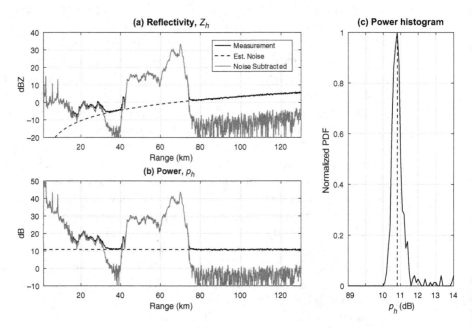

Figure 7.12 CSU-CHILL S-band observations at 301° azimuth and 4.5° elevation. Precipitation is present up to a range of 75 km. For ranges less than 18 km, surface clutter contributes to the measured power. From 32 to 40 km, the echo power is small, and the measurement is dominated by noise. After 75 km, the measurements are only noise. The dashed line indicates the estimated noise power from the data (from panel [c]). Note that the noise equivalent reflectivity increases with range in panel (a). The gray "Noise subtracted" curve shows the reflectivity (panel [a]) and power (panel [b]; arbitrary units) after the estimated noise is subtracted from the measured power (the power is estimated from $N = 512$ pulses).

Clutter targets that are moving (planes, birds, insects, smoke plumes, etc.) may be difficult to identify with single-polarization systems. When observed with a dual polarization radar, most hydrometeor types exhibit a high degree of correlation between the scattering amplitudes at horizontal and vertical polarizations. This is a consequence of the similarity of shapes and uniform fall behavior governed by aerodynamics and hydrostatic (for raindrops) forces. On the other hand, for most nonmeteorological targets composed of irregular objects with different sizes, shapes, and composition (buildings, trees) and varying flight behavior (birds, insects), this correlation is reduced. This generally makes them easily recognizable, in particular through analysis of the copolar correlation, as discussed in Section 7.2.2.

Figure 7.14 shows an example of probability density functions (PDFs) of Z_h, Z_{dr}, and ρ_{hv} for weather and clutter echoes. Similarly, Figure 7.15 presents the statistical distributions of textures of Z_h, Z_{dr}, and Φ_{dp}, providing information about the spatial variability of the radar observations. The distribution of the radar variables (and relative textures) for weather echoes is generally much narrower than the distribution for nonmeteorological targets. The two PDFs (weather and clutter) typically show a

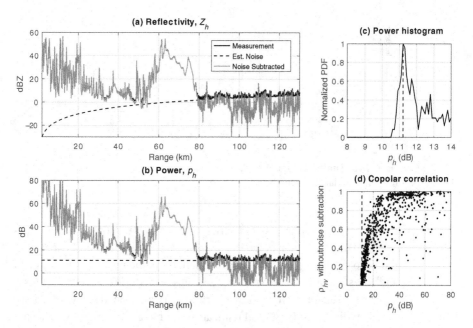

Figure 7.13 CSU-CHILL S-band observations, fixed pointing at 345° azimuth and 0.5° elevation. Precipitation extends from 50 to 80 km. Substantial surface clutter is observed up to 50 km, and weaker clutter contributions are observed along the entire radial. Beyond 80 km, the echo power is small, and the measurements are dominated by noise. The dashed line indicates the estimated noise power from the measurements (from panel [c]). The gray "Noise subtracted" curves are the estimate signal power after noise subtraction ($N = 512$ is used for the estimates). In panel (d) the noise-only observations also have the lowest copolar correlation. This relationship provides an effective parameter for data quality control as well identifying noise-only measurements.

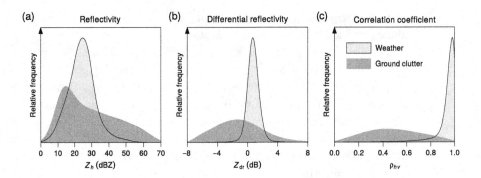

Figure 7.14 PDF of radar variables for ground clutter and weather echoes during an event observed by a C-band radar in Italy. Distributions of Z_h (a), Z_{dr} (b), and ρ_{hv} (c). Note the clear separation between the distributions of the radar variables for weather echoes and ground clutter. Note, the particular polarimetric response of nonmeteorological echoes may differ for radar sites at disparate locations.

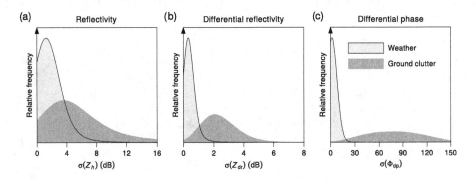

Figure 7.15 Similar to Figure 7.14, but for textures (standard deviation over a local spatial region) of radar variables Z_h, Z_{dr}, and Φ_{dp}.

moderate overlap for the reflectivity and its texture (panel [a] in Figs. 7.14 and 7.15). Using the dual polarization radar variables allow a clear discrimination of the echo type. In particular, the correlation coefficient ρ_{hv} (panel [c] of Fig. 7.14) and the standard deviation of Z_{dr} and Φ_{dp} (panels [b] and [c] of Figs. 7.15) show a marked separation (smallest overlap) between the density functions, providing robust indications for the classification of radar observations. The occurrence of anomalous propagation may sometimes lead to new and stronger echoes from ground clutter, which can still be successfully identified by the low ρ_{hv} and high standard deviation of Φ_{dp}.

Classification algorithms based on fuzzy logic (discussed in detail in Section 8.3) require the definition of membership functions for each class and radar variable. Although these are often defined for various hydrometeor types using electromagnetic scattering simulations, for nonmeteorological targets, it may be convenient to directly use the observed probability density functions like the ones illustrated in Figures 7.14 and 7.15. This empirical approach allows users to take into account the distinctive

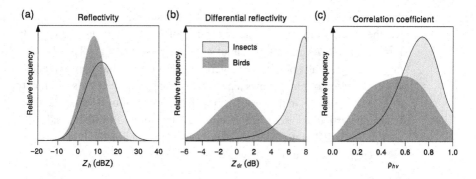

Figure 7.16 PDFs of radar variables Z_h (a), Z_{dr} (b), and ρ_{hv} (c) for birds and insects. The distributions refer to PPI scan observations collected with the NEXRAD KTLX S-band radar in Oklahoma. Insect and bird distributions respectively refer to July 19, 2013, during the day (17:35 UTC, corresponding to 12:35 local time) and March 3, 2018 at night (04:51 UTC, corresponding to 22:51 local time on the previous day) [202]. In general, echoes from birds are more frequent at night, when migration mainly occurs, whereas insects are more often observed during the day. Note that the Z_{dr} distribution of insects is truncated at ~8 dB because of the limited range of values (−7.9 dB, 7.9 dB) reported by the WSR-88D radars. Other studies report Z_{dr} values from insects of up to 10 dB [203].

characteristics of individual radar sites. In fact, different sources of clutter signals (terrain, buildings, sea, etc.) may result in markedly different polarimetric signatures.

Besides the previously mentioned sources of clutter, another important source is birds and insects. It is important to distinguish between the two because whereas insects are in general carried by the wind, birds are capable of flying across the wind. The obvious implication for radar applications is that insects are good tracers of the wind velocity, allowing the estimation of clear-air wind profiles, whereas birds are generally a source of contamination for Doppler measurements. Gauthreaux et al. [201] compared estimates of wind velocity and direction from velocity azimuth display (VAD) products with corresponding radiosonde data. The results indicated that despite birds' tendency to migrate following winds, differences between the birds' trajectory and the wind direction of more than 120° may be observed. Similarly, the difference in absolute velocity could be as large as 15 m s^{-1}.

Both birds' and insects' echoes may be erroneously interpreted as precipitation. Reflectivity from birds and insects is highly variable, depending on their radar cross-section (RCS; typically falling in the Mie scattering regime for birds) and concentration. Dual polarization observations offer a chance to distinguish between insects and birds. In fact, birds are physically larger, less coordinated and more randomly distributed than insects; therefore, their expected correlation coefficient is lower. Insects generally have highly positive Z_{dr} (up to 10 dB [203]), whereas birds may show both positive and negative Z_{dr} values. This is illustrated in Figure 7.16 for radar observations of birds (dark gray) and insects (light gray) with the S-band NEXRAD KTLX radar in Oklahoma. The correlation coefficient of insects and birds in Figure 7.16 (average of ~0.7 and ~0.5, respectively) is significantly lower than the typical values observed in precipitation ($\rho_{hv} > 0.9$). The differential reflectivity of insects, with an average value of $\sim$7 dB, is larger than the typical values associated with weather observations, with the exception of layers of plate-like crystals (see, e.g., Fig. 4.21). However, this potential ambiguity is easily resolved by observing that insects are generally not present at temperatures colder than approximately 10°C. The Z_{dr} distribution of birds shows a possible overlap with weather scatterers (panel [b] of Fig. 7.14), although in this case, the much lower ρ_{hv} of birds allows a clear separation. For these particular cases, the reflectivity distributions are similar and centered around a Z_h of $\sim$10 dBZ, illustrating the significance of dual polarization radars to distinguish between the two types of scatterers.

7.3.6   Ground-Clutter Filtering

Echoes from mountains, structures, and vegetation all contribute to the ground-clutter echoes observed by weather radar when scanning at low-elevation angles. As the elevation angle increases, these surface-bound elements are no longer within the antenna's beamwidth, and therefore their contribution to the overall echo power is significantly reduced, and in most cases, it is nonexistent. An increase of only a few degrees in the radar's elevation will largely negate any adverse contamination from ground clutter. Even so, observations near the surface are highly desirable for many applications, such as in measuring near-surface winds and estimating rainfall. As the range from the radar increases, small changes in elevation result in a significant change in the height of the observation volume above the surface (Section 2.4.5). Radar observations at the lowest elevations include important atmospheric observations but are also most affected by ground clutter. Ground-clutter detection methods are used to identify contaminated observations. Ground-clutter filtering techniques are used to separate the unwanted ground echoes from the precipitation echoes of interest.

In Section 7.3.5, it was shown that the dual polarization properties of ground clutter differ from the dual polarization properties of precipitation. Precipitation also has a spatial consistency (being a distributed volume of scatterers) that is less typical of ground clutter. Ground-clutter identification, using the dual polarization observations, is an effective method of detecting prevalent ground clutter in weather radar observations [204]. The detection of ground clutter does not suppress it when it exists in the same observation volume as precipitation. In the absence of precipitation, ground-clutter maps can also be generated from observations in clear conditions. For other sources of clutter, such as highways or wind turbines (where the moving vehicles or turbine blades are difficult to filter), clutter maps can also be generated. The clutter maps can be used to identify contaminated observation volumes during precipitation events. The identification or detection of clutter typically results in the observation volume being ignored in subsequent algorithms (e.g., rainfall estimation). In these cases, the surrounding "good" observation volumes are used to infer precipitation within the clutter-contaminated volumes. Note the clutter maps and clutter-detection results may be selectively applied to different applications.

In the case where both ground clutter and precipitation occupy a common volume, signal-processing techniques are required to separate the ground clutter's echo from that of the precipitation. Ground-clutter filters (GCFs) have been implemented in a number of ways, but all rely on a common feature of this clutter: it does not move; it is fixed, stationary. (Vegetation is an exception, where the leaves and branches can sway in the wind, but it is near-zero enough to be easily filtered.) If the radar does not move and the ground clutter does not move, then the Doppler velocity (and therefore Doppler frequency) will be zero. A simple high-pass filter on the received signal removes the zero-frequency ground-clutter signal. Although this filter method works well, the precipitation's Doppler spectrum can also be present at near-zero Doppler velocities. The high-pass filter cannot discriminate between clutter and precipitation and therefore, it removes both. This effect can be seen in filtered weather radar data (even with more advanced filtering techniques, which are discussed next) where the reflectivity field, which should be uniform based on the meteorological conditions, shows a drop in power consistent with the zero-Doppler velocity line.

A GCF can be applied to all observation volumes, or it can be selectively applied based on predetermined clutter maps, a clutter-detection algorithm, or simply by using the radar's elevation angle. An example of ground-clutter filtering for a PPI scan is shown in Figure 7.17 with and without the GCF enabled (panels [d] and [c] respectively). For low-elevation angles, the GCF provides a substantial improvement in the detection and identification of precipitation in the presence of ground clutter. The GCF also improves the quality of the other radar variables by reducing or removing biases caused by the ground clutter. It is worth noting that a clutter filter allows the retention of the information about precipitation where there is a superposition of weather and clutter (unless the clutter signal is overwhelming), whereas a data-masking procedure (e.g., using a threshold on ρ_{hv}, as illustrated in Fig. 7.8) implies the loss of information for a particular range volume.

A Gaussian Doppler spectrum is commonly used as a model for weather radar signals:

$$S(v) = \frac{P}{\sigma_v \sqrt{2\pi}} e^{-\frac{(v-\bar{v})^2}{2\sigma_v^2}}, \qquad (7.26)$$

where P is the signal's power, σ_v is the Doppler spectrum width, $\bar{v}$ is the mean Doppler velocity, and v is the independent variable – the spectrum's Doppler velocity. Precipitation and ground clutter can simultaneously exist in the same volume. The contributions of each are statistically independent and therfore, the radar's received signal can be modeled as the sum of their spectra and noise:

$$S_x = S_p + S_c + S_n. \qquad (7.27)$$

The Doppler spectra of precipitation (S_p) and ground clutter (S_c) are Gaussian, whereas noise (S_n) has a constant (flat) spectrum. Eq. (7.27) can be expanded as

$$S_x(v) = \frac{P_p}{\sigma_{vp}\sqrt{2\pi}} e^{-\frac{(v-\bar{v}_p)^2}{2\sigma_{vp}^2}} + \frac{P_c}{\sigma_{vc}\sqrt{2\pi}} e^{-\frac{v^2}{2\sigma_{vc}^2}} + \frac{P_n}{2v_a}. \qquad (7.28)$$

The precipitation spectrum's power, velocity, and spectrum width are P_p, $\bar{v}_p$, and σ_{vp}, respectively. The clutter spectrum's power and spectrum width are P_c and σ_{vc},

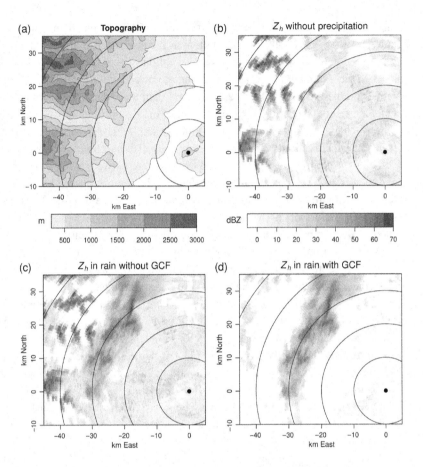

Figure 7.17 Illustration of ground-clutter filtering in complex orography. Topography (a) and average unfiltered reflectivity in the absence of precipitation at $2°$ elevation from a C-band radar in the Alpine region (Torino, Italy) (b). Reflectivity observations at $2°$ elevation of a summer storm near the mountains with (d) and without (c) ground-clutter filtering.

respectively (the mean velocity of the clutter is assumed to be zero). The noise-power spectrum has power of P_n, recalling from eq. (5.12) that the mean time domain power is the integral of the spectrum's power.

Ground clutter is stationary and therefore nominally has a mean velocity of zero. In the Doppler spectrum shown in Figure 7.18, the ground clutter is easily identified. With this prior knowledge of the ground clutter's velocity, a simple GCF would remove the power at zero Doppler velocity. The zero-Doppler velocity signal components are the mean of the signal (with $k = 0$ in eq. [6.41]) following

$$S_x[v = 0] = \frac{1}{N^2} \left| \sum_{n=0}^{N-1} x[n] \right|^2. \tag{7.29}$$

To remove the zero-Doppler power $S_x[v = 0]$ from the signal, the IQ signal's mean can be subtracted as $x[n] - \bar{x}$, where $\bar{x}$ is the coherent average of the N IQ samples: $\bar{x} = \frac{1}{N}\sum x[n]$. From eq. (7.29) (and eq. [5.10] for $X[0]$), it is clear that a stationary scatterer results in a non-zero mean, or a bias, in the complex-valued signal. A strong clutter signal is also a dominant point scatterer, which violates the typical weather radar assumption of a uniform distribution of homogeneous scatterers within the radar volume. Ground clutter filtering through subtraction of the zero-Doppler velocity power has two drawbacks: (1) it removes all zero-Doppler velocity power, including any part of the precipitation spectrum that is at zero-Doppler velocity, and (2) ground clutter may have slight nonzero velocity components as a result of the motion of the antenna or the motion of trees and vegetation from wind. The window functions that are used in the estimation of the Doppler spectrum can also introduce an artificial spectral broadening (Section 6.2).

As the antenna scans, its pattern acts as a weighting function on the amplitude and phase of the precipitation scatterers. This, in effect, implements a window on the data. A window applied in the time domain acts as convolution in the spectral domain, which broadens the estimated spectrum. The faster the antenna scans, the broader the spectral features of the clutter become.

The ground clutter can be approximated by a narrow Gaussian spectrum centered around a mean Doppler velocity of $0\,\mathrm{m\,s^{-1}}$ (the spectral width of the filter is a parameter, but it is typically around $0.2–0.3\,\mathrm{m\,s^{-1}}$). The precipitation's mean Doppler velocity and spectrum width must be estimated. The measured spectrum's Doppler velocities that are identified as clutter can be replaced with estimates of what the precipitation's spectral power are using a model parameterized by the precipitation's Doppler velocity and spectrum width. Examples of algorithms that adopt a similar technique include Gaussian model adaptive processing (GMAP) [205] and GMAP in the time domain [206].

Figure 7.18 shows the average spectrum observed by the CSU-CHILL S-band radar, which includes precipitation, ground clutter, and noise. The spectra of these three components are estimated and fit to the observations, which are shown in panel (a) of Figure 7.18. The portion of the ground-clutter Doppler spectrum that is to be filtered from the data is indicated by the circles. The GCF replaces the clutter points of the Doppler spectrum with the precipitation model's estimated spectrum for these Doppler velocities. The estimated noise power is subtracted from the remaining spectrum to give the ground-clutter-filtered, noise-subtracted spectrum in panel (b) of Figure 7.18. The biasing effect of the ground clutter (and noise) is removed. The filtered Doppler spectrum is then used to estimate the precipitation's radar variables. Both the horizontal and vertical polarizations' observations are filtered in the same way (although the power level for the three components may be different between the polarizations). The same mean Doppler velocity and Doppler spectrum width are used for both polarizations to minimize any biases in the dual-polarization radar variables that could occur from applying different GCF parameters.

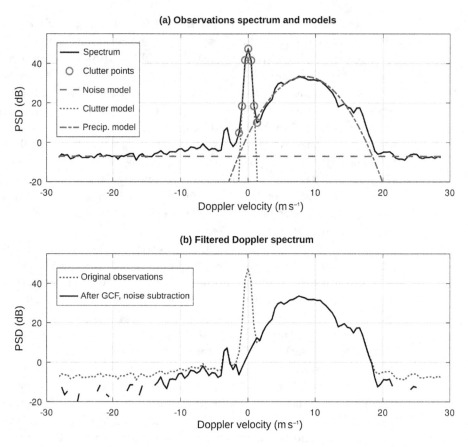

(a) Observations spectrum and models

(b) Filtered Doppler spectrum

Figure 7.18 Spectrum estimates of precipitation with ground clutter from the CSU-CHILL S-band radar on June 26, 2014, at an azimuth of $245°$, elevation of $0.5°$, and range of 61.65 km. The black line is the average of 17 spectra, each estimated with $N = 128$ samples and a Hann window. The Doppler spectrum's points that contain significant contributions from ground clutter are marked by circles in panel (a). The models from eq. (7.28) for the noise, ground clutter, and precipitation spectra are shown fitted to the observations in panel (a). After ground-clutter filtering and noise subtraction, the resulting precipitation is shown in panel (b) as a solid black line. Compare this to the original observation, shown as the dotted line.

7.3.7 Three-Body Scattering

When an electromagnetic wave propagates through a collection of scatterers, the wave may scatter many times, leading to the phenomenon known as *multiple scattering*. The effects of multiple scattering are generally negligible for common centimeter-wavelength radars, whereas they become more important at higher frequencies, such as those used by cloud radars [207]. This section discusses a specific example of three-body scattering that may affect radar observations of severe weather.

Three-body scattering is a widely accepted explanation for the "hail-spike" signature that is often observed in the presence of large hail. The hail scatterers all exist

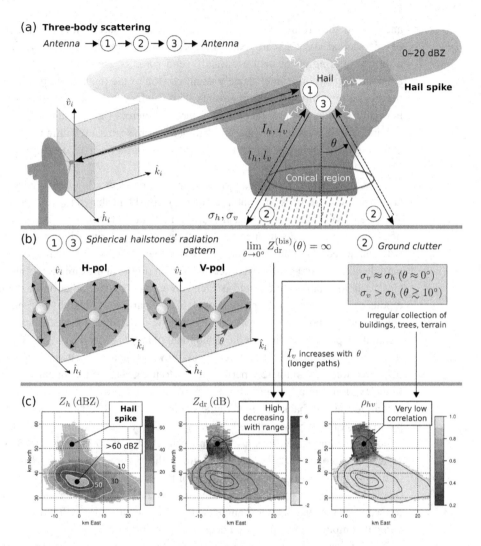

Figure 7.19 Illustration of the schematic geometry (a), main physical mechanisms (b) and observational evidence (c) of three-body scattering caused by large hail. The observations in panel (c) are from a C-band PPI scan at 15° elevation through an intense hailstorm.

within a limited range, but their echoes appear at ranges beyond where the hail occurs. This is because the radar maps the range with the echo's arrival time (using the speed of light as the conversion factor). When the signal scatters multiple times, the total distance that the echo travels during the multiple scattering events is interpreted as if it was a single scattering event, which results in an echo that appears in a farther radar observation volume.

Consider the schematic illustration in panel (a) of Figure 7.19, representing a radar observing a hail-bearing storm at a low-elevation angle. The wave transmitted by the radar is scattered by the hailstones in all directions, including back to the radar and toward the ground, which subsequently scatters part of this energy. A portion of this second event's scattered power propagates back into the hail region, where a third scattering event may eventually send some of the energy back to the radar antenna. Clearly, only an extremely small portion of the transmitted power can reach the radar after multiple scattering events. This is why a three-body scattering signature normally implies the presence of very large scatterers (hail) in the radar resolution volume. The echoes in the hail spike originates from scattering within a conical region defined by the incidence angle θ (echoes at the farther ranges correspond to larger θ values).

In a single-polarization radar, the hail spike appears as a low-reflectivity protuberance behind the storm's core and may be erroneously attributed to weak precipitation. With dual polarization systems, the different scattering behavior of the orthogonal polarized waves results in specific signatures of the polarimetric variables, typically with enhanced Z_{dr} and low ρ_{hv} values. In the Rayleigh approximation, spherical particles illuminated by the radar act like a point dipole in the direction of the incident electric field. The scattered field is then represented by the radiation of the oriented dipole, with an amplitude proportional to the particles' volume. For horizontally and vertically polarized waves, the scattering pattern assumes the characteristic toroidal shape, whose slices in the orthogonal planes $\hat{k}_i - \hat{v}_i$ and $\hat{h}_i - \hat{v}_i$ are illustrated in panel (b) of Figure 7.19. It is immediately obvious that the scattered radiation in the $\pm\hat{v}_i$ direction (approximately perpendicular to the ground if the antenna is looking at low-elevation angles) is null for the vertical polarization, whereas it is the same magnitude as the backscattered radiation for the horizontal polarization. If a dual polarization receiver was placed on the ground in the area delimited by the conical region shown in Fig. 7.19(a), it would measure the bistatic $Z_{dr}^{(bis)} = I_h/I_v$ [208], where I_h and I_v denote the power scattered along the θ direction. Although this bistatic differential reflectivity would be infinite for the example of the idealized infinitesimal dipole when $\theta = 0°$ (note, however, that in general for Mie scattering, $I_h \gg I_v > 0$), the scattered power at higher θ angles within the conical region will have an increasing vertical component, resulting in a decrease of the bistatic Z_{dr}.

In addition to scattering by the hail particles, the subsequent propagation effects and backscattering by ground clutter should be considered. Following the notation used by Picca and Ryzhkov [209], the observed Z_{dr} in the hail spike can be written as a combination of three terms:

$$Z_{dr} \sim Z_{dr}^{(bis)} \frac{\sigma_h}{\sigma_v} \frac{l_h}{l_v} = \frac{I_h}{I_v} \frac{\sigma_h}{\sigma_v} \frac{l_h}{l_v}, \tag{7.30}$$

where $\sigma_{h,v}$ and $l_{h,v}$, respectively, denote the scattering amplitudes of ground clutter and the two-way attenuation through the precipitation layer between the hail core and the ground. Near vertical incidence, the radar cross-section of the ground clutter is similar for the horizontally and vertically polarized radiation ($\sigma_h \approx \sigma_v$), and consequently, it has no effect on the ratio of the radiated powers. Similarly, differential attenuation through the rain layer (l_h/l_v) is zero at vertical incidence. Therefore, for small θ angles, the observed Z_{dr} at the radar receiver is the result of the different radiation patterns of the hail particles. With larger θ angles, in addition to increasing I_v, both the relative scattering amplitude from ground clutter and the signal attenuation in the rain medium increase for the vertical radiation component. In particular, using a statistical model for ground clutter, Hubbert and Bringi [210] have shown that the vertical backscattered power from ground clutter becomes larger than the corresponding horizontal power when the incident angle $\theta > 10°$. For this reason, Z_{dr} is generally very large near the hail shaft but shows a decreasing trend with farther ranges in the hail spike (panel [c] in Fig. 7.19, middle graph). At the same time, multiple scattering (including scattering by the ground) independently decorrelates the signals, drastically reducing the correlation coefficient (panel [c] of Fig. 7.19, right panel), which can be effectively used for the identification of these echo artifacts.

7.4   Calibration

The calibration of weather radars is essential to ensure accurate conclusions and results for qualitative and quantitative applications. The term *calibration* normally implies both the reflectivity calibration (measured power accuracy) and the position calibration (pointing accuracy of the antenna). In fact, any bias in either the received power or in the antenna position will introduce a systematic error in the observations. The antenna pointing accuracy can be verified using measurements from either fixed targets or the sun. For the calibration of reflectivity (absolute calibration), the goal is to achieve an accuracy of better than 1 dB. In dual polarization systems, the presence of a second channel (i.e., the vertical polarization) requires an additional effort. Although both channels may be independently calibrated (each one within a 1-dB accuracy), this is not sufficient to ensure calibration of the differential reflectivity to a target accuracy of 0.2 dB, which is generally required for quantitative applications such as rainfall estimation. Therefore, techniques for the direct calibration of differential reflectivity have also been developed.

Phase measurements don't require calibration. However, it is a good practice to routinely estimate the system differential phase shift ϕ_{sys}, for example, by considering the first few range bins in weather along each beam. Setting the ϕ_{sys} value close to a preset value (some people use $0°$) may simplify the calculation of K_{dp} and prevent aliasing.

This section discusses the role of antenna quality, pointing verification, and end-to-end calibration procedures characterizing the radar system as a whole, for reflectivity and differential reflectivity. For a more comprehensive discussion of calibration techniques, including a description of subsystem verification and internal calibration loops, the reader is referred to Chandrasekar et al. [211].

In addition to one-time calibrations as a means to measure the end-to-end system performance, the calibration techniques can also be used periodically to assess changes in system performance. Radars, like all other mechanical or electronic devices, are subject to changes in component characteristics over time. By regularly assessing the radar's calibration, the change in performance can be tracked to

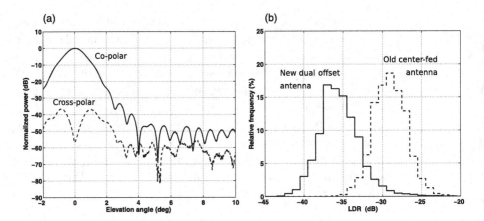

Figure 7.20 (a) Antenna copolar (solid line) and cross-polar (dashed line) pattern for the CSU-CHILL's new dual offset Gregorian antenna. (b) LDR in stratiform precipitation observed with the old CSU-CHILL center-fed parabolic antenna (dashed lines, data collected on May 18, 2007) and the new dual offset antenna (solid lines, data collected on June 5, 2008). Adapted from [194], ©American Meteorological Society. Used with permission

identify changes that may indicate the need for maintenance or replacement. Periodic calibration also allows these changes to be compensated for in the radar data to maintain the accuracy of the observations.

7.4.1   Antenna Quality

At the most basic level, a weather radar consists of a transmitter and a receiver, with the antenna serving as its interface to the environment. It is then no exaggeration to say that the quality of data from a polarimetric radar is dependent on the electrical performance of the antenna.

Knowledge of the antenna copolar and cross-polar patterns is important to understand the effects of potential biases and the achievable accuracy on the dual polarization radar variables of a specific radar system. Characterization of the minimum observable LDR is a practical and concise metric to establish the quality of a dual polarization antenna's cross-polar isolation. For dual polarization radars that do not estimate LDR, the antenna's cross-polar isolation also affects the accuracy of the other dual polarization estimates: Z_{dr}, ρ_{hv}, and Φ_{dp}. The antenna's LDR limit (the lowest LDR the antenna can measure) and the antenna's integrated cross-polar ratio (ICPR$_2$; eq. [2.48]) are equivalent:

$$LDR_{limit} = ICPR_2. \qquad (7.31)$$

For STSR mode, $LDR_{limit} < -25$ dB is desired to ensure negligible Z_{dr} bias (within ± 0.1 dB in rain) [20].

Dual offset antennas may attain very low sidelobe levels and high system cross-polarization isolation [194] (panel [a] of Fig. 7.20). Consider the observations in panel (b) of Figure 7.20, showing distributions of LDR collected by the CSU-CHILL radar with the old center-fed antenna (dashed lines) and the new dual offset Gregorian antenna (solid lines), with the LDR reaching values as low as -40 dB. The analysis of LDR in stratiform rain provides a way to empirically quantify the system cross-channel isolation, which is mainly dictated by the antenna performance. In fact, due to the narrow canting-angle distribution of raindrops, the cross-polarization signal is expected to be very low, and the measured LDR can be interpreted as the system's lower bound. This is particularly true for the case of drizzle, where the scatterers are nearly spherical droplets.

Most operational weather radars around the world use center-fed parabolic antennas and cannot meet the performance demonstrated by the CSU-CHILL's dual offset Gregorian antenna. For such antennas, the mechanical struts that hold the feed horn in place cause a blockage that deteriorates the side-lobe performance and reduces the system isolation by enhancing the cross-polarization returns (Section 2.6.3). Knowledge of the LDR's lower limit for an operational radar is important for characterizing the overall system performance and understanding potential biases. Even if most operational systems use the STSR mode and do not measure LDR routinely, a subset have the capability to do so by means of an electrical switch to route all the transmitter's power to a single polarization channel.

When operating in STSR mode, biases in the polarimetric variables can arise as a result of both the cross-polar antenna patterns and the nonzero mean canting angle of precipitation particles along the propagation path. In general, the precipitation particles can be modeled to have orientations following a Gaussian distribution with a mean canting angle of zero (i.e., oriented in the horizontal plane). However, the mean canting angle may differ from zero in particular situations, such as in the presence of strong electric fields in ice-crystal regions of thunderstorms, which can result in a mean canting angle that is closer to vertical. In this case, the cross-polar induced errors may show up as radial streaks in Φ_{dp} and Z_{dr} within and beyond the regions of the canted ice crystals (see discussion in Section 8.5.3).

7.4.2   Pointing Calibration

The radar has been adopted as a workhorse instrument for operational meteorology and atmospheric research because it provides rapid coverage of a large area. For most applications, these observations are only useful if the radar's observations can be accurately localized to a position. This would imply that the reported antenna pointing (and range) are accurate. For radars that are part of a larger network (see Chapter 10), the accuracy of the position and pointing is even more important to ensure high-quality observation products from the network (e.g., multi-Doppler estimates).

This is a one-time time check after the radar system is installed or maintained. For most operational systems, that would imply that the position would only need to be validated once during the radar's life, but positioner electronics and radar components do occasionally require replacement. For mobile radar systems that are deployed

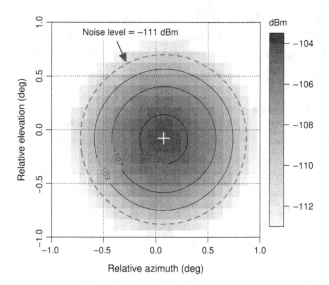

Figure 7.21 Observed power during a dedicated sun raster scan with CSU-CHILL S-band radar. The scan spanned elevation in the range of 11–14° and azimuth in the range of 228–234°. On the x- and y-axes, the relative position with respect to the sun is shown. The deviation from the observed power peak (marked by the white cross symbol) is less than 0.1° both in elevation and azimuth, indicating a good pointing accuracy.

seasonally or as part of research efforts, calibration of the radar's pointing is a more frequent activity.

The process is relatively straightforward. Today, the radar's position is determined using modern global navigation satellite systems (GNSSs) (e.g., global positioning system [GPS]), which is a simple and highly accurate process. With the radar's position and accurate timekeeping (also available from GNSS), the sun's position in the sky, in terms of azimuth and elevation angle relative to the radar, can be calculated. The radar observes the region around the sun's location. The sun results in an increased radiometric noise power measured by the radar receiver. If the radar scans are fine enough with respect to the antenna beamwidth (e.g., $\theta_1/10$ to $\theta_1/5$), the peak of the radiometric noise is observed when the antenna's boresight is aligned with the sun (Fig. 7.21). Any errors between the commanded positioner angle and the peak of the solar noise are attributed to position bias errors that can be corrected. After correction, ideally, the commanded positioner angle should align with the peak of the sun's noise power. Note that the sun's position is constantly changing, and therefore each radar measurement throughout the scan must account for this relative change.

When using the sun for position calibration, higher-elevation angles are typically preferable to lower-elevation angles because of atmospheric refraction effects, which can distort the path between the sun and radar and introduce slight errors. By using multiple solar position estimates throughout the day, the azimuth and elevation plane can be corrected. The leveling and the orientation of a reference azimuth (typically north) are the sources of errors in the position knowledge after an installation. A radar whose pointing is calibrated should have an accuracy better than half of the antenna's beamwidth to ensure that the reported position is within the antenna's main lobe. Using the sun as a position calibration source, accuracy to better than tenths of a degree can be achieved.

The antenna's half-power beamwidth (HPBW; θ_{HPBW}) indicates the azimuth and elevation extent of the V_6 radar volume. The antenna's HPBW is defined for either the transmit or receive antenna. Recall that the two-way effect of the antenna is -6 dB because of the G^2 in the weather radar equation (eq. [2.39]). (Note that for the solar calibration, because the antenna is only receiving the solar noise and not relying on the transmitted signal, only the one-way effect of the antenna applies.) The main lobe of the antenna pattern's gain, relative to the boresight's peak gain at $\theta = 0$, can be approximated as a parabola using a logarithmic scale (i.e., the pattern's boresight-relative gain is in decibels):

$$G_n \text{ (dB)} = a\theta^2, \tag{7.32}$$

where a is a scale factor for the particular antenna. The mainlobe of the antenna's one-way pattern, such as on transmit or receive, is characterized by its HPBW where the gain is 3 dB below the boresight's gain:

$$-3 \text{ (dB)} = a_1(\theta_{HPBW}/2)^2 \tag{7.33}$$

which gives the constant for the one-way relative gain:

$$a_1 = -12/\theta_{HPBW}^2. \tag{7.34}$$

In the radar equation, the receive and transmit antenna gains are multiplied (on a log scale, they are added) and the product results in a combined pattern that is 6 dB below the boresight gain at single antenna's HPBW:

$$-6 \text{ (dB)} = a_2(\theta_{HPBW}/2)^2, \tag{7.35}$$

where the two-way pattern's constant is

$$a_2 = -24/\theta_{HPBW}^2. \tag{7.36}$$

For a monostatic radar system, it is clear that $a_2 = 2a_1$.

The desired reflectivity accuracy is typically 1 dB for weather radar applications. For the reflectivity estimate of a precipitation volume to be accurate to within 1 dB, and if the only source of error is assumed to be the pointing accuracy, then the tolerable pointing error θ_{err} can be estimated as follows:

$$-1 \text{ (dB)} = a_2\theta_{err}^2 \tag{7.37a}$$

$$-1 \text{ (dB)} = -24\theta_{err}^2/\theta_{HPBW}^2 \tag{7.37b}$$

$$\theta_{err}/\theta_{HPBW} \approx 0.2, \tag{7.37c}$$

or the pointing error can be as much as 20 percent of the antenna's HPBW for a 1-dB reflectivity error. Typically, a portion of the reflectivity error is allowed for the estimator bias and standard deviation. For 0.5 dB of reflectivity error allowed for pointing, this results in a tolerable pointing error of 10 percent of the HPBW, or approximately 0.1° for 1° HPBW weather radar antennas. The effect of pointing on reflectivity error is really only an issue when comparing to other georeferenced observations (like a network of radars, rain gauges, etc.) and where there are significant gradients in the weather system.

Besides dedicated raster scans, the sun's observations can also be exploited during routine operational scans for monitoring the stability of the antenna pointing accuracy [212, 213]. For typical volume scan strategies, the solar signals can be on the order of 10 or more per day. The cumulative analysis of several days is then sufficient to detect possible biases (or changes) in the elevation and azimuth pointing.

7.4.3   Reflectivity Calibration

The radar resolution volume is determined by the range resolution and beam's width at a selected range (a function of the transmitted waveform and receiver filter, as well as the antenna pattern). Now consider a single target within the volume (i.e., eq. [2.10] with $N = 1$) that is a calibration target. The equivalent reflectivity factor can be calculated for the calibration target with a known RCS. The calculated (true) reflectivity can be compared against actual radar observations of the calibration target to estimate biases in the radar's measured reflectivity.

For a single target within the radar volume and the target's RCS is known, then the radar constant C, which includes the effects of all of the radar system's components, can be estimated directly from the observed signal. Note that for a point scatterer, the antenna gain (which is azimuth and elevation sensitive) will affect the estimated radar constant. Thus, for point targets, the receiver power is dependent on the pointing direction of the transmitting and receiving antennas.

Using this technique, a "sphere calibration" can be performed, providing an end-to-end system calibration to accurately estimate the system's radar constant. For a perfectly conducting metal sphere whose diameter is large with respect to the radar's wavelength (i.e., the sphere is in the optical scattering regime), the RCS is

$$\sigma = \pi r_s^2, \tag{7.38}$$

where r_s is the metal sphere's radius. For a metal sphere located in the antenna's far-field at range r, and along the boresight direction in the center of the range resolution, the point target radar equation (eq. [2.36c]) and sphere's RCS from eq. (7.38) can be combined as:

$$P_r = \frac{\lambda^2 P_t G^2 r_s^2}{64\pi^2 r^4}. \tag{7.39}$$

For calibration, a correction to the radar system's parameters may be necessary to match the calculated power to the measured power. This can be implemented with a separate calibration factor, or modification of an existing system parameter. In eq. (7.39), G represents the combination of the antenna gain with waveguide losses (transmit and receive paths) and radome losses factored in:

$$G^2 = G_0^2 / (l_{tx} \, l_{rx} \, 2 l_{radome}), \tag{7.40}$$

(a)   (b)

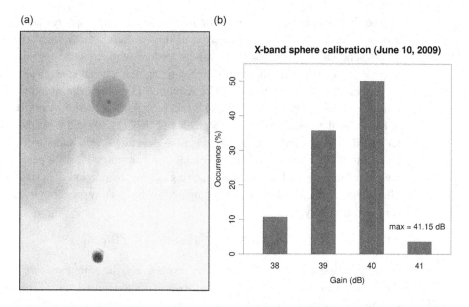

Figure 7.22 Metal sphere calibration performed with a tethered balloon for an X-band radar in Italy (a). The peak gain (b) calculated from the received power using eq. (7.41) needs to be compared with the corresponding value derived from the system specifications (antenna gain, waveguide and radome losses).

where G_0 is the antenna's one-way gain; l_{tx} and l_{rx} are the transmit and receive path losses, respectively; and l_{radome} is the one-way radome loss. Equation (7.39) can be rearranged to calculated the calibrated antenna gain as

$$G = 8\pi \frac{r^2}{\lambda r_s} \sqrt{\frac{P_r}{P_t}}, \qquad (7.41)$$

from which the radar constant can be directly calculated using eq. (2.39d).

Figure 7.22 shows an example of a calibration experiment performed with a tethered balloon carrying a Styrofoam sphere wrapped with aluminum foil (panel [a]). During this experiment, the sphere was suspended at a height of approximately 250 m above the ground, and the radar performed azimuth scans closely spaced in elevation. In order to maximize the chances of having the sphere close to the center of the along-range filter response, data collection was carried out with range over-sampling. The histogram in panel (b) shows the gain distribution for an experiment conducted using a sphere with a 20-cm diameter. The obtained peak gain is consistent with the system specifications (antenna gain $G_0 = 42.2$ dB, one-way radome loss of $10\log_{10}(l_{radome}) \sim 0.3$ dB, and total waveguide losses of $10\log_{10}(l_{tx}\, l_{rx}) \sim 1.9$ dB). In fact, from eq. (7.40), the calculated antenna gain G in decibels would be $(41.15 + 1.9/2 + 0.3)$ dB $= 42.4$ dB, which is very close to the nominal antenna gain $G_0 = 42.2$ dB, demonstrating the overall good radar calibration.

The absolute calibration using point targets such as metal spheres or corner reflectors with known RCSs is the preferred method to verify the system's performance (end-to-end calibration). However, this procedure requires some logistic effort and interruption of the operational scan. Recognizing that a metal sphere calibration may be infrequent, other techniques can be exploited to monitor the stability of the calibration over time. One such technique is relative calibration adjustment (RCA), which is based on the clutter-reflectivity statistics [214, 215]. The CDF of the daily clutter distribution is used to assess the possible deviation of the radar calibration with respect to a given reference value (baseline). In particular, the 95th percentile of the clutter-reflectivity CDF for the baseline (e.g., consider a period of several days right after a metal sphere calibration) is compared with the corresponding 95th percentile for the current day:

$$\text{RCA (dB)} = Z^{95}(\text{baseline}) - Z^{95}(\text{daily}), \tag{7.42}$$

where $Z^{95}(\text{baseline})$ and $Z^{95}(\text{daily})$ are, respectively, the 95th percentiles (in dB) for the baseline and for a specific daily distribution of clutter reflectivity. The RCA value is then the reflectivity adjustment needed to obtain agreement with the reference calibration, with positive (negative) values indicating underestimation (overestimation) relative to the baseline.

An alternative end-to-end method for absolute calibration exploits returns from the rain medium, based on the self-consistency principle (Section 9.1.1). It is worth noting that the self-consistency approach, because of microphysical rain variability and propagation effects, is intended as a monitoring tool rather than as a replacement for a calibration procedure like the one provided by the metal sphere. Basically, the radar variables Z_h, Z_{dr}, and K_{dp} lie in a constrained three-dimensional space (Fig. 9.7), so it is possible to express K_{dp} as a function of Z_h and Z_{dr} as in eq. (9.22). Then, assuming Z_{dr} is properly calibrated (refer to next section), it is possible to estimate the eventual reflectivity bias ΔZ_h by comparing the observed and the retrieved differential phase shifts along a range segment r_1–r_2 in rain:

$$\Delta\Phi_{dp}{}^{obs} = \Phi_{dp}(r_2) - \Phi_{dp}(r_1) \tag{7.43a}$$

$$\Delta\Phi_{dp}{}^{ret} = 2\int_{r_1}^{r_2} a Z_h{}^b Z_{dr}{}^c \, dr \tag{7.43b}$$

$$\Delta Z_h \text{ (dB)} = 10\log_{10}\left(\Delta\Phi_{dp}{}^{ret}/\Delta\Phi_{dp}{}^{obs}\right). \tag{7.43c}$$

This method has the advantage of not requiring a dedicated scan, but it relies on important assumptions. First, the coefficients a, b, and c in eq. (7.43b) depend on the assumed drop-shape model (Section 3.3), in addition to wavelength and temperature (especially at the X band). Second, care must be taken in the selection of the data, avoiding contamination by nonrain scatterers (e.g., clutter, melting snow) and ensuring that the rain along the path r_1–r_2 is properly chosen to minimize the Φ_{dp} measurement errors. Finally, observations should be corrected in advance for attenuation and differential attenuation, in particular for measurements at the C and X bands.

In Chapter 9, it will be shown that in cases where the absolute calibration is already ensured by some other technique, the triplet of observations Z_h, Z_{dr}, and K_{dp} can be used to estimate the mean shape of raindrops (Fig. 9.9).

An Interesting Data-Quality Experiment for Wet Radome

The majority of operational weather radars have their antenna covered by a radome. The radome attenuation under dry conditions is caused by the transmission loss through the radome wall. The dry radome attenuation is on the order of a few tenths of a decibel and is normally precisely measured by the manufacturer. When it rains at the radar location, additional signal attenuation is caused by the water on the surface of the radome. This attenuation may increase with the rainfall rate, affecting the radar system's calibration.

The effect of water on the radome surface has been extensively studied in the past, mainly because of the relevance of attenuation for the performance of communication systems at high frequencies. The interest of the weather radar community rose more recently, with both theoretical and experimental studies focusing on the main frequency bands used by weather radar systems. It is known that the attenuation due to the wet radome scales with operating frequency (similar to path-integrated attenuation). Thus, while at the S band the two-way loss due to a wet radome is generally negligible, it can reach a few decibels at the C band [216] and several decibels at the X band [217]. The radome attenuation mainly depends on the thickness of the water film flowing on its surface, which in turn is a function of the rain rate. (For laminar flow on a hemispherical radome, as a first approximation, the water layer thickness is proportional to the cube root of the rain rate.) However, on many radomes with water-resistant coating, the water forms beads instead of a film.

Figure 7.23 illustrates a simple experimental setup for measuring the attenuation induced by a wet radome. During any radar calibration experiment, such as using a metal sphere, a corner reflector, or a standard gain horn, artificial rain produced with a water hose is used to spray the radome surface while collecting measurements with the antenna pointing at the calibration target. As a result, it is possible to measure the reduced received power caused by the wet radome. The dip in the signal power is observed when the radome is wetted. This is followed by a rise in signal power after the water is turned off during the time it takes the radome to dry out. Although the measured power loss depends on the radar frequency and the rain intensity, other conditions also control the amount of attenuation. The type of hydrophobic coating and the radome structure are among the factors that may affect the excess attenuation in rain that is experienced by a specific radar system.

Although real-time correction procedures may be devised [218, 219], these are typically affected by a relevant degree of uncertainty. It is nevertheless important to know the general radome performance under heavy rain at the radar site and recognize when measurements are affected to adopt the proper mitigation techniques, the most fundamental one being the use of differential phase shift measurements for rainfall estimation (K_{dp} is completely immune to the radome attenuation effects).

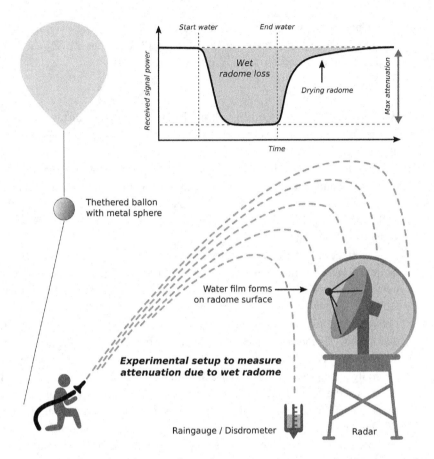

Figure 7.23 Illustration of an experimental setup to measure attenuation due to wet radome.

7.4.4   Differential Reflectivity Calibration

One viable option to assess the calibration of differential reflectivity is provided by the symmetry of raindrops when viewed from below. As discussed in Section 4.10, the expected Z_{dr} in rain when the antenna elevation is approaching $90°$ is 0 dB ($Z_h = Z_v$). This calibration technique generally requires a specific PPI scan at vertical incidence [157]. In fact, by rotating the antenna in azimuth, the polarization plane is also rotated (Fig. 2.2), which should reduce the impact of azimuthal dependencies such as those caused by the possible alignment of raindrops in strong wind shear or ground-clutter contamination through side lobes and back lobes. The average Z_{dr} over the full $360°$ azimuth domain will then reflect the system bias between the horizontal and vertical channels.

Vertical pointing scans are not always feasible, as a result of either hardware limitations or strict volume-scan requirements. Alternative calibration techniques include Z_h–Z_{dr} statistics in drizzle or very light rain observed during operational scans at low-elevation angles. In fact, drizzle is dominated by small, nearly spherical drops, implying that Z_{dr} should approach 0 dB for low reflectivity values. In practice, Z_{dr} is affected by the drop size distribution (DSD) variability even in light rain. For example, the appearance in the radar resolution volume of a few large, oblate drops grown by collision-coalescence or a modification of the DSD by size sorting (with a consequent increase in Z_{dr} and decrease in Z_h; see Sections 3.5.4 and 8.4) may lead to $Z_{dr} >$ 0 dB even at low reflectivity values. This implies a difficulty in the choice of an appropriate reference value for Z_{dr} and reflectivity thresholds for selecting data to use for calibration. As an example, for the calibration of the Meteo France radar network, a differential reflectivity reference value of 0.2 dB for light rain is compared with the median Z_{dr} in the reflectivity interval 20–22 dBZ [220] (see, e.g., Z_{dr} vs. Z_h plot in Fig. 9.7). In the selection of the data, it is important to exclude observations with a low SNR or a low correlation coefficient to avoid contamination from ground clutter. Typical threshold values are SNR> 20 dB and $\rho_{hv} > 0.99$.

In order to cope with the relatively large Z_{dr} variability in drizzle or light rain, dry aggregated snow observations above the melting layer can also be considered as calibration targets [156]. Dry snow is characterized by a small intrinsic differential reflectivity because of its very low density. Considering a typical density lower than $\sim 0.1 \, \text{g cm}^{-3}$ and an aspect ratio larger than ~ 0.5, $Z_{dr} \leq 0.3$ dB (see Fig. 4.21). For larger aggregates, the density is further reduced (eq. [3.74]), and Z_{dr} usually drops below 0.2 dB when $Z_h > 30$ dBZ. The accuracy of the calibration can also be improved by considering observations at high-elevation angles, which have the effect of damping the value of Z_{dr}. For example, an observed $Z_{dr} = 0.2$ dB at horizontal incidence would reduce to $Z_{dr} \approx 0.05$ dB at $60°$ elevation (the highest possible elevation of the WSR-88D radar) according to eq. (4.85), allowing a calibration of Z_{dr} within ~ 0.1 dB.

7.5   Selected Problems

1. If the signal power is $|s|^2 = 2$ and the noise power is $|w|^2 = 1$:

   a) what is the observed power?

   b) What is the bias in the estimate signal power, in decibels, if the estimated noise power $|\hat{w}|^2 = 1.2$ is subtracted instead of the true noise power?

2. What is the standard deviation of noise power estimated with $N = 32$, $N = 64$ and $N = 512$ samples?

3. Convective storm observations made by the US National Weather Service Radar WSR-88DP (2.7 GHz) commonly use the volume coverage patterns (VCP 12) scanning mode. The lowest PPI elevation is $0.5°$ and the highest PPI elevation in VCP 12 is $19.5°$. Assume, a scan rate of approximately $12° \text{s}^{-1}$ with a PRF of 860 Hz for the low elevations. For the highest elevation scans, assume a scan-rate of approximately $28° \text{s}^{-1}$ with a PRF of 1280 Hz. If the pulses are integrated for $1°$ in azimuth (note, round the number of pulses to an integer number) and the precipitation's spectrum width is $\sigma_v = 2 \, \text{m s}^{-1}$ with $\rho_{hv} = 0.99$:

a) Compute the standard deviations of: reflectivity, differential reflectivity, velocity, spectrum width, differential phase and copolar correlation assuming high-SNR observations.

b) Recalculate the standard deviations of the six standard dual-polarization radar variables assuming an SNR of 6 dB at both polarizations.

4. NCP is commonly used as a simple observation filter for assessing signal quality but NCP is dependent on the observations spectrum width.

a) For observations with SNR > 10 dB and $\sigma_v > 2\,\mathrm{m\,s}^{-1}$, what is the lower-limit for NCP assuming a large number of samples at S-band (2.7 GHz) and X-band (9.4 GHz)?

b) With an NCP threshold of 0.25 and $N = 64$ samples, what is the approximate confidence level that the filtered data only contains noise?

c) How would you adjust the NCP threshold, and what are the impacts on the data quality, if the radar frequency changes from S-band to C-band or X-band?

5. A C-band weather radar (5.6 GHz) is operating at a PRF of 1 kHz with an unambiguous range of 150 km. The radar is observing clear skies within its unambiguous range, but a 70 dBZ hailstorm is at a range of 170 km.

a) What reflectivity levels will be observed on the radar display and at what ranges?

b) With a minimum detectable reflectivity of 10 dBZ at 10 km, what path integrated attenuation would be required so the hailstorm is not detected?

c) The radar can operated in either STSR or ATSR mode. Is there any advantage of one mode compared to the other when observing second-trip echoes? State all the assumptions and explain the reasoning.

6. Noise estimating and subtraction improve the estimation of differential reflectivity if done correctly. This can be a non-trivial process, and errors in the noise estimation can introduce additional biases in the Z_{dr} estimates at low SNR. Assume the mean noise power levels are the same for both polarizations, therefore the polarizations' SNRs are dependent on Z_{dr}. For the follow cases, what is the maximum error in the noise power estimates (of both polarization) that can be tolerated to ensure the Z_{dr} estimator's noise power bias is within ± 0.1 dB of the intrinsic differential reflectivity, and how many noise samples must be averaged so its standard deviation meets the Z_{dr} bias's requirement:

a) For observations at 0 dB SNR with a intrinsic Z_{dr} of -1 dB.

b) For observations at -6 dB SNR with a intrinsic Z_{dr} of -1 dB.

c) For observations at 0 dB SNR with a intrinsic Z_{dr} of 4 dB.

d) For observations at -6 dB SNR with a intrinsic Z_{dr} of 4 dB.

8   Radar Observations and Classification

Dual polarization technology has greatly expanded the range of meteorological applications relying on weather radar observations. Dual polarization adds the capability to retrieve information about particles, specifically their dimension, orientation, concentration and the degree of mixing between different hydrometeor types. This additional information can help determine the characteristics of precipitation systems, evaluate hazardous conditions, and anticipate the evolution of the storm.

This chapter discusses some of the most relevant and common patterns in polarimetric radar observations of weather systems and relates them to the underlying precipitation processes. The microphysical interpretation is not only fundamental to the identification of the precipitation type but also reveals important processes in the cloud. A simple example is the vertical columns of Z_{dr}, whose detection can provide information about the location and intensity of the updraft within the storm.

Attempts to synthesize the large amount of available radar information have led to the development of automatic procedures for the identification of the dominant hydrometeor type within the radar resolution volume. Each type of hydrometeor has a specific signature with a range of expected values for the radar variables. Early attempts to exploit the additional information provided by the measurement of differential reflectivity date back to the 1980s, when Aydin et al. [113] proposed a technique for the identification of hail using a linear dual polarization S-band radar. Since then, great progress has been made in the development of techniques exploiting the full available set of polarimetric radar variables, which typically consist of reflectivity (Z_h), differential reflectivity (Z_{dr}), correlation coefficient (ρ_{hv}), and specific differential phase shift (K_{dp}). Hydrometeor classification techniques have been successfully applied to polarimetric weather radars from S band to X band for a wide range of hydrometeor types. This chapter illustrates the most widely used supervised methods for particle classification, which are based on fuzzy logic. Fuzzy logic provides a simple and robust method for coping with the overlap in possible polarimetric variable values. New and emerging techniques are also discussed, which adapt and learn from data, relying on either unsupervised or semisupervised approaches. Hydrometeor classifications are also important for improving the accuracy of rain-rate estimates and for the identification of meteorological phenomena (such as in fast-developing extreme weather or aircraft icing conditions).

Finally, this chapter also focuses on specific polarimetric signatures in warm rain and discuss recent advancements in the observation of ice-phase processes.

8.1   Stratiform Precipitation

Stratiform precipitation is generally characterized by extensive horizontal development and weak vertical motions, light to moderate rainfall/snowfall intensity, and long duration. Dual polarization observations of the three-dimensional structure of stratiform precipitation add a new level of information for cloud physicists, including more accurate representations of microphysical processes in numerical models, and for operational applications. One notable example is the implementation of real-time tools for monitoring the melting-layer evolution. The identification and short-term forecasting of surface precipitation types (rain/snow) also benefit from proper interpretation of the cloud microphysics.

8.1.1   The Melting Layer – the "Bright Band" (Observations and Detection)

One of the most notable dual polarization radar signatures is the one associated with the melting layer. The main physical mechanisms involved in the transition from solid to liquid precipitation across the melting layer were illustrated in Chapter 3, together with the corresponding behavior of the radar observations. The identification of the melting layer has important applications, in particular for quantitative precipitation estimation, attenuation correction, and hydrometeor classification. Radar observations have been used for a long time to infer the main characteristics of the melting layer. Before the advent of dual polarization, the melting layer could be identified based on the "bright-band" signature, a layer of enhanced reflectivity that appeared as a bright ring on the radar display for a plan position indicator (PPI) conical scan. This is illustrated in Figure 8.1. The radar's beam at a given elevation angle goes through three different stratiform precipitation regions: the rain region, the melting layer, and the ice region. Because the reflectivity values within the melting layer are higher than the corresponding values below (in rain) and above (in ice), the projected observations on the surface will show a characteristic ring with stronger reflectivity. When multiple elevations are combined in a single product using the maximum value, a series of concentric rings will show up as illustrated in the figure, with the rings closer to the radar corresponding to the higher elevations.

A well-defined bright band in the reflectivity field may not always be seen in connection with the melting layer. The appearance and intensity of the bright band may in fact depend on a number of factors, including the distribution of the snowflakes aloft and the degree of riming. The combined use of Z_{dr} and ρ_{hv} (and linear depolarization ratio [LDR] if available) in addition to Z_h provides the basis for more robust melting-layer detection algorithms. As discussed in Section 3.6.1, the characteristic signature of the melting layer includes an increase of Z_{dr}, typically right below the reflectivity maximum. However, the most unambiguous and robust indication of the melting layer is normally provided by the drop in the correlation coefficient resulting from the diversity of the hydrometeor shapes within the resolution volume, which may include a varying mixture of ice crystals and aggregates, partially frozen particles, and raindrops.

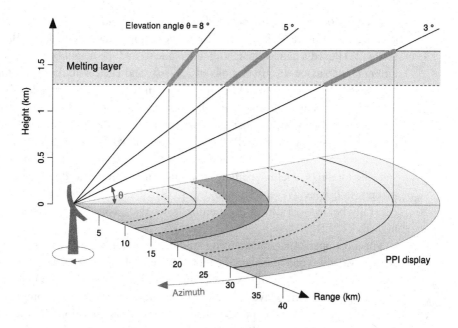

Figure 8.1 Conceptual diagram of melting-layer observation with a radar volume scan consisting of three elevations. The radar beams at lower elevations pass through a larger portion of the melting layer, resulting in large and wide rings on the PPI display, as opposed to the narrower and closer rings corresponding to the higher elevations.

Differential phase shift measurements are normally not readily used for the identification of the melting layer because of the complex interactions affecting both forward propagation and backscattering within a population characterized by the presence of large melting particles. On the one hand, small, oriented crystals melt to nearly spherical droplets (K_{dp} decreases), whereas large, nearly spherical aggregates melt and become oblate (K_{dp} increases), although those particles do not contribute as much to Φ_{dp} as they do for Z_h and Z_{dr} (sixth-moment weighting). On the other hand, large, partially melted particles may cause a relevant phase shift upon backscattering. In fact, recalling that the observed differential phase shift $\Psi_{dp} = \Phi_{dp} + \delta_{co}$, it has been shown that the magnitude of δ_{co} due to wet snowflakes can dominate the total phase shift across the melting layer in some instances. This behavior may be particularly enhanced at the C band and S band, where δ_{co} may reach several tens of degrees [97], whereas X-band observations are generally less affected by the backscatter differential phase in the melting layer. Despite these difficulties related to the use of differential phase-shift observations for the identification of the melting layer, the retrieval of δ_{co} is a current research topic that may provide useful complementary information for the characterization of microphysical processes producing large, wet snowflakes, such as accretion and aggregation.

Focusing on the more obvious dual polarization signatures of Z_h, Z_{dr}, and ρ_{hv}, multiple algorithms have been proposed for the detection of the melting layer. Brandes and Ikeda [221] used a pattern-matching technique to find the height of the freezing level through comparison of observations of Z_h, ρ_{hv}, and LDR with model profiles based on statistics obtained from field campaigns in different climatic regions. Their study used measurements collected with the National Center for Atmospheric Research (NCAR) S-pol radar using alternate transmission and simultaneous reception, which allowed the estimation of LDR. The signal depolarization arises from nonspherical particles whose main axis is not aligned with the transmitted polarization. In comparison with raindrops, melting particles manifest a more pronounced oscillatory behavior as they fall, which results in a higher LDR. In fact, whereas the LDR in rain is generally below -25 dB, in the melting layer, the S-band LDR can reach values as high as -15 dB.

For most operational radars using simultaneous transmission and reception, LDR is not available, and the identification of the melting layer greatly relies on the measurements of ρ_{hv}, with possible contributions from Z_h and Z_{dr}. Figure 8.2 shows an example of melting-layer detection for operational volume scanning radars. The scatterplot in panel (a) contains all measurements at elevations between 4.4° and 10° matching specific criteria for their ρ_{hv}, Z_h, and Z_{dr}. First, all values of ρ_{hv} falling within a given interval (0.90–0.97 for S band) are identified. Next, for these selected points, the values of Z_h and Z_{dr} are checked within a given height interval to account for the possible altitude mismatch between the signatures in the radar variables. For example, the observations in Figure 8.2 show the peak Z_h values at a higher altitude (farther range) with respect to both the lowest ρ_{hv} values and the peak Z_{dr}, which are typically found closer to the middle and bottom part of the melting layer. Because the algorithm is applied to PPI scans, the values of Z_h and Z_{dr} are considered along segments with varying extension, depending on the elevation. If the maximum values of both Z_h and Z_{dr} fall within a given predefined interval (30–47 dBZ for Z_h, 0.8–2.5 dB for Z_{dr} at S band [222]), the point is marked as the melting layer. The intervals for ρ_{hv}, Z_h, and Z_{dr} may be adjusted for the operating frequency band and climatic region [223].

Application of simple frequentist statistics to the points identified as part the melting layer allows the estimation of its lower and upper boundaries. The melting layer's thickness is related to the microphysical composition of precipitation aloft (large snowflake aggregates take longer to melt completely, leading to a thicker melting layer). Considering the points in the height-azimuth domain as in panel (a) of Figure 8.2, the quantiles of the distribution are estimated using a running window in the azimuth with a given sector width (e.g., $\pm10°$). The 20th and 80th quantiles can then be taken to represent the bottom (dashed thick line) and top (solid thick line) of the melting layer, respectively [222]. The analysis of the distribution for a given azimuth sector can also allow for filtering of possible misclassifications. Although the distribution is expected to be unimodal, in practice, multimodal distributions may arise. For example, the presence of pristine crystals aloft in the ice region can show a polarimetric signature similar to the melting layer, or in other instances,

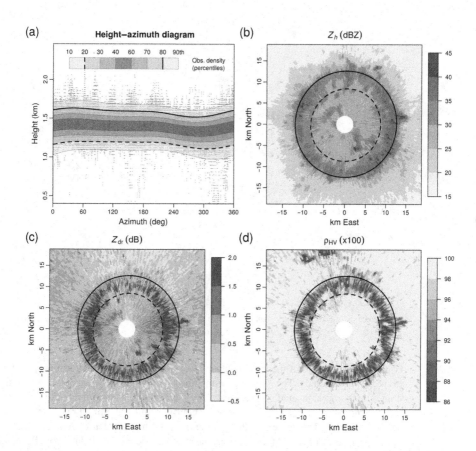

Figure 8.2 Example of melting-layer identification for a C-band radar in Italy.
(a) Height-versus-azimuth plot of all observations at elevations between 4.4° and 10° matching
the criteria for melting-layer detection (see text for details). The dashed and solid lines
correspond to the 20th and 80th percentiles of the observations, representing, respectively, the
bottom and top of the melting layer. Reflectivity (b), differential reflectivity (c), and correlation
coefficient (d) at 7.4° elevation are shown. A median filter in range was applied to the radar
variables to reduce the measurement noise.

residual ground-clutter can contaminate estimates at the lower elevations. In these
cases, a selective filtering of the secondary modes of the distribution, before the
estimation of the quantiles, may help to avoid a biased estimation of the melting-layer
boundaries.

Similar approaches can also be applied to individual PPIs in order to provide a
more spatially detailed estimation when the freezing level is not uniform over the radar
domain. However, the use of the correlation coefficient at very low-elevation angles
for melting-layer estimation at far distances may be difficult because of possible side-
lobe contamination from ground clutter and nonuniform beam filling (Section 7.3.3).
A combined approach that also includes surface observations or thermodynamic output from sounding observations or numerical models can help to improve the detection

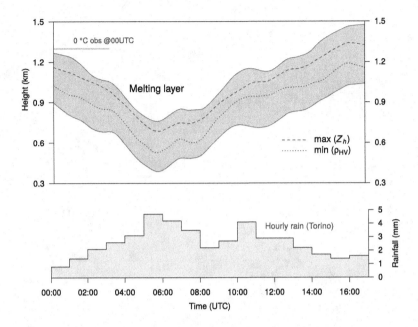

Figure 8.3 Time series of the melting layer's vertical extent during a stratiform precipitation event in northern Italy. The depth of the melting layer ranges between 300 and 500 m. The minimum of ρ_{hv} is located on average ≈ 150 m below the maximum reflectivity of the bright-band peak. The lower plot shows the hourly cumulated rainfall measured by a rain gauges in Torino. The descent of the melting layer during the intense rainfall around 06 UTC is indicative of the "melting effect," that is, the atmospheric cooling induced by the melting process (see text for details).

of the melting layer, especially in cases with a low-altitude freezing level or strong advection with significant horizontal gradients of temperature.

Dual polarization radar observations are fundamental in the detection of the fine-scale evolution of the melting layer. Figure 8.3 shows a time series of the melting-layer altitude and its depth using C-band radar observations in the Alpine region. The radar detection of the melting layer (gray band) is in agreement with the sounding observations at 00 UTC, but during the following hours, a consistent downward propagation is observed. The descent of the melting layer between 02 UTC and 12 UTC is associated with the higher intensity of the precipitation at the surface, as illustrated by the rain-gauge hourly measurements. This is a manifestation of the "melting effect," that is, the cooling of the ambient air as a result of the latent heat of fusion absorbed during the melting process. This effect is often difficult to forecast, given its relatively small magnitude compared to other processes, such as advection and evaporation [224]. The example in Figure 8.3 shows the potential for accurate monitoring of the rain/snow transition altitude with dual polarization radar. This figure also highlights the heights of the maximum reflectivity (the bright-band peak; dashed line) and the minimum of the correlation coefficient (dotted line). As discussed in Section 3.6.1, the lowest ρ_{hv} is generally observed below the reflectivity peak. In this example, the altitude of the minimum ρ_{hv} is approximately 150 m below the maximum reflectivity, in agreement with other studies. In particular, Baldini and Gorgucci [225] studied the dual polarization characteristics in the melting layer from vertical-looking observations, allowing them to establish that the peak in the second-order moments of Z_{dr} or Φ_{dp} (which are essentially controlled by the correlation coefficient ρ_{hv}) are found on average 150 m below the peak of reflectivity.

8.1.2   Orographic Precipitation

When moist air is forced upward over a mountain slope, precipitation may be enhanced as a result of both warm-rain and cold-rain processes. Figure 8.4 shows a conceptual diagram of the main microphysical processes involved in orographic precipitation. The diagram also highlights some issues specifically related to radar observations in complex orography. Ground-clutter contamination resulting from the sloped terrain intersecting the main lobe or side lobes of the radar beam may limit the possibility of observing precipitation processes close to the surface. The backscattered power from ground clutter has a small dependence on the radar wavelength in the microwave region [226], whereas the weather echoes power is proportional to λ^2 (eq. [2.39a]). It follows that the relative contamination due to ground clutter is smaller for radars operating at higher frequencies, such as the X band and Ku band.

Another important limitation for radar observations in complex orography is posed by the blockage of the beam by the mountain reliefs. Dual polarization radar, through the measurement of the differential phase shift, provides a unique opportunity to mitigate biases from partial beam blocking. In fact, phase measurements are insensitive to the power loss affecting reflectivity measurements. Considering the inverse

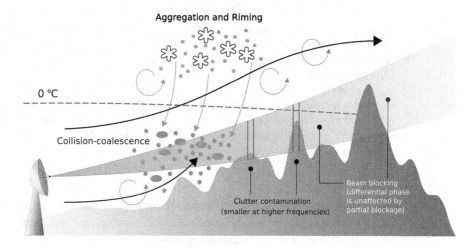

Figure 8.4 Schematic illustration of radar observation challenges and main microphysical processes in the rain and ice regions of orographic precipitation.

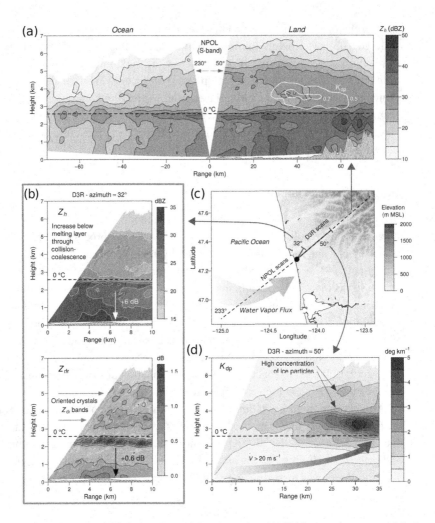

Figure 8.5 Radar RHI observations of the November 13, 2015 event that occurred during the OLYMPEX campaign [88], showing warm-rain and cold-rain processes in orographic precipitation. (a) S-band reflectivity from the NASA NPOL radar. The scan on the left (over the ocean) is at 02:00 UTC, and the scan on the right is at 02:12 UTC. (b) Z_h and Z_{dr} observation at 02:10 UTC from the NASA Ku-band D3R radar. (c) Map showing the local topography and directions of the vertical radar scans illustrated in panels (a), (b), and (d) K_{dp} observations from the Ku-band radar at 02:13 UTC. Note the larger K_{dp} values (up to $\sim 5\ °\text{km}^{-1}$) in comparison with the NPOL observations in panel (a) ($< 1\ °\text{km}^{-1}$), reflecting the frequency scaling between the Ku and S bands.

wavelength dependence of the differential phase-shift observations, higher-frequency radars provide better sensitivity to estimate Φ_{dp} which allows precipitation detection and estimation even in weak-rainfall regimes.

Figure 8.5 shows range-height indicator (RHI) scans collected in a stratiform precipitation event during the Olympic Mountains Experiment (OLYMPEX) [88]. Panel

(a) shows a southwest-to-northeast combined reflectivity RHI from the National Aeronautic and Space Administration (NASA) mobile S-band radar (NPOL). The combination of two separate RHI scans collected 12 minutes apart over the ocean (left) and over land (right) illustrate the general morphology of a typical landfalling midlatitude storm. As discussed by Zagrodnik et al. [227] and McCurdie et al. [228] in their analysis of terrain-enhanced precipitation processes during the OLYMPEX campaign, the radar observations show a higher frequency of occurrence of large reflectivity values over land compared to ocean, in particular in the ice layer around 4–6 km height. This secondary reflectivity peak above the bright band is an indication of the orographic enhancement over complex terrain. The turbulent overturning associated with the forced upward flow along the mountain slope may favor particle growth through riming and aggregation, as schematically depicted in Figure 8.4. In particular, Houze and Medina [229] highlight how "the turbulence embodied in this orographically induced cellularity allows a quick response of the precipitation fallout to the orography since aggregation and riming of ice particles in the turbulent layer produce heavier, more rapidly falling precipitation particles." This explains the strong correlation between the enhanced precipitation at the surface and the upslope flow over the windward side of the mountain range. In fact, a more steady but weaker updraft would advect the stronger precipitation farther downwind.

It is interesting to note how, besides cold-rain processes favoring particle growth above the melting layer, warm-rain microphysical processes are also taking place in the rain layer below. Panels (b) and (d) of Figure 8.5 show dual polarization observations collected by the NASA D3R dual frequency radar. In particular, panel (b) presents detailed observations of the reflectivity and differential reflectivity at the Ku band over the first inland hills (32° azimuth). These observations, focusing on the closer 10-km range, document the precipitation enhancement near the surface through collision-coalescence (see Sections 3.5.1 and 3.5.4). The overall increase of Z_h and Z_{dr} below the melting layer is, respectively, approximately 6 dB and 0.6 dB around the 5-km range, denoting a very efficient particle-growth mechanism associated with the upslope flow over the hilly terrain. Rainfall enhancement in the proximity of the surface can also be seen in the S-band scan over land, around the 30-km range (panel [a]; reflectivity increases below the melting layer from approximately 35 to 40 dBZ), and in the corresponding K_{dp} observations from the Ku-band radar (panel [d]; note the 35-km range displayed), with values increasing from approximately 0.4 to 1.0 °km^{-1} below the melting layer.

The D3R differential phase observations show an extended region of enhanced K_{dp} aloft, between 2.5 and 4 km height, and a secondary peak at higher altitude (around 5 km height). The stronger K_{dp} signal above the melting layer is also appreciable in the S-band radar observations (white contour in panel [a]), although with a much lower magnitude owing to the characteristic frequency scaling (the peak K_{dp} at the Ku band is over 5 °km^{-1} around 3.5 km height, whereas at the S band, K_{dp} barely reaches 1 °km^{-1}). These enhanced K_{dp} signatures aloft, discussed more thoroughly in Section 8.5, in general indicate a large concentration of ice particles, either oriented crystals or early aggregates with relatively high density and an oblate shape [230].

In this particular event, the secondary K_{dp} peak at the Ku band around 5 km height ($T \approx -14°C$ according to the local sounding launched from the radar site at 03 UTC) appears roughly co-located with patches of positive Z_{dr}, indicating a sustained dendritic growth in the cloud region, where the difference between the saturation vapor pressure for water and ice is the greatest (recalling Fig. 3.18). On the other hand, the stronger K_{dp} signature below (2.5–4 km, corresponding to temperatures between $0°$ and $-8°C$) may be an indication of active riming. When supercooled liquid droplets are carried aloft by the turbulent flow, riming takes place, and the increased particle density enhances the differential phase shift [231]. At the same time, secondary ice production through rime splintering or collisions between faster-falling rimed particles and slower dendritic crystals acts to increase the concentration of the ice particles. The abundance of newly formed, small, oriented ice crystals has little effect on the reflectivity, which is mostly influenced by the largest aggregates, but may determine a significant increase in the specific differential phase [86].

This case illustrates how the dual polarization radar variables can be used to interpret something about the concentration, shape, and density of the hydrometeors, distribution. In order to get a more comprehensive picture of the precipitation processes, it is important to consider multiple dual polarization radar variables at the same time and examine their characteristic spatial patterns. This concept is the basis of the advanced hydrometeor classification techniques discussed in Section 8.3.

8.2   Convective Precipitation

Figure 8.6 shows radar observations collected by the Sioux Falls NEXRAD radar in South Dakota during the tornado outbreak of June 16–18, 2014. The area denoted by the letter A on the western side is a supercell that produced a tornado (indicated by an open white circle) traveling northwestward for approximately 5 km between 03:44 UTC and 04:02 UTC. The time of the 0.8° elevation PPI in panels (a) and (b) is 03:39 UTC. The vertical profiles in panels (c) and (d) are synthesized range-height profiles along the 270° azimuth which are generated by interpolating the PPI scans within the volume coverage pattern (03:39–03:43 UTC). The letter B denotes another major convective structure, whereas two minor cells are visible in the southeastern sector. One thing that stands out in the low-level PPIs is the spatial decoupling of the Z_h and Z_{dr} peak locations, especially for the major storm structures A and B. This particular behavior is often observed in convective precipitation, for ordinary cells as well as multi-cell and supercell storms, and is related to the microphysical processes induced by the dynamic circulation in the storm's environment. In Figure 8.6, only the southeastern cell at the 38-km range shows roughly co-located maxima of reflectivity and differential reflectivity, whereas a clear spatial lag is observed for the other cells. The relative displacement of the Z_h and Z_{dr} peaks tends to occur near the inflow region of the storm and is indicative of the particular circulation at the storm's boundary.

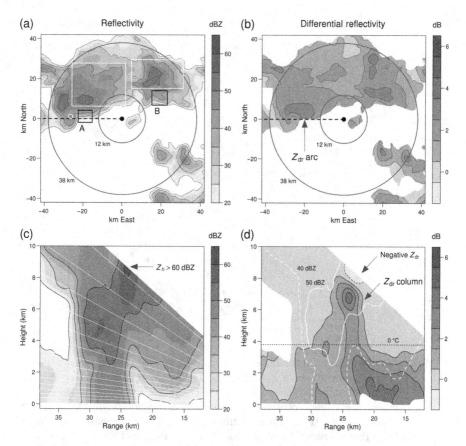

Figure 8.6 A supercell storm in southeastern South Dakota observed by the Sioux Falls NEXRAD radar on June 18, 2014. PPI at 03:39 UTC of reflectivity (a) and differential reflectivity (b) at 0.8° elevation. The white open circle marks the initial location of the tornado at 03:44 UTC. Corresponding vertical cuts obtained through interpolation of the volume scan observations along the westward direction (270° azimuth) are shown in panels [c] and [d] for the 12- to 38-km range segment. The white lines in panel (c) denote the actual elevations of the volume scan, ranging between 0.5° and 19.5°.

8.2.1 Size Sorting Induced by Wind Shear at the Storm's Boundary

One obvious distinction between widespread stratiform precipitation and convective rain is the fact that convective storms generally have a cellular shape and are surrounded by drier air. The discontinuity between precipitation and dry air at the storm's boundary may produce border effects, which show up in radar observations as regions of enhanced differential reflectivity. These regions are often associated with the growth of new convective cells, as schematically represented in the $Z_h - Z_{dr}$ observation space in Figure 3.21.

The distribution in the $Z_h - Z_{dr}$ plane of observations from two distinct regions of the storm, defined by the white and black rectangles in panel (a) of Figure 8.6,

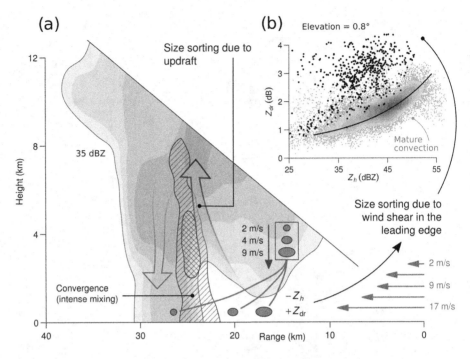

Figure 8.7 Simplified two-dimensional representation of size sorting in a convective storm (gray-shaded contours of Z_h), using real dual polarization and Doppler observations from the case illustrated in panels (c) and (d) of Figure 8.6. The vertical storm-relative wind profile on the right is derived by combining the storm motion with Doppler velocity observations on the range-height vertical plane up to a 20-km range. The enhanced Z_{dr} in the lower layers between the 10- and 20-km range (panel [d] of Fig. 8.6) results from size sorting due to wind shear, whereas the columnar positive Z_{dr} between the 20- and 25-km range is related with size sorting inside the updraft, represented here by the upward-pointing arrow. The hatched area shows a deep convergence zone, indicating that intense mixing is taking place from the surface to middle levels. This region encompasses values of radial convergence higher than 0.005 and 0.01 s^{-1}, respectively, for the outer and inner regions (with peak divergence up to 0.03 s^{-1}, corresponding to a variation of 30 m s^{-1} over a length of 1 km). The scatterplot in panel (b) displays Z_h, Z_{dr} observations on the low-level PPI corresponding to the small (black) and large (white) rectangles in panel (a) of Figure 8.6. The black dots in the scatterplot correspond to the smaller black rectangle in the leading edge of the convective storm in panel (a) of Figure 8.6, whereas the gray dots with density shading and the regression line correspond to the larger white rectangle representative of mature convection in panel (a) of Figure 8.6.

is represented in the scatterplot in panel (b) of Figure 8.7. The larger areas (white rectangles) comprise the "older" portions of the storm (mature convection), which show the characteristic monotonically increasing average relation between Z_{dr} and Z_h expected in rain (see, e.g., Fig. 9.7 in Chapter 9). On the other hand, the smaller areas on the southern boundary of cell A and B (black rectangles) are representative of an anomalous drop size distribution (DSD) characterized by relatively few large drops and a consistent lack of small drops. In the case of supercell storms, as a consequence of the unique internal circulation, the Z_{dr} structure typically takes the shape of an arc, hence the notation "Z_{dr} arc." A similar behavior, but without the peculiar arc shape, can be observed in non-supercell storms like cell B.

As discussed in Chapter 3, the particular modification of the DSD with a decrease of the small-droplet concentration may result from either evaporation in the lower layers or from size sorting. When these particular observations (high Z_{dr} and low to moderate Z_h) are located near the storm's boundary, size sorting is the most likely mechanism governing the modification of the DSD. Although size sorting is generally a transient effect (Section 3.5.4), updraft and wind shear, like that typically present in the convective storm's inflow region, may sustain it for a longer time [232]. Wind shear affects the circulation in the three-dimensional storm's environment, but for the sake of illustration, Figure 8.7 shows a simplified two-dimensional conceptual range-height diagram of size sorting, superimposed on the observations of cell A in Figure 8.6. The increasing wind speed in the lower atmospheric layer near the surface, associated with the sustained inflow, produces a varying response on drops of different sizes. Larger drops or partially melted graupel/hail particles fall faster and are only weakly affected by the horizontal flow, whereas smaller drops remain longer in the sheared flow and are advected farther downstream toward the storm's core. The resulting deprivation of smaller drops below the overhanging echo structure explains the decreasing reflectivity toward the surface (lowering number concentration of raindrops) and the concurrent large Z_{dr} values observed between the 10- and 20-km range in panel (d) of Figure 8.6. Some of these drops are further fed into the updraft, where they can still experience size sorting (the smallest particles are carried aloft into the upper region of the storm) in addition to growth by collision-coalescence, as illustrated in more detail in the diagram in Figure 3.25.

8.2.2 Strong Updraft Detection: Z_{dr} columns

Figure 8.7 also shows the interpolated radial convergence from the observed Doppler velocity. A deep convergence zone is clearly visible, extending from the surface, near the storm's right boundary, to the middle levels. This region of strong convergence is also characterized by a very high Doppler spectrum width, in excess of 10 m s^{-1}, indicating intense mixing at the updraft–downdraft interface [233] (refer to Fig. 3.27 for the flow inside a supercell storm).

Doppler observations are routinely used to detect low-level convergence, which may indicate the presence of an updraft, and can be used as an indication of an intensifying storm. However, the convergence signature may not show up in Doppler velocity data when converging winds are aligned perpendicular to the radial beam. Multiple-Doppler analysis is needed to reliably retrieve the three-dimensional wind field, as illustrated in the example in Chapter 3 (Fig. 3.26). But even in dense radar networks, multiple-Doppler coverage may be limited, suggesting that alternative radar signatures of intense updrafts may be valuable for both research and operational applications.

Size sorting may often be maintained by wind shear in the inflow region of the storm, identifying a potential region of cell growth. Other combinations of dynamics and microphysical processes may lead to similar dual polarization observations, implying that these two-dimensional signatures in low-level PPIs cannot be taken as an unambiguous indication of the storm's intensification. A clearer marker of strong updraft and cell development is available from the three-dimensional analysis of the differential reflectivity field, as illustrated for the supercell in Figures 8.6 and 8.7. Besides supercell storms, where Z_{dr} columns may extend to extremely high altitudes (approximately 4 km above the environmental freezing level in panel [d] of Fig. 8.6), in ordinary convection, Z_{dr} values in excess of 2 dB up to approximately 2 km above the freezing are commonly observed, as in the Florida cell example in Figure 3.26, with cloud top around 8 km. An extended analysis conducted over England with an X-band radar during the Convective Precipitation Experiment (COPE) has shown a frequent occurrence of Z_{dr} columns also in shallow convection with moderate peak updraft intensity ($7-9$ m s^{-1}) and 4- to 8-km cloud top heights [102]. The study focused in particular on the relation between the presence of a Z_{dr} column in the convective cells and the intensity of precipitation near the surface. It was found that the cells exhibiting a Z_{dr} column tended to produce higher rainfall intensities. In particular, the median rainfall rate was on average >20 mm h^{-1} for cells with Z_{dr} columns and ~ 10 mm h^{-1} for general convection.

Again considering the event in Figure 8.6, Z_{dr} columns were already present in cell A about 1 hour before it reached supercell status. The $10°$ elevation PPI represented in panel (a) of Figure 8.8 shows the location of the positive Z_{dr} regions above the environmental freezing level (dotted line) at 02:50 UTC. Panels (b) and (c) show the vertical sections corresponding to azimuth cuts (S1 and S2) through two distinct Z_{dr} columns. The Z_{dr} core along section S2 (panel [c]) is roughly co-located with the peak reflectivity and corresponds to the more mature region of the storm. The Z_{dr} peak along S1 (panel [b]) is instead marking a new intensifying updraft on the southwestern flank of the main storm. In this case, the Z_{dr} region with values greater than 2 dB extends from the weak-reflectivity low-level layers up to approximately 6 km in height, where reflectivity reaches 50–55 dBZ.

The use of Z_{dr} columns as a proxy for updraft detection in deep convection is gaining increasing relevance for both diagnostic and prognostic applications. Snyder et al. [234] proposed an automated algorithm for the detection of Z_{dr} columns for application to the WSR-88D radar network in the United States. The identification is based on the estimation of the column depth, defined by the contiguous region where $Z_{dr} > 1$ dB above the environmental 0°C level. Owing to the deep significance for the storm dynamics, the information provided by the location and morphology of the Z_{dr} columns is also prone to be exploited in numerical models through data assimilation.

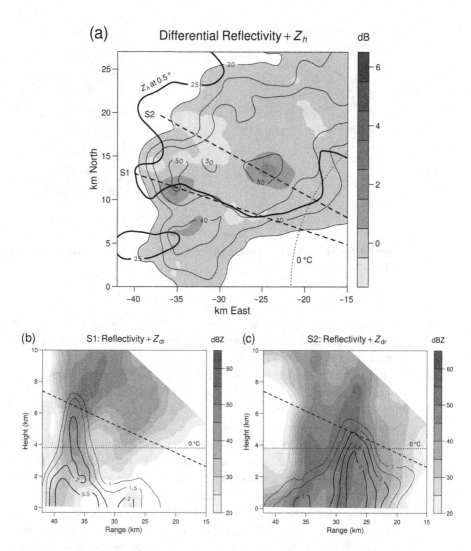

Figure 8.8 The convective cell A in Figure 8.6, shown at an earlier time (02:50 UTC). (a) High-elevation PPI (10°) of Z_{dr} with contours of reflectivity between 20 and 50 dBZ. The thick black line shows the 25-dBZ reflectivity contour at the lowest elevation (0.5°). The range-height plots in panels (b) and (c) show, respectively, the vertical cuts through azimuth 288° (S1) and 298° (S2). The differential reflectivity contours for $Z_{dr} \geq 1$ dB are overplotted on the reflectivity (gray shading). The slanted dashed line indicates the elevation of 10° matching panel (a).

8.2.3   Strong Updraft Detection: K_{dp} Columns

In addition to Z_{dr} columnar enhancements, regions of high K_{dp} extending well above the 0°C level are also often observed in deep convective storms. Figure 8.9 shows the same vertical sections as in Figure 8.8, but in this case with K_{dp} contours overplotted on the reflectivity field. Similar to Z_{dr}, the information provided by the identification

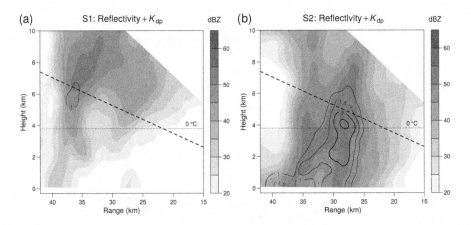

Figure 8.9 Similar to panels (b) and (c) from Figure 8.8, but with contours of K_{dp} overplotted on the reflectivity field.

of K_{dp} columns can be useful to identify intense updrafts in convective cells, relying on the interpretation that these contiguous positive signatures underlie the presence of supercooled liquid drops carried aloft in rapidly ascending air.

Recalling the definition of K_{dp} in terms of the DSD (eq. [4.68]), it is clear that for a well-defined K_{dp} signature to appear (in particular at low frequencies such as the S band), a large concentration of oblate drops is needed. On the other hand, because Z_{dr} is independent of the number concentration, a small number of oblate particles is sufficient to produce a well-marked signal. This is illustrated in Figure 8.10 with a realistic simulation of two different DSDs at the S band. The DSD in the big box consists of very sparse but relatively large drops (median volume diameter $D_0 = 3.4$ mm), producing a rainfall rate of just 0.2 mm h^{-1}, an accordingly negligible K_{dp}, and a very large Z_{dr}. The second DSD depicted in the small 1-m^3 box is representative of heavy rain, with a large relative concentration of medium-size raindrops ($D_0 = 2.2$ mm), which results in a high rainfall rate in excess of 50 mm h^{-1}, reflectivity of approximately 50 dBZ, and a markedly positive Z_{dr} and K_{dp}.

It is clear that Z_{dr} columns have a higher chance of being observed during the early stage of developing updrafts in both shallow and deep convection, hence the interest for use in prognostic applications. In general, with further development of the convective cell, a K_{dp} column may become apparent, reflecting the increase in water content above the environmental freezing level as a consequence of a prolonged updraft. In the mature stage of deep convection, K_{dp} columns may be especially useful, in comparison with Z_{dr} columns. In fact, the formation of hail aloft can mask the continued presence of supercooled drops, suppressing the positive Z_{dr} signal, while K_{dp} would still be sensitive to the presence of liquid particles. Because differential phase observations are not affected by attenuation, K_{dp} can be used to provide a robust indication of the updraft location in the mature stage. Van Lier-Walqui et al. [235] analyzed four deep convective events during the Mid-Latitude Continental

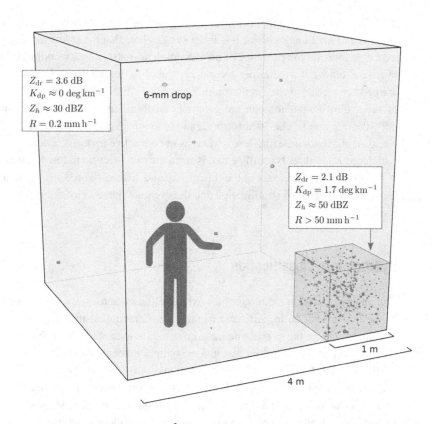

Figure 8.10 The larger cube (64 m³) illustrates a possible DSD realization in the leading edge of convection. The marked scarcity of drops results in a negligible K_{dp} and rainfall rate, weak or moderate reflectivity, and a very large Z_{dr} at the S band. The human silhouette helps to portray, with realistic proportions, the typical situation experienced when a storm is approaching and large but very sparse drops start falling. This may occur in the inflow region as a result of the wind shear effect illustrated in Figure 8.7, but it also may occur as a result of the transient effect of sedimentation (refer to Section 3.5.4) during the onset of rainfall below the cloud base. In the smaller cube (1 m³), instead, a DSD typical of heavy precipitation in the storm's core is represented, with all radar variables attaining large values. All drops are scaled 20:1 with respect to the cubic boxes and silhouette.

Convective Clouds Experiment (MC3E) field campaign, finding that a consistent volume of Z_{dr} above the melting level may appear 30–60 minutes before a corresponding increase of the K_{dp} volume. This seems to confirm the significance of the Z_{dr} columnar enhancement as an early observational indication of updraft intensification. The later appearance of the enhanced K_{dp} aloft may also be attributed to droplet shedding from wet hailstones [103], as illustrated in Figure 3.31, in addition to lofting of supercooled drops.

As discussed in the orographic precipitation example (Fig. 8.5), K_{dp} may also attain relatively large values above the 0°C level in stratiform precipitation, but in that case, the differential phase signal is primarily originating from ice particles such as dendrite crystals and rimed aggregates. It is interesting to note that the stratiform K_{dp} signature may also indicate the presence of an ascending air mass, a weak updraft on the order of several tenths of a meter per second, typical of mesoscale forcing associated with a warm-front passage (see discussion in Section 8.5). From a strict morphological point of view, the two phenomena are generally easily distinguishable when considering only the K_{dp} field. The stratiform signature is on average weaker and completely confined above the freezing level, whereas in convective updrafts, a contiguous region with large K_{dp} values typically extends from the rain layer up to few kilometers above $0°C$. A contextual analysis of the reflectivity and differential reflectivity fields, with a focus on the degree of stratification and the eventual presence of the bright band, may provide clarity.

8.3   Hydrometeor Classification

The dual polarization radar variables exhibit different sensitivity to the characteristics of the hydrometeors, in particular to their size, shape, density, orientation, and concentration. Given the complementary information available about the microphysical composition of the sampling resolution volume, it seems natural to devise a way to synthesize the dual polarization observations to assess the dominant microphysical species observed by the radar. For real-time operations in particular, even a well-trained meteorologist may not be in a position to carefully scrutinize the three-dimensional dual polarization radar observations collected with a typical temporal sampling of ~5 minutes or less. For this reason, an objective, automatic procedure is needed. The search for objective methods to partition the parameter space for the purpose of microphysical classification kicked off with the development of dual polarization weather radar in the 1980s. Hall et al. [236] demonstrated how the availability of differential reflectivity observations could lead to the identification of hydrometeor types and other nonmeteorological targets. With the specific purpose of hail detection, Aydin et al. [113] proposed a method to discriminate between rain and hail using reflectivity and differential reflectivity measurements. This method was based on a hard partition of the Z_h, Z_{dr} parameter space. The same concept was later extended to other hydrometeor types and leveraged additional dual polarization variables [237]. The main drawback of the early Boolean classifiers was the inability to deal with the overlapping decision boundaries in the parameter space. In addition, these methods could not account for uncertainty in the radar measurements. The most significant evolution of these early approaches came from the fuzzy-logic classifiers [238, 239]. The fuzzy-logic techniques allowed users to tackle the hard-boundary limitations, in favor of soft boundaries. These techniques have become widely used, and hydrometeor classification is currently available with most weather radars as part of the standard meteorological products for systems operating at the S, C, and X bands. The application of hydrometeor classification at higher frequencies is a current topic of research and beyond the scope of this book.

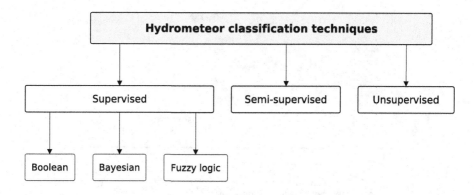

Figure 8.11 Representation of different hydrometeor classification techniques.

Fuzzy-logic methods, together with Boolean and Bayesian methods, belong to a wider class of supervised classification techniques. Other, more recent classification approaches use either semi-supervised or unsupervised techniques. A general representation of the hydrometeor classification techniques is presented in Figure 8.11. The next section focuses on the main architecture of fuzzy-logic-based algorithms, which are currently the most popular supervised methods. Section 8.3.2 discusses more recent approaches, including both semi-supervised and unsupervised methods.

8.3.1   Fuzzy-Logic Classifiers

The dual polarization radar variable signatures obtained from different hydrometeor species are generally not mutually exclusive. For example, the radar observations of reflectivity and differential reflectivity typically show a large overlap for drizzle and small, randomly oriented ice aggregates in the portion of the parameter space delimited by $Z_h \leq 25$ dBZ and $-0.5 < Z_{dr} < 0.5$ dB. Fuzzy logic has proven to be a suitable tool to deal with this classification problem, being specifically devised to address the concept of transferring from hydrometeor space to observation space, which is not a one-to-one mapping.

The fuzzy-logic hydrometeor classification algorithm follows the classical flow from fuzzification to inference, aggregation, and defuzzification, as depicted in Figure 8.12. The fuzzy sets are the hydrometeor classes, whereas the fuzzy rule base relies on the definition of proper membership functions to relate the input radar variables x to the hydrometeor class k. One of the important elements of the development of a hydrometeor classification algorithm, including those based on fuzzy logic, is the choice of the number (N_c) and type of classes. The number of classes can either be defined a priori, based on the expected discernible capacity of the dual polarization information, or determined by means of some objective processing of real observations. In practice, it is common to deal with a number of classes ranging between 5 and 10. For example, the NEXRAD hydrometeor classification product

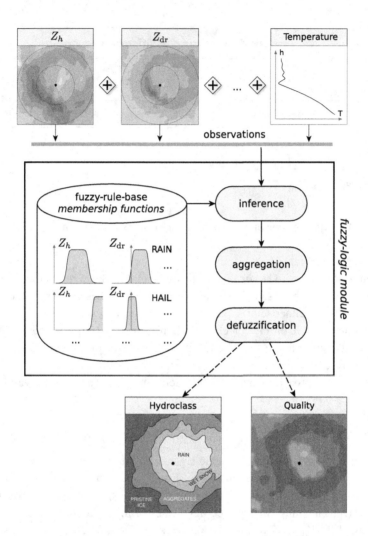

Figure 8.12 Diagram of a fuzzy-logic algorithm for radar hydrometeor classification. The observations typically include measurements of Z_h, Z_{dr}, ρ_{hv}, K_{dp}, and LDR (when available), besides additional parameters such as the signal-to-noise ratio (SNR) and spatial textures of dual polarization moments. To improve the classification, one or more temperature vertical profiles (either from sounding or numerical model forecasts) are often considered. The output of the fuzzy-logic system is the hydrometeor classification map. An associated quality measure can also be retrieved to aid in the interpretation of the classification results.

aims at discerning among 10 possible echoes: biological, clutter, ice crystals, dry snow, wet snow, rain, heavy rain, big drops, graupel, and hail.

In fuzzy-logic systems, the proper choice of the fuzzy rule base and the corresponding inference mechanism is critical for the success of its classification. In particular, the set of membership functions is the main component of the fuzzification phase, representing the propagation and scattering response of the different hydrometeor types. The value of the membership functions ranges between 0 and 1 and represents the degree of truth that an element belongs to a set (hydrometeor class). The membership functions can be represented by arbitrary functions, such as triangular, trapezoidal, and sigmoid functions. In weather radar applications, the beta functions [238, 240] and the trapezoidal functions [239] have been among the most used. The beta function for a given class k is defined as

$$\text{MBF}_k(x) = \frac{1}{1 + \left(\frac{x - m_k}{a_k}\right)^{2b_k}}, \tag{8.1}$$

where x is the value of the input variable, m_k is the location parameter, a_k is the spread, and b_k is the slope. To better approximate the actual shape of the variables' distribution, asymmetric functions can be used, setting different spread (a_k) and slope (b_k) parameter values for the left ($x \leq m_k$) and right ($x > m_k$) tails of the function. The membership degree $\text{MBF}_k(x)$ quantifies the grade of membership of the element x to the hydrometeor class k. The value of the coefficients that define the membership functions (m_k, a_k, and b_k for the beta function) is usually assessed through specific electromagnetic scattering simulations at the radar frequency of interest. As an example, Table 8.1 provides the MBF parameters that were used in the application illustrated in fig. 8.13.

In the inference phase, the individual membership functions are evaluated for the current observations, giving the strength of the patial truth for the corresponding radar variables. For example, an observation of $Z_h = 30$ dBZ will lead to a confidence of

Table 8.1 Values of the parameters used to define the S-band membership functions (eq. 8.1) for the example in fig. 8.13. Note that in some case an asymmetric MBF is used with different spread parameter a for the left ($x \leq m_k$) and right ($x > m_k$) tails. The functions f_1, g_1, f_2, g_2 are used to account for the expected physical correlation between Z_{dr} and Z_h in rain: $f_1(x) = 0.4 + 0.08\,x$, $g_1(x) = 0.5 + 0.02\,x$, $f_2(x) = -0.4 + 0.055\,x$, $g_2(x) = 0.1 + 0.025\,x$, with $x = \max\{Z_h \text{ (dBZ)}; 7 \text{ dBZ}\}$.

	Z_h			Z_{dr}			ρ_{hv}			K_{dp}		
Class	m	a	b	m	a	b	m	a	b	m	a	b
Large drops	40	15.	8	$f_1(x)$	$g_1(x)$	2	0.99	0.03	2	0.3	0.3/1.0	2
Drizzle	0	25	8	0.5	0.7	2	1.00	0.02	2	0.01	0.2	3
Rain	34	14	8	$f_2(x)$	$g_2(x)$	2	0.99	0.03	2	0.3	0.3/1.0	2
Heavy rain	50	10	8	$f_2(x)$	$g_2(x)$	2	0.99	0.05	2	0.5	0.2/10.0	2
Rain plus hail	65	15	4	0.5	1.5	3	0.95	0.07	2	5.0	5.0	8
Hail	65	13	4	−2.0	3.0	3	1.00	0.06	2	0.0	1.0	2
Graupel	42	14	5	1.2	2.0	2	1.00	0.06	2	0.8	1.5	2
Wet ice	20	25	4	1.5	1.7	3	0.88	0.08	2	1.0	1.2	2
Dry ice	22.5	15	5	0.2	1.0	2	1.00	0.05	2	0.2	0.3	2
Crystals	0	22.5	5	1.5	3.5	2	0.98	0.06	2	0.5	0.7	2
Dendrites	20	15	5	1.0	0.5/1.5	2	1.00	0.07	3	0.8	0.4/2.5	2

~1 for rain and ~0 for hail. Once the membership functions are evaluated for every radar variable in the inference phase, rules are needed to combine these values (the strength of the individual partial truths) for any given hydrometeor class. In fact, the evaluation of an individual membership function is a piece of information that represents a constraint. More constraints mean more information. The combination of all the constraints is the objective of the aggregation phase. The aggregation is often implemented in a fuzzy-logic system by means of a combination of linguistic propositions in the form of concatenated conditional IF-THEN statements (IF A_1 THEN B_1, AND IF A_2 THEN B_2, AND ...). Another approach, an alternative to the conditional linguistic combination, is to model the fuzzy rules as follows:

$$S_k = \prod_{i=1}^{N_v} \mathrm{MBF}_k(x_i). \tag{8.2}$$

In the multiplicative rule (eq. [8.2]) x_i is the value of the input variable, k is the hydrometeor class, N_v is the number of radar variables, and S_k is the resulting confidence associated with the class k. The output class can then be retrieved in the defuzzification phase by selecting the class for which S_k is maximum. It is important to note that in addition to the radar variables, the inputs often include the temperature corresponding to the radar resolution volume. This is typically estimated either from a nearby atmospheric sounding or using the output from a numerical weather-prediction model.

An alternative and widely used formulation of the aggregation phase relies on the weighted sum of the individual membership functions for the input variables:

$$S_k = \frac{\sum_{i=1}^{N_v} [w_{k,i} \cdot \mathrm{MBF}_k(x_i)]}{\sum_{i=1}^{N_v} w_{k,i}}, \tag{8.3}$$

where $w_{k,i}$ represents the weights associated with the input variable i and hydrometeor class k. Again, the retrieved class is the one with the maximum S_k. With respect to the multiplicative approach, the additive operator has the advantage of mitigating the effect of measurement errors. For example, inputs values may be unreliable due to partial beam filling, side-lobe effects, and low SNRs. These inputs may lead to low confidence for a specific radar variable, implying a corresponding low overall confidence S_k as a result of the the product in eq. (8.2). The use of the additive method (eq. [8.3]) instead allows to retain the complementary information provided by the remaining input variables. The weights $w_{k,i}$ can be all set equal to 1 as default, but then adjusted based on the relative confidence in the different radar variables for a specific radar system.

Lim et al. [241] proposed a hybrid scheme where the additive inference rule is used for the radar variables Z_{dr}, K_{dp}, and ρ_{hv}, whereas the product rule is used for Z_h and T. In such a scheme, the reflectivity and temperature are strong constraints, with the advantage of reducing most misclassifications due to overlapping dual polarization radar membership functions. As a consequence, the multiplicative membership function of T introduces very sharp and often unrealistic transitions across precipitation phase changes. Bechini and Chandrasekar [242] adopted a parabolic weighting function for the temperature in the additive inference. This allows the use of the temperature information to minimize misclassifications in regions far from phase transitions and to rely more on the radar variables to discriminate between different particle types near the melting layer.

Besides the output classification map, it is important to devise a metric for the quality of the classification. Considering the aggregation methods in eqs. (8.2) and (8.3), the quality (or strength) of the classification is related to the confidence S_k of the selected particle class (max confidence among all hydrometeor classes). The difference between the confidence of the first- and second-ranked identification may also provide useful information for the interpretation of the classification results [243]. If the difference between the highest (first choice) and the second-highest (second choice) confidence decreases below a given threshold (e.g., 25 percent, as suggested in [243]), a user may be warned about a questionable classification. A close ranking may also indicate that the number and type of classes were not properly devised in the algorithm development, with two or more classes not easily discernible in practice.

8.3.2   Semi-Supervised and Unsupervised Methods

In their original paper, Liu and Chandrasekar [238] devised the development of a neuro-fuzzy hydrometeor classifier as an evolution of the fuzzy-logic system with static membership functions. This method was conceived to provide the ability to learn from data and dynamically adapt the membership functions. The learn-from-data concept has been increasingly explored over the years, together with new theoretical models proposed as an alternative to, or in connection with, fuzzy-logic systems. For example, Roberto et al. [244] presented a classification method using a support vector machine supervised learning model, where the output of a standard fuzzy-logic classifier drives the training phase of the algorithm. All supervised models generally include a training component to fine-tune the system, based on actual observations of different hydrometeor species or electromagnetic scattering simulations under different microphysical scenarios. Further evolution of these models is partially hindered by the limited availability of in situ observations and the complexity of simulating the covariance scattering matrix for many frozen or mixed-phase particles.

Bechini and Chandrasekar [242] proposed a semi-supervised technique to take advantage of the information provided by the spatial coherence and the self-aggregation propensity of atmospheric processes and their signatures in the polarimetric radar observations. The algorithm is designed to operate in the two-dimensional observation space defined by the range and radial components (either a PPI or RHI scan) to exploit the spatial and temporal contiguity of the observations. After a preliminary bin-based fuzzy-logic classification, a K-means clustering technique is modified to incorporate spatial contiguity and weak microphysical constraints. The microphysical constraints basically involve the relative position in the vertical profile of different hydrometeors (e.g., the technique is more likely to detect rain below snow than vice versa) and is implemented through a penalty term in the minimization of a cost function. The K-means clustering analysis, owing to the spatial constraints, results in a low-noise classification, where the cluster prototypes may have drifted significantly with respect to the initial values of the centroids obtained from the bin-based classification. The subsequent application of a connected component-labeling algorithm results in the identification of relatively few connected regions. At this point, the distribution of the radar variables within each connected region is used to perform a more robust classification, exploiting a full statistical sample as opposed to the single observation used in the classical bin-based approach. In practice, a comparison between the probability density function (PDF) of the radar variables within a connected region and the theoretical membership functions is used to assign a final class and obtain a region-based classification. Other semi-supervised methods have been proposed by Wen et al. [245] and Besic et al. [246], who developed a procedure where the microphysical characterization obtained through scattering simulations is combined in an iterative process with the statistics obtained from the observations through clustering analysis.

Figure 8.13 shows an example of radar observations through a convective storm, with the region-based classification resulting from the semi-supervised approach [242]. The clear separation between different hydrometeor types facilitates the interpretation of the classification map, highlighting the main larger-scale features in the storm. Small-scale features may be undetected at far ranges, where the decreased radar resolution hinders the self-aggregation propensity of multiple radar bins with similar characteristics. In Figure 8.13, a stratiform region is identified closer to the radar (0–30-km range), with a well-defined melting layer (wet snow class, in dark gray). In the 30–60-km range, the nature of the precipitation is highly convective, with large regions of graupel and hail aloft, and a rain/hail mixture closer to the surface. In particular, the rain+hail region around the 50–57-km range (hatched dark-gray area with black lines) is the result of a strong reflectivity signal (large particles), low differential reflectivity (mainly affected by large and roughly spherical hail particles), and high K_{dp} values (sensitive to rain and unaffected by hail). The wet snow layer is identified mainly by virtue of the relatively high Z_h and Z_{dr} and low correlation coefficient ρ_{hv}. Also, it is worth noting the small region of large drops in the 36–40-km range (hatched light-gray area with black lines) likely originates from the melting of graupel particles above.

Grazioli et al. [247] proposed an unsupervised approach for hydrometeor classification, using an agglomerative hierarchical clustering algorithm with spatial constraints to aggregate the observations based on both the similarity of the polarimetric radar variables and the spatial smoothness. The learn-from-data approach is extended in this case to establish the number of particle classes, which is a very important step for any classification method. Considering the trade-off between compactness and separability of the clusters, an optimal solution was obtained with seven clusters ($N_c = 7$) for observations collected with an X-band radar during two measurement campaigns in

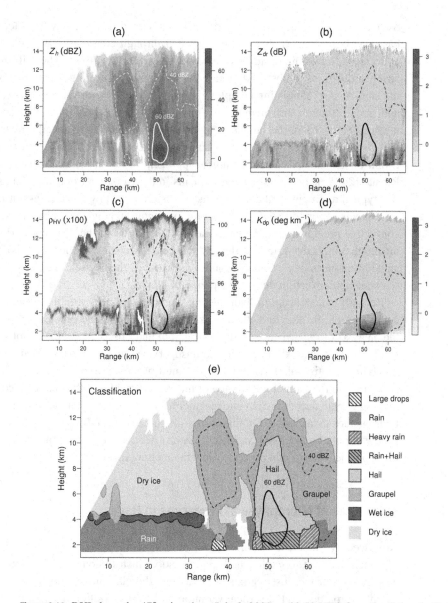

Figure 8.13 RHI along the 47° azimuth on July 2, 2008, at 22:58 UTC from the Colorado State University–Chicago Illinois (CSU-CHILL) radar. Panels (a) through (d) show the radar variables Z_h, Z_{dr}, ρ_{hv}, and K_{dp}. The hydrometeor classification map in panel (e) is the result of the semi-supervised approach that combines fuzzy logic with cluster analysis and spatial constraints [242] to obtain a clean segregation between the different particle types. The reflectivity contours at the 40-dBZ (dashed line) and 60-dBZ (thick solid line) levels are overplotted in all panels to aid interpretation. For ease of visualization some classes from table 8.1 are rendered with the same gray code, i.e. rain includes drizzle, and dry ice includes crystals and dendrites.

Switzerland and France. The clusters are later assigned to specific hydrometeor classes based on human interpretation, then verified through comparison with observations collected by a two-dimensional video disdrometer on the ground. The shift from scattering simulations to observation-based unsupervised clustering is in part justified by the challenge of simulating complex ice-phase particles with varying shape, density, and size. Although this method falls under the general category of "unsupervised" methods (Fig. 8.11), the human assignment of the clusters to specific hydrometeor types still introduces a necessary "supervised" component for the practical application of the method. Using a similar approach based on agglomerative clustering, Ribaud [248] analyzed radar measurements in the Brazilian tropical region, showing that typical microphysical distributions are characterized by five classes during stratiform precipitation (drizzle/light rain, rain, wet snow, aggregates, and ice crystals). For convective precipitation, six classes are identified during the wet season (light rain, moderate rain, heavy rain, low-density graupel, aggregates, and ice crystals), whereas an addition class (high-density graupel) is found in the dry season, likely associated with deeper convection during this period of the year.

Recently, the concept of hydrometeor classification are evolving toward the more complex identification and characterization of microphysical processes. To this aim, the analysis of radar data at a single range bin and observation time is not sufficient, as discussed previously in relation to unsupervised and semi-supervised approaches involving a statistical characterization of multiple observations in space. The spatial structure and the temporal evolution of the radar variables both need to be considered in order to gain deeper insight into the microphysical processes underlying a particular hydrometeor distribution. On this matter, the next section discusses how the vertical profile of polarimetric radar variables may be effectively used to identify the dominant microphysical warm rain process below the melting layer.

8.4   Signatures of Microphysical Processes in Rain

Advanced hydrometeor classification algorithms go beyond the traditional bin-based approach and aim to exploit both the statistical distribution and the relative spatial arrangement of the observations to strengthen the classification using physical constraints. In particular, the vertical variation of dual polarization radar variables in rain is well characterized in relation to the dominant warm-rain process. Based on the theoretical background illustrated in Chapter 3 and the examples discussed in this chapter, Figure 8.14 summarizes the typical behavior of the radar variables Z_h and Z_{dr} below the melting layer in response to specific microphysical warm-rain processes. A rain layer where collision-coalescence is the dominant process will show increasing Z_h and Z_{dr} with the distance h below the melting layer (refer to the simplified example in Fig. 3.15), whereas the opposite behavior (decrease of both Z_h and Z_{dr}) is observed when breakup takes over. Size sorting implies opposite vertical gradients for Z_h and Z_{dr}, as discussed in relation to Figures 8.6 and 8.7, with increasing Z_{dr} and decreasing Z_h toward the surface. Evaporation in general implies a similar radar response, the

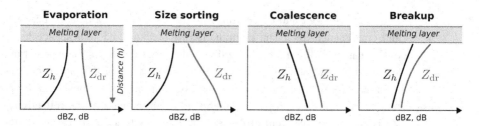

Figure 8.14 Conceptual diagram of how warm-rain processes affect the vertical profiles of reflectivity and differential reflectivity below the melting layer.

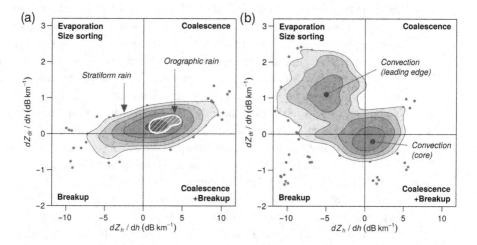

Figure 8.15 Identification of the dominant warm-rain process in the space defined by the vertical gradients of Z_h and Z_{dr} below the melting layer, according to the conceptual diagram in Figure 8.14. (a) Observations in stratiform and orographic rain for the cases illustrated in Figure 8.2 and panel (b) of Figure 8.5, respectively. The points represent data from stratiform precipitation, with superimposed gray shading denoting the density of observations (darker shading corresponds to higher density); the hatched white area comprises all data from the RHI in panel (b) of Figure 8.5 (orographic rain). (b) Same analysis as panel (a) but for the convective case in Figure 8.6. The bimodal distribution reflects the different dominant microphysical processes on the leading edge (size sorting) and in the core of the storm (a combination of coalescence and breakup).

major difference being a smaller change in both Z_h and Z_{dr} in comparison with size sorting. The preferential size reduction of the smaller drops due to evaporation may lead to a barely detectable change in Z_{dr} in some instances (Section 3.5.3).

These general tendencies can be exploited to identify the dominant warm-rain process from real polarimetric radar observations. Figure 8.15 shows some of the observations discussed in previous examples in a space defined by the vertical gradients of reflectivity (x-axis) and differential reflectivity (y-axis). The partition resulting from drawing the zero gradient lines identifies four quadrants, corresponding to the warm-rain processes in Figure 8.14. Observations falling in the lower-right quadrant (increasing Z_h and decreasing Z_{dr}) are attributed to a balance between breakup and coalescence [69]. The combination of breakup and coalescence has the effect of decreasing both tails of the DSD (coalescence decreases the concentration of the small drops, and breakup decreases the concentration of the big drops), resulting in a larger shape factor μ for a gamma distribution [232]. The increased concentration of medium-size drops may then lead to an increase of the reflectivity, whereas Z_{dr} (which is independent of the concentration) will decrease, being mostly affected by the removal of the larger drops from the distribution. For the stratiform case of Figure 8.2, within a range of ~10 km from the radar collision-coalescence is the dominant process, as highlighted by the highest concentration of the observed gradients in the upper-right quadrant of panel (a) of Figure 8.15. Despite some spread as a result of measurement errors and the uncertainty in the gradients' estimation, this type of analysis allows the identification of the most significant warm-rain process over a specific region. Besides coalescence, other mechanisms, such as evaporation and breakup, may prevail locally in this case. The observations during orographic precipitation along the Ku-band RHI presented in panel (b) of Figure 8.5 lie entirely within the white hatched area, showing a very well-defined characteristic of this orographically enhanced precipitation event.

For convective precipitation, the spread in the Z_h, Z_{dr} gradient space is generally larger, as expected because of the enhanced dynamics and possible contamination by partially melted particles. Notwithstanding, the example in panel (b) of Figure 8.15 clearly shows the prevailing nature of the precipitation processes across the storm depicted in panels (c) and (d) of Figure 8.6. As previously discussed, along this vertical section, size sorting is prevalent in the leading edge of the storm (closer to the radar, between the 12- and 22-km range), whereas a combination of coalescence and breakup appears to dominate in the core of the storm. This explains the bimodal distribution of the observations' vertical gradients in panel (b) of Figure 8.15, with two separate density peaks centered around $(-5, 1)$ dB km^{-1} (size sorting) and $(1, -0.2)$ dB km^{-1} (coalescence and breakup). It should be noted that in regions of intense updraft, the polarimetric tendencies depicted in Figure 8.14 may get perturbed.

8.5   Ice-Phase Process Studies

Polarimetric radar observations in rain have been studied since the introduction of dual polarization technology in the 1980s, supported by accurate electromagnetic scattering simulations. In fact, despite some uncertainty in the exact characterization of the drop shape as a result of oscillations and canting-angle variability (Section 3.3), the theoretical radar variables calculated for a generic distribution of raindrops can be described quite accurately and match actual radar observations. Hydrometeors in the solid or mixed phase are more difficult to represent in theoretical simulations because of the wide range of complex shapes, varying density, and different particle sizes. This complexity hindered a comprehensive understanding of dual polarization radar observations in the ice region of clouds. Until the beginning of the 2000s, the number of studies focused on observations of ice particles with polarimetric radars were limited. Some studies focused on the radar observations of differential reflectivity and the LDR, often assisted by the analysis of in situ aircraft measurements. Other works also considered the specific differential phase shift K_{dp} observations of ice particles. More recently, a new impetus has been given to studies of ice-phase processes, partly owing to the ease of advanced techniques to simulate electromagnetic scattering by arbitrarily shaped particles, like the discrete dipole approximation method [249]. This section focuses on dual polarization radar signatures in the ice region of stratified layers, and discusses how the radar variables may be affected by electrification in thunderstorms.

8.5.1   Stratified Layers – Z_{dr} and K_{dp} Enhancement Aloft

As discussed in Chapter 3, ice crystals initially form by nucleation and may further grow by deposition, riming, and aggregation. Then, depending on the height of the freezing level, ice particles may either melt and reach the ground in the form of rain or may cause snowfall to occur at the surface. The growth of ice crystals by vapor deposition is especially efficient around the $-15°C$ level, where the difference between the saturation vapor pressure with respect to water and ice is maximum (Fig. 3.18). Dual polarization radar observations in stratiform precipitation often show characteristic signatures around this altitude, with enhanced layers of Z_{dr} and K_{dp}.

The Z_{dr} enhancement in the ice region, like the bands between the 3- and 5-km height in panel (b) of Figure 8.5, is attributed to the presence of horizontally oriented ice crystals, which tend to fall with their major axis aligned horizontally. Although ice particles are in general smaller than raindrops, their aspect ratio (the dimension of the minor axis divided by the dimension of the major axis) can be much smaller, reaching values as low as 0.05 for plate crystals [250]. Because Z_{dr} is independent of the particles' concentration, even a relatively small number of oriented crystals can result in large Z_{dr} values. Wolde and Vali [251] analyzed in situ measurements from microphysical probes and W-band airborne polarimetric observations, reporting Z_{dr} up to 7 dB in cloud regions populated with hexagonal plates and stellar crystals. These extremely high values of differential reflectivity arise from the combination of the small aspect ratios and high volume fractions of ice for hexagonal plates, as illustrated by the theoretical calculations synthesized in Figure 4.21. Similar Z_{dr} signatures are often observed, in particular in the cloud layer between $-10°$ and $-20°C$, where plate-like crystals and dendrites represent the most common crystal habit (Fig. 3.18). In comparison with solid plates, dendrite crystals have a lower bulk density, often resulting in moderate Z_{dr} in the range of 1–3 dB [251, 252]. In addition, considering the dilution effect of aggregation on Z_{dr}; it is common to observe higher Z_{dr} values in the upper portion of the cloud, where the reflectivity is lower, indicating aggregation has not started.

The appearance of early aggregates quickly decreases the observed differential reflectivity, because Z_{dr} is reflectivity weighted, masking the contribution of the lower reflectivity pristine crystals. Keat and Westbrook [253] developed a technique

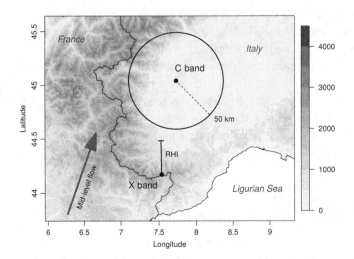

Figure 8.16 Diagram of the observations in northwestern Italy for the X-band RHI in Figure 8.17 and the space-time-averaged vertical profiles at the C band (Fig. 8.18). The arrow on the left indicates the direction of the moist flow from the sea.

to retrieve the relative contribution of pristine oriented ice crystals from the radar signal (compared with aggregates). In addition to Z_{dr}, the technique considers the correlation coefficient ρ_{hv}, which is sensitive to the diversity of shapes within the radar resolution volume. When only pristine ice crystals are present, ρ_{hv} will be close to 1. The same happens when the radar sampling volume is filled with nearly spherical aggregates. With a combination of oriented crystals and aggregates, $\rho_{hv} < 1$. Under realistic assumptions, each pair of Z_{dr} and ρ_{hv} measurements corresponds to an estimate of the relative contribution of oriented crystals. At the same time, the intrinsic differential reflectivity of pristine crystals can be estimated. The method was applied to observations with the S-band Chilbolton radar in southern England, showing an intrinsic Z_{dr} of 5–9 dB associated with oriented ice crystals. The analysis also revealed that pristine ice crystals were likely nucleated around $-14°$ to $-15°C$, grew by vapor deposition, and later aggregated.

Kennedy and Rutledge [254] focused on the analysis of the differential phase shift in winter precipitation at the S band, revealing regions of enhanced K_{dp} near the $-15°C$ isotherm. They suggested that these K_{dp} signatures, on the order of a few tenths of a degree per kilometer at the S band (see, e.g., panel [a] of Fig. 8.5), could be related to a vigorous growth of dendritic crystals. Bechini et al. [83] demonstrated the same, considering observations at both the C and X bands in northwestern Italy Fig. 8.16 is a map for the experiment, and the observations are shown in Figs. 8.17 and 8.18. One advantage of working with shorter-wavelength radars is the higher sensitivity of differential phase measurements (see, e.g., the Ku-band RHI in panel [d] of Fig. 8.5), making those instruments especially useful for ice process studies. The RHI along an Alpine valley in Figure 8.17 shows a prominent K_{dp} peak around 6 km in height, near the $-15°C$ levels, as inferred from a nearby sounding. During this event, a front passage caused localized convective activity during the early morning. The subsequent occlusion of the frontal system set synoptic conditions for widespread rain, which tended to intensify during the course of the day with the stronger advection of moist air from the Mediterranean Sea. The C-band radar located approximately 100 km north (downwind) of the X-band system also observed similar differential phase signatures. The average vertical profiles over a range of 50 km are illustrated in Figure 8.18 for three different 1-hour periods. For all cases, both Z_h and Z_{dr} show similar qualitative profiles, with the characteristic bright-band peak below the freezing level and a secondary Z_{dr} peak aloft. When considering, K_{dp}, a markedly different signature is evident between the morning weak rainfall (the average reflectivity below the melting layer is less than 25 dBZ) and the afternoon intensifying precipitation ($Z_h \sim 30\text{–}35$ dBZ), denoting a change in the precipitation growth process. The 11 UTC measurements of K_{dp} show a rather uniform vertical profile extending up to about 6 km in altitude, with average values barely above $0 \ °\text{km}^{-1}$. In contrast, the evening profiles (18 UTC and 20 UTC) show a deeper vertical extension (up to approximately 8 km in altitude) and a clear K_{dp} peak aloft, located slightly below the corresponding peak of differential reflectivity.

These dual polarization signatures are often observed in stratified layers, in different climatic regions, and with radar systems operating at different frequencies, as evidenced by the observations in Figure 8.5 (S band and Ku band), Figure 8.17 (X band), and Figure 8.18 (C band). The K_{dp} signature is typically found close (in space and time) to the Z_{dr} peak. When considering space-time averages, the resulting profiles may easily show both signatures, as in Figure 8.18, although instantaneous profiles of Z_{dr} and K_{dp} in the dendritic growth region may present an anticorrelated behavior. In particular, Griffin et al. [255] recently considered several winter precipitation events observed with the WSR-88D S-band radars to investigate their microphysical evolution through the analysis of dual polarization data. Their results indicate that the maximum Z_{dr} tend to occur in regions of weak reflectivity ($Z_h \sim 10$ dBZ) and low K_{dp}, whereas the maximum K_{dp} is observed within regions of low Z_{dr}.

The enhancement of Z_{dr} is more common in stratiform ice processes than the K_{dp} positive signature and tends to occur at higher altitudes. This typically indicates the onset of the growth of planar ice crystals in the upper part of the cloud, where the particle concentration is not sufficient to initiate the aggregation process [256]. In particular, in water subsaturation conditions, planar crystal growth may result in a stronger Z_{dr} signal (up to 6–8 dB) [257] but negligible K_{dp}. However, it is important to note that although Z_{dr} provides useful qualitative information to reveal the presence of pristine oriented ice crystals aloft, K_{dp} can bring additional quantitative information, given its relation with both the shape and the concentration of the ice particles. Kennedy and Rutledge [254], in their analysis of winter events, noted a correlation between the K_{dp} signatures aloft and the snowfall intensification at the surface. This correlation is not unique to snowfall events but is a more general characteristic of stratiform precipitation. Figure 8.19 shows the scatterplots between K_{dp} and Z_h in stratiform and convective rainfall events in northern Italy, observed with the C-band

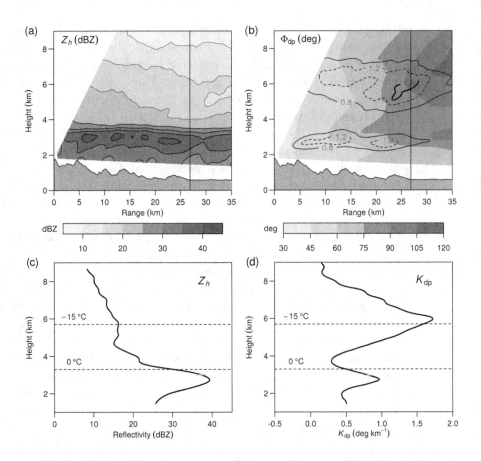

Figure 8.17 RHI scan from Col de Tende, Italy (1800 m mean sea level [MSL]) at 17:43 UTC on August 14, 2010, along the 359° azimuth, showing reflectivity (a) and filtered differential phase shift with overplotted contours of K_{dp} (b). The gray area from 0 km to 1-2 km height represents the topographic profile of the Alpine valley. Panels (c) and (d) show the vertical profiles of Z_h and K_{dp} at the range of 27 km, respectively. The horizontal dashed lines mark the location of the 0° and −15°C temperature levels as inferred from the nearby radiosounding of Cuneo Levaldigi.

radar on the Torino hill (740 m MSL). Only events with a freezing level above 1500 m MSL were selected in order to ensure observations of rain close to the surface. Panel (a) shows the scatterplot between daily average values of K_{dp} and Z_h at the lowest vertical level, illustrating the expected correlation based on the theoretical behavior of the dual polarization variables in rain (see Chapter 9). In panel (b), the average K_{dp} aloft (around −15°C) is considered instead. Although the scatterplot shows no correlation for convective events, stratiform rainfall is clearly characterized by a significant correlation between K_{dp} aloft in the dendritic growth region and the average surface precipitation intensity. This statistical relation may be useful for practical applications in rainfall and snowfall estimation, especially at far ranges or in complex orography where the radar visibility at low-elevation angles is limited.

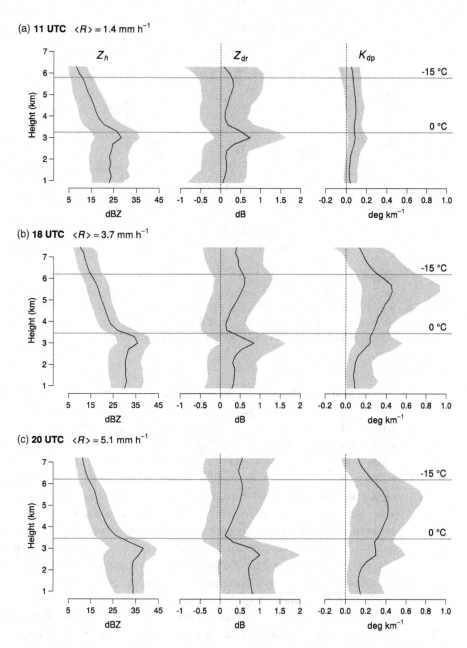

Figure 8.18 Hourly-averaged vertical profiles of radar variables at 11 UTC (a), 18 UTC (b), and 20 UTC (c) from the C-band Bric Della Croce radar (740 m MSL) over the 50-km range depicted in Figure 8.16. The shaded area encompasses the values between the 10th and 90th percentiles and the average rainfall rate ($< R >$) is estimated from radar observations at the lowest level using eq. 9.71.

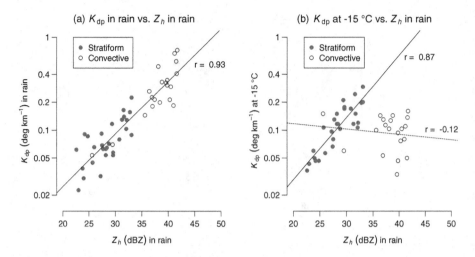

Figure 8.19 Daily average K_{dp} and Z_h during 48 rainfall events in northern Italy, observed from the C-band Bric Della Croce radar (740 m MSL). (a) Scatterplot between K_{dp} and Z_h at the lowest level in rain (900 m MSL). (b) Similar to panel (a), but K_{dp} in the ice region at elevations around the $-15°$C level. Note that for stratiform precipitation, K_{dp} aloft is still highly correlated ($r = 0.87$) with reflectivity in the rain layer below.

Although an accurate microphysical explanation of these polarimetric signatures is still an object of current research, recent studies have revealed a significant relation with the cloud vertical motions. In particular, Moisseev et al. [230], using coincident RHI scans and vertical-pointing Doppler radar observations, have shown that the positive K_{dp} signatures in winter storms tend to appear coincident with vertical ascent.

This confirms the suggestion of Kennedy and Rutledge [110] that the maintenance of active depositional growth requires upward air motions to provide a continuous water vapor supply. It follows that the observation of positive K_{dp} layers at altitudes between $-10°$ and $-20°$C may provide an indication of weak updrafts associated with mesoscale forcing. The intensity of upward vertical motions in stratiform precipitation systems is only a few tenths of a meter per second, but may be sufficient to activate the growth of supercooled drops in the ice portion of the cloud. This is illustrated in Figure 8.20 (adapted from fig. 5 in [80]). The line separating the gray and white regions indicates the threshold vertical velocity w^* above which both liquid drops and ice crystals grow in the coexistence regime (see Section 3.5.6) at a temperature of $-15°$C. For typical values of the integral ice-particle radius, given by the product of number concentration N_i and average radius $\bar{r}_i$, w^* is on the order of 0.1 m s^{-1}. Depositional growth is favored in ascending air masses, where both ice crystals and supercooled liquid drops may rapidly grow simultaneously. In such a supersaturated environment, the relative abundance of supercooled drops results in further growth by light riming, which, similar to aggregation, tends to dilute the anisotropy of the ice particles and lessen the differential reflectivity [257]. The combined effect of

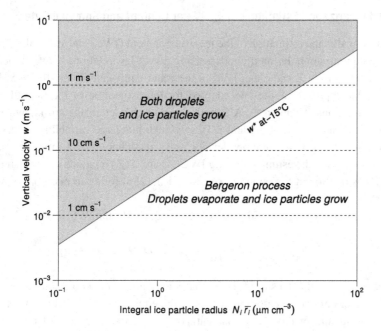

Figure 8.20 Particle-growth processes in an ascending air mass with vertical velocity w. In the gray-shaded region ($w > w^*$), both droplets and ice particles grow simultaneously. The threshold vertical velocity w^* is calculated as a function of the integral radius of the ice particles (product of number concentration N_i and average radius $\bar{r}_i$), for a pressure of 500 hPa and a temperature of $-15°C$, based on eq. 22 in Korolev and Mazin [80]. When $w < w^*$, the Wegener–Bergeron–Findeisen process takes place, with ice growth at the expense of the evaporating liquid drops.

early aggregation and light riming may explain the observed decrease of Z_{dr} when K_{dp} reaches its peak value.

In addition to the role of depositional growth of dendritic crystals, other processes have been considered to explain these polarimetric signatures. It was recently suggested that a seeder-feeder mechanism may account for PSDs with sufficient concentrations to generate the observed specific differential phase shifts [230]. In this view, the appearance of the K_{dp} layer would indicate the onset of aggregation. Ice multiplication is another process that may increase the concentration of ice particles and explain the observed high K_{dp} values. In particular, when the K_{dp} enhancement is observed at higher temperatures (i.e., lower altitudes), as in the example in panel (d) of Figure 8.5 or panel (c) of Figure 8.18, secondary ice production of columnar crystals through rime splintering may play a role in providing a large concentration of oriented ice particles [86].

We anticipate that in the near future, a larger availability of polarimetric radar observations in different climatic regions (in particular, using higher-frequencies, such as X band, and in conjunction with in situ sensors) will foster a deeper understanding of precipitation-growth mechanisms in the ice phase.

8.5.2 Dual Polarization Estimation of Ice Water Content and Snowfall Rate

Traditionally, the estimation of the ice water content (IWC) and snowfall rate has been pursued with nonpolarimetric radars using power-law relations between the observed radar reflectivity Z and either IWC or the water equivalent snowfall rate S, following a similar approach as the one adopted for the estimation of the rainfall rate using $R(Z)$ relations (Chapter 9). Although the rain characteristics are known to a great level of detail (Sections 3.2 and 3.3), snow exhibits large variability in terms of size distribution, shape, orientation, density, and growth habits. This introduces additional uncertainties in the estimation using $IWC(Z)$ and $S(Z)$ relations as compared with the $R(Z)$ estimation of the rain rate (see table 1 in [258] for a summary of the main $Z(S)$ relations).

Analogously to the liquid water content (eq. [3.4]), the IWC is defined as

$$\text{IWC} = \frac{\pi}{6} \times 10^{-3} \int_{D_{\min}}^{D_{\max}} \rho_{\text{snow}}(D)D^3 N(D)dD \quad [\text{g m}^{-3}], \quad (8.4)$$

where ρ_{snow} is the density of the ice particle in grams per cubic centimeter (g cm^{-3}), N is the number concentration, and D is the equivalent spherical diameter of the ice particle in millimeters. Note that with respect to the definition of the liquid water content W (eq. [3.4]), now the density is inside the integral because it may vary for different ice particles. The particle's density is often approximated as a function of its size (eq. [3.74]). The water equivalent snowfall rate is

$$S = 0.6 \times 10^{-3} \int_{D_{\min}}^{D_{\max}} \frac{\rho_{\text{snow}}(D)}{\rho_w} D^3 V_{ts}(D)N(D)dD \quad [\text{mm h}^{-1}], \quad (8.5)$$

where V_{ts} is the terminal velocity of snow and ρ_w is the density of water.

A relation to estimate the ice particles' IWC from K_{dp} can be introduced based on the discussion in Chapter 4 (eq. [4.68]), as follows [259]:

$$\text{IWC}(K_{\text{dp}}) = c\frac{K_{\text{dp}}}{f}, \quad (8.6)$$

for frequency f in GHz and a scaling coefficient c. This simple linear relation between K_{dp} and IWC arises from the fact that in general, for a given population of ice particles observed by the radar, the aspect ratio is not functionally related to the particles' size (unlike for raindrops). This implies that for pristine ice crystals with high density, the term $\text{Re}(p_h - p_v)$ (which includes the shape and dielectric properties of the scatterer) can be taken out of the integral in eq. (4.68), leading to a linear relation between K_{dp} and IWC that is independent of the PSD but depends critically on the average aspect ratio of the ice particles (Fig. 4.16). If, by some other means, an estimate of the mean aspect ratio is available, then the relation in eq. (8.6) may prove useful for the estimation of the IWC. From a practical point of view, the presence of low-density irregular snowflakes or crystal aggregates will bias K_{dp} toward low values, preventing its use for quantitative applications.

In principle, the IWC and snowfall rate can be parameterized in terms of K_{dp} and reflectivity [260]. Under Rayleigh scattering approximation, and assuming an exponential size distribution for the snowflakes, both the IWC (in $\mathrm{g\,m^{-3}}$) and the water equivalent snowfall rate S (in $\mathrm{mm\,h^{-1}}$) can be expressed as a function of the reflectivity and specific differential phase:

$$\bullet \qquad F_{\mathrm{ice}}(K_{\mathrm{dp}}, Z_h) = c \left(\frac{K_{\mathrm{dp}}}{f} \right)^b Z_h{}^a, \qquad\qquad (8.7)$$

where F_{ice} denotes either IWC or S.

For the estimation of the ICW ($F_{\mathrm{ice}} = \mathrm{IWC}$ in eq. [8.7]), the exponents of K_{dp} and Z_h (coefficients b, a) can be shown to be related to the density–size relation for snowflakes as $b = (3 + \beta)/3$, $a = -\beta/3$, where β represents the exponent of the density-size relation in eq. (3.74) (see appendix in [260]). Once a given density–size relation is assumed, most of the variability in IWC estimation arises from the uncertainty about the shape and orientation of the snowflakes and is contained in the multiplicative coefficient c.

For the estimation of the water equivalent snowfall rate ($F_{\mathrm{ice}} = S$ in eq. [8.7]), from a theoretical point of view, the exponents b and a are additionally affected by the fall velocity of the snow particles through the relations $b = (3 + \beta - \delta)/3$, $a = (\delta - \beta)/3$, where δ is the exponent of a velocity–size relation like eq. (3.72). Note that the fall velocity of snow aggregates is generally the order of $1\ \mathrm{m\,s^{-1}}$ and, unlike for raindrops, is a weak function of the snowflake size. The value of δ typically varies in the range 0.1–0.2, eventually leading to minor differences in the exponents b and a when using the parameterization in eq. (8.7) to estimate S as opposed to IWC. Similar to the estimation of IWC, the coefficient c includes the dependence on the particles' shape and orientation, but in addition, c is also sensitive to the altitude of radar observations because the terminal fall speed is affected by the atmospheric pressure.

In order to derive usable relations, f in eq. (8.7) should be replaced with the actual radar frequency, and the coefficients (in particular the multiplicative coefficient c) should be estimated for a specific region or measurement campaign through analysis of in situ observations such as disdrometer data. As an example, assuming $b/a = 0.65$ and $\sigma = 0°$, the following relations for the S band at 350 m MSL were obtained [260]:

$$S(K_{\mathrm{dp}}, Z_h) = 1.48\, K_{\mathrm{dp}}^{0.615} Z_h{}^{0.33} \qquad\qquad (8.8a)$$

$$\mathrm{IWC}(K_{\mathrm{dp}}, Z_h) = 0.71\, K_{\mathrm{dp}}^{0.66} Z_h{}^{0.28}. \qquad\qquad (8.8b)$$

Similar relations can also be derived for the C and X bands. The use of dual polarization relations reduces the uncertainty in IWC and snowfall rate estimation (spatial averaging of the radar observations also reduce measurement uncertainties). Like rainfall rate estimation, extensive validation is needed to demonstrate the selected coefficients and that the inclusion of the radar's K_{dp} measurements improves the accuracy of S and IWC. With respect to reflectivity-only estimators of S and IWC, the combination of K_{dp} and Z_h allows for more stable estimates through all of the ice layer below the dendritic growth region. In fact, $S(Z)$ and $\mathrm{IWC}(Z)$ relations are highly sensitive to the measurement altitude because of the typical steep vertical gradient of reflectivity (increase downward as a result of aggregation). The opposite vertical variation of K_{dp} (decrease downward; see, e.g., Figs. 8.17 and 8.19) provides a compensation effect in $S(K_{dp}, Z_h)$ and $IWC(K_{dp}, Z_h)$ relations.

8.5.3   Cloud Electrification—Signatures of K_{dp} and LDR

Ice crystals, like raindrops, tend to fall with their major axis oriented horizontally as a result of aerodynamic forces. When an intense electric field is present, however, particles can deviate from their aerodynamically induced orientation and become aligned parallel to the electric field [261–263]. Electric fields in thunderstorms are typically on the order of 100–200 kV m^{-1}, but even larger values can be found before breakdown which results in lightning. Smaller particles, with their lower inertia and surface area, experience lower aerodynamic forces and therefore smaller electric fields are required to change their preferential orientation. It has been calculated that an electric field of 100 kV m^{-1} is capable of aligning plates with a major axis of less than 0.6 mm and columns with a height of $\sim$2 mm [263]. Larger particles need stronger electric fields to be aligned (e.g., 200 kV m^{-1} for plates of 1.7 mm). It follows that the effect on the radar reflectivity and differential reflectivity is negligible during cloud electrification because of their strong dependence on the larger particles, unless the electric field reaches very strong values and/or the PSD in the cloud is especially narrow and dominated by small ice particles. When negative Z_{dr} is detected in the upper portion of thunderstorms, this is more often due to the presence of graupel particles. Instead, given the specific differential phase shift's dependence on the particle concentration and relative insensitivity to large, more spherical aggregates, K_{dp} will be more easily affected by the canting of small ice crystals in response to the cloud's electric field. Panel (a) of Figure 8.21 shows a time series of K_{dp} values collected near the cloud top of a Florida thunderstorm with the NCAR CP-2 dual frequency radar (S band and X band). The radar system is operated in fixed-pointing mode, with the antenna held at constant elevation and azimuth angles, allowing collection of data for a fixed location with high temporal resolution. The time series of the S-band K_{dp} shows a periodic pattern with a decreasing trend lasting several seconds, followed by a sudden increase. These fluctuations are interpreted as being the result of a progressive alignment of the small ice crystals with the electric field, as illustrated in the conceptual diagram in panel (b) of Figure 8.21. After the lightning strike, the electric field disappears, and the particles' orientation quickly returns to nearly horizontal, with positive values of K_{dp}. The recovery time for horizontal realignment in the Florida thunderstorms was estimated to be less than 1 s.

K_{dp} represents the local gradients of the path-integrated differential phase (Φ_{dp}); therefore, its value is related to the atmospheric composition of a small cloud portion extending over a few range gates (where the derivative of Φ_{dp} is calculated). The X-band system of the CP-2 radar transmits on the vertical polarization only but receives both the copolar (V) and the cross-polar (H) signals. The time series of the X-band LDR, estimated from the ratio of the cross-polar to the copolar reflectivity, is also reported in panel (a) of Figure 8.21. For linear (horizontal or vertical) transmit

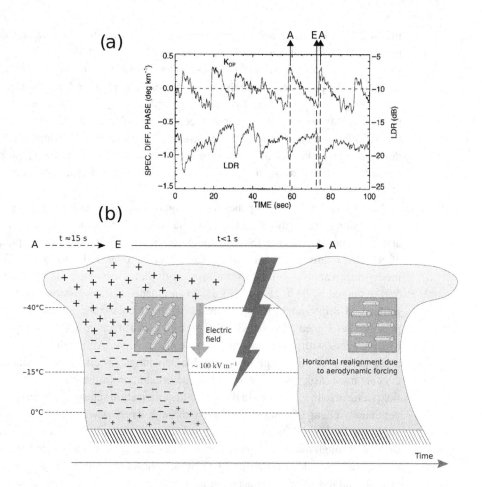

Figure 8.21 (a) Time series of K_{dp} and LDR near the top of a Florida thunderstorm on September 20, 1991, at 23:43:36 UTC. The letters A and E, respectively, denote the time periods where aerodynamic forcing and the electric field have the greatest impact on the particles' alignment. Adapted from [263]. (b) Conceptual diagram explaining the polarimetric observations in panel (a). The ice crystals' orientation starts to deviate from the horizontal as the electric field builds up in the cloud. The maximum alignment with the electric field is reached just before the lightning discharge (E), followed by a quick reorientation in the horizontal plane induced by the dominant atmospheric forces (A).

pulses, the maximum depolarization is expected when the particles are oriented at $45°$, whereas the LDR should be minimum for particles that are horizontally or vertically aligned. This explains the anticorrelated behavior of LDR with respect to K_{dp} in panel (a). Assuming that the ice particles are nearly horizontally oriented right after a lightning strike (i.e., no electric field), K_{dp} is positive, and the LDR is at a local minimum. As the electric field builds up, K_{dp} decreases and can reach negative values as the ice crystals orientation projects more into the vertical plane. At the same time, the LDR increases as a result of the growing magnitude of the cross-polar return (refer to

Fig. 4.24). Even when crystal canting occurs over a limited range along the radar beam, the LDR will show a characteristic streak beyond that region (similar to Φ_{dp}) because of the scatterers' cross-polar forward scattering response (some of the single polarization's transmitted power is depolarized and converted into the orthogonal polarization by the scatterers). This power then propagates and scatters at farther ranges as copolar echoes, but these echoes are measured by the radar as an increased LDR because of the initial depolarization. Ultimately, the detection of the small crystals' orientation is mainly a matter of propagation characteristics, and as such, K_{dp} and LDR are the most useful indicators of cloud electrification.

The K_{dp} observations in panel (a) of Figure 8.21 were collected using alternate transmission of the orthogonal H and V channels. When using simultaneous transmit and simultaneous receive (STSR mode; see Section 2.2), the same depolarization effects that manifest as LDR streaks can affect the other radar variables. The STSR mode radar variable estimators assumes negligible cross-polar signals (see Section 5.5). As noted by Scott et al. [142], Z_{dr} in simultaneous transmission mode may show artifacts of differential phase as a result of the depolarizing effects induced by the presence of aligned particles. The origin of these artifacts resides in the presence of hydrometeors with a nonzero net mean canting angle. Although the mean canting angle is zero for rain, implying negligible depolarization effects, a population of ice crystals may easily assume a mean canted orientation in the presence of an electric field (refer also to Fig. 4.23). In this case, the polarization state of the propagating wave continually changes as the wave travels through a region with canted particles. This applies to both polarization states (H and V) and affects the signal both ways along the path through the depolarizing region. The consequence is a bias in the estimation of Z_{dr} (and the other radar variables), which shows up within the region of canted crystals and then remains constant at farther ranges, resulting in a radial streak in PPI or RHI plots. The differential phase shift Φ_{dp} is also affected by cross-coupling in the STSR mode, although the slope of the radial profile remains similar to the ATSR mode for regions without significant depolarization. In addition, while the sign and magnitude of the radial streak in Φ_{dp} only depend on the mean canting angle in the depolarizing region, for Z_{dr} the sign and magnitude may change depending on both the mean canting angle and the value of the system differential phase on transmission. This implies that the relevance of the Z_{dr} streaks may greatly vary between different systems operating at the same frequency.

8.6   Selected Problems

1. The vertical cut in panel (c) of Figure 8.6 is obtained from a volume scan consisting of 14 PPIs. The elevation angle θ, beam height, and value of the S-band polarimetric parameters at the range of 28 km are listed in the following table:

θ (°)	T (°C)	Z_h (dBZ)	Z_{dr} (dB)	ρ_{hv}	K_{dp} (°km^{-1})
0.5	22.8	48.4	1.7	0.982	0.94
0.9	20.9	46.5	1.2	0.980	0.83
1.3	18.7	45.9	1.2	0.981	0.81
1.8	16.6	48.3	1.3	0.983	0.93
2.4	15.4	48.5	0.9	0.983	1.2
3.1	15.8	48.8	0.9	0.981	1.65
4.0	14.4	49.5	1.1	0.937	1.96
5.1	10.5	52.1	0.9	0.967	2.19
6.4	5.1	51.9	1.3	0.942	2.32
8.0	−1.1	56.0	0.9	0.971	2.14
10.0	−8.6	56.5	−0.4	0.974	1.61
12.5	−16.5	51.6	−0.6	0.984	1.12
15.6	−28.2	49.4	−0.9	0.981	0.44
19.5	−43.0	44.2	−0.3	0.975	−0.48

Perform a fuzzy logic hydrometeor classification on these observations, based on the membership functions defined in Table 8.1. Proceed following these steps:

a) Calculate the value of the beta membership functions (Table 8.1) for the radar observations at the first elevation ($\theta = 0.5°$).

b) Apply the additive rule (eq. [8.3]) assuming identical weights for all the radar parameters ($w = 1$), and find the most likely outcome of the hydrometeor classification for $\theta = 0.5°$.

c) Repeat steps 1 and 1 for all the other levels.

d) Perform the classification again (a-c), but include the temperature information making proper assumptions about the temperature membership functions (e.g. use a beta function with $m = 30, a = 30, b = 4$ for liquid particles, etc.) and apply either the additive (eq. [8.3]) or the multiplicative rule (eq. [8.2]) for the temperature. How does the classification change?

e) Perform the classification again including temperature, but do not consider K_{dp} observations; that is, use only temperature, Z_h, Z_{dr}, and ρ_{hv}. Is there any change in the classification results? If so, how would you explain it?

2. Consider the case of observation from a thin melting layer as shown in Fig. 8.1, and assume the thickness of the melting layer is 400 m. With a uniform storm structure observed as a RHI scan:

a) Sketch the observed melting layer structure as a function of the distance from the radar. Use the hint from Figure 8.1 that the closer observations will be made at higher elevation angles and faraway observations are made at smaller elevation angles.

b) If the intrinsic Z_{dr} (i.e. the Z_{dr} at the elevation of 0°) at 1 km height is 2 dB, sketch the range variation of Z_{dr} caused by the elevation angle dependence. Hint: use eq. (4.85) for the elevation dependence of Z_{dr}.

3. $Z - Z_{dr}$ scattergram can be used for several applications and provide insight into the microphysical processes of precipitation and their hydrometeor types.

a) Download data publicly available from the CSU-CHILL website, the US national Weather Service Observations of WSR-88D, or other observations accessible to you through your local weather service. Visually examine the displays of the radar observations and locate the region of storms for both the convective and stratiform precipitation.

b) Plot the Z_h versus Z_{dr} scattergram and compare it to the diagram on Figure 8.7b.

    c) Plot the same Z_h versus Z_{dr} diagram in the ice region aloft.

    d) Plot the Z_h versus Z_{dr} diagram in the melting layer.

    e) Interpret the results of the three different precipitation phases (ice, mixed-phase, and rain).

4. The concepts used for hydrometeor classification evolved over time, with more observations being added to the process through the years. Consider the following questions that would have been research topics during the advancements of these techniques:

    a) Assume you have a simple task of distinguishing between water and ice. Use the radar observations downloaded for problem 3. With only the Z_{dr} observations, make a simple rule to distinguish between water and ice and create a two-output rule diagram. This simple ability to distinguish between water and ice is what propelled dual-polarization radars to the forefront.

    b) Using your simple classification, draw histograms of Z_h and Z_{dr}, for water and ice. These are the membership functions used in the literature. Also draw a two-dimensional membership function.

    c) From the microphysical properties and constraints, we know with certainty at high altitudes above melting layer, the precipitation is ice. However, if you made a decision just based on reflectivity rules, you will notice errors. From the histograms, compute the probability of error in the classification.

    d) Now, include the altitude in your decision process with a simple rule, namely, 250 m above melting layer is ice and 250 m below the melting layer is water. Inside the 500 m band of the melting layer, assume there is no change in the classification (or make a random choice). The altitude of the melting layer can be estimated using radar observation to detect the melting layer or from the sounding information. Try both methods for determining the melting layer altitude and look for differences. Document and quantify how the hydrometeor classification changes when using altitude information.

9 Rainfall Estimation and Attenuation Correction

One of the most common applications using weather radar is rainfall estimation. The advantage of monitoring and quantifying precipitation over large areas with a single instrument makes radar a unique tool for use in hydrology. As a matter of fact, the spatial and temporal resolution achievable by a weather radar cannot be beaten by any practical rain-gauge network. Even so, the radar estimates rely on remote-sensing observations, implying that their accuracy may be influenced by geometric and physical factors, including (but not limited to) the height of the observation volume above the ground, the variability of the drop size distribution (DSD) in rain, the signal attenuation occurring between the radar and the precipitation target, and the presence of frozen or mixed-phase particles. The advent of dual polarization contributed to overcoming or mitigating many of the factors that traditionally have limited single-polarization systems. In particular, dual polarization observations led to great improvements in the area of attenuation correction and rainfall estimation. This fostered the development of new applications, such as hydrometeor classification (Chapter 8). The sensitivity of dual polarization observations to the shape, density, orientation, phase, and concentration of particles within the radar resolution volume allows the user to discern between different particle types. In addition to providing specific classification products for a range of applications (e.g., meteorology, transportation, aviation, agriculture), the identification of the dominant hydrometeor type within the radar resolution volume is a key processing step for the subsequent application of rainfall-estimation algorithms.

Correction for signal attenuation is the other fundamental processing step that is required prior to quantitative rainfall estimation. Although attenuation may be neglected in many instances at the S band, for higher frequencies (C band and above), attenuation can lead to underestimation of reflectivity with corresponding inaccuracies in the rainfall estimates. Hydrometeor classification, attenuation correction, and rainfall estimation are strongly interconnected topics and are integral parts of modern implementations of the radar rainfall-estimation processing systems through generalized blended algorithms.

Reflectivity and differential reflectivity values are often represented on a logarithmic scale for ease of visualization (dBZ and dB, respectively). However, in this chapter just as in the rest of the book, the notations Z_h and Z_{dr} indicate the radar variables in linear scale, that is, the reflectivity in units of millimeters to the sixth power per cubic meter ($\text{mm}^6\,\text{m}^{-3}$) and Z_{dr} is a unitless ratio, unless where explicitly noted.

9.1   Dual Polarization Variables in Rain (Z_h, Z_{dr}, K_{dp})

The relevant radar observables for the estimation of the rainfall rate are Z_h, Z_{dr}, and K_{dp}. In the Rayleigh regime, the radar cross-section of spherical particles is proportional to the sixth power of the diameter, as shown in eq. (4.34). For the general case of nonspherical particles, the dual polarization variables can be written in terms of the raindrop size distribution $N(D)$, that is, the number of particles per unit volume and per interval of diameter, and the backscatter and forward-scatter amplitudes:

$$Z_{h,v} = \frac{4\lambda^4}{\pi^4 |K_w|^2} \int |s_{hh,vv}|^2 N(D) dD$$

$$= \frac{\lambda^4}{\pi^5 |K_w|^2} \int \sigma_{h,v}(\lambda, D) N(D) dD \tag{9.1}$$

$$Z_{dr} = \frac{\int \sigma_h(\lambda, D) N(D) dD}{\int \sigma_v(\lambda, D) N(D) dD} \tag{9.2}$$

$$K_{dp} = \lambda \int \text{Re}[f_h(\lambda, D) - f_v(\lambda, D)] N(D) dD. \tag{9.3}$$

In these equations, D is the equivalent diameter, $\sigma_{h,v}$ represents the radar cross-sections at h or v polarization, and $\sigma_{h,v} = 4\pi |s_{hh,vv}|^2$, where $s_{hh,vv}$ represents the scattering amplitudes (eq. [4.50]). The radar cross-section (eq. [4.34]) and the forward-scattering amplitudes, $f_{h,v}$, are a function of the particle's diameter and the radar's wavelength λ. The dielectric factor of water (see Section 4.2.1) is $|K_w|^2 \approx 0.93$.

Raindrops have a shape that can be well approximated by an oblate spheroid (Chapter 3). In the case of spheroids, the radar cross-sections can be expressed based on the theory of Gans, which is a generalization of the Rayleigh scattering theory for spheres to more broadly include oblate or prolate spheroids. Under the Rayleigh–Gans approximation, it is possible to separate the scattering contribution of the drop's volume and the drop's axis ratio (due to its oblateness) in the scattering amplitude (eqs. [4.40, 4.41]), leading to the following expressions for the horizontal and vertical reflectivity:

$$Z_h = \frac{1}{9|K_w|^2} \int D^6 N(D) |p_h|^2 dD \tag{9.4}$$

$$Z_v = \frac{1}{9|K_w|^2} \int D^6 N(D) |p_v|^2 dD \tag{9.5}$$

where p_h and p_v are polarization factors (eqs. [4.46, 4.47]) which depend on the drop's axis ratio and the dielectric factor. The reflectivities in eqs. (9.4) and (9.5) have the standard units of $\text{mm}^6 \, \text{m}^{-3}$, as a result of $N(D)$ being expressed in inverse cubic meters per millimiter ($\text{m}^{-3} \, \text{mm}^{-1}$]) and D in millimiters . Following from eq. (4.68), the specific differential phase shift is commonly expressed in units of degrees per unit kilometer:

$$K_{dp} = 10^{-6} \left(\frac{180}{\lambda} \right) \left(\frac{\pi}{6} \right) \int \text{Re}(p_h - p_v) D^3 N(D) dD \; [\text{deg km}^{-1}], \tag{9.6}$$

where λ is the wavelength in meters. From eq. (9.6), it is clear that K_{dp} is inversely proportional to the radar wavelength (scaling linearly with the frequency) and it is proportional to the product of the drop's volume and a function of its axis ratio.

These analytical expressions are useful for direct calculation of the radar variables, for example, from a DSD measured with a disdrometer. However, it is important to remember that these relations are only valid under the Rayleigh–Gans approximation. For any combination of DSDs and radar frequencies involving non-Rayleigh scatterers, numerical methods are needed to calculate the radar variables.

The rainfall rate is given by (meter, kilogram, second [MKS] system):

$$R = \frac{\pi}{6} \int V_t(D) D^3 N(D) dD \quad [\text{m s}^{-1}], \tag{9.7}$$

or in the conventional units of millimeters per hour (mm h^{-1}):

$$R = 0.6\pi \times 10^{-3} \int V_t(D) D^3 N(D) dD \quad [\text{mm h}^{-1}], \tag{9.8}$$

where $V_t(D)$ is in meters per second, D in millimiters, and $N(D)$ in inverse cubic meters per millimiter. Using the power-law formulation for the terminal velocity of raindrops from eq. (3.34):

$$V_t(D) = 3.78 D^{0.67} \quad [\text{m s}^{-1}], \tag{9.9}$$

eq. (9.8) then becomes the same as eq. (3.35). Similarly, the water content is expressed by the third moment of the DSD, as in eq. (3.4).

9.1.1 DSD, Radar Moments, and Self-Consistency

Under the assumption of a DSD following an exponential form (eq. [3.1]), from eqs. (9.4) and (9.5), it follows that the differential reflectivity is only a function of the distribution's slope parameter Λ:

$$
\begin{aligned}
Z_{dr} &= \frac{\int D^6 N_0\, e^{-\Lambda D} |p_h|^2 dD}{\int D^6 N_0\, e^{-\Lambda D} |p_v|^2 dD} \\[2mm]
&= \frac{\int D^6 e^{-\Lambda D} \left| \frac{(\epsilon_r - 1)}{1 + \frac{1}{2}(1 - L_z)(\epsilon_r - 1)} \right|^2 dD}{\int D^6 e^{-\Lambda D} \left| \frac{(\epsilon_r - 1)}{1 + L_z(\epsilon_r - 1)} \right|^2 dD},
\end{aligned}
\tag{9.10}
$$

where L_z is the vertical depolarization factor (see Section 4.4) and ϵ_r is the raindrop's dielectric factor. For any given pair of Z_h and Z_{dr} observations, it would then be possible to completely specify the DSD, retrieving both parameters (Λ and N_0). After Λ is calculated from Z_{dr}, the intercept parameter N_0 can be calculated from either Z_h or Z_v. In this simplified ideal case, any other statistical moment of the DSD (i.e., K_{dp}, R, W) could be expressed as a function of Z_h and Z_{dr}.

Recalling eqs. (3.9) and (3.10), the mass-weighted mean diameter D_m and the median volume diameter D_0 are strictly related to the slope parameter Λ. Figure 9.1 shows Λ, D_m, and D_0 as a function of Z_{dr}. This illustrates the distinctive relation

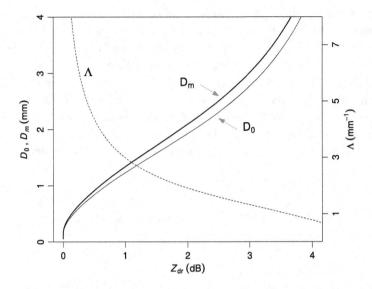

Figure 9.1 Mass-weighted mean diameter D_m and median volume diameter D_0 as a function of differential reflectivity at the S band. The values are simulated based on an exponential DSD with $D_{max} = 8$ mm and a canting-angle Gaussian distribution with zero mean and $7°$ standard deviation. The value of the slope parameter Λ is also plotted with a dashed line (right axis).

between the characteristic diameters of the DSD and the radar differential reflectivity, which forms the basis of the dual polarization rainfall estimators described in Section 9.5.2. The relations shown are obtained for the S band, using an exponential DSD, with $D_{max} = 8$ mm and canting angle following a Gaussian distribution with a $7°$ standard deviation. The D_m–Z_{dr} and D_0–Z_{dr} relations are defined under these assumptions but may show differences depending on the specific radar operating frequency (refer to the discussion later in Section 9.2). Note that although D_0 and D_m can be used almost interchangeably in most instances (with proper parameterizations), D_m has found increased usage for being a moment of the DSD.

In order to elaborate more on the relevant relations between the DSD characteristics and the radar observations, we consider DSDs measured by disdrometer during several rainy days. The radar moments Z_h and Z_{dr} are calculated from gamma fits to the observed DSDs using electromagnetic scattering simulations and are plotted in Figure 9.2. The different fitting curves for $\log_{10}(N_w) = 2.2, 3.2, 4.2$ (N_w in $mm^{-1}m^{-3}$) in the left panel show that for a given value of reflectivity, the higher Z_{dr} (corresponding to higher D_m) is "compensated" by a smaller N_w. From the definition of N_w (eq. [3.24]), it follows that the observation of a smaller Z_{dr} for a given reflectivity corresponds to a higher water content (or rainfall rate). As an example, for $Z_h = 45$ dBZ, the rainfall rate increases from ∼5–10 mm h^{-1} to ∼30–40 mm h^{-1}, with N_w between 2.2 and 4.2, whereas Z_{dr} decreases from 3.3 to 1.0 dB. This is a consequence of Z_{dr} being related to the reflectivity-weighted axis ratio (eqs. [9.11–9.14]), whose

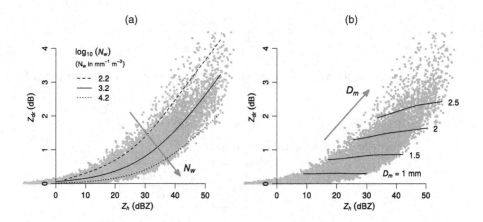

Figure 9.2 Scatterplot of Z_{dr} vs. Z_h, based on gamma fits to observed DSDs during 5 days, with N_w (a) and D_m (b) contours overplotted. The radar moments are calculated using electromagnetic (EM) scattering simulations at 3 GHz (S band).

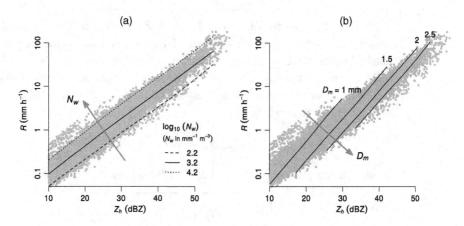

Figure 9.3 The same observations as Figure 9.2, but for rainfall rate R (log scale) versus reflectivity Z_h.

microphysical implication is that the distributions with low Z_{dr} are dominated by a large number of small drops, whereas the distributions with high Z_{dr} have a longer right tail and fewer small droplets. Panel (b) in Figure 9.2 shows the corresponding isolines of D_m, demonstrating the close relation with Z_{dr} already illustrated in Figure 9.1. Figure 9.3 uses the same data as Figure 9.2, but shown on the Z_h–$\log_{10}(R)$ plane. This is useful to stress the characteristic increase of N_w with the rain rate R for a given reflectivity and the corresponding decrease of the mass-weighted mean diameter D_m.

Without any a priori hypothesis about the drop-shape model, Z_{dr} can also be shown to be explicitly related to the reflectivity-weighted axis ratio. Using the Rayleigh–Gans formulation in eqs. (9.4) and (9.5), the ratio of the squared absolute value of the polarization factors is very closely approximated by the following [267]:

$$\frac{|p_h|^2}{|p_v|^2} = \frac{|1 + L_z(\epsilon_r - 1)|^2}{\left|1 + \frac{1}{2}(1 - L_z)(\epsilon_r - 1)\right|^2} \approx r^{-7/3}, \tag{9.11}$$

where r is the axis ratio of the raindrops ($r = b/a$). Substituting into eq. (9.10) leads to

$$(Z_{dr})^{-1} = \frac{\int r^{7/3} D^6 N(D)|p_h|^2 dD}{\int D^6 N(D)|p_h|^2 dD}. \tag{9.12}$$

The reflectivity-weighted axis ratio is

$$\bar{r}_z = \frac{\int r D^6 N(D) dD}{\int D^6 N(D) dD}, \tag{9.13}$$

which is connected to Z_{dr} for typical DSDs as $\bar{r}_z^{7/3} \approx \overline{r_z^{7/3}}$, and therefore, accurate estimator of $\bar{r}_z$ is [267]:

$$\bar{r}_z = Z_{dr}^{-3/7} = 10^{-0.043[Z_{dr}(dB)]}. \tag{9.14}$$

Building on this theoretical relation between Z_{dr} and the reflectivity-weighted axis ratio, the assumption of a drop-shape model indirectly leads to the previously illustrated relations between Z_{dr} and the characteristics drop diameters D_m and D_0.

Next, consider the physical nature of the relation between the reflectivity and the rainfall rate. For a three-parameter gamma DSD, the reflectivity of spherical drops under the Rayleigh approximation is expressed using the normalized DSD (eq. [3.26]):

$$Z = N_w f(\mu) D_m^{-\mu} \int D^{6+\mu} e^{-(4+\mu)\frac{D}{D_m}} dD \tag{9.15}$$

Solving the integral with the help of eq. (3.17) leads to

$$Z = N_w f(\mu) \Gamma(7 + \mu) \frac{D_m^7}{(4 + \mu)^{7+\mu}}. \tag{9.16}$$

Analogously, the rainfall rate (eq. [3.35]) can be expressed by

$$R = (7.12 \times 10^{-3}) N_w f(\mu) \Gamma(4.67 + \mu) \frac{D_m^{4.67}}{(4 + \mu)^{4.67+\mu}}. \tag{9.17}$$

It follows that the ratio Z/R is exclusively a function of D_m and μ:

$$\frac{Z}{R} = \frac{1}{(7.12 \times 10^{-3})} \frac{\Gamma(7 + \mu)}{\Gamma(4.67 + \mu)} \left(\frac{D_m}{4 + \mu}\right)^{2.33}. \tag{9.18}$$

Without loss of generality, the integration limits for Z and R have been assumed as $(0, \infty)$ in the previous equations. Corresponding expressions for the case of a finite upper limit (D_{max}) can be obtained considering the tables of the incomplete gamma function.

Using either simulated or observed DSDs, it can be easily demonstrated that although the ratio Z/R is a function of both D_m and μ, D_m plays the major role,

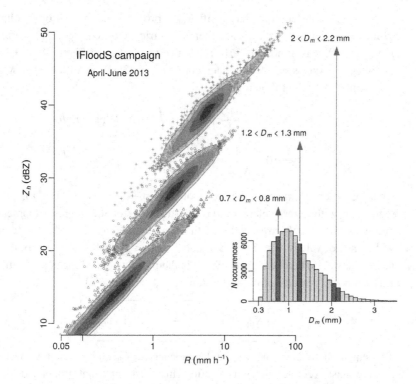

Figure 9.4 $Z–R$ scatter plots with smoothed observation densities (gray levels) for the disdrometer observations collected during the IFloodS campaign. The horizontal reflectivity is calculated from the Rayleigh–Gans approximation (eq. [9.4]). The three different plots correspond to classes of the mass-weighted mean diameter D_m centered on 0.75, 1.25, and 2.1 mm, as highlighted in dark gray in the histogram showing the overall distribution of the estimated D_m during the campaign.

whereas the dependence on μ is weaker [268]. To further illustrate the stratification based on D_m, Figure 9.4 shows scatterplots of Z versus R for three classes of the mass-weighted mean diameter centered on 0.75, 1.25, and 2.1 mm. The reflectivity and rainfall rate are calculated from a large data set of disdrometer observations and using the Rayleigh–Gans approximation (eq. [9.4]) for the integration of Z_h. The histogram in the lower-right corner of Figure 9.4 shows the distribution of the mass-weighted mean diameter during approximately 2 months of measurements, with the three classes of D_m highlighted in dark gray. The scatterplot shows a tight relation with a nearly unit slope in the log-log plane for the different D_m classes, matching the prediction from eq. (9.18). This implies that an estimate of the drops' mass-weighted mean diameter is needed to evaluate the rainfall rate accurately from the radar measurement of reflectivity. As indicated in Figure 9.1, this estimate may be obtained from the differential reflectivity observations. This is the basis for the first multiparameter rainfall estimation algorithm of the form $R(Z_h, Z_{dr})$, proposed in the 1970s and described later in Section 9.5.2.

The specific differential phase shift K_{dp} is proportional to the drops' volume and is a function of the drops' shapes and their dielectric factor (eq. [9.6]). The term $\mathrm{Re}(p_h - p_v)$ in eq. (9.6) is almost linearly related to $(1 - r)$ from Figure 4.16. Considering the integral expression of the water content (eq. [3.4]), it follows that K_{dp} can be approximated by the following:

$$K_{dp} \approx 10^{-6} \left(\frac{180}{\lambda}\right) C \left(\frac{\pi}{6}\right) \int D^3 N(D)(1 - r)dD$$

$$= 10^{-3} \left(\frac{180}{\lambda}\right) C \left(\frac{W}{\rho_w}\right) \left[1 - \frac{\int r D^3 N(D)dD}{\int D^3 N(D)dD}\right],$$

(9.19)

where C is an estimate of the slope of $\mathrm{Re}(p_h - p_v)$ versus $(1 - r)$. From linear regression with a zero intercept and optimization for the range of diameters 0.5–5 mm, $C \approx 3.5$, mostly irrespective of the rain temperature and radar frequency over a wide range of values ($0 < T < 30°C$ and $3 < f < 36$ GHz). The ratio of the integrals inside the square brackets in eq. (9.19) is the mass-weighted mean axis ratio ($\bar{r}_m$), so K_{dp} can be expressed as follows (assuming $\rho_w = 1 \, \mathrm{g\,cm}^{-3}$):

$$\boxed{K_{dp} \approx 10^{-3} \left(\frac{180}{\lambda}\right) CW(1 - \bar{r}_m).}$$

(9.20)

This equation concisely shows the dependence of K_{dp} on both the total water content and the mass-weighted mean axis ratio. This is an important property of the specific differential phase, different from the differential reflectivity. In fact, Z_{dr} is independent of the drops' concentration and is tightly related to the reflectivity-weighted axis ratio (eq. [9.14]). Few large drops within a DSD dominated by small, nearly spherical droplets may lead to a high Z_{dr} as a consequence of the sixth-moment weighting. K_{dp} would hardly change from $0 \, °\mathrm{km}^{-1}$ instead, as long as the total water content remains small (refer to Figure 8.10). On the other hand, a narrow DSD with a very large number of drops in the 0.5- to 1.0-mm range would imply a significant K_{dp} (and water content), with Z_{dr} only slightly above 0 dB.

If the linear fit of Pruppacher and Beard [37] drop shape ratio ($r = 1.03 - 0.062D$) is used in eq. (9.20), K_{dp} can also be expressed as a function of the water content and the mass-weighted mean diameter (noting that r_m is related to D_m via the drop shape ratio):

$$\boxed{K_{dp} \approx 10^{-3} \left(\frac{180}{\lambda}\right) CW \, 0.062 \, (D_m - 0.5); \quad D_m > 0.5 \text{ mm.}}$$

(9.21)

Figure 9.5 shows the ratio K_{dp}/W versus D_m from scattering simulations at different frequencies, stressing the effect of the drop's shape on the specific differential phase shift. The linearity of the ratio K_{dp}/W is remarkable, especially for the S and X bands. For frequencies higher than 10 GHz, the linear approximation ceases to be valid for the range of typical D_m due to the effects of Mie scattering (the drop sizes becomes large with respect to the wavelength).

A similar analogy can be noticed between eq. (9.21), where the ratio of K_{dp} to W results in a linear function of the mass-weighted mean diameter D_m, and eq. (9.18),

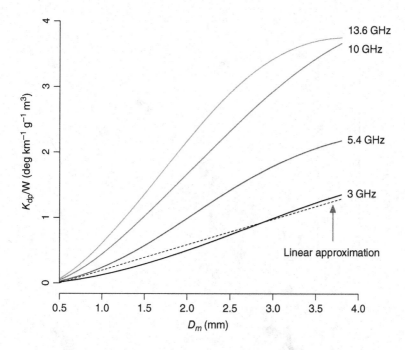

Figure 9.5 Scattering simulations of the ratio of K_{dp} to W as a function of the mass-weighted mean diameter, with scattering calculations based on exponential DSDs. The normalization by W illustrates the effect of the mean drop shape on K_{dp}. The dashed line indicates the linear approximation in eq. (9.21) applied at the frequency of 3 GHz.

which shows the ratio of the reflectivity to the rainfall rate is linearly proportional to $D_m^{2.33}$. This analogy can be appreciated by looking at Figure 9.6, which illustrates the stratification of K_{dp} versus W into ranges of D_m, for the same observations as in Figure 9.4.

A remarkably tight relation is observed between the two moments calculated from real observations. The K_{dp}–W scatter plots show a lower dispersion in comparison with the Z–R plots in Figure 9.4, especially for the higher D_m. This different behavior can be understood by again considering eqs. (9.18) and (9.21). Whereas Z/R, in addition to D_m, is also dependent on the shape parameter μ, the ratio K_{dp}/W only depends on the mass-weighted mean diameter D_m. The lower dispersion in Figure 9.6 thus shows a fundamental property of the K_{dp}–W relation, that is, the insensitivity to the shape of the DSD.

Equations (9.12)–(9.21) illustrate some relevant relations between the radar variables and the properties of the DSD. Now, the relation between the three radar variables Z_h, Z_{dr}, and K_{dp} is considered in greater detail. Following the hypothesis of an exponential DSD, eq. (9.10) showed that it would be possible to uniquely relate any measurement of the DSD, including K_{dp} to the values of Z_h and Z_{dr}. In practice, even if K_{dp} can not be uniquely specified from Z_h and Z_{dr} for an arbitrary DSD, the three radar variables (Z_h, Z_{dr}, and K_{dp}) are basically constrained over a limited portion of

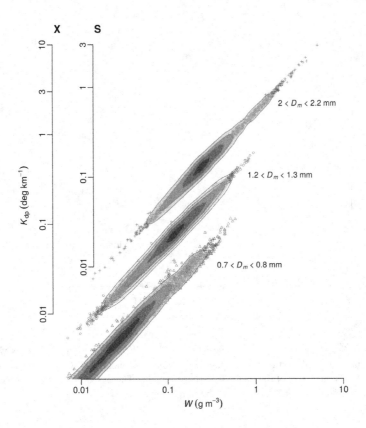

Figure 9.6 K_{dp}–W log-log plots with smoothed observation densities (gray levels) considering the same data set of disdrometer observations as in Figure 9.4. K_{dp} is calculated according the Rayleigh–Gans approximation (eq. [9.6]) from the DSD observations, with y-axes reporting the values corresponding to the S band and X band. The three different groups correspond to mass-weighted mean diameter D_m classes centered on 0.75, 1.25, and 2.1 mm.

the parameter space. This implies that it is, in principle, possible to retrieve one radar variable from measurements of the other two with a certain level of accuracy. This is the fundamental basis of the self-consistency principle introduced by Scarchilli et al. [269]. The inherent consistency among the radar variables is fundamental for several applications, including rainfall estimation, attenuation correction, hydrometeor classification, and radar calibration. Assuming a gamma DSD with varying parameters, the resulting radar moments show a distribution in the three-dimensional parameter space (Z_h, Z_{dr}, K_{dp}) like the one depicted in Figure 9.7.

The specific differential phase can then be expressed as a function of reflectivity and differential reflectivity:

$$K_{dp} = c\, Z_h^b\, Z_{dr}^a.$$

(9.22)

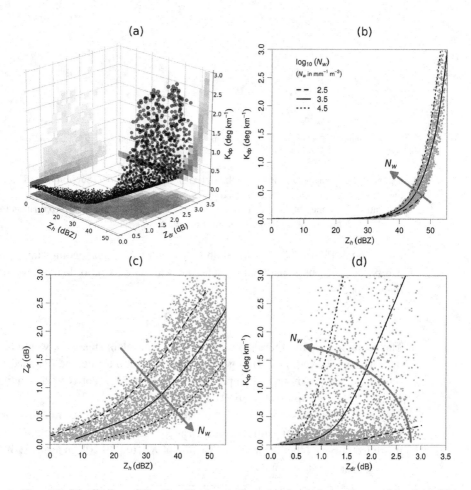

Figure 9.7 Distribution of Z_h, Z_{dr}, K_{dp} in rain from simulated observations at S-band (a). The corresponding 2D scatterplots between the pairs of parameters K_{dp}-Z_h (b), Z_{dr}-Z_h (c) and K_{dp}-Z_{dr} (d) illustrate the typical functional relations in rain.

The values of the three coefficients are typically assessed through nonlinear regression analysis over simulated DSDs representative of the rainfall's natural variability . For example, the values for S band and C band reported in [269] are respectively: $c = 1.05\,10^{-4}$, $b = 0.96$, $a = -2.6$ and $c = 1.46\,10^{-4}$, $b = 0.98$, $a = -2.0$. The comparison between the observed K_{dp} and the corresponding computed values using eq. (9.22) would then allow for an assessment of the calibration of the radar (Section 7.4), provided that the differential reflectivity is unbiased.

9.2   A Detailed Look at Rainfall Physics: DSD Retrievals

A drop-shape model establishes the relation between the drop diameter and its axis ratio (Section 3.3). The shape of raindrops according to the Beard–Chuang model are

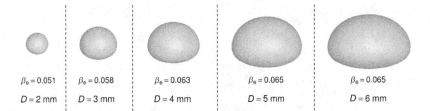

$\beta_e = 0.051$ $\beta_e = 0.058$ $\beta_e = 0.063$ $\beta_e = 0.065$ $\beta_e = 0.065$

$D = 2$ mm $D = 3$ mm $D = 4$ mm $D = 5$ mm $D = 6$ mm

Figure 9.8 Drop shapes according to the Beard and Chuang model (Section 3.3). β_e indicates the equivalent slope that would be used in the linear model ($b/a = 1.03 - \beta D$) to get the same axis ratio as the one obtained from the oblate spheroid polynomial fit approximation in eq. (3.31), depicted by the dashed lines in Figure 3.8.

illustrated in Figure 9.8 for several drop diameters. A corresponding simplified shape can be obtained by considering a spheroid with an axis ratio defined by the simple linear model:

$$r = \frac{b}{a} = 1.03 - \beta D. \tag{9.23}$$

The coefficient β_e in Figure 9.8 represents the equivalent slope that would be used in the linear model (eq. [9.23]) to get the same axis ratio as the one obtained from the more complex polynomial fit approximation (eq. [3.31]). For typical raindrop sizes, β_e varies between approximately 0.05 and 0.07.

A useful application of the self-consistency principle is the estimation of the drop-shape model directly from the radar measurements [270]. Considering eq. (9.16) for the reflectivity and the linear approximation of K_{dp} (eq. [9.20]) in terms of W and r_m, the ratio K_{dp}/Z_h is given by

$$\boxed{\frac{K_{dp}}{Z_h} \approx g(\mu)\frac{(1 - \bar{r}_m)}{D_m^3},} \tag{9.24}$$

where $g(\mu)$ is a weak function of of the DSD shape parameter μ, similar to eq. (9.18). Note that N_w canceled out in eq. (9.24) using eq. (3.27) to express W as a function of D_m and N_w. Given the tight relation between Z_{dr} and D_m (Fig. 9.1), it follows that from a theoretical perspective (i.e., not considering measurement errors), the variability in the three-dimensional parameter space (Z_h, Z_{dr}, K_{dp}) can be related to the actual shape of the raindrops represented by $\bar{r}_m$. In fact, by plotting the ratio K_{dp}/Z_h as a function of Z_{dr} for different drop-shape relations, the points in this 2D space appear tightly aligned along distinct curves, depending on the underlying drop shape model. Figure 9.9 considers the radar variables calculated from observed DSDs, with gray bands encompassing 95 percent of the observations, for six drop-shape models (see Section 3.3). The relevance of this particular formulation of the self-consistency principle is that it distinctly shows how it is theoretically possible to estimate the dominant drop-shape model from a set of real radar observations. This is also the basis for the DSD retrieval method proposed by Gorgucci et al. [271], where the empirical relations between the radar variables and the DSD parameters are parameterized in terms of β.

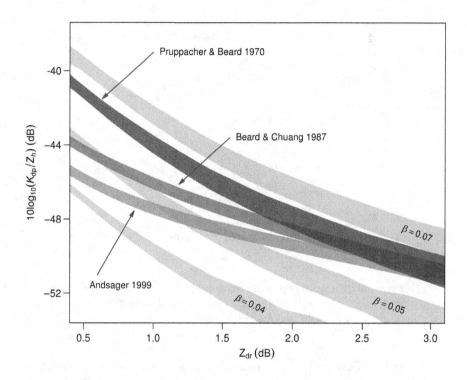

Figure 9.9 Plot of K_{dp}/Z_h ratio as a function of Z_{dr} for different drop-shape models, using the same disdrometer observations as in Figure 9.4. The radar moments are calculated according to the Rayleigh–Gans approximation from $\sim 10^5$ 1-minute measured DSDs, with the extent of each gray band including 95 percent of the observations. In addition to the drop-shape models of Pruppacher and Beard [37], Beard and Chuang [38], and Andsager [42], the results adopting a generic linear model of the form $r = 1.03 - \beta D$ are plotted for β values 0.04, 0.05, and 0.07 mm^{-1}.

Gorgucci [272] proposed a technique to estimate the mean shape of raindrops from the triplet of radar observations Z_h, Z_{dr}, and K_{dp}. According to the Pruppacher and Beard [37] linear fit to wind-tunnel data (eq. [3.30]), $\beta = 0.062$ in eq. (9.23). The 0.062 slope also provides a good approximation to the equilibrium shape model [38], as shown in Figure 9.8. However, raindrops falling in the real atmosphere may show axis ratios that deviate from equilibrium as a result of canting and oscillations. Therefore, an "effective" β can be estimated to account for the characteristics of different rain regimes. Using simulated gamma DSDs over a wide range of the three radar variables, the following relation is obtained for S-band radar observations [272]:

$$\beta = 2.08 \, Z_h^{-0.365} \, Z_{dr}^{\ 0.965} \, K_{dp}^{0.38}. \tag{9.25}$$

This relation can be used to retrieve the mean drop-shape model from the triplet of radar observations. As an example, Figure 9.10 shows the distribution of β values

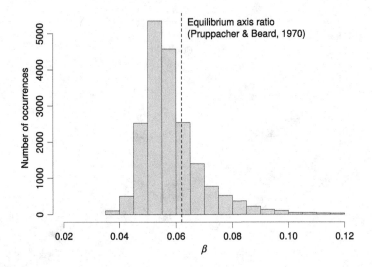

Figure 9.10 Histogram of β values computed from S-band radar observations of the squall line represented in Figure 9.15. The values are calculated in the precipitation regions where $Z_h > 35$ dBZ, $Z_{dr} > 0.2$ dB, and $K_{dp} > 0.3\ °\mathrm{km}^{-1}$.

retrieved using eq. (9.25) from S-band observations through a squall line. Once the drop shape is evaluated using eq (9.25), Gorgucci et al. [271] went on to derive the parameters of the gamma DSD in terms of the triplet of dual polarization radar observations (Z_h, Z_{dr}, K_{dp}) and the value of β.

Because of practical difficulties in dealing with all three radar variables and their relative observation errors, in addition to possible deviations from theoretical expectations, it is common to adopt a simpler approach to estimate the parameters of the DSD. Recalling that $\Lambda D_m = 4 + \mu$ (eq. [3.14]), for the case of an exponential DSD ($\mu = 0$), the mass-weighted drop diameter D_m could be readily estimated from Z_{dr} because the intercept parameter N_0 cancels out in eq. (9.10). The same applies for the median volume diameter, for which $\Lambda D_0 = 3.67 + \mu$ (eq. [3.13]). For more generic DSDs with varying μ, scattering simulations are used to derive polynomial expressions for the retrieval of both the median volume diameter and the mass-weighted mean diameter from Z_{dr} (in dB). A commonly used approximation for D_0 at the S band was derived from retrievals over a large DSD data set collected by the National Aeronautics and Space Administration (NASA), including disdrometer data from Huntsville, Alabama, and Oklahoma during the Mid-Latitude Continental Convective Clouds Experiment (MC3E) field campaign [273]:

$$D_0\ (\mathrm{mm}) = 0.802 + 0.790\,Z_{dr}(\mathrm{dB}) - 0.084\,Z_{dr}^2(\mathrm{dB}) + 0.022\,Z_{dr}^3(\mathrm{dB}).\qquad (9.26)$$

Considering the IFloodS DSD data set, which includes both convective and stratiform precipitation, it is possible to derive empirical relations between D_0 and Z_{dr} for

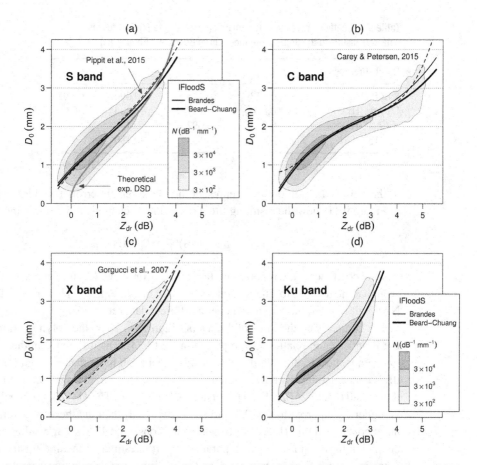

Figure 9.11 Density plots of D_0 vs. Z_{dr} for S (a), C (b), X (c) and Ku band (d) from over 40000 disdrometer observations during the IFloodS campaign. The scattering simulations to retrieve Z_{dr} at different frequency bands are performed assuming the Beard and Chuang axis ratio (eq. [3.31]) and a Gaussian distribution of the canting angles with zero mean and $7°$ standard deviation. The thick solid line represents the third-order polynomial fit, whose coefficients are reported in Table 9.1, and the thin solid line represents the corresponding fit using the Brandes et al. [43] approximation for the drop-shape model (eq. [3.33]). The dashed lines indicate relevant models from the literature: Pippit et al. for the S band [273] (third-order polynomial), Carey and Petersen for the C band [274] (fourth-order polynomial, with $D_{max} = 3D_0$), and Gorgucci et al. for the X band [275] (power-law relation, with $\beta = 0.06$).

different radar frequency bands. Figure 9.11 shows the density plots of D_0 versus Z_{dr} for the IFloodS data set at the S, C, X, and Ku bands. The differential reflectivity is calculated using scattering simulations from the DSD parameters estimated with the method of moments [268]. Assumptions for the simulations include the Beard and Chuang approximation (eq. [3.31]) for the axis ratio and a Gaussian distribution for the canting angle with zero mean and a standard deviation of $7°$. Also, following the same approach as in Carey and Petersen [274], a Gaussian white noise has been added

Table 9.1 Coefficients of the polynomial fit (eq. [9.27]) to $D_0(Z_{dr})$ for the IFloodS data set, assuming the drop-shape model from Beard and Chuang (eq. [3.31])

Band (Frequency)	a	b	c	d
S (2.8 GHz)	0.853	0.725	−0.096	0.024
C (5.6 GHz)	0.820	0.877	−0.204	0.024
X (9.4 GHz)	0.853	0.692	−0.193	0.047
Ku (13.6 GHz)	0.839	0.655	−0.180	0.067

to the retrieved Z_{dr} in order to represent the measurement error. The thick solid lines in Figure 9.11 show the resulting fitted polynomial relations of the following form:

$$D_0 \text{ (mm)} = a + b\, Z_{dr}(\text{dB}) + c\, Z_{dr}^2(\text{dB}) + d\, Z_{dr}^3(\text{dB}), \qquad (9.27)$$

and the coefficients are reported in Table 9.1 for the S, C, X, and Ku bands. It can be seen that the relation for the S band (thick solid line in Fig. 9.11) is very similar to the NASA fit (dashed line) obtained over a different data set.

For comparison, the thin solid lines in Figure 9.11 show the polynomial fits using the Brandes et al. relation [43] reported in eq. (3.33). The difference in the estimated D_0 with respect to the Beard and Chuang model is no more than 0.1 mm for $Z_{dr} < 3$ dB. Other relations from the literature have been added to the plot (dashed lines) for the S band (eq. [9.26]), C band [274], and X band [275]. The D_0–Z_{dr} fitted polynomial is higher (by approximately 0.3 mm for $Z_{dr} < 4$ dB) than the theoretical relation for an exponential DSD (Fig. 9.1) because most observed DSDs during this particular campaign have a positive shape parameter μ (the estimated median for the IFloodS data set is 3.5). For DSDs with $\mu > 0$, the tail of the distribution decreases faster. Thus, for the same characteristic diameter (either D_m or D_0), Z_{dr} is lower. On the other hand, for large median volume diameter (and Z_{dr}), μ can be negative, implying that the theoretical relation for the exponential distribution overestimates D_0 and D_m.

The C-band relation in Figure 9.11 shows a notable difference with respect to the S-band results (at the S band, the scatterers basically follow the Rayleigh–Gans theory). In fact, when D_0 reaches $\sim$2 mm at the C band, the right tail of the raindrops' distribution starts to be significantly affected by the resonance peak for drops with diameters around 6 mm (Fig. 4.11), and Z_{dr} increases much faster.

The mass-weighted mean diameter D_m can also be modeled using a polynomial relation like eq. (9.27):

$$D_m \text{ (mm)} = a + b\, Z_{dr}(\text{dB}) + c\, Z_{dr}^2(\text{dB}) + d\, Z_{dr}^3(\text{dB}), \qquad (9.28)$$

with the proper coefficients for the IFloodS data set reported in Table 9.2. Relying on the estimate of D_0 or D_m, and assuming a fixed value for the shape parameter μ, the N_w parameter can be easily calculated using the relation with the moments of the DSD in Section 3.2.3. In particular, from eq. (9.16), N_w can be expressed in terms of D_m as follows:

Table 9.2 Like Table 9.1, but for the polynomial fit to $D_m(Z_{dr})$

Band (Frequency)	a	b	c	d
S (2.8 GHz)	0.871	0.752	−0.058	0.015
C (5.6 GHz)	0.831	0.943	−0.205	0.023
X (9.4 GHz)	0.871	0.722	−0.178	0.045
Ku (13.6 GHz)	0.855	0.679	−0.158	0.065

$$N_w = \frac{Z_h}{D_m{}^7} \frac{(4+\mu)^{7+\mu}}{f(\mu)\Gamma(7+\mu)}. \tag{9.29}$$

In practice, instead of assuming a prescribed value of μ in eq. (9.29), empirical relations can be obtained through statistical analysis of observed DSDs, and specified in terms of either D_m or D_0. For example, retrievals on the same NASA data set used to derive eq. (9.26) results in the following relation for the S band:

$$N_w = 20.96\, Z_h D_0{}^{-7.7}. \tag{9.30}$$

9.3 Attenuation in Precipitation

For light rainfall rates, Z–R relations and dual polarization techniques typically produce similar results for observations at the S and C bands. However, dual polarization techniques have demonstrated clear advantages in moderate to heavy rain rates, particularly those where the observations may include hail contamination (e.g., convective systems). When considering rainfall-rate retrievals using radars operating at C-band and higher frequencies, estimates using the Z–R relation are subject to further biases due to attenuation. It is necessary to correct for attenuation prior to estimating the rainfall rate to minimize any biasing effects. This section reviews the fundamental notions of attenuation, whereas the focus of Section 9.4 is on the techniques to correct for attenuation using dual polarization measurements.

The microwave frequencies used by weather radar suffer little attenuation by atmospheric gases. This small portion of loss is generally accounted for in the radar equation through the addition of a constant specific attenuation term. The radar equation to compute the reflectivity from the received power P_r (dBW) and range r (km) (eq. [2.39c]) can then be updated to

$$10\log_{10}(Z) = P_r + 20\log_{10}(r) - C_{dB} + C_{gas}r, \tag{9.31}$$

where C_{gas} represents the two-way atmospheric gas specific attenuation (dB km^{-1}), and $C_{dB} = 10\log_{10}(C)$ is the radar constant in dB (eq. [2.39d]). The atmospheric gas attenuation term has little impact for the S band through the X band (typical values are 0.016 dB km^{-1} at the S band, 0.019 dB km^{-1} at the C band, and 0.024 dB km^{-1} at the X band) and is primarily caused by oxygen molecules. However, for frequencies higher than 10 GHz, the increasing absorption caused by the water vapor may produce an order-of-magnitude-higher attenuation ($\sim$0.1 dB km^{-1}), leading up to 3–5-dB of total attenuation at the 30–40-km range for Ka-band systems.

The attenuation introduced by precipitation is much higher than the attenuation from atmospheric gas. The path-integrated attenuation depends on the radar wavelength, the composition of the precipitation particles (dimension, shape, orientation, phase), and their distribution along the radar's beam between the antenna and the target resolution volume. As such, the precipitation's total attenuation cannot be simply accounted for with a constant term in the radar equation and instead, a range-dependent estimate and integration along the beam's path must be evaluated.

9.3.1   Absorption and Scattering

The total power loss suffered by an incident wave is due to both scattering and absorption by the particles along the propagation path, as discussed in Chapter 4. Recall that the extinction cross-section represents the total power loss as

$$\sigma_e = \sigma_a + \sigma_s, \tag{9.32}$$

where σ_a is the absorption cross-section, and σ_s is the total scattering cross-section. The attenuation depends on the radar frequency and the phase of the particle through its complex relative permittivity. The relative permittivity of water is a function of wavelength and temperature (Fig. 4.1). Specifically, the imaginary component ϵ_r'' is inversely proportional to temperature for typical weather radar frequencies, causing higher attenuation at lower temperatures. The extinction cross-section as a function of drop diameter and temperature is shown in Figure 9.12 for several frequency bands.

In the Rayleigh scattering regime, all frequencies between the S and Ka bands show a monotonically increasing extinction cross-section with decreasing temperature. Such increase can reach almost 6 dB when the temperature drops from 30 to $-15°C$ (supercooled drops). At the W band, the extinction cross-section also shows similar behavior, but only for temperatures well above freezing. For temperatures lower than $\sim$6°C, the trend is reversed, with the extinction cross-section decreasing with colder drops. In the Mie and optical scattering regimes, the extinction cross-section, σ_e, is dominated by the total scattering cross-section, σ_s, rather than the absorption cross-section, σ_a. As the particle size enters the Mie scattering regime, the extinction varies less with temperature because the total scattering cross-section is only weakly dependent on temperature.

The attenuated (measured) received power P_m can be expressed in terms of the unattenuated (intrinsic) power P_i and the attenuation coefficient k_L, using the Beer–Lambert law:

$$P_m = P_i \exp\left(-2\int_0^r k_L(s)ds\right), \tag{9.33}$$

where the factor 2 takes into account the two-way propagation of the radar's wave. The attenuation coefficient has units of inverse length (m^{-1}) and is the result of

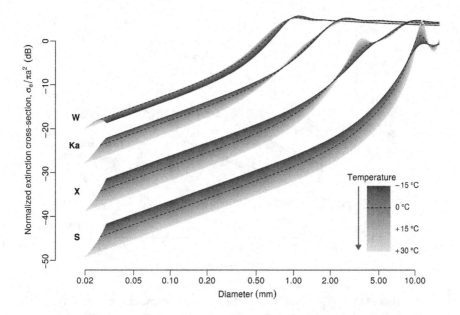

Figure 9.12 Normalized extinction cross-section for spherical drops at different frequency bands as a function of diameter and temperature (gray shading). Note that water droplets may also be found in clouds at subfreezing temperatures, down to $-15°C$ (Section 3.5).

the weighted sum of the extinction cross-section of the particles illuminated by the radar beam:

$$k_L = \int \sigma_e(D) N(D) dD, \qquad (9.34)$$

where $\sigma_e(D)$ is in meters squared and $N(D) dD$ is the number of particles per cubic meter. It is generally more convenient to work using a logarithmic scale for attenuation (i.e., the attenuation is in decibels), the two-way path-integrated attenuation at a given range r is written as follows:

$$\text{PIA}(r) = 10\log_{10}\frac{P_m}{P_i} = -2\int_0^r A(s)ds, \qquad (9.35)$$

where the specific attenuation, in decibels per kilometer (dB km^{-1}), is defined as

$$\begin{aligned} A &= (10\log_{10}e) \times 10^3 \, k_L \\ &= 4.343 \times 10^3 \, k_L. \end{aligned} \qquad (9.36)$$

From eq. (9.34), the specific attenuation is related to the extinction cross-section of the DSD following

$$\boxed{A = 4.343 \times 10^3 \int \sigma_e(D) N(D) dD.} \qquad (9.37)$$

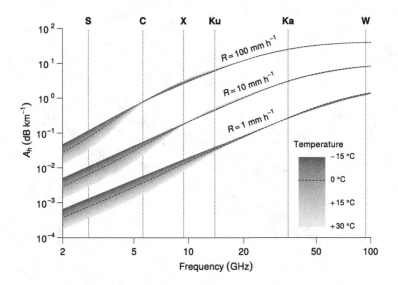

Figure 9.13 Specific attenuation of the horizontal polarization is plotted as a function of the radar frequency and temperature (gray shading). The values are calculated using scattering simulations based on an exponential DSD ($N_0 = 8000$ m^{-3}mm^{-1}) and increasing rainfall rates (1, 10, 100 mm h^{-1}), assuming the Beard–Chuang axis ratio approximation for the shape of the raindrops (eq. [3.31]). Note the attenuation increases with colder temperatures for the lower-frequency range. This suggests how super cooled drops and partially melted large ice particles can cause substantial attenuation, even for frequencies that are traditionally considered to be non-attenuating.

Recall that in Section 4.1.6, the path-integrated attenuation was derived from the attenuation constant α that acts on the wave's electric field. Here, the attenuation coefficient k_L acts on the wave's power density. Both have the same effect on the radar's signal.

Precipitation particles may not be spherical, and therefore, the attenuation may vary depending on the polarization. For rain, given the oblate shape of the drops, the specific attenuation for the horizontal polarization A_h is higher than the corresponding attenuation of the vertical polarization A_v. Figure 9.13 shows A_h as a function of frequency relevant to weather radar. The sensitivity to temperature is also illustrated (gray shading). The difference between A_h and A_v results in a positive specific differential attenuation in rain ($A_{dp} = A_h - A_v$), which is represented in Figure 9.14.

The extinction cross-section is proportional to the volume of the particles (eq. [4.38]), implying that specific attenuation in rain is directly related to the total liquid water mass within the resolution volume, including both the small spherical cloud droplets and the larger raindrops (hence the possible use of specific attenuation for rainfall estimation, as discussed in the next section). It is worth noting that because of the isotropy of cloud and drizzle droplets, these particles do not contribute to differential attenuation.

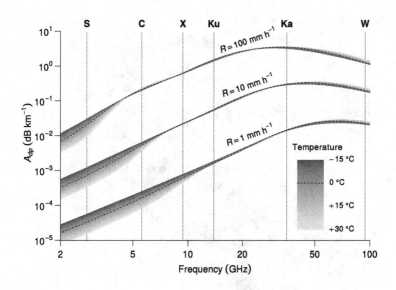

Figure 9.14 Same as in Figure 9.13, but for the specific differential attenuation. Note that the y-axis is shifted to lower values by one order of magnitude.

9.3.2 Impact of Attenuation

Attenuation clearly has an impact on all quantitative applications at frequencies above the S band (sometimes even at the S band over long squall lines). Attenuation may also affect the qualitative interpretation of the radar return, resulting in an apparent elongation of the rain structures in the cross-beam direction. Figure 9.15 shows S-band reflectivity observations of a squall line propagating from the northwest, with the stronger precipitation cores at the leading front and a trailing stratiform region behind. The apparent spatial extent of the rain cells and the location of the storm's core may change depending on the radar wavelength and the characteristics of the precipitation. At the C band, the echo is generally unaltered in the light stratiform region of precipitation systems, whereas attenuation may be significant for severe convection.

Considering S-band observations of reflectivity and differential reflectivity (which are not much affected by attenuation), it is possible to retrieve the DSD parameters as described in Section 9.2. From the retrieved DSD, all radar parameters (including variables which are not observed like attenuation) can be simulated at any operating frequency. To evaluate the impact of attenuation, the observations in Figure 9.15 along the radial at 136° azimuth are used to estimate the DSD parameters D_0 and N_w shown in Figure 9.16. N_w ranges between 10 and 10^5 mm^{-1} m^{-3}, whereas D_0 reaches values in excess of 2.5 mm on the leading edge of the squall line, at around the 70-km range.

The corresponding simulated range profiles of the radar reflectivity at the S band and the C band across the squall line are presented in Figure 9.17. The value of intrinsic (unattenuated) reflectivity (gray solid lines) is similar for the two frequencies, although reflectivity at the C band is in general slightly lower than the

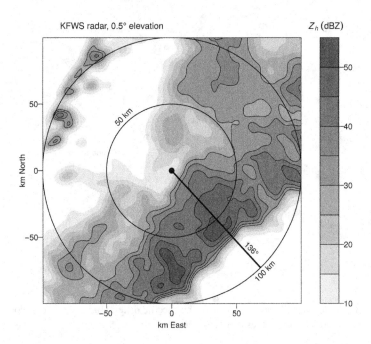

Figure 9.15 Plan position indicator (PPI) of reflectivity at the S band observed on May 12, 2014 at 22:20 UTC by the Dallas–Fort Worth NEXRAD radar (contour lines are plotted every 5 dB, starting at 30 dBZ). A squall line has just passed over the radar and moves southeastward.

corresponding reflectivity at the S band. In fact, referring to the backscattering cross-section of water particles in Figure 4.6, particles with diameter $D < 6$ mm at the C band (corresponding to size parameter $k_0 a < 0.33$) show a downward deviation from the Rayleigh approximation. One notable exception from this trend occurs around the range of 70 km, where Mie scattering resonance effects are induced by the presence of large drops (Fig. 9.16). Here, the C-band horizontal polarization's intrinsic reflectivity is approximately 1.5 dB higher than the corresponding S-band value. The simulated observations across the squall line (thick solid lines in Fig. 9.17) show the increasing impact of attenuation through intense rainfall. The PIA reaches approximately 3 dB at the S band and 17 dB at the C band. This example shows that when the radar beam propagates through extended regions of heavy rain, attenuation also needs to be accounted for at the S band to ensure accurate reflectivity estimates.

In Figure 9.18, the simulated Φ_{dp} profiles along the same radial as in Figure 9.15 demonstrate how path-integrated attenuation is related to the differential phase shift. The amount of the differential phase shift scales with frequency according to eqs. (4.68) and (4.69). It is important to remember that a weather radar does not measure Φ_{dp} directly. The received differential phase shift signal is given by $\Psi_{dp} = \Phi_{dp} + \delta_{co}$, where δ_{co} represents the phase difference between the horizontal and vertical polarizations due to backscattering by particles within the radar resolution

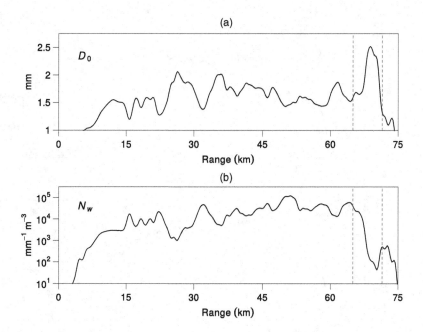

Figure 9.16 Range profiles of retrieved DSD parameters. The median volume diameter D_0 (a) and normalized intercept parameter N_w (b) are estimated along the radial at $136°$ azimuth in Figure 9.15. The parameters are estimated from the NEXRAD observations of reflectivity (Fig. 9.15) and differential reflectivity (not shown) using eqs. (9.28) and (9.30). The dashed vertical lines mark the range interval corresponding to the inset in Figure 9.18.

volume (Section 4.7.2). Although the contribution of δ_{co} in rain is generally negligible at the S band, for higher frequencies, the impact of backscatter phase shift can be a significant contribution to Ψ_{dp} observations and therefore affect the estimation of K_{dp}. This occurs in the Mie scattering regime when nonspherical particles are large relative to the radar wavelength. The insets in the middle panel of Figure 9.18 show the contribution of δ_{co} to Ψ_{dp} at the C band and the X band, on the far end of the rain profile corresponding to the leading edge of the squall line. This portion of the storm is often characterized by a relative abundance of big raindrops (see, e.g., Section 8.2.1), as indicated by the large median volume diameter in the range segment marked by the dashed vertical lines in Figure 9.16. At the C band, the resonance scattering regime starts with the appearance of raindrops with a diameter of $\sim$6 mm in the radar resolution volume. Given the fact that most observed DSDs do not contain such large drops, the phase shift due to backscattering is also often negligible at the C band, and Ψ_{dp} is expected to show a monotonic increase with range (not considering the measurement noise). When large drops ($D > 6$ mm) are present, this often occurs over small regions of intense rainfall; that is, the impact of backscattering affects several range gates. The consequence is that the contribution of δ_{co} typically appears as a bump in the differential phase profile, similar to the simulated effect illustrated in Figure 9.18. Compared with the C band, the resonance region for the X band starts at smaller drop sizes ($\sim$3 mm), so the effect of δ_{co} on the observed

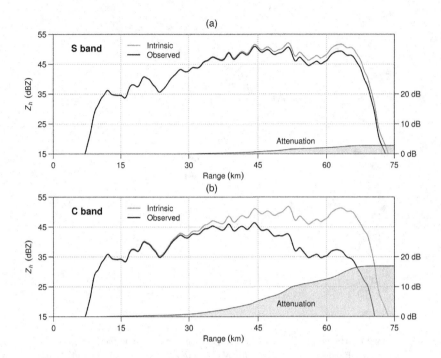

Figure 9.17 Range profiles of simulated reflectivity at the S band (a) and the C band (b), along the radial at 136° azimuth in Figure 9.15. The difference between the intrinsic (true) and observed reflectivity is the path integrated attenuation, which is shown as the shaded area. The attenuation curves correspond to the vertical axis on the right-hand side of the graphs.

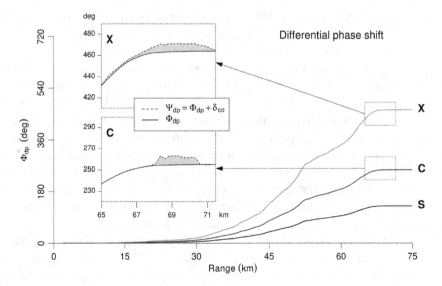

Figure 9.18 Range profiles of simulated Φ_{dp} along the radial at 136° azimuth in Figure 9.15 for different radar frequencies. The contribution of the backscatter differential phase δ_{co} to the total differential phase shift may show local peaks of several degrees at the C band and the X band (δ_{co} is shown as the gray area on top of the Φ_{dp} line in the dashed rectangle), whereas it is generally negligible at the S band (less than 0.1° in the same region).

differential phase Ψ_{dp} is more ubiquitous. For this reason, and also considering the frequency scaling of Φ_{dp} (the component of the measured differential phase shift due to forward propagation through precipitation), the relative impact of δ_{co} on the measured differential phase shift is generally weaker, and similarly, the estimation of K_{dp} is only slightly affected at the X band.

The possible effects induced by the differential phase shift due to backscattering, in addition to measurements noise, need to be addressed in order to get a "clean" differential phase shift profile to be used in quantitative applications (see discussion in Section 6.3). These applications include rainfall estimation based on the specific differential phase (Sections 9.5.3 and 9.5.4) and attenuation correction (Section 9.4).

9.4   Attenuation Correction for Precipitation Radar

Attenuation is due to the path-integrated extinction cross-section σ_e (eq. [9.37]), which determines the power loss suffered by the incident wave as a result of absorption and scattering. Recall that for Rayleigh scattering, $\sigma_e \approx \sigma_a$. Before the advent of dual polarization, attenuation correction was especially challenging. In fact, early attempts relied on iterative procedures and the estimation of specific attenuation from the observed equivalent reflectivity using power-law relations [276]. However, these unconstrained approaches are known to be unstable because of errors in the reflectivity measurements and the large variability of the attenuation–reflectivity relations. Improvements were gathered from the adoption of constrained approaches, where the path-integrated attenuation at a given far range is constrained by some auxiliary information [277]. These methods led to the development of range-profiling algorithms for rainfall-rate retrieval from spaceborne radars, based on the constraint provided by the path-integrated attenuation given by the surface echo measurement.

The measurement of the differential phase shift in dual polarization systems allowed the development of new methods for attenuation correction. In the following sections, we examine three methods with an increasing degree of complexity: the simple one-parameter method based on the differential phase shift, Φ_{dp} -constrained methods (the rain-profiling algorithm, based on both Φ_{dp} and Z_h observations), and a modified version of the rain-profiling algorithm that relies on additional measurements of Z_{dr}. Several of factors contribute to the success of any operational implementation of attenuation correction. In particular, the presence of residual ground clutter, hail, or large raindrops that results in backscatter differential phase shift (δ_{co}) may alter the performance of the attenuation correction algorithm. This underlines the need for a prior hydrometeor classification for partitioning the radar observation space (Section 8.3).

9.4.1   Simple Φ_{dp} -Based Methods

Equation (9.6) showed that under the Rayleigh–Gans approximation, K_{dp} is proportional to the product of the third moment of the DSD as well as a function of the axis ratio. Considering an equilibrium drop-shape model, the term $\mathrm{Re}(p_h - p_v)$ in eq. (9.6) can be shown to be approximately linearly related to the drop diameter for $D > 0.5$ mm (recalling the discussion of eq. [9.20]). The specific differential phase is therefore related to the fourth moment of the DSD. This makes K_{dp} an excellent candidate for the estimation of both the rainfall rate (Section 9.5.3) and the specific attenuation. In fact, the relation between the specific attenuation at the horizontal polarization (A_h, in dB km^{-1}) and K_{dp} (in deg km^{-1}) can be expressed as

$$A_h = \alpha K_{dp}^b, \qquad (9.38)$$

where the exponent $b = 0.85$ at the S band, and the parameter α depends on the radar frequency, the temperature of the raindrops, and the drop-shape model. Attenuation is generally negligible at the S band, whereas it becomes increasingly relevant at higher frequencies. For frequencies between $\sim$5 and $\sim$20 GHz, the coefficient b becomes very close to unity, and the relation between specific attenuation and K_{dp} can be approximated by a linear equation ($b = 1$):

$$A_h = \alpha K_{dp}. \qquad (9.39)$$

Recalling the definition of path-integrated attenuation (eq. [9.35]), the intrinsic reflectivity can then be calculated from the observed reflectivity in logarithmic scale and the specific attenuation:

$$Z_h(r) = Z_h'(r) + 2 \int_0^r A_h(s) ds \text{ [dBZ].} \qquad (9.40)$$

Given the definition of K_{dp} ($\frac{1}{2} \frac{d\Phi_{dp}}{dr}$) and using eq. (9.39), it follows that the attenuation-corrected reflectivity at a given range r can be estimated from the observed attenuated reflectivity (Z_h') and the differential phase shift:

$$Z_h(r) = Z_h'(r) + \alpha[\Phi_{dp}(r) - \Phi_{dp}(0)] \text{ [dBZ],} \qquad (9.41)$$

where $\Phi_{dp}(0)$ is the so-called system differential phase (ϕ_{sys}). Similarly, the specific differential attenuation A_{dp} can be expressed in terms of K_{dp} :

$$A_{dp} = \beta K_{dp}^b, \qquad (9.42)$$

where again, as a first-order approximation, the exponent b can be set to unity, and the corrected differential reflectivity Z_{dr} may be retrieved from the observed values (Z_{dr}') in logarithmic scale (dB):

$$Z_{dr}(r) = Z_{dr}' + \beta[\Phi_{dp}(r) - \Phi_{dp}(0)] \text{ [dB].} \qquad (9.43)$$

Note that the multiplicative coefficient β in eqs. (9.42) and (9.43) should not be confused with the axis ratio slope in eq. (9.23). The relations in eqs. (9.39) and (9.42) generally provide robust estimators of attenuation and differential attenuation for practical implementations. However, in addition to the sensitivity to temperature, the value of the α and β parameters may vary significantly depending on the DSD.

Table 9.3 Typical values of the α and β coefficients for the specific attenuation and specific differential attenuation using the linear relations of the form $A_h = \alpha K_{dp}$ and $A_{dp} = \beta K_{dp}$

Band (Frequency)	α Value			β Value		
	0°C	10°C	20°C	0°C	10°C	20°C
S (2.8 GHz)	0.022	0.017	0.013	0.0054	0.0039	0.0030
C (5.6 GHz)	0.12	0.11	0.10	0.033	0.033	0.032
X (9.4 GHz)	0.30	0.32	0.34	0.057	0.057	0.058
Ku (13.6 GHz)	0.47	0.50	0.53	0.097	0.092	0.090

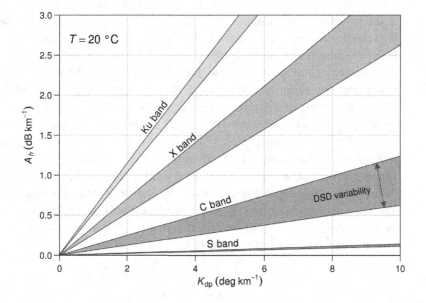

Figure 9.19 Relation between horizontal specific attenuation A_h and K_{dp} at different radar wavelengths and 20°C, as inferred from scattering simulations of observed DSDs from the IFloodS disdrometer data set. The width of each gray band encompasses $\sim$80% of the data. The least-squares fit values of α are summarized in Table 9.3 for different temperatures.

In particular, the value of α is related to the median volume diameter D_0, and the presence of big raindrops can lead to large attenuation.

Figure 9.19 shows the typical $A_h(K_{dp})$ relation for radar wavelengths between 3 GHz (S band) and 13.6 GHz (Ku band) at 20°C. The width of the shaded bands illustrates the degree of uncertainty on the slope α in relation to the variability of the DSD. The temperature dependence of α (eq. [9.39]) and β (eq. [9.42] with $b = 1$) can be observed in Table 9.3, where the typical value of the coefficients in the various frequency bands is reported for $T = 0°C$, $T = 10°C$, and $T = 20°C$.

In addition to Φ_{dp}, Z_{dr} can be used as a constraint to estimate the total differential attenuation [278]. The estimation of the path-integrated A_{dp} is done by measuring the magnitude of the negative Z_{dr} in low-intensity precipitation regions at far ranges where Z_{dr} is expected to be close to 0 dB. However, the estimation of the differential attenuation from Z_{dr} on the far end of the precipitation region is generally complicated for real-time applications and involves assumptions about the intrinsic (unattenuated) Z_{dr} (it is assumed to be $Z_{dr} = 0$ dB) and difficulties related to measurements in regions characterized by a low signal-to-noise ratio (SNR).

9.4.2 Φ_{dp}-Constrained Method (Rain-Profiling Algorithm)

The rain-profiling algorithms (RPAs) were initially developed for downward-looking spaceborne radars, to provide an estimate of the rainfall rate along the slanted beam in the presence of attenuation. These methods rely on an auxiliary constraint (e.g., the echo power from the surface) to normalize the measured echoes along the beam. In fact, a comparison of the backscattered echo from the surface in the presence and absence of precipitation allows for the estimation of the total path-integrated attenuation. By using the surface echo, the procedure compensates for the weakness of the simple unconstrained iterative schemes, which are unstable because of measurements errors, calibration offsets, and uncertainty in the DSD. The Hitschfeld and Bordan [276] solution provides the basis for the attenuation-correction algorithm, which relies on the attenuated reflectivity and differential phase shift observations. A similar concept can be implemented for ground-based radars by applying a constraint on the total path-integrated attenuation estimated from the differential phase [279]. The main assumption of the Φ_{dp}-constrained method is that the specific attenuation (in units of dB km^{-1}) can be expressed both as an analytical function of the linear reflectivity (using a power-law relation) and as a linear function of K_{dp}:

$$A_h(r) = a Z_h(r)^b \tag{9.44}$$

$$A_h(r) = \alpha K_{dp}(r). \tag{9.45}$$

Although eqs. (9.44) and (9.45) are written for the horizontal reflectivity, analogous relations can be derived for the vertical reflectivity, with a proper set of empirical coefficients a, b, and α. In eq. (9.44), the parameter a may show a significant variability for a given frequency, depending on the temperature and the normalized intercept parameter N_w of the rain DSD, whereas the exponent b is not sensitive to temperature and can be considered stable if N_w is nearly constant along the radar beam (this assumption will not be needed when using the generalization of the RPA in Section 9.4.3).

As noted in Section 9.4.1, the formulation in eq. (9.45) holds with excellent accuracy for the C, X, and Ku bands. Recalling the Beer–Lambert law (eq. [9.33]) and the relation (eq. [9.36]) between the attenuation coefficient k_L in linear units (m^{-1}) and the specific attenuation in decibels per kilometer (dB km^{-1}), the measured (attenuated) reflectivity can be specified in terms of the specific attenuation A_h and the intrinsic (unattenuated) reflectivity:

$$Z'_h(r) = Z_h(r) \exp\left(-0.46 \int_0^r A_h(s)ds\right). \qquad (9.46)$$

Note that Z'_h represents the observed (attenuated) reflectivity, and Z_h indicates the intrinsic (unattenuated) reflectivity.

After substituting eq. (9.44) into eq. (9.46) and taking the derivative with respect to range r, it can be shown after some manipulation (see [21] for a complete derivation) that the multiplicative coefficient a cancels out, and the solution for the specific attenuation is given by

$$A_h(r) = A_h(r_{max}) \frac{[Z'_h(r)]^b}{[Z'_h(r_{max})]^b + A_h(r_{max})I(r, r_{max})} \qquad (9.47)$$

with

$$I(r, r_{max}) = 0.46b \int_r^{r_{max}} [Z'_h(s)]^b ds, \qquad (9.48)$$

where r is any range closer than the reference range r_{max} ($r < r_{max}$). The specific attenuation at range r can then be calculated from the observed reflectivity profile $Z'_h(r)$ and the attenuation at r_{max}. The reference range is typically chosen at the end of a rain cell (Fig. 9.20).

The information from the measurement of Φ_{dp} is used to determine the value of $A_h(r_{max})$ in eq. (9.47). Using the expression in eq. (9.45) for the specific attenuation, the measurement of Φ_{dp} at the initial (r_{min}) and final (r_{max}) range of the rain cell is used to determine the value of $A_h(r_{max})$ in eq. (9.47). In fact, integrating A_h between r_{min} and r_{max} gives the following:

$$\int_{r_{min}}^{r_{max}} A_h(s)ds = \alpha \int_{r_{min}}^{r_{max}} K_{dp}(s)ds$$

$$= \frac{\alpha}{2}\left[\Phi_{dp}(r_{max}) - \Phi_{dp}(r_{min})\right] \qquad (9.49)$$

$$= \frac{\alpha}{2}\Delta\Phi_{dp}(r_{min}, r_{max}).$$

Considering the specific attenuation expression (eq. [9.44]), eq. (9.49) can be solved for $A_h(r_{max})$ (see [21] for details) as follows:

$$A_h(r_{max}) = \frac{[Z'_h(r_{max})]^b}{I(r_{min}, r_{max})}\left(10^{0.1(b\alpha)\Delta\Phi_{dp}} - 1\right). \qquad (9.50)$$

Substituting into eq. (9.47) leads to the customary expression for specific attenuation in the RPA:

$$A_h(r) = \frac{[Z'_h(r)]^b \left(10^{0.1(b\alpha)\Delta\Phi_{dp}(r_{min}, r_{max})} - 1\right)}{I(r_{min}, r_{max}) + \left(10^{0.1(b\alpha)\Delta\Phi_{dp}(r_{min}, r_{max})} - 1\right)I(r, r_{max})}, \qquad (9.51)$$

where the value of the coefficient α is given in Table 9.3, $b \approx 0.7$ at the S band, and $b \approx 0.8$ at the C band and X band [279]. It is worth noting that eq. (9.51) requires the full profile of the measured reflectivity inside the rain cell but only the value of

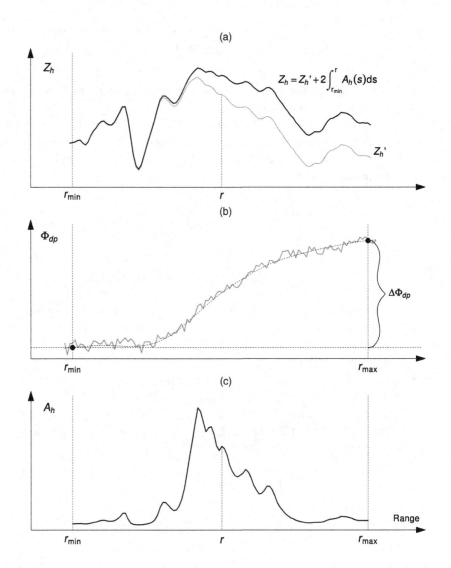

(a)

$$Z_h = Z_h' + 2\int_{r_{\min}}^{r} A_h(s)ds$$

(b)

(c)

Figure 9.20 Illustration of the Φ_{dp}-constrained method (RPA) with X-band radar observations through convective rain. The input observations, represented in gray, are the measured (attenuated) reflectivity $Z_h'(r)$ (panel [a]) and the differential phase shift $\Phi_{dp}(r)$ (panel [b]). For the retrieval of the specific attenuation $A_h(r)$ (panel [c]), only the Φ_{dp} increase between $r_{\min}$ and $r_{\max}$ is needed ($\Delta\Phi_{dp}(r_{\min}, r_{\max})$). From the estimation of $A_h(r)$, the unattenuated reflectivity can be retrieved (thick black line in panel [a]).

Φ_{dp} at the boundaries to constrain the cumulative attenuation. Filtering the noise in the differential phase measurements at the two boundaries is therefore important to provide a robust estimation of the constraining $\Delta\Phi_{dp}(r_{\min}, r_{\max})$. It is also important to recall that Φ_{dp} is the intrinsic differential phase shift, whereas the radar measures $\Psi_{dp} = \Phi_{dp} + \delta_{co}$, where δ_{co} is the phase shift due to backscattering. Because δ_{co} is not accounted for in this formulation, the integration boundaries along the radar beam should be carefully selected. As opposed to Φ_{dp}, δ_{co} is not cumulative along the range. It is therefore important to make sure that the backscattering differential phase shift is negligible at the boundaries r_{min} and r_{max}. A departure of Ψ_{dp} from Φ_{dp} due to δ_{co} inside the rain segment will not affect the performance of the method. In practice, however, a nonnegligible backscattering differential phase shift is often observed at the leading edge of a rain cell (see, e.g., the Ψ_{dp} inset in Fig. 9.18). In the updraft region, the relative abundance of large drops due to size sorting can lead to very large Z_{dr} values (Section 3.6.2) and nonnegligible δ_{co} (Section 4.7.2). In this case, it is important to rely on a careful estimation of the system differential phase shift and apply a filter to remove the biases from δ_{co} at the boundaries of the rain segment (Section 6.3). The same applies to the simple attenuation-correction method discussed in the previous section (eq. [9.41]). Once the specific attenuation is retrieved from eq. (9.51), the intrinsic (unattenuated) reflectivity can be calculated using eq. (9.40) with appropriate consideration of the limits of integration from r_{min} to r.

In addition to attenuation correction, the method can be exploited and applied to rainfall estimation. The specific attenuation in rain is directly related to the total liquid water mass within the resolution volume (eqs. [4.38, 9.37]) and is almost linearly related to the rain rate [47]. It follows that once the specific attenuation is retrieved from the range profile, the rainfall rate can be estimated using a relation of the form $R(A_h) = c(\lambda, T) A_h^{b}$ [280]. Although the coefficient b is roughly constant and close to 1.0, it is important to note that the multiplicative parameter c is dependent on both the radar wavelength (frequency scaling of Φ_{dp}) and the temperature (Fig. 9.13).

An important property of both the constrained method and the Φ_{dp}-based method is that the estimates of specific attenuation does not depend on the radar calibration. This is obvious for eq. (9.45) because K_{dp} is immune to radar system calibration errors. For the Φ_{dp}-constrained method, the reflectivity observations are used to modulate the attenuation progression along the rain profile, and it can be easily verified that the estimation of A_h does not change if a constant offset is added to $Z'_h(r)$ in eq. (9.51).

The formulation in eq. (9.51) depends on the parameters b (exponent in eq. [9.44]) and α (multiplicative constant in eq. [9.45]). Whereas the exponent b varies within a relatively small range (e.g., Delrieu et al. [281] reported 0.76-0.84 at the X band), the α coefficient has been shown to vary due to both the temperature dependence and the sensitivity to the drop-shape model. Carey [282] reports a factor-of-two variation for α at the C band (between 0.05 and 0.11), whereas Park et al. [283] show that α can assume values between 0.139 and 0.335 at the X band.

When applying the RPA method or the simpler Φ_{dp}-based relation (eq. [9.39]), it is important to set the value of the coefficients as accurately as possible. Tuning these parameters could be done by relying on auxiliary information about the environmental temperature and drop-shape statistics for a specific region and precipitation type. However, this auxiliary information may not be available, or it may not be easy to incorporate in real-time processing. Self-consistency can be used to dynamically adjust the value of the α parameter in eq. (9.51), exclusively relying on the radar's observations. The self-consistent rain-profiling algorithm (SC-RPA) is constrained using the difference between the observed and self-consistency based Φ_{dp} [284]. The basic idea is to exploit the full profile of Φ_{dp} instead of only the two boundary values (i.e., $\Phi_{dp}(r_{min})$ and $\Phi_{dp}(r_{max})$). The parameter α is initially set to a reasonable value within a predefined range ($\alpha_{min} < \alpha < \alpha_{max}$). Then, an iteration is performed to find the optimal value that minimizes the error function:

$$\text{err} = \sum_{j=min}^{max} |\Psi_{dp}^{filt}(r_j) - \Phi_{dp}^{calc}(r_j; \alpha)|, \qquad (9.52)$$

where r_j ranges between r_{min} and r_{max}, Ψ_{dp}^{filt} is a smooth-filtered profile of the observed differential phase (eq. [6.3]), and Φ_{dp}^{calc} is calculated using the attenuation profile retrieved from eq. (9.51):

$$\Phi_{dp}^{calc}(r_j; \alpha) = 2 \int_{r_{min}}^{r} \frac{A_h(s; \alpha)}{\alpha} ds \qquad (9.53)$$

The optimal α is, in practice, the one that allows for the best match between the observed and calculated Φ_{dp} profiles within the rain-cell segment. This procedure works well especially for heavy rainfall, when $\Delta\Phi_{dp}$ is much higher than the measurement noise. The filtering of the observed Ψ_{dp} is especially important for removing the backscatter phase shift δ_{co}, whose contribution to the observations may cause a departure from the corresponding calculated differential phase. Considering that δ_{co} is also most likely to occur in heavy rain, the success of the minimization procedure greatly depends on the quality of the Ψ_{dp} filtering.

9.4.3   Dual-Polarization Rain Profiling with Differential Reflectivity

The rain-profiling method has been generalized by Lim and Chandrasekar [146] to overcome one of the major limitations of the standard RPA algorithm, namely, the assumption of constant N_w along the rain segment (r_{min}, r_{max}). The algorithm has been developed with X-band applications in mind, but it could be easily adapted for use with radars operating in other frequency bands. At the X band, the contribution of δ_{co} to Ψ_{dp} in moderate to heavy rain is more ubiquitous (starts at $D_0 \approx 1.5$ mm) and varies more gradually than at the C band, where the backscattering phase shift typically affects short-range segments with characteristic "bumps" in the observations when D_0 exceeds 2.5 mm (Fig. 9.18).

It has been shown in Section 9.2 how the availability of both the reflectivity and the differential reflectivity allows the retrieval of the DSD parameters D_0 and N_w. Considering the additional measurements of Z_{dr}, it is possible to derive a new formulation of a general dual polarization rain-profiling algorithm (DRPA) to improve the performance of the RPA (which is based on Φ_{dp} and Z_h alone). The DRPA method adds the parameterization of the specific differential attenuation A_{dp}, in terms of K_{dp}, to the parameterization of A_h from eq. (9.45):

$$A_h(r) = \alpha K_{dp}(r)$$
$$A_{dp}(r) = \beta K_{dp}(r). \qquad (9.54)$$

In addition, eq. (9.44) is replaced by new expressions for the attenuation at the horizontal and vertical polarizations:

$$A_h(r) = a_1 [Z_h(r)]^{b_1} [Z_{dr}(r)]^{c_1}$$
$$A_v(r) = a_2 [Z_v(r)]^{b_2} [Z_{dr}(r)]^{c_2}. \tag{9.55}$$

Using eqs. (9.54) and (9.55) leads to the following expressions for the Φ_{dp}-constrained estimation of the specific attenuation at the horizontal and vertical polarizations:

$$A_h(r) = \frac{[Z_h'(r)]^{b_1} [Z_{dr}'(r)]^{c_1} \left(10^{0.1\alpha(b_1+\beta c_1)\Delta\Phi_{dp}(r_{min}, r_{max})} - 1\right)}{I_h(r_{min}, r_{max}) + \left(10^{0.1\alpha(b_1+\beta c_1)\Delta\Phi_{dp}(r_{min}, r_{max})} - 1\right) I_h(r, r_{max})}$$

$$A_v(r) = \frac{[Z_h'(r)]^{b_2} [Z_{dr}'(r)]^{c_2} \left(10^{0.1\alpha(b_2+\frac{\beta}{1-\beta}c_2)\Delta\Phi_{dp}(r_{min}, r_{max})} - 1\right)}{I_v(r_{min}, r_{max}) + \left(10^{0.1\alpha(b_2+\frac{\beta}{1-\beta}c_2)\Delta\Phi_{dp}(r_{min}, r_{max})} - 1\right) I_v(r, r_{max})}, \tag{9.56}$$

where

$$I_h(r_i, r_j) = 0.46 \left(b_1 + \beta c_1\right) \int_{r_i}^{r_j} [Z_h'(s)]^{b_1} [Z_{dr}'(s)]^{c_1} ds$$

$$I_v(r_i, r_j) = 0.46 \left(b_2 + \frac{\beta}{1-\beta} c_2\right) \int_{r_i}^{r_j} [Z_h'(s)]^{b_2} [Z_{dr}'(s)]^{c_2} ds, \tag{9.57}$$

and the multiplicative coefficients a_1 and a_2 canceled out, as in the standard RPA (eq. [9.51]). The other coefficients in eq. (9.55) can be calculated using scattering simulations at the X band (typical values are $b_1 = 1.01$, $c_1 = -3.36$ and $b_2 = 1.01$, $c_2 = -2.63$).

A self-consistent extension is devised (SC-DRPA) to optimize the value of the two parameters α and β. In this case, there are two error functions to minimize, one for each independent specific attenuation estimation (horizontal and vertical polarizations):

$$err_1 = \sum_{j=min}^{max} |\Psi_{dp}(r_j) - \Phi_{dp}^{calc_1}(r_j; \alpha; \beta)|$$

$$err_2 = \sum_{j=min}^{max} |\Psi_{dp}(r_j) - \Phi_{dp}^{calc_2}(r_j; \alpha; \beta)| \tag{9.58}$$

One notable difference with respect to the SC-RPA method is that in this case, by exploiting the availability of Z_{dr} there is no need to filter the observed Ψ_{dp} profile to remove the backscatter differential phase shift (the actual observed Ψ_{dp} appears in eq. [9.58], without the "filt" superscript as in eq. [9.52]). Instead, δ_{co} can be estimated from Z_{dr} using, e.g. eq. (4.71), and included in the calculated differential phase expressions:

$$\Phi_{dp}^{calc_1}(r;\alpha;\beta) = 2 \int_{r_{min}}^{r} \frac{A_h(s;\alpha;\beta)}{\alpha} ds + \delta_{co}(r)$$

$$\Phi_{dp}^{calc_2}(r;\alpha;\beta) = 2 \int_{r_{min}}^{r} \frac{A_v(s;\alpha;\beta)}{\alpha^{\frac{\beta}{1-\beta}}} ds + \delta_{co}(r).$$

(9.59)

An alternate expression for δ_{co} is given in [146] as:

$$\delta_{co} = 0; \ Z_{dr} < 1.25$$
$$\delta_{co} = -11.5 + 9.35 Z_{dr}; \ Z_{dr} \geq 1.25,$$

(9.60)

with Z_{dr} in linear units and δ_{co} in degrees. Note that eq. (9.60) is very close to the best-fit relation (eq. [4.71]) for $Z_{dr} > 1.5$ dB.

The simultaneous minimization of the error functions (eq. [9.58]) does not lead to a unique solution for the pair of parameters α and β. In fact, in the α–β plane, the error functions present a valley shape, with the minima aligned along a line that can be shown as a function $\beta(\alpha)$. To solve this underdetermined problem, a further constraint is needed. The selection of the proper pair of parameters among all available options can be accomplished by exploiting the fact that reflectivity and differential reflectivity are located in a limited region in the Z_h–Z_{dr} plane (Fig. 9.7). With this constraint, the range of possible α, β pairs is substantially reduced, and the mean values obtained from the two minimization processes can be used as the optimal values.

Although the procedure for attenuation estimation with fixed parameters (eq. [9.56]) is unaffected by reflectivity bias, the optimized self-consistent procedure assumes that the reflectivity and differential reflectivity are unbiased. In fact, an offset in either Z_h or Z_{dr} may lead to the selection of unsuitable coefficients, given the role of the bounded Z_h–Z_{dr} domain. A bias in Z_{dr} may also affect the estimation of δ_{co} in eq. (9.59). For application of the SC-DRPA as an alternative to the SC-RPA, it is therefore fundamental to ensure the proper calibration of the radar system for reflectivity and differential reflectivity.

9.5   Rainfall-Rate Estimation

Reflectivity–rainfall rate (Z–R) relationships were derived early on when radar started to be used for monitor weather and are still widely used today for rainfall-rate estimation (these estimators do not rely on dual polarization observations). The advent of dual polarization radars brought the possibility of retrieving the essential properties of the rain medium, which can be exploited to improve the estimation of the rainfall rate. Starting from conventional $R(Z_h)$ estimators, the next sections revise the fundamental one-parameter and two-parameter rainfall estimators based on the additional observations of Z_{dr} and K_{dp}. For each rainfall-estimation algorithm, scattering simulations are used to derive convenient parameterizations that are valid for the most common weather radar frequencies. The drop-shape model used to simulate the radar variables is from Brandes et al. [43], described in eq. (3.33). For the calculation of the rain rate, the terminal fall velocity relation (eq. [3.37]) is used [43]. It is important to note that although scattering simulations are widely used to derive the parameters of the rainfall relations, the results are sensitive to a number of specific assumptions regarding the rain microphysics, as well as the specific input DSDs. For demonstration here, disdrometer observations collected during the IFloodS campaign are used, which are representative of a wide range of rainfall conditions.

The rainfall rates, R, are estimated from the radar's observations (namely, Z, Z_{dr}, and K_{dp}). To form this relationship, details about the rainfall's microphysical processes must be assumed. The DSD, the rainfall terminal velocity, and the drop-shape ratio all have an impact on the rainfall rate, as well as the radar observations. For the various rainfall rate estimators discussed in this section, typical values for the coefficients are provided to relate the radar variables to the rainfall rate. Other sources in the literature may provide different values, but the overall impact on the results is minor. Regional or seasonal variations of these coefficients may also be found in the literature. These typical values provide a reference point from which coefficients can be refined if local characteristics of the rainfall process are available. Localized optimization can offer improved estimates compared with the coefficients given here, but they then may be less suitable for other regions. The errors from the dual polarization models are generally much smaller than the typical errors associated with rainfall estimation from single-polarization radars using a Z–R relation. The respective coefficients are provided to demonstrate the relationships between the models, and the frequencies, but also as typical values that may be adopted immediately.

It is also noted that attempts have also been made to estimate rainfall from attenuation measurements. In particular, the nearly linear relation between rainfall (R) and attenuation (A) implies that $R(A)$ relations are weakly affected by the variability of the drop size distribution. This led to the development of methods to estimate path-integrated rainfall from radar observations [47]. However, these methods require either targets with a known cross-sectional area (two-way methods) or separate transmitting and receiving antennas placed at the beginning and at the end of the path (one-way methods), making them not practical for application to operational weather radar systems. Other attenuation-based rainfall methods, as mentioned in Section 9.4.2, are derived through differential phase observations.

9.5.1 $R(Z_h)$ Algorithm

In their 1947 paper, Marshall, Langille, and Palmer [285] stated that "a fortuitous relation between the precipitation quantity important to radar ($\sum D^6/V$) and one that is more relevant to meteorology (rate of rainfall) may prove useful in practical and research applications." That relation was of the following form:

$$\boxed{Z(R) = a\, R^b,}\qquad(9.61)$$

and did actually prove very useful, having been the basis for rainfall estimation by radar for more than half a century. The coefficients a and b were 190 and 1.72, respectively, in the 1947 paper and later modified in the more renowned 1948 paper [23]. The more widely adopted coefficients are $a = 200$ and $b = 1.6$, which are typically suitable for stratiform rain. The value of the exponent in the power-law expression is related to the ratio of the different moments of the DSD (Z_h and R). With $c = a^{-\frac{1}{b}}$, eq. (9.61) can be written as

$$R(Z) = c\, Z^{\frac{1}{b}}. \tag{9.62}$$

In practice, the coefficients a and b in eq. (9.61) show considerable variability [286]. In particular, the multiplicative constant a is related to N_w, the intercept parameter of the normalized gamma DSD. The reflectivity in eq. (9.61) and (9.62) is intentionally written without a subscript. In fact, this power-law relation can be applied to either the horizontal or vertical reflectivity observations. For example, the scattering simulations at the S band for the IFloodS data generate the relation $R(Z) = 0.02\, Z^{0.67}$ (coefficients $a = 340$, $b = 1.4$) for the horizontal polarization and $R(Z) = 0.014\, Z^{0.74}$ ($a = 340$, $b = 1.5$) for the vertical polarization. From eqs. (9.16) and (9.17), it can be shown that the Z–R relation can also be written as follows [21]:

$$\frac{Z}{N_w} = \tilde{a}(\mu)\left(\frac{R}{N_w}\right)^b, \tag{9.63}$$

where $b = 7/4.67 \approx 1.5$ is a constant, in fair agreement with the previous empirically retrieved exponents, and $\tilde{a}$ is a weak function of μ. From the previous expression, the multiplicative coefficient a in eq. (9.61) can be written to explicitly show the inverse relation with N_w:

$$a = \tilde{a}(\mu)N_w^{1-b} \approx \tilde{a}(\mu)N_w^{-0.5} \tag{9.64}$$

Single-polarization radars have traditionally used horizontal polarization, which provides a higher sensitivity, given the oblate shape of raindrops. For the S band, as illustrated by the Z_h–Z_{dr} scatterplots (e.g., Fig. 9.7), the horizontal reflectivity can be up to ~3 dB higher than the corresponding reflectivity on the vertical channel. For shorter wavelengths, as a result of Mie scattering effects, the difference can be even larger. On the other hand, for attenuating frequencies, the vertical channel in rain suffers less path attenuation, which needs a proportionally smaller correction (reduced uncertainty). Considering the C-band profile in Figures 9.17 and 9.18, for which the total phase shift is ~250°, the simulated path-integrated attenuation is ~17 dB on the horizontal polarization and ~14.5 dB on the vertical polarization. The resulting differential attenuation (~2.5 dB) typically shows in the observation of negative Z_{dr} streaks beyond the storm's core. In the precipitation regions at farther ranges, where the magnitude of the differential path-integrated attenuation is larger than the intrinsic Z_{dr} of rain, the vertical reflectivity signal may then appear stronger (and easier to detect above the noise level) than the horizontal reflectivity. This shows that there is a balance between a stronger signal arising from the scattering properties

of the DSD within the radar resolution volume (horizontal polarization) and less attenuation due to propagation through the rain medium (vertical polarization).

Measurements at horizontal and vertical polarizations in the simultaneous transmit,
simultaneous receive (STSR) mode may also be combined to improve the sensitivity
of reflectivity estimates. In fact, because the uncorrelated noise vanishes when consid-
ering the off-diagonal elements of the covariance matrix, using $|\sum_i H_i V_i^*|$ in place of
$\sum_i |H_i|^2$ or $\sum |V_i|^2$ (where H_i and V_i represent the complex voltage amplitudes) for
estimating the copolar echo may lead to enhanced detection capability [287].

The application of the Z–R relation typically involves a threshold at 53–55 dBZ
to avoid hail contamination. The R–Z algorithm may be useful after a substantial
average of the DSD, but it has limitations for the local and instantaneous estimations
required in many radar applications. Panel (a) of Figure 9.21 shows the scatterplot of
the rainfall-rate estimates applying eq. (9.62) with the optimized coefficients $a = 340$
and $b = 1.5$ on the reflectivity derived from DSD measurements collected during the
IFloodS campaign. The deviation from the true R (calculated using eq. [9.8] from the
same DSD observations) represents the parametric error, that is, the inaccuracy due to
the specific form of the expression in eq. (9.62). Note that this error estimate does not
include the contribution arising from a suboptimal choice of the coefficients a and b
in the Z–R relation.

9.5.2 $R(Z_h, Z_{dr})$ Algorithm

As shown in Section 9.1.1, for the case of an exponential distribution, Z_{dr} is only
related to the slope parameter Λ, whereas Z_h depends on both N_w and Λ. The differential reflectivity was the most studied dual polarization measurement during the
1970s and 1980s [288], and the initial aim was to develop a dual polarization rainfall
estimator focusing on the radar reflectivity at the horizontal polarization and the differential reflectivity in logarithmic scale, with expressions of the following form [289]:

$$R(Z_h, Z_{dr}) = c\, Z_h^b \left[10 \log_{10}(Z_{dr})\right]^a .\tag{9.65}$$

For an exponential size distribution, Figure 9.1 shows that D_m is a monotonically
increasing function of Z_{dr} for any given value of the maximum diameter D_{max}. From
eq. (9.18) R and Z are related though the mass-weighted mean diameter as

$$\frac{R}{Z} \sim D_m^{-2.33}.\tag{9.66}$$

It follows that the exponent of Z_{dr} in the expression in eq. (9.65) must be negative.
However, the negative exponent implies a high sensitivity to Z_{dr} measurement errors,
especially when $Z_{dr} \sim 0$ dB. A more robust estimator can be obtained by considering
reflectivity (either at horizontal or vertical polarization) and differential reflectivity in
linear scale ($Z_{dr} = 1$ for spherical droplets):

$$\boxed{R(Z, Z_{dr}) = c\, Z^b\, Z_{dr}^a.}\tag{9.67}$$

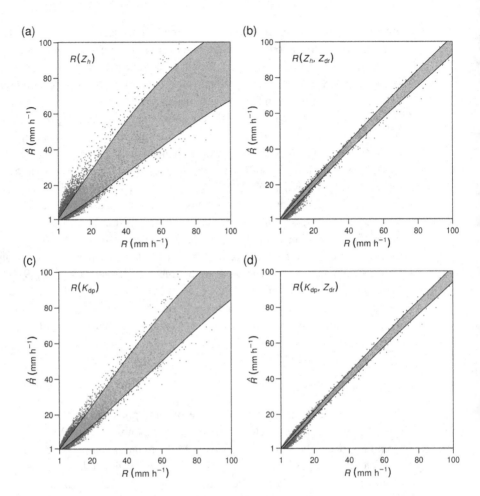

Figure 9.21 Comparison of different radar rainfall estimators: $R(Z_h)$ (a), $R(Z_h, Z_{dr})$ (b), $R(K_{dp})$ (c) and $R(K_{dp}, Z_{dr})$ (d). R is the "true" rainfall rate calculated from the disdrometer DSD observations collected during the IFloodS campaign, whereas $\widehat{R}$ is the rain rate estimated using the radar dual polarization relations in Tables 9.4 and 9.5 at the S band. The $R(Z_h)$ relation in eq. (9.62) uses the coefficients $c = 0.020$, $b = 0.67$. The gray area marks the region between the 10th and 90th percentiles, that is, the region including 80% of the data.

The role of the Z_{dr} component in eq. (9.67) can be seen as a "correction" factor to the standard reflectivity-based estimate, which takes into account the natural variability of the mass-weighted mean diameter D_m (see discussion in Section 9.1.1). In fact, given the large difference between the moments of the DSD, in a distribution dominated by large drops, the $R(Z_h)$ algorithm would lead to strong overestimation of the rain rate as indicated by eq. (9.66). The reverse would happen with a DSD dominated by small drops. The introduction of Z_{dr}, which is tightly related to the mass-weighted mean diameter, allows more accurate estimates for widely varying DSDs. Consider eq. (9.67) in logarithmic form (Z in dBZ, Z_{dr} in dB):

$$10\log_{10}(R) = 10\log_{10}(c) + b\, Z(\text{dBZ}) + a\, Z_{dr}(\text{dB}); \qquad (9.68)$$

Table 9.4 Typical coefficients of the $R(Z_h, Z_{dr})$ (eq. [9.72]) and $R(Z_v, Z_{dr})$ relations, obtained through nonlinear regression over simulated radar variables and rainfall rate

Band (Frequency)	$R(Z_h, Z_{dr})$			$R(Z_v, Z_{dr})$		
	c	b	a	c	b	a
S (2.8 GHz)	4.6×10^{-3}	0.96	-3.7	4.6×10^{-3}	0.96	-2.7
C (5.6 GHz)	4.5×10^{-3}	0.94	-2.5	4.5×10^{-3}	0.94	-1.6
X (9.4 GHz)	6.8×10^{-3}	0.99	-5.3	6.8×10^{-3}	0.99	-4.3
Ku (13.6 GHz)	6.3×10^{-3}	0.98	-5.6	6.3×10^{-3}	0.98	-4.7

it is easy to see how a change in Z_{dr} results in a shift of the intercept parameter in the $\log(Z)$–$\log(R)$ plane (refer, e.g., to Fig. 9.4 with the x- and y-axes flipped).

The coefficients of eq. (9.67) are commonly determined using nonlinear regression on a set of values for the rain rate (R) and radar variables (Z_h and Z_{dr}). To this aim, surface measurements of the rain DSD are fundamental to providing the detailed microscale structure of precipitation. The DSD observations can be used directly or they can be used to estimate the physical range of parameters for specific analytic forms (e.g., the gamma distribution), from which new synthetic DSDs can be generated.

While the rainfall rate R is directly calculated from either the measured or synthetic DSD using eq. (9.8), scattering simulations are needed to retrieve the radar variables for the appropriate radar frequency. When the raindrop shape is not directly measured (e.g., many disdrometers only provide information about the drops' equivalent diameter) or when parameteric DSDs are used, a drop-shape model and canting distribution need to be assumed to simulate the dual polarization radar variables. Table 9.4 provides a list of coefficients for eq. (9.67) at different radar frequency bands under realistic assumptions about the drop-shape model (eq. [3.33]), terminal fall velocity (eq. [3.37]), and canting angle following a Gaussian distribution with a $7°$ standard deviation. The two separate parameterizations are obtained through non-linear regression between the rainfall rate and the respective radar variables. The dramatic reduction of the $\widehat{R}$–R dispersion in the scatterplot in panel (b) of Figure 9.21 clearly illustrates the improvement provided by the two-parameter relation in eq. (9.67) compared to the single-polarization estimator $R(Z_h)$ (panel [a]). The $R(Z_h, Z_{dr})$ and $R(Z_v, Z_{dr})$ relations (coefficients in Table 9.4) only differ in the exponent of Z_{dr}. Considering that the horizontal reflectivity can be expressed as $Z_h = Z_v Z_{dr}$ (linear scale), the relation in eq. (9.67) is written for either Z_h or Z_v as follows:

$$R = c\, Z_h^b\, Z_{dr}^a = c\, (Z_v Z_{dr})^b\, Z_{dr}^a = c\, Z_v^b\, Z_{dr}^{(a+b)}. \tag{9.69}$$

Considering that $b \approx 1$, it follows that the difference between the two parameterizations reflects the decreased magnitude of the Z_{dr} exponent in the $R(Z_v, Z_{dr})$ relation. A practical implication is that by using Z_v instead of Z_h, the resulting rain-rate estimation is slightly less susceptible to the Z_{dr} measurement error (see eq. [9.76] in Section 9.5.5).

9.5.3 $R(K_{dp})$ Algorithm

As noted in Section 9.1.1, K_{dp} is approximately proportional to the fourth moment of the DSD. Considering that R is the 3.67th moment of the DSD (eq. [3.35]), it follows that K_{dp} is an excellent candidate for rainfall estimation. The relationship between the rainfall rate and the specific differential phase can then be written in the following form:

$$R(K_{dp}) = c\, K_{dp}^{b}$$

(9.70)

The scaling with frequency of K_{dp} (eq. [9.6]) implies that the coefficient c in eq. (9.70) depends on the radar wavelength, whereas the exponent b is nearly constant ($b \approx 0.85$). In particular, for Rayleigh scatterers, (eq. [9.70]) can be approximated by a simple function of the radar frequency f (GHz) as follows:

$$R(K_{dp}) \approx 130 \left(\frac{K_{dp}}{f}\right)^{0.85}.$$

(9.71)

As a result of the increasing Mie scattering effects at higher frequencies, the relation in eq. (9.71) is only usable up to approximately 13 GHz (Ku band).

The estimator in eq. (9.71) is the only single-parameter relation for rainfall estimation, besides the standard $R(Z_h)$ relation. As such, it has the advantage of being easy to implement. In addition, it is affected by a single source of measurement error, as compared with the multiparameter estimators $R(Z_h, Z_{dr})$ and $R(K_{dp}, Z_{dr})$ (errors are discussed in Section 9.5.5).

In comparison with the reflectivity-based rainfall estimator in eq. (9.62), the $R(K_{dp})$ relation is denoted by a reduced parametric error, as illustrated in panel (c) of Figure 9.21. The other notable advantage of this algorithm derives from the fact that K_{dp}, being a phase measurement, is immune to calibration errors and is unaffected by attenuation. On the other hand, for the K_{dp} observations to be successfully used in eq. (9.70), the slope of the differential phase shift must be estimated, which has its own challenges, as discussed in Section 9.5.5.

9.5.4 $R(K_{dp}, Z_{dr})$ Algorithm

Similar to the $R(Z_h, Z_{dr})$ algorithm, in this case, the aim is to improve the K_{dp}-based rainfall rate estimate by introducing a "correction" factor from the differential reflectivity. The revised rainfall rate estimator takes the following form:

$$R(K_{dp}, Z_{dr}) = c\, K_{dp}^{b}\, Z_{dr}^{a}.$$

(9.72)

Given the reduced fraction of the K_{dp}-to-R DSD moments (4/3.67) with respect to the Z_h-to-R moments (6/3.67), the impact of Z_{dr} is smaller in this case, as suggested by the smaller magnitude of the exponent a (Table 9.5) in comparison with the corresponding exponent of Z_{dr} in the $R(Z_h, Z_{dr})$ algorithm (Table 9.4). The relation in eq. (9.72) is characterized by a low parametric error, as shown in panel (d) of Figure 9.21.

Table 9.5 Typical coefficients of the $R(K_{dp})$ (eq. [9.70]) and $R(K_{dp}, Z_{dr})$ (eq. [9.72]) relations. For the $R(K_{dp})$ relation, the exponent is fixed to a nominal value ($b = 0.85$).

Band (Frequency)	$R(K_{dp})$	$R(K_{dp}, Z_{dr})$		
	c	c	b	a
S (2.8 GHz)	44	110	0.97	−2.0
C (5.6 GHz)	23	33	0.91	−0.7
X (9.4 GHz)	16	34	0.99	−1.7
Ku (13.6 GHz)	12	25	0.98	−1.9

9.5.5 Impact of Measurement Errors

The uncertainty in the rainfall rate estimates are directly impacted by the radar variables' measurement error (Chapter 7). To evaluate the impact of these measurement errors, a zero-mean Gaussian noise can be added to the simulated radar variables that were previously used to evaluate the parametric relations in Figure 9.21. To this end, the following standard deviations for the radar variables are used: 1 dB for Z_h, 0.1 dB for Z_{dr}, and 0.25 $^\circ$km^{-1} for K_{dp}. Note that higher accuracy is sought for Z_{dr} for rainfall estimation compared to other applications such as hydrometor classification or melting layer detection. It is also important to note that whereas the standard deviations of Z_h and Z_{dr} refer to the actual gate (single-point) measurement (i.e., 150-m gate spacing), for, K_{dp} the measurement error is given by eq. (7.3h), which is repeated here:

$$\text{SD}(\hat{K}_{dp}) \approx \sqrt{\frac{3}{M}} \frac{\text{SD}(\hat{\Psi}_{dp})}{L}; \quad M \gg 1. \tag{9.73}$$

According to eq. (9.73), if the standard deviation of the measurement of the differential phase shift $\hat{\Psi}_{dp}$ is 2°, K_{dp} would need to be estimated along an $L = 3$ km range segment in order to get $\text{SD}(\hat{K}_{dp}) = 0.25$ $^\circ$km^{-1}, with $M = 20$ range gates and a 0.15-km range resolution.

The scatterplots in Figure 9.22 illustrate the impact of the given measurement errors on the radar variables at the S band, in addition to the error arising from the specific parametric forms (Fig. 9.21). Figure 9.23 shows scatterplots at the X band for the two estimators involving K_{dp} observations. Although the dispersion is clearly larger for all four relations when compared to the case with only parametric error, there are substantial differences in the total error structure for the Z_h-based and the K_{dp}-based algorithms.

First consider the two single-parameter relations $R(Z_h)$ and $R(K_{dp})$. The radar received power is evaluated on a logarithmic scale, implying that any given measurement error (e.g., $\text{SD}(10\log_{10}(\hat{Z})) = 1$ dB) corresponds to a constant relative error in the linear reflectivity factor Z. It follows from eq. (9.62) that

$$\frac{\text{SD}(\hat{R})}{R} = \frac{1}{b}\left(\frac{\text{SD}(\hat{Z})}{Z}\right), \tag{9.74}$$

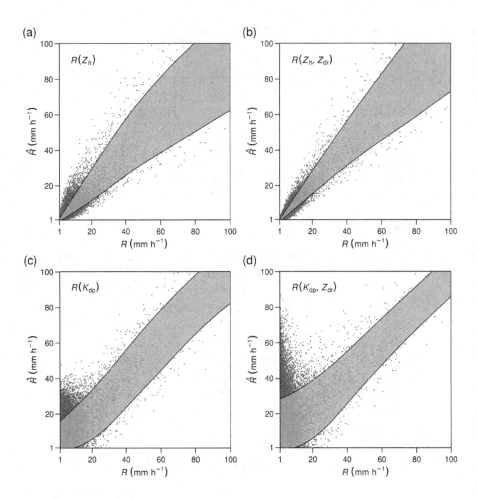

Figure 9.22 Like Figure 9.21, but with synthetic Gaussian measurement error added to the radar parameters (1 dB for Z_h, 0.1 dB for Z_{dr}, and 0.25 $^\circ$km^{-1} for K_{dp}).

where b is the exponent of R in eq. (9.61) and SD(R) represents only the uncertainty due to measurement errors (the total error is the sum of the parametric and measurement errors). The relative error in the rainfall rate is proportional to the relative error in the linear reflectivity. For example, a 1-dB accuracy on Z_h (25 percent relative accuracy) corresponds to approximately 15 percent accuracy on the rainfall rate for the Marshall–Palmer relation ($b = 1.6$). Now consider that the phase of the signal is measured in linear scale ($^\circ$); hence, the absolute error in K_{dp} is constant, whereas the relative error decreases with increasing K_{dp}. From eqs. (9.70) and (9.73):

$$\frac{SD(\hat{R})}{R} = b \left(\frac{SD(\hat{K}_{dp})}{K_{dp}} \right)$$

$$\approx b \frac{SD(\hat{\Psi}_{dp})}{L} \sqrt{\frac{3}{M}} \left(\frac{c}{R} \right),$$

(9.75)

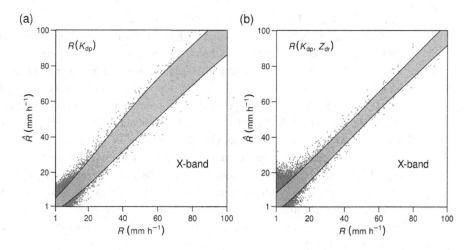

Figure 9.23 Like panels (c) and (d) in Figure 9.22 but for the X band.

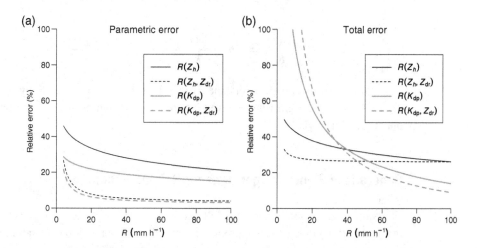

Figure 9.24 Normalized standard deviation (relative error) of different rainfall-rate relations at S band, for the IFloodS data set. (a) The relative error due to the parametric form of the rainfall estimators (corresponding to the scatter in Fig. 9.21). (b) The total error due to both the parameterization error and the measurement error (corresponding to the scatter in Fig. 9.22). The measurement error is simulated assuming a zero-mean Gaussian distribution with $SD(Z_h) = 1$ dB, $SD(Z_{dr}) = 0.1$ dB, and $SD(K_{dp}) = 0.25$ °km^{-1}.

implying that the relative measurement error in the rain-rate estimation using $R(K_{dp})$ varies with $1/R$. From these considerations, it follows that the measurement error in K_{dp} has a larger impact for the estimation of weak rainfall intensities in comparison with the reflectivity-based estimator, whereas the inverse is true for heavy rainfall. This can be clearly seen in Figure 9.24, which shows the parametric relative error (panel [a]) and the total relative error (panel [b]) with increasing R for the four rain-rate relations at S band. Neglecting any consideration of path attenuation, there is a clear trade-off between $R(Z_h)$ and $R(K_{dp})$ relations (black solid lines and gray solid lines, respectively). At the S band, this trade-off occurs at approximately 30-40 mm h^{-1}, whereas for C-band and X-band radars, the preference of applying an $R(K_{dp})$ relation, as opposed to an $R(Z_h)$ relation, generally starts at around 10-20 mm h^{-1} and 5-10 mm h^{-1}, respectively. This can also be appreciated by noting the reduced scatter in the $R(K_{dp})$ plot at the X band (panel [a] of Fig. 9.23) in comparison with the corresponding plot at the S band (panel [c] of Fig. 9.22).

In the absence of measurement errors, the two multiparameter relations of eqs. (9.67) and (9.72) provide much better accuracy with respect to their corresponding single-parameter relations (see Fig. 9.22). However, with two radar parameters involved, the measurement errors' effect on R may grow larger. According to error-propagation theory, the variance formulas for the two-parameter estimators are given by the following:

$$\frac{SD(R)}{R} = \sqrt{b^2 \frac{(SD(Z_h))^2}{Z_h{}^2} + a^2 \frac{(SD(Z_{dr}))^2}{Z_{dr}{}^2}} \qquad (9.76)$$

$$\frac{SD(R)}{R} = \sqrt{b^2 \frac{(SD(K_{dp}))^2}{K_{dp}{}^2} + a^2 \frac{(SD(Z_{dr}))^2}{Z_{dr}{}^2}}. \qquad (9.77)$$

In addition to measurement errors, the Z_h and Z_{dr} values can be affected by systematic errors (bias), which must be removed beforehand through proper calibration procedures (Section 7.4).

9.6   Generalized Blended Algorithms for Rainfall Estimation

The various dual polarization rainfall estimators have specific error structures that make certain methods more appropriate than others depending on the rainfall conditions. It is important to note, however, that before application of any given rainfall estimator to real observations, it's important to make sure that the radar resolution volume is filled with rain (as opposed to ice, mixed phases, or clutter). In fact, the estimators $R(Z_h)$, $R(Z_h, Z_{dr})$, and $R(K_{dp}, Z_{dr})$ (eqs. [9.62, 9.65, 9.72] respectively) are designed based on the assumption that the radar is sampling a population of raindrops following a prescribed shape model. The only exception is given by the $R(K_{dp})$ estimator because K_{dp} is insensitive to nearly spherical ice particles, and therefore, this relation can be applied in the presence of a mixture of ice and liquid particles, such as hail mixed with rain in convective storms (Section 3.6.3).

In order to ensure a physically consistent application of rainfall estimators, it appears natural to link the choice of the algorithm to a hydrometeor classification of the radar measurements (Section 8.3). The combination of these two fundamental applications of dual polarization radar observations leads to the development of blended algorithms for rainfall estimation driven by hydrometeor classification

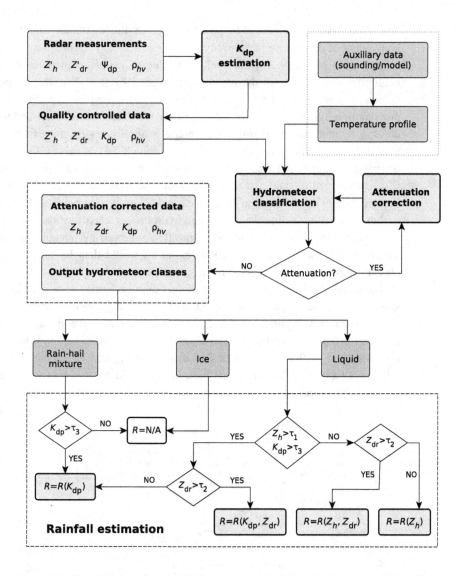

Figure 9.25 Flow diagram of a rainfall blended algorithm driven by hydrometeor classification. The primed variables (Z'_h and Z'_{dr}) denote measured (attenuated) observations. The typical thresholds for the S band are $\tau_1 = 38$ dBZ, $\tau_2 = 0.5$ dB, and $\tau_3 = 0.3$ $^\circ$km^{-1} [290].

[290–292]. Figure 9.25 shows a flow diagram of a generalized blended algorithm, including data preparation (e.g., quality control and K_{dp} estimation), a core processing phase (hydrometeor classification and attenuation correction), and the final rainfall estimation. The diagram shows a loop control structure involving hydrometeor classification and attenuation correction. Indeed, these are highly interconnected topics. In principle, hydrometeor classification algorithms (discussed in Section 8.3) should be applied after observations are corrected for attenuation. However, path attenuation may greatly vary depending on the particular distribution of the precipitation type along the radar beam. In particular, given the extremely small imaginary component of the relative permittivity of ice (Fig. 4.2) and considering the extinction cross-section in eq. (4.38), the attenuation due to ice-phase particles is generally negligible compared with the attenuation induced by rain. It is thus important to avoid applying an attenuation-correction procedure devised for rain to regions of the precipitation system containing solid hydrometeors. In fact, it is not uncommon to observe an increase in the differential phase in the ice region, for example, due to dendritic ice growth in stratiform systems (Section 8.5.1). If these differential phase observations were used for attenuation correction following a simple K_{dp}-based or Φ_{dp}-constrained method assuming a rain parameterization, the consequence would be a consistent overestimation of both Z_h and Z_{dr} in the ice portion of the cloud. It is thus reasonable to first devise a basic hydrometeor classification relying heavily on auxiliary information (temperature profile), followed by attenuation correction. The resulting attenuation-corrected observations would then be used for a refined classification relying more on the radar variables.

The thresholds driving the selection of the rainfall relation can be established using simulations at the frequency of interest, and specifically considering the accuracy of the K_{dp} extimates. As mentioned in Section 9.5.5, in practical applications, it is quite common to achieve an estimation error for K_{dp} of ≈ 0.2–$0.3\,^{\circ}\text{km}^{-1}$. It follows that a threshold of $\tau_3 = 0.3\ ^{\circ}\text{km}^{-1}$ in the flow diagram in Figure 9.25 is generally adequate. For the S band, Cifelli et al. [290] indicate threshold values of $\tau_1 = 38$ dBZ and $\tau_2 = 0.5$ dB, respectively, for Z_h and Z_{dr}.

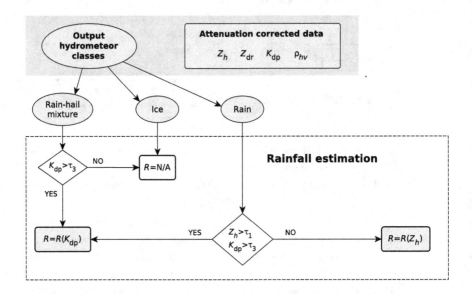

Figure 9.26 A simplified version of the rainfall-estimation process for the X band. As a result of frequency scaling, the threshold on reflectivity is reduced to $\tau_1 \approx 30$ dBZ, whereas the threshold on K_{dp} is unchanged ($\tau_3 = 0.3^{\circ}\text{km}^{-1}$).

The generalized blended algorithm illustrated in Figure 9.25 can also be applied to radars operating at higher frequencies. Considering the widespread use of X-band radar networks, it is useful to consider a simplified version of the blended algorithm for robust operational applications. Although the first part of the blended algorithm is basically unchanged (the details of the hydrometeor classification may change to account for the varying significance of the radar variables), the final part (rainfall estimation) has been modified to rely exclusively on the observation of the differential phase shift and reflectivity. This is illustrated in the diagram in Figure 9.26, showing how the choice of the rainfall relation is simplified to $R(K_{dp})$ for heavy rain and $R(Z_h)$ for weak rainfall intensity.

9.7   Selected Problems

1. Assume a raindrop distribution with $N_0 = 8000 \text{ m}^{-3}\text{mm}^{-1}$, $D_0 = 2$ mm, and $D_{max} = 8$ mm. Compute the reflectivity in dBZ, and the rainfall rate.
2. A scanning weather radar measures a volume of precipitation with a rainfall rate of $R = 7.6 \text{ mm h}^{-1}$ and an intrinsic reflectivity of $Z = 37.1$ dBZ.
   a) Define three different $Z - R$ relations that can fit this radar observation.
   b) Consider the widely used $Z - R$ relation from the literature: $Z(R) = a\, R^b$, with $a = 200$, $b = 1.6$. What is the error in rainfall rate estimation using this standard relation? What is the source of the error compared to the three relations identified in the previous part?
   c) Find the drop median diameter assuming an exponential distribution. (Hint: refer to eq. [9.18], with $\mu = 0$.)
3. If you have a way of providing D_0 measurements from a rain event characterized by an exponential DSD ($\mu = 0$), explain why the rainfall estimate will be free of parameterization error. You need to do this by writing the rainfall rate as an equation in terms of Z and D_0. Look at the standard equations provided to you for guidance. Hint: Get independent expressions for R and Z, and take the ratio.
4. Consider the following two pairs of observations: Z_h =42 dBZ, Z_{dr} =1.3 dB, and Z_h =42 dBZ, Z_{dr} =0.9 dB. Which pair will produce the larger rainfall rate? Explain why.
5. Consider the polynomial approximations for D_0-Z_{dr} relations in table 9.1. What is the expected D_0 value corresponding to an observation of Z_{dr} =3 dB, respectively, at the S and C bands? How would you explain the difference?
6. Consider the S, C, and X bands that are commonly used for weather radar.
   a) Which band is subject to the highest attenuation in rain?
   b) How does temperature affect attenuation in rain at the different frequencies?
7. Imagine an intense squall line passing over an X-band radar. Which would be the most suitable rainfall estimation algorithm to use in this situation?
8. The vertically integrated liquid (VIL) is a widely used radar product found in applications for severe thunderstorm warning and as a hail size indicator. VIL is defined as:

$$\text{VIL} = \int_{h_{\text{base}}}^{h_{\text{top}}} W(h)\,dh$$

where W is the liquid water content and h is the height. For single polarization radar, a widely used estimate of the liquid water content is $W = 3.44 \times 10^{-3}\, Z_h^{4/7}$, with Z_h in linear units ($\text{mm}^6\,\text{m}^{-3}$) and W in $\text{g}\,\text{m}^{-3}$. When dual-polarization measurements are available, K_{dp}-based relations allow improved water content estimation both in rain and rain/hail mixtures (Chapter 9). Analogously to eq. (9.71) for rainfall estimation, a relation for W can be formulated in terms of K_{dp} and the radar frequency f in gigahertz as follows:

$$W_{\text{liquid}}(K_{\text{dp}}) = 3.565 \left(\frac{K_{\text{dp}}}{f} \right)^{0.77}$$

For graupel and hail particles you may consider the reflectivity-based relation:

$$W_{\text{graupel}}(Z_h) = 2.438 \times 10^{-3}\, Z_h^{0.56}$$

Using the table from Chapter 8, problem 1, calculate the standard VIL (reflectivity-based) and a polarimetric VIL at the range of 28 km, proceeding as follows:

a) Calculate the height of the radar beam at 28 km range for all elevations (eq. [2.24]), and derive the proper vertical thickness dh to be used in the discretized version of the VIL equation above.

b) Assume hydrometeors in liquid phase up to 4500 m and graupel/hail at higher altitudes. Apply the proper relations to calculate the liquid water content at each altitude.

c) Integrate W from the lowest to the highest elevation to calculate the VIL value in kilograms per square meter ($\text{kg}\,\text{m}^{-2}$).

d) How do the single polarization and the dual polarization VIL estimates compare?

10   Weather Radar Networks

The modern weather radar layout commonly consists of a network of radars to cover a wide geographic area, often with many located in the coastal zone to observe storms coming from the sea. The traditional design of a single weather radar characterizes its performance using the weather radar equation (eq. [2.38b]). The single radar's maximum range and unambiguous Doppler velocity are traded off following eq. (5.37). The radar's design and operational configuration must balance how finely or coarsely the space is sampled (e.g., how many plan position indicators [PPIs] are positioned at different elevations), the desired measurement accuracy, and the required revisit rate for the volume scan (e.g., all elevations must complete their scans every 10 minutes). With a network of radars, the spatial and temporal coverage can be divided among many different radar systems. Therefore, the operational and design constraints of a radar network have additional parameters available to optimize the performance of each weather radar as part of a larger observing system.

For a single radar to effectively cover a large area, the radar's design must increase in size and complexity. The subsystems of the radar continue to evolve and extend their performance limits. Although these are, and will continue to be, important technological advancements for weather radar systems, the advancements in the way the radars are operated, and the way their performance is assessed, are incremental. There is a point where a network of smaller radars is more effective than a very big single radar. Spatially distributed networks of weather radars open a new observation paradigm. The weather observation network are designed and characterized using fundamentally new performance metrics compared to those traditionally used for a single radar.

Numerous technology developments for individual radars are relieving some of the weather radar performance limitations of the past. For example, phased arrays can more rapidly observe the radar's vicinity without sacrificing data quality. There are also advances in signal-processing techniques that can be used to bend the range–velocity ambiguity limitation (see Section 6.4.1). Even with these technological advancements, the single radar's coverage near the surface is limited as the range increases because of Earth's curvature (Fig. 10.1). Velocity is still observed in the radial direction, and two-dimensional (2D) or three-dimensional (3D) velocity fields must be inferred. The detection sensitivity is a function of the radar's range. The spatial resolution is limited by the antenna's beamwidth and the range. Considering this, most of these performance and operational gains are small compared with the

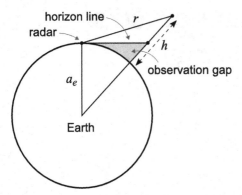

Figure 10.1 A conceptual diagram for the height of the beam above the surface due to Earth's curvature.

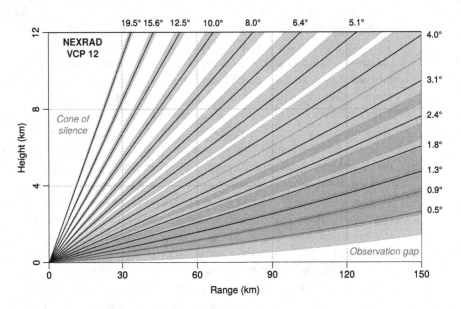

Figure 10.2 The NEXRAD volume coverage pattern (VCP) 12 includes 14 elevations; it takes approximately 4.1 minutes to complete this volume scan. VCP 12 was developed for improved detection and storm tracking [293]. Note, the effect of Earth's curvature, and the resulting observation gap, is included in the beam's height above the surface. Similarly, the maximum elevation angle is limited to 19.5°, resulting in a cone volume above the radar that is not observed.

measurement advantages that are available to a network of radars. In addition, all of the improvements available to the individual radar can also benefit the radar network.

The traditional approach of a single radar providing long-range observations becomes less attractive as the area to cover increases. For large areas (e.g., the continental United States), a network of distributed radars is the only option for effective coverage. Although a collection of WSR-88D radars comprises the NEXRAD network, in some instances, the individual radars within the network may not provide optimal coverage for all regions. As shown in Figure 10.2 for one of the volume coverage patterns (VCPs) executed by NEXRAD radars, the lowest altitude the radar can observe exceeds 1000 m aboveground at ranges beyond 100 km.

Consider the parameter space of trade-offs to determine the optimal number of radars to cover an area. On one extreme, one radar system could be used to cover the entire region. On the other extreme, a radar system could be dedicated for each observation volume (the size of the radar volume could also be open to debate). For large areas (e.g., national weather monitoring), either of these two extremes is not practical. The actual implementation of a radar or radar network is typically determined by a number of factors, including performance (these metrics are covered in detail later in the chapter), cost (both operations and the initial investment), availability of locations to deploy radars, and the ability to aggregate and utilize the data collected. This last item, the ability to aggregate and utilize the data, is the area of fastest growth and potentially the biggest benefit from a network of radars. A dense radar network, like the X-band radar network represented in Figure 10.3, can also allow for applications that would be impossible with a single radar (e.g., 3D wind-field retrievals) and provide additional solutions to cope with the limitations of short-wavelength systems, such as path attenuation, as discussed in Section 10.5.

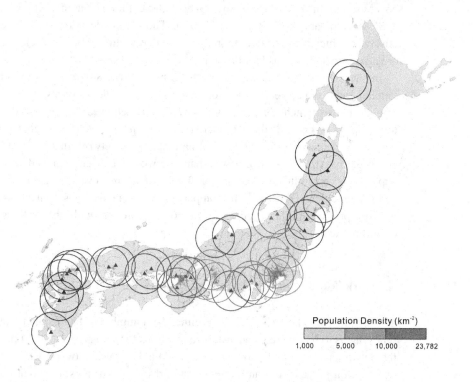

Figure 10.3 The Ministry of Land, Infrastructure, Transport and Tourism (MLIT) Japan X-band weather radar coverage map (2019). The observation range indicated by the circles is 80 km. Note the denser radar coverage in urban regions with higher population density. Image courtesy of Takeshi Maesaka.

10.1   Extending Radar Coverage through Networks

Radar networks can be broadly characterized into three scales: continental, regional, and urban. The composition of the radar networks also fall into two categories: homogeneous (e.g., NEXRAD) and heterogeneous (e.g., Italy, China, India). The first and most obvious use for a radar network is to expand the observation area. Implementation of such a network does not require any communication between the radars; their data are independently collected and archived. The user implements the data fusion by inferring meteorological phenomena from the available sets of observations.

The radar beam may be blocked at low elevations, either partially or completely, by trees, buildings, or mountains. Eventually, the radar's beam is too high above the ground and cannot observe the lower parts of the troposphere as a consequence of Earth's curvature. Figure 10.1 illustrates the beam's height above the surface for a radar at $0°$ elevation. Even with a clear field of view, at a range of 150 km, the beam center is 1324 m aboveground (using $a_e = 4/3 \cdot 6371$ km; see Section 2.4.5).

A single radar system's observational coverage is determined by its scanning strategy. There is a balance between the volume coverage repetition interval and the accuracy of the radar observations, which limits the scanning speed and fractional coverage within the radar volume. One of the commonly used VCPs executed by WSR-88D radars is VCP 12 (Fig. 10.2). The lower elevations have overlapping coverage, which then becomes more sparse with elevation. Above $20°$ elevation, there are no observations. This blind zone is sometimes referred to as the "cone of silence".

In the continental United States, the S-band WSR-88D system provides overlapping radar coverage for monitoring weather over the majority of the territory. The altitude at which complete radar coverage is achieved depends on the region's topography. In general, there is increased overlap in regions of higher population density. In Japan, besides the C-band national radar network that provides uniform coverage over the country, a new X-band network has been deployed with specific application to rainfall estimation for flood monitoring over the large urban areas and major cities (Fig. 10.3). In comparison with standard national networks, this network provides higher spatial and temporal resolution, with observations updated every minute.

10.2   Network Topologies for Radar Operations

The layout of a radar network is determined by a number of factors, including topography, areas to be covered, and available locations for siting a radar. The operations of the networks are divided into two classes: (1) a data-push network, where the radars operate in a predefined scan sequence and provide the data as it becomes available, and (2) a data-pull network, where the users specify the data and coverage of interest and the radar network is coordinated and prioritized.

The radar systems in a network can be operated cooperatively, where the radars are controlled together as a collective rather than only opportunistically (each radar

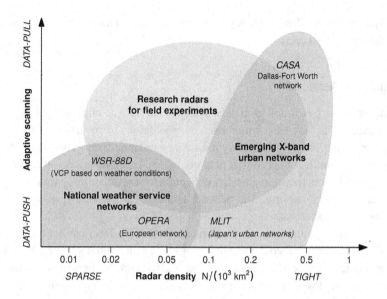

Figure 10.4 Examples of different types of radar networks. The network's geometry (sparse to tight) and the radar coordination strategy (data-push to data-pull) are shown according to a broad definition at the time of writing this book. In the lower-left corner of the diagram, National Weather Service (NWS) networks are typically sparse and adopt fixed scan strategies, whereas tight networks of X-band systems are more likely to use adaptive scanning strategies (e.g., the CASA network [3, 295], shown in the upper-right corner).

operates without knowledge or feedback from other radars). Also, there is no requirement that the radars in the network be homogeneous. A network can be composed of diverse operating frequencies and system capabilities. Consider the fusion of the NEXRAD S-band, TDWR C-band, and local X-band radar systems, which can provide multiscale (and multitemporal) coverage.

The network's layout, whether it be a tight network or a sparse network, refers to the physical geometry, but it also typically correlates to how the network is used. The diagram in Figure 10.4 shows different radar network which are broadly classified by the spatial density of the nodes and their scan flexibility. Most national radar networks are sparse (density on the order of 0.02–0.05 radars per 1000 km^2) and adopt fixed scan strategies (data-push networks). The WSR-88D radars in the United States can be considered an example of a moderately adaptive scanning network because the VCP is changed by operators according to the weather conditions. New emerging networks of X-band radars are typically much tighter with respect to long-range S-band and C-band systems, reaching a density on the order of 0.1 radars per 1000 km^2 in urban regions (see, e.g., the MLIT network in fig. 10.3). In the top-right corner of the diagram in Figure 10.4, the X-band CASA network in the Dallas–Fort Worth metroplex [294] is both tight (density of $\sim$ 0.3 radars per 1000 km^2) and operates based on a collaborative and adaptive approach (data-pull network). Data-pull networks don't necessarily need to be tight. An example of a sparse data-pull system is a research radar (or a combination of radars) tracking a severe storm based on reflectivity profiles. The concepts of data-push and data-pull networks are discussed in the next subsections.

10.2.1   Data-Push Networks

A network can be formed from radar systems that have no other relationship or interaction except that they exist within each other's regions of coverage. This opportunistic network is the simplest form of synthesizing a radar network. The data from the radar systems within the opportunistic network are collected in an open-loop fashion, with no feedback from the aggregate data collection to coordinate observations. The biggest advantage of the opportunistic network is that it can provide "gap-filling" coverage between radar systems.

A common example is for local radar systems to leverage national networks to provide observational coverage outside of the regional radar's range, as well as large-scale context for the short-range regional radars data. In many cases where research radars are deployed for targeted observations, an overlapping national network provides valuable information for planning the operations of the research radar, intercomparison of data, and the larger context of the atmospheric system.

In opportunistic radar networks, the placement and operation of the radar systems are not designed with these other radars in mind. The combination of multiple radars' observations is done when it is convenient, but it is not critical to the objectives of the radars. For the example of a research radar, if its objective (the research it is conducting) does not rely on other radar systems' observations, it could be opportunistically combined with observations from another radar or another radar network. Television station radar, air-surveillance radar (ASR), and terminal Doppler weather radar (TDWR) can be used to complement existing radar networks. Conversely, data from a larger network, such as NEXRAD, can be used to supplement a dedicated radar, such as an airport's ASR.

When radar systems are destined to be part of a network, the radar systems' architecture, radar locations, and operations are designed to fill particular roles in the network. For the data-push network, each radar operates independently but within a well-defined set of parameters such that when its data are combined with the data from other radars in the network, the performance is characterized, and the errors are understood and bound. For mobile radars that are deployed temporarily to provide coverage of radar gaps, these deployments and the integration of its observations are undertaken in a coordinated way. The gap-filling radar enhances the overall network, and the observation scan patterns are typically defined in a manner that maximizes the quality of the network's data product instead of the data product of any individual radar.

The mission of a radar system in a coordinated network is to supplement the network itself. In an opportunistic network, the individual radar's mission is separate from and not dependent on the other systems. Although this distinction may be trivial in some scenarios, in others, the radars' locations and operations can highly depend on the objective of the coordinated system.

10.2.2   Data-Pull Networks

A radar network can be designed and deployed so that all radars contribute to a collective operational strategy in which observations from one radar feed back to the operation of the other radars.

The "distributed collaborative adaptive sensing (DCAS)" operating concept adapts the scanning strategy of the radars with overlapping coverage within the network based on the location and evolution of severe weather [295–297]. Rather than a fixed volume scan strategy (e.g., NEXRAD's VCP 12) that covers 360° regardless of where the weather of interest is located and repeats with a fixed period that could be 5–10 minutes, the DCAS can interleave and operate multiple radars within the network to focus the scanning on the regions of interest. This adaptive scanning strategy can still provide the full volume scan (if desired) but can also greatly increase the rate of observations in sectors where hazardous weather (e.g., tornadoes) is evolving, improving revisit times to every minute or better as required by users.

Phased-array radar systems can potentially provide another degree of flexibility to fulfill multiple roles. Consider air-traffic control applications, which rely on weather radar information and short-term local forecasting to ensure the safety of aircraft. In high-traffic airspaces, a dedicated system of weather radar (e.g., the TDWR) is used to provide detailed and dedicated weather information. In other instances, the same ASRs used to track aircraft are outfitted with weather signal processors to provide weather information in the absence of a dedicated TDWR (and with insufficient NEXRAD coverage). This dual use approach provides an effective means of applying one radar system to fulfill multiple purposes. With multiple users of the radar systems, a data-pull configuration is needed for each user to define their needs and allow the radar system to determine the optimal scan strategy to achieve its goals.

10.3   Implications for the Radar Equation in a Network

Up to this point, the discussion has focused on using the observations from an individual radar to observe meteorological phenomena and infer the microphysical properties. As shown by the weather radar equation (eq. [2.39c]), the sensitivity (the minimum reflectivity the radar can detect) of the radar is primarily determined by the radar constant and the range from the radar. For cases where the sensitivity is sufficient, the radar's maximum range may be constrained by measurement accuracy requirements (in terms of the error bounds of the estimated radar variables).

The opportunity to implement a weather radar network allows the radar design problem to be reframed. The location and technical requirements are different between a single radar and a network of radars. The economics and capabilities of a single radar versus a network of radars can be approached with different philosophies to achieve the optimal performance. Although the performance of any individual radar does not change when used in a network, the inclusion of the radar's observations in the larger system adds additional capabilities and opens additional avenues for design optimization. The network optimization primarily considers the following radar performance parameters:

1. Minimum sensitivity: detection capability
2. Observation volume size (beamwidth): spatial resolution
3. Beam height: coverage
4. Number of overlapping radars: network's data accuracy

10.3.1   Network Geometry

The network optimization problem frequently starts as a geometry problem for positioning the network's radar systems. To simplify the problem, consider a network of N identical radar systems on a flat plane that are equally distributed along a circumference with radius $r_{cell} = r_{max}/M$ (the network cell range) around a central point that will be the origin O (i.e., the center of the network cell). The flat-plane assumption is not quite accurate for Earth's surface, but the error introduced by this approximation can be neglected for the moment. r_{max} is the maximum range of each radar within the network, and r_{cell} represents the distance between a radar node and the center of the network cell. Each of the N radars in the network is equally spaced along the circumference of a circle whose radius is given by r_{max}/M. M is called the *overlap ratio*. If $M = 1$, the radar's range extends to the center of the network. With $M < 1$, there is a coverage gap at the center of the circle about the origin, whereas for $M > 1$, the range of the single radar is larger than the range of the network cell.

The smallest N-sided polygon inscribable within a circumference of radius r_{max}/M is a triangle. Panel (a) of Figure 10.5 shows the minimal possible network cell composed of three radars for two particular overlap values, $M = 1$ (dashed circumference) and $M = \sqrt{3}$ (solid circumference). As shown by Junyent and Chandrasekar [298], this network configuration is more efficient in comparison with elemental network cells composed of higher-order polygons (squares, hexagons).

Adopting the network geometry from Junyent and Chandrasekar [298], the angular separation between the radars and the center is

$$\theta_s = \frac{\pi}{2} - \frac{\pi}{N}, \tag{10.1}$$

where $\theta_s = 30°$ for the triangular network cell in panel (a) of Figure 10.5a, and the separation distance between neighboring radars is

$$r_{sep} = 2\frac{r_{max}}{M} \sin\left(\frac{\pi}{N}\right). \tag{10.2}$$

The size of a radar's beam is $b_s = r\,\theta_{HPBW}$, where r is the range from the radar, and θ_{HPBW} is the 3-dB antenna beamwidth in radians. Within the network's observation space, the mean beam size is calculated as (eq. 25 in [298]):

$$\overline{b}_s^{(N)} = \frac{b_{s,max}}{3M \sin\theta_s} \left[\cos^2\theta_s \ln\frac{1+\sin\theta_s}{\cos\theta_s} + \sin\theta_s\right], \tag{10.3}$$

where $b_{s,max}$ is the radar's beam size at range r_{max}.

Assuming for simplicity that the radar is located on the ground (i.e., $h_0 = 0$ in Fig. 2.8), the minimum beam height b_h is defined as the aboveground height of a line tangent to Earth's surface at the radar location:

$$b_h = \sqrt{a_e^2 + r^2} - a_e, \qquad (10.4)$$

where r is the range from the radar, and a_e is Earth's effective radius (Section 2.4.5). This corresponds to the aboveground altitude of the radar beam's lower 3-dB line, with the antenna pointed at an elevation equal to half of its one-way 3-dB beamwidth; for example, an antenna with a $1°$ beamwidth pointing at $0.5°$ elevation. The mean minimum beam height over the network cell can then be calculated as follows (eq. 21 in [298]):

$$\overline{b}_h^{(N)} = \frac{b_{h,\max}}{M^2} \frac{1 + 2\cos^2\theta_s}{6}, \qquad (10.5)$$

where $b_{h,\max}$ is the radar's minimum beam height at the range $r_{\max}$. The mean detection sensitivity is as follows (eq. 29 in [298]):

$$\overline{Z}_{\min}^{(N)} = \frac{Z_{\max}}{M^2} \frac{1 + 2\cos^2\theta_s}{6}, \qquad (10.6)$$

where $Z_{\max}$ is the radar's minimum sensitivity at the range $r_{\max}$, $Z_{\min}(r_{\max})$. Note that in these equations, if we want to keep the network size constant and increase the overlap ratio M, we also need to increase the maximum range of the single radars $r_{\max}$, as shown in panel (a) of Figure 10.5.

In a real environment, the network grid cannot be perfectly regular following the idealized framework for many reasons including terrain limitations and logistics issues. Over the flat terrain of Oklahoma, the CASA IP1 network (panel [b] in

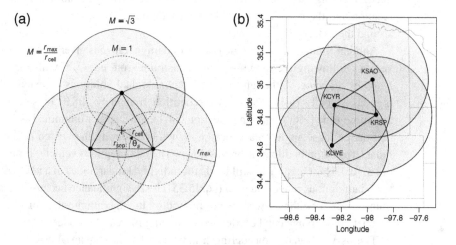

Figure 10.5 (a) Radar network cell diagram with $N = 3$ radar nodes. M is the overlap ratio (see text for details). (b) CASA IP1 X-band network in Oklahoma, with 40-km range rings.

Fig. 10.5) is an example of a small network of X-band radar systems resulting from the combination of two triangular elemental cells.

The network geometry can also significantly deviate from the pure geometric considerations illustrated previously, depending on the specific scope and application of the radar observations. For example, radars have the potential to benefit the population through severe weather early-warning systems in densely populated urban environments. In this case, the geometry of the network can be altered by the population density in order to add observations from multiple systems, as well as a degree of redundancy, where most people live. Examples of network designs that are influenced by population density include the X-band network in Japan (Fig. 10.3) and the CASA Dallas–Fort Worth urban test bed [299].

10.3.2   Network Sensitivity

For a single-radar system, sensitivity versus range is the typical driving performance parameter. As discussed in Chapter 2, increasing the sensitivity can be accomplished by increasing the antenna gain or transmitter power. The range effect, where sensitivity decreases as a function of r^2, has a large impact on defining the radar's performance.

Starting with eq. (2.61), assume a homogeneous network of radars with the same performance specifications. By substituting the antenna gain with the beamwidth using eq. (2.44), the minimum sensitivity is related to the radar's performance parameters as

$$Z_{\min} \propto \frac{\lambda^2 r^2 \theta_{\text{HPBW}}^2}{P_t}. \tag{10.7}$$

From this, the sensitivity of the radar is improved ($Z_{\min}$ is made smaller) by increasing the transmitter's power (P_t) or by decreasing the wavelength (λ), decreasing the range between the radar and target (r), or decreasing the antenna's beamwidth (θ_{HPBW}). The maximum range at which the radar system is effective may be influenced by the radar's location and the resulting beam blockage from the terrain, vegetation, and structures around the radar.

Another limitation for a single radar's maximum range is given by the unambiguous velocity following the range–velocity constraint from eq. (5.37), also known as the *Doppler dilemma*. Although there are technological solutions to extend the unambiguous velocity (using dual pulse-repetition frequency [PRF] or staggered-pulse-repetition time [PRT] sampling; Section 6.4.1), the Doppler dilemma is still an important driver for a single radar's design. When considering the design of a network of radars, the Doppler dilemma breaks down. The unambiguous velocity limit (eq. [5.36] or eq. [6.4.1] for dual PRF) still holds for individual radar nodes in a network, whereas the unambiguous range equation (eq. [5.35]) loses significance because the maximum range covered by the network is not limited by the maximum range of a single radar. The network's range can be extended by adding more radar nodes.

This is why, when considering a network of radars as an alternative to a single radar, the solutions and capabilities that can be achieved are fundamentally altered. If a single radar can be replaced by multiple radars, in the aggregate, the network of radars can achieve the same minimum sensitivities with multiple smaller and lower-power radar systems. To quantify the network's performance compared to a single radar, consider the improvement factor for the sensitivity, defined as the inverse ratio between the network detection sensitivity (eq. [10.6]) and the corresponding mean sensitivity of a single radar ($Z_{max}/2$):

$$I_Z = \overline{Z}_{min}(\text{dBZ}) - \overline{Z}_{min}^{(N)}(\text{dBZ}) = 10 \log_{10}\left(\frac{3M^2}{1 + 2\cos^2\theta_{sep}}\right). \qquad (10.8)$$

This sensitivity improvement factor compares a single network elemental cell with a single radar covering the same domain. Clearly, combining multiple elemental cells to build a wider network leads to increasingly larger improvement factors.

In addition to detection sensitivity and geometric improvements (namely, the mean beam size and beam height), the network also brings with it a number of other advantages not possible with the single-radar system:

- Doppler velocity estimation from multiple directions are available to directly measure 2D or 3D wind vectors
- Multiple observations from different viewing angles reduce the impact of beam blockages and path-integrated attenuation
- Potential for increased temporal sampling through multiple radars observing the same volume, improving severe weather (e.g., tornadoes or hail) detection lead times
- Improved measurement accuracy through an increased the number of independent observations of a volume
- Inherent fault tolerance

10.4 Gridded Data

Gridding of radar data is not specific to radar networks but it is a necessary process when combining observations from multiple radars. The term *gridding* refers to the following process: The radar naturally samples in the spherical coordinate system. For combining data from multiple sensors, common sampling points are needed. A Cartesian coordinate system is a natural system for combining multiple data sets. It has the advantage of mathematically simple operations for working with vectors (e.g., velocity), easy integration with numerical models, and uniform sampling within the domain, which simplifies the interpretation and combination of results from multiple sensors. Several gridding techniques with varying levels of complexity are typically available in common radar software packages, including the nearest-neighbor approach, bilinear interpolation, Cressman analysis [300], and Barnes analysis [301, 302]. The choice of the technique is mainly dictated by the available computational resources and by the variable being represented (e.g., scalar variables such as reflectivity, vector variables such as Doppler velocity, or categorical variables such as hydrometeor classification). Additionally, the volume of radar observations is collected over a variable period of time, which depends on the individual radar's scan strategy. The simplest interpolation assumes that there is negligible motion during the observation period. Although the period of time is finite, storm motion can result in a skewing of observations due to the time elapsed over the volume scan. Techniques can be used to try to estimate and compensate for the motion of the meteorological systems to time-align the observations from multiple elevations and multiple radars within the network prior to spatial interpolation [303].

In some instances, it may be more appropriate to grid model data or other sensor data onto a radar's spherical coordinate system for a direct comparison, such as when mapping one radar's observations onto another radar's for direct comparison of results. Conversion onto an existing radar's spherical coordinate system results in an overall reduced error (error from one gridding rather than two).

As a first step for any regridding activity, a mapping must be established between two different coordinate systems (see also the discussion about radar view geometry in Section 2.4.5). This can be done in either direction; for example, to convert from polar to Cartesian coordinates:

$$
\begin{bmatrix} x \\ y \\ z \end{bmatrix} = \begin{bmatrix} r \cos(\pi/2 - \phi) \cos \theta \\ r \sin(\pi/2 - \phi) \cos \theta \\ r \sin \theta \end{bmatrix},
\tag{10.9}
$$

and conversion from Cartesian to polar coordinates is calculated as follows:

$$
\begin{bmatrix} r \\ \phi \\ \theta \end{bmatrix} = \begin{bmatrix} \sqrt{x^2 + y^2 + z^2} \\ \tan^{-1}(y/x) \\ \sin^{-1}(z/r) \end{bmatrix}.
\tag{10.10}
$$

Note that the angle value returned by $\tan^{-1}$ is ambiguous for x and y spanning both positive and negative values. The function "atan2," available in most programming languages (also see the Appendix), can be used to return a correct and unambiguous value for the angle ϕ. These formulas are used to project onto a Cartesian space resting on a plane tangent to Earth at the radar location (flat-plane assumption). For reprojecting the radar observations onto the surface (with x and y becoming arc distances from the radar along two orthogonal great circle paths), Earth's equivalent radius a_e needs to be taken into account, using the equations in Section 2.4.5.

The transformation provided by eq. (10.9) is useful for projecting single-radar observations onto a Cartesian plane, without interpolation. For example, the PPI in Figure 1.3 shows a typical plot of reflectivity that preserves the native polar geometry. On the other hand, the reverse transformation (eq. [10.10]) is used when observations on a regular Cartesian grid are needed, for example, for the elaboration of network composites.

The gridding operation provides interpolation from the radar's spherical sampling points to the desired grid points, and as a result of the interpolation, the data undergo a degree of filtering. The characteristics of the interpolation filter have an impact on the overall data quality. Because of the modification of the radar data, it is important for the user of the data to understand the basic operation and limitations to properly inform the interpretation of the results. For example, gaps in radar data coverage in combination with a vertical gradient in the wind field can induce a bias in the interpolation of Doppler observations onto a Cartesian grid [304]. Similarly, strong gradients in Z_{dr} resulting from precipitation variability or phase changes (think of rain/hail transitions) may pose a challenge for gridding onto a Cartesian domain. A viable solution in this case consists of first applying interpolation separately on the two reflectivity fields (Z_h and Z_v) in dBZ, whose spatial variation is often close to linear, and then obtaining Z_{dr} from their difference.

Although specific applications may need special solutions, the goal here is to provide a discussion of some common techniques for interpolating radar data onto a grid and the resulting effects on the data. Considerations and techniques for gridding and interpolating radar data can be explored in far more detail than covered here. For further study, the interested reader may refer to the available literature (e.g., linear interpolation methods [305, 306], the Cressman weighting scheme [300, 307], and the Barnes exponential weighting scheme [301, 302, 308]).

Gridding is an important tool for maximizing the effectiveness of observations from multiple radars. The simplest method of gridding is the nearest-neighbor approach. For each point on the grid, the value of the closest radar measurement is assigned. Assuming a higher density of radar samples than the grid's resolution, the nearest-neighbor approach provides reasonable results. Common alternatives to the nearest neighbor are bilinear interpolation and the Cressman or Barnes analysis techniques, which allow for weighted interpolation. The Cressman and Barnes analysis techniques interpolate the existing data in between samples while trying to match the interpolated field where samples exist. This should yield similar data if the grid points are the same as that of the original data but has the benefit of using a region of data where gaps exist in the sampled data set. Other techniques also exist for gridding, including statistical [309] and variational methods [310]. Techniques specifically conceived for network applications can be designed to account for the range-dependent radar beam geometry; vertical gaps between radar scans; the lack of time synchronization; and varying beam resolutions between radar nodes in the network, storm movement, and terrain blockage [303].

Alternate solutions exist to interpolate data at constant altitudes and fill some of the gaps between the scanning elevations. Techniques range from computationally complex multivariate interpolations with physical constraints to more simple region-of-influence averages, to a nearest-neighbor search. Each of these techniques has its place, and the appropriate choice largely depends on the needs of the end user. To characterize the fine-scale structure of precipitation features, a nearest-neighbor or region-of-influence interpolation technique can lead to incorrect conclusions because of the inherent spatial filtering effects, but the resulting fields are frequently more than adequate for estimating the accumulated rainfall within a region.

One of the biggest challenges of gridding radar observations is the relatively sparse sampling within a volume. Although the radar samples are relatively dense in range and azimuth, the sampling in elevation is typically sparse and limited in its maximum elevation. This leads to a "cone of silence" above the radar that remains unsampled (see Figure 10.2).

10.4.1   Bilinear Interpolation

Given a regular Cartesian grid, for any grid point (x, y, z), the corresponding coordinates in the spherical domain (r, ϕ, θ) can be calculated from eq. (10.10). For simplicity, consider the 2D Cartesian (dashed lines) and polar (solid lines) domains represented in panel (a) of Figure 10.6 (assume $\theta = 0$ for the moment). Gridding using a nearest-neighbor approach would simply imply using the radar observation at P_{21} for the estimation of an observation field f at the Cartesian grid point marked by the symbol "+":

$$f(C_{x,y}) = f(P_{\phi,r}) = f(P_{21}),   (10.11)$$

where C is used to indicate a location in the Cartesian domain, P is used to indicate the same location in radar-centric polar coordinates, and mapping between the two is obtained through eqs. (10.9) and (10.10) (with $\theta = 0$, $z = 0$).

Because in general $C_{x,y} \neq P_{i,j}$ for any i, j (where i and j indicate the discrete azimuth and range bin location, respectively), a better estimation of the radar moments at $C_{x,y}$ can be obtained by considering the four adjacent observations on the polar

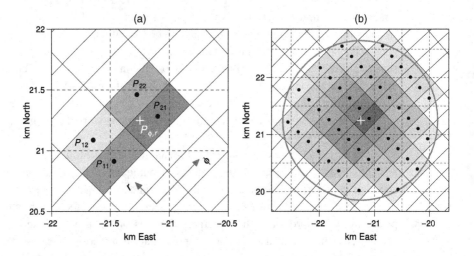

Figure 10.6 Geometry for bilinear interpolation (a) and Cressman analysis (b) for typical radar observations on a polar domain (solid lines) with a range resolution of 250 m and $1°$ azimuth resolution. The Cartesian grid (dashed lines) has 500-m resolution, the gray shading provides a qualitative indication of the weight given to radar observations (darker is higher), and the circle in panel (b) denotes a radius of influence of 1.4 km.

grid (i.e., the gray-shaded bins in Fig. 10.6). An improved estimate at $C_{x,y}$ can be calculated using bilinear interpolation, which consists of applying a first linear interpolation in one direction (e.g., azimuth) and then a second linear interpolation in the other direction (range). Note that as a result of the product of two linear functions, the bilinear interpolation is not linear but quadratic in the sample location. The value at $C_{x,y}$ is retrieved following:

$$f(C_{x,y}) = f(P_{\phi,r})$$
$$= \frac{1}{(r_2 - r_1)(\phi_2 - \phi_1)} \begin{bmatrix} \phi_2 - \phi & \phi - \phi_1 \end{bmatrix} \begin{bmatrix} f(P_{11}) & f(P_{12}) \\ f(P_{21}) & f(P_{22}) \end{bmatrix} \begin{bmatrix} r_2 - r \\ r - r_1 \end{bmatrix}.$$
$$(10.12)$$

Note that with this formulation, observations spaced at Δr (range resolution) will be given the same weight as observations spaced at $\Delta \phi$ (azimuth resolution). This limitation is overcome in the Barnes analysis, where the weight function is split into separate directions (radial distance and angle for 2D interpolation), with specific smoothing parameters (refer to the discussion in the next subsection).

The same procedure can also be used for range-height indicator (RHI) scans, considering the pertinent coordinate mapping. Three-dimensional volume scans can be constructed from either a series of PPI scans or a series of RHI scans, although the former method is by far the most commonly used scan strategy. In either case, the sampling in one angular direction is contiguous, following the uniform movement of the antenna (e.g., the azimuthal direction for PPIs), whereas the other angular direction (elevation for PPIs) is typically sampled at not-equally-spaced angles and noncontiguous times. Although the 2D interpolation scheme illustrated previously can easily be extended to a 3D domain (trilinear interpolation), for example, to create constant altitude PPI (CAPPI) products at different altitudes, for the interpolation of volume scans, it is common practice to use the bilinear interpolation over 2D sweeps, followed by linear interpolation between adjacent scans. For a polar volume composed of PPI scans, this second step may include only vertical interpolation (between two observations on the adjacent tilts that are below and above the grid cell) or it may include both vertical and horizontal interpolation (adding two observations at the same height, which translates to different ranges due to the elevation from the two adjacent tilts below and above the grid cell). For example, Zhang et al. [311] adopt a scheme where the horizontal interpolation is only included when adjacent tilts are more than $1°$ apart. The additional horizontal interpolation may help to reduce possible artifacts; for example, the bright band may introduce ring-shaped discontinuities arising from the different sampling in azimuth (uniform) and in elevation (nonuniform, with varying gaps between adjacent elevation sweeps). On the other hand, horizontal interpolation is not recommended when the field's variability in the horizontal direction is large (such as for convective cases). The weights should then be adapted based on the actual spatial scales of precipitation, for example, relying on objective methods for the identification of the bright band (see Section 8.1.1) to distinguish between stratiform and convective precipitation.

10.4.2   Cressman and Barnes Analysis

As shown in eq. (10.12), bilinear interpolation estimates the appropriate value at the Cartesian grid point through a weighted average of the four nearest radar bins. A weighted average can be obtained in a number of other ways, with varying levels of sophistication. Here, empirical objective analysis methods developed for meteorological applications are discussed for gridding and interpolation.

Cressman proposed one of the earliest techniques for interpolating meteorological observations (specifically atmospheric soundings) onto a grid [300]. This technique is still widely used today for interpolation of radar observations because of its simplicity. For a selected point to be estimated, the Cressman technique uses a radius of influence to determine the value for the selected point. Within the region of influence, closer points (points with smaller radii) have a greater influence than points farther away. The maximum radius of the region of influence can also be adjusted with the range from the radar to accommodate the varying density of observations.

Within the region of influence each point has a weighting factor

$$w_{ik} = \begin{cases} (r_m^2 - d_{ik}^2)/(r_m^2 + d_{ik}^2) & d_{ik} \leq r_m \\ 0 & \text{otherwise,} \end{cases} \tag{10.13}$$

where r_m is the maximum radius of influence, and $d_{ik} = \sqrt{(x_k - x_i)^2 + (y_k - y_i)^2}$ is the Euclidean distance from the interpolation point (x_i, y_i) on the Cartesian grid to the sample point (x_k, y_k) in the polar domain.

The interpolated field g is estimated from the observation field f by the weighted sum of the values within the region of influence:

$$g(x_i, y_i) = \frac{\sum_k w_{ik} f(x_k, y_k)}{\sum_k w_{ik}}. \tag{10.14}$$

Similar to the Cressman circular region of influence, the Barnes analysis interpolation technique uses a Gaussian interpolation window [301] with a weighting function:

$$w_{ik} = \exp(-d_{ik}^2/k_d), \tag{10.15}$$

where k_d is the fall-off range parameter. In the adaptive Barnes scheme [302], the isotropic fall-off range parameter k_d is replaced by a smoothing parameter split into radial and angular components. For the interpolation of two-dimensional fields (PPI), the weighting function takes the following form:

$$w_{ik} = \exp(-r_{ik}^2/k_r - \phi_{ik}^2/k_\phi), \tag{10.16}$$

where r_{ik} is the radial distance between the ith analysis location and the kth observation location, ϕ_{ik} is the azimuthal difference between the ith analysis point and the kth observation, and k_r and k_ϕ are the smoothing parameters in the radial and azimuthal directions. This definition of the weighting function explicitly considers the sampling geometry of the radar, providing an automatic adaption to data density.

The interpolated field is first estimated from the observation field f by the weighted sum of the values within the region of influence, as in eq. (10.14). This initial interpolated field (g_0) provides the first guess for a subsequent correction step to represent smaller scales in the analysis. This is accomplished by comparing the interpolated field g_0 to the observation field f at the location of the observations:

$$g_1(x_i, y_i) = g_0(x_i, y_i) + \frac{\sum_k \left[f(x_k, y_k) - g_0(x_k, y_k) \right] w'_{ik}}{\sum_k w'_{ik}}, \qquad (10.17)$$

where the modified weighting function

$$w'_{ik} = \exp\left[-r_{ik}^2/(\gamma k_r) - \phi_{ik}^2/(\gamma k_\phi) \right] = w_{ik}^{1/\gamma} \qquad (10.18)$$

incorporates decreased smoothing parameters (scaled by γ, with $0 < \gamma < 1$) and, consequently, steeper fall-off to force fast convergence. The parameter γ is typically set to a value around 0.5. In eq. (10.17), $g_0(x_k, y_k)$ is the value of the first-pass analysis at the polar sample points and can be obtained through bilinear interpolation.

Note that the original Barnes analysis scheme [301] did not include the γ parameter and required more passes to reach the same level of convergence as the modified two-pass scheme including γ (eq. [10.17]) [312]. This second pass is important for practical applications because it restores reliable peak values in the analysis. Think, for example, of threshold-based products such as hail-detection maps (reflectivity threshold at 55 dBZ). If an observed peak reflectivity (e.g., 56 dBZ) is smoothed too much in the first pass, the interpolated value may fall below the predefined threshold and miss a detection. The two-pass Barnes analysis scheme achieve the two fundamental functions of any gridding operation, interpolation and filtering, while retaining peak values close to the observations in the final analysis.

The Cressman and Barnes analysis schemes can be easily extended to 3D observations (volume scans). However, as discussed for bilinear interpolation, it is often convenient to perform the interpolation between adjacent scans in a second step.

10.5   Network-Based Applications

The availability of multidirectional observations in a dense radar network allows the development of new applications or the improvement of existing ones, including wind-field retrievals, rainfall estimation, network calibration, and attenuation correction. In particular, applications like quantitative precipitation estimation (PQE) and wind-field retrievals are among the most studied and widely used applications for operational weather radars. Another common application is network calibration, which consists of exploiting the enormous amount of data collected by overlapping radars during precipitation to derive the relative calibration alignment and track eventual deviations with high accuracy.

More recent applications include multiple-radar retrieval of reflectivity and specific attenuation, which is specifically linked to the development of weather radar networks operating at higher frequencies, such as the X band. The concept was initially studied by Srivastava and Tian [313], who used an analytical method to estimate specific attenuation and reflectivity in a dual radar configuration. Chandrasekar and Lim [314] later developed a new procedure in a networked radar environment, based on the solution of the integral equation for reflectivity, in a manner similar to that used in the rain-profiling algorithm with a differential phase constraint (Section 9.4.2) but relying only on reflectivity observations (Section 10.5.4). Other advanced applications include retrieval of the drop size distribution (DSD) by exploiting the redundancy of radar observations in a network [315].

10.5.1   Improved Rainfall Estimation

One of the main limitations of long-range single-radar QPE is the radar observation's height above the ground, which steadily increases with range (eq. [2.24]). A network of radars enables observation of precipitation closer to the ground, reducing the impact of microphysical processes, and the resulting variation in QPE estimates, between the height of the radar resolution volume and the surface (see, e.g., Section 8.4 for warm-rain processes). In addition, a network of shorter-range radars, as opposed to a single long-range system, has the advantage of providing higher-resolution observations within the network's domain (eq. [10.3]). This is very important, given the spatial variability of the DSD, for providing accurate high-resolution rainfall estimation data (e.g., to a hydrological model for flash-flood prediction), which is particularly important over complex terrain or urban areas. The DSD not only varies locally in space but also over time (see Figure 3.6 as an example). To increase the time resolution of the observations, radar networks can provide solutions that go beyond any individual radar's system design. In particular, adaptive scanning strategies in data-pull networks allow for the concentration of the network observation capability over specific regions of interest, thus substantially increasing the frequency a selected storm is observe [316]. In fact, convective precipitation systems may evolve very rapidly, showing substantial change in radar signatures on the scale of tens of seconds. For these fast-evolving storms, the increased temporal sampling may allow a dramatic improvement of the QPE quality.

With an adaptive and collaborative scanning approach, the same scattering region is observed almost simultaneously by multiple radars from multiple directions. The different views of a storm imply different paths through the atmosphere and precipitation in between, with corresponding different attenuation suffered by the individual radars. This allows for new reflectivity and specific attenuation retrieval methods (Section 10.5.4), which also results in higher accuracy of the rainfall estimates.

Polarimetry brings additional value to radar network QPE applications. The use of $R(K_{dp})$ in particular has the notable advantage of being less sensitive to DSD variations compared with other estimators relying on the power terms, such as Z_h and Z_{dr}. At short radar wavelengths such as the X band, $R(K_{dp})$ relations can also be applied in light-rainfall regimes (Section 9.5.5). Furthermore, fusion of independent K_{dp} estimates at the same location, for example, based on the best estimation quality of

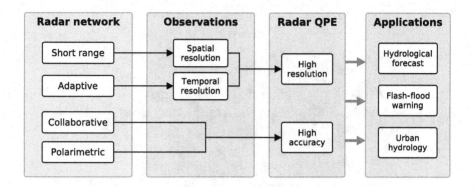

Figure 10.7 Salient features of a radar network for improved QPE applications.

K_{dp} [175], can more easily be achieved compared with power estimates because K_{dp} is immune to attenuation and radar calibration. Considering the challenges of absolute calibration across all nodes of a network, and the variable effects of attenuation, the use of differential phase observations for rainfall estimation can be considered a great advantage of QPE in a network. Overall, the combination of short-range dual-polarization radars in an adaptive and collaborative sensing strategy for the lower atmosphere results in accurate and fine-resolution estimates of precipitation for hydrological applications (Fig. 10.7).

10.5.2   Multiple-Doppler Velocity Estimation

In a Cartesian coordinate system, the wind velocity is expressed by the three scalars u, v, and w, which represent the velocity components in the east, north, and vertical directions, respectively. Radar measures the radial Doppler velocity v_r along the radar's pointing angle in the radar spherical coordinate system with the coordinates, range (r), elevation (θ), and azimuth (ϕ). The precipitation particles fall with a terminal velocity V_t, subject to the gravitational force (Table 3.1). The Doppler velocity v_r is then the projection of the vector v (with components u, v, $w + V_t$) onto the radar's line of sight:

$$v_r = u \sin \phi \cos \theta + v \cos \phi \cos \theta + (w + V_t) \sin \theta. \qquad (10.19)$$

Equation (10.19) can be written in vector form as follows:

$$v_r = a^\mathsf{T} v, \qquad (10.20)$$

where a is the unit vector in the radial direction from the radar to the observation point:

$$a = \begin{bmatrix} \sin \phi \cos \theta & \cos \phi \cos \theta & \sin \theta \end{bmatrix}^\mathsf{T}. \qquad (10.21)$$

Using the standard convention, azimuth is zero at north and increases in the clockwise direction; elevation is zero parallel to the ground and increases to 90° at the

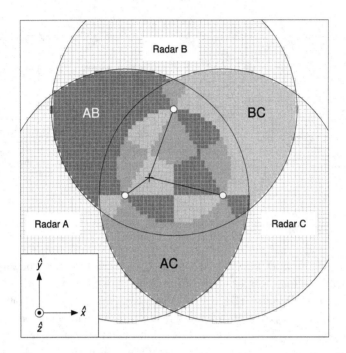

Figure 10.8 Multi-Doppler diagram shows the radar locations and Cartesian grid points for a network of three radars with overlap ratio $M = 2.2$. The gray colors identify the regions of the best dual Doppler pairs, as denoted by the overplotted letters (e.g., the darker gray is where the best dual Doppler estimate is obtained using radars A and B.

zenith. The sign of the Doppler velocity denotes whether the scatterers are moving away from or toward the radar. Following Section 5.3.2, the convention is that a positive v_r is moving away from the radar, and a negative v_r moves toward the radar.

Now consider the case when multiple radars measure the velocity of the scatterers, as shown in Figure 10.8. To solve for the velocity vector (which has three unknowns), three radial velocity measurements that are not all in the same plane are required (i.e., they should not all have an elevation of zero). With three or more radars, the point's velocity vector v can be estimated from the radial velocity measurements of the N radars $(v_{r1}, v_{r2}, \ldots, v_{rN})$ using the relative observation geometry vectors $(a_1, a_2, \ldots, a_N)$. In fact, for the matrix A constructed of linearly independent rows (the radars are in different locations, and their radials are not coincident), as follows:

$$ A = \begin{bmatrix} a_1^{\mathrm{T}} \\ a_2^{\mathrm{T}} \\ \vdots \\ a_N^{\mathrm{T}} \end{bmatrix} \tag{10.22} $$

the velocity vector v is obtained from

$$ v = A^{\dagger} v_r, \tag{10.23} $$

where $A^\dagger$ is the pseudo-inverse of A:

$$A^\dagger = A^T(AA^T)^{-1}, \tag{10.24}$$

and v_r is a vector of the measured radial velocities from radars 1, 2, ..., N:

$$v_r = \begin{bmatrix} v_{r1} & v_{r2} & \cdots & v_{rN} \end{bmatrix}^T. \tag{10.25}$$

Typically, radar velocity estimates are primarily within the horizontal plane because of the low-elevation angles of the radars' scans, causing the vector a to have a negligible third component ($\sin\theta \approx 0$). Consider, for example, the estimation of the wind vector at the location denoted by the "+" symbol in Figure 10.8. If the three radars are scanning at low-elevation angles, it is evident that using observations from radars B and C is preferable to the other two possible pairs (AB or AC) because of the quasi-orthogonality of these radar beams. In such a case, the radial velocity observation from radar A would not add significant information (being the radar beam almost aligned with the beam of radar B). The radar's observations may have small angles relative to one or two of the dimensions (see eq. [10.21]). As a results, small radial velocity measurement errors can have a large multiplier effect from the inversion in (eq. [10.23]) for wind directions that are nearly orthogonal to the radar's radial. While a variety of techniques to mitigate these errors can be considered, a robust solution is to select the appropriate observations to include in the multi-Doppler estimates.

In dual Doppler retrieval, the vertical velocity component ($w + V_T$) can be treated as unrecoverable in a first instance, and the transform matrix A in eq. (10.23) becomes square (two rows for the radars and two columns for the first two components of vector a) and invertible, leading to a simple solution for the horizontal wind field:

$$v = A^{-1}v_r. \tag{10.26}$$

Inverting matrix A, we get

$$u = \alpha_1 v_{r1} - \alpha_2 v_{r2} \tag{10.27a}$$

$$v = \beta_2 v_{r2} - \beta_1 v_{r1}, \tag{10.27b}$$

where

$$\alpha_1 = \frac{\cos\phi_2}{\det A}; \ \alpha_2 = \frac{\cos\phi_1}{\det A}; \ \beta_1 = \frac{\sin\phi_2}{\det A}; \ \beta_2 = \frac{\sin\phi_1}{\det A}, \tag{10.28a}$$

and

$$\det A = (\sin\phi_1 \cos\phi_2 - \sin\phi_2 \cos\phi_1)\cos\theta_1 \cos\theta_2$$
$$\approx \sin\phi_1 \cos\phi_2 - \sin\phi_2 \cos\phi_1; \text{ for } \theta_1 \approx 0 \text{ and } \theta_2 \approx 0. \tag{10.28b}$$

The accuracy of the velocity retrieval depends only on the radar network geometry relative to the precipitation system being observed. In particular, assuming that the variance of the radial velocity is the same for all radars, that is, $\sigma^2(v_r) = \sigma^2(v_{r1}) = \sigma^2(v_{r2})$, the normalized variance for the dual Doppler retrieval is obtained by simple trigonometric manipulation:

$$\sigma_N^2(v) = \frac{\sigma^2(u) + \sigma^2(v)}{2\,\sigma^2(v_r)} = \frac{1}{\sin^2(\phi_1 - \phi_2)} \frac{\cos^2\theta_1 + \cos^2\theta_2}{2\cos^2\theta_1\cos^2\theta_2}$$

$$\approx \frac{1}{\sin^2(\phi_1 - \phi_2)}; \text{ for } \theta_1 \approx 0 \text{ and } \theta_2 \approx 0. \tag{10.29}$$

Equation (10.29) shows that the retrieval error tends to infinity when the angle subtended by the two Doppler radar beams on the horizontal plane approaches $0°$ or $\pm 180°$. In general, a subtended angle that lies between a $30°$ and $150°$ angle (which corresponds to a normalized standard error $\sigma_N(v) < 2$) is adequate for retrieving the horizontal wind velocity [317].

The accuracy estimate in eq. (10.29) can be used to evaluate the impact of the two radar's observation geometry on the wind velocity retrieval. The best pairs in Figure 10.8 have been identified by minimizing the variance of the three possible dual Doppler pairs in the radar network's overlapping regions at every Cartesian grid point. The optimization of the dual radar selection for wind retrieval in a network can be extended to include other factors, such as higher-elevation angles and the range from the radar (which deteriorates both beam resolution and sensitivity). The problem can then be stated in general terms through the definition of an objective function, whose minimization results in the best dual Doppler pair according to the specified criteria [318, 319].

Given the difficulty of retrieving the vertical velocity component for typical ground-based weather radar networks, it is common to rely on an additional constraint represented by the principle of mass conservation. The measured horizontal velocity is combined with the anelastic mass conservation equation to derive the vertical velocity component:

$$\frac{\partial(\rho u)}{\partial x} + \frac{\partial(\rho v)}{\partial y} + \frac{\partial(\rho w)}{\partial z} = 0. \tag{10.30}$$

The air density ρ is spatially varying but is assumed to only change with height (i.e., along $\hat{z}$). It is important to note that as the radar's elevation increases from zero, the vertical velocity (which includes both the fall speed of the precipitation and the vertical velocity of air) contributes to the observed radial velocity. Using empirical relationships, the terminal fall speed of precipitating particles can be estimated from the radar reflectivity (see, e.g., [49]).

The practical solution of eq. (10.30) in the framework of multi-Doppler wind retrieval has been a research topic for years. While a brief introduction was provided here, solutions range from simple, nonsimultaneous iterative schemes (which first derive the horizontal wind components, then the vertical component, and iterate in the Cartesian coordinate system) [320] to simultaneous solutions using variational formalism [321] or minimization techniques that preserve the radial nature of the wind observations [322]. Wind-analysis methods based on data assimilation in cloud models with physical constraints (see, e.g., [323]) are also gaining increasing attention as a result of expanding computational resources.

10.5.3   Network Calibration

When a network of radars is used to observe weather phenomena, it is important that all systems are calibrated to a common reference level. End-to-end absolute calibration of individual radars is often performed sporadically, so it is possible for a system's calibration to deviate from that of neighboring nodes in the network at some point.

Observations routinely collected during precipitation events provide the basis for checking the relative calibration alignment for all radars in a network that has as significant overlap ratio (tight network). Consider the intersection between two PPI scans, as illustrated in Figure 10.9. Leaving aside the different geometry of the radar resolution volumes (size, orientation), in an idealized scenario without measurement uncertainties or calibration errors, the observations corresponding to the common volumes at the intersection of the cones should match. In a real-world scenario, measurement uncertainties will introduce a dispersion about a regression line whose intercept value will give the relative bias between the two radars.

Although in theory, all intersecting volumes between the scans of two radars could be considered, it is preferable to apply some constraint on the selection of the observations to limit the impact of measurement uncertainties and observation geometry. Panel (a) of Figure 10.10 shows the intersections between a low-level scan of the C-band Monte Settepani radar in Italy and four scans from the C-band Bric Della Croce radar, located 95 km to the northwest. The example of daily inter-calibration illustrated in panel (b) of Figure 10.10 was obtained by selecting observations for which the height mismatch between the two radars' resolution volumes (considering

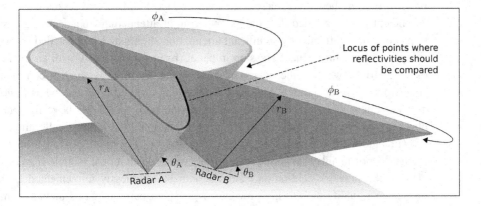

Figure 10.9 The intersection between two conical scans (thick parabolic line) at generic elevations θ_A and θ_B identifies the pairs of polar coordinates $(\theta_{A,B}, \phi_{A,B}, r_{A,B})$ for use in network calibration.

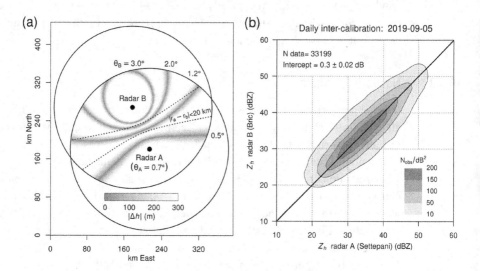

Figure 10.10 (a) Surface projection of the intersections between the 0.7° elevation of the Settepani radar (radar A, 1400 m mean sea level [MSL]) and four low-level scans (0.5°–3.0° elevation) of the Bric radar (radar B, 740 m MSL) in Italy. Gray shading indicates the magnitude of the beam-height difference Δh between the two conical scans, whereas the dotted lines delimit the area where the range from the two radars differs by less than 20 km. (b) Example of daily Z_h–Z_h calibration scatterplot using observations where $|\Delta h| < 100$ m (to ensure good matching in the vertical) and $|r_A - r_B| < 20$ km (to ensure a similar size of resolution volumes). Only measurements with $\rho_{hv} > 0.94$ are used to avoid contamination by the melting layer, large precipitation gradients, or other artifacts.

the altitude of the beam center) is within 100 m and the range from the two radars differs by less than 20 km. These constraints ensure good matching in the vertical direction (measurements close to the ideal intersection of the conical scans) and a similar size of the resolution volumes. In addition, to avoid the impact of strong precipitation gradients (e.g., across the melting layer), a threshold on the correlation coefficient is also applied. For radars operating at attenuating frequencies, observations potentially affected by significant attenuation should also be excluded from the analysis. The result of the regression in panel (b) of Figure 10.10, with an intercept of −0.3 dB, shows that the radars are well aligned (within the current community standard of 1 dB accuracy). Although the residual standard error can be as large as a few decibels (the spread around the regression line), given the large number of observations typically available ($>10^4$ in this example), the relative calibration can be estimated with very high accuracy.

Because of its inherent simplicity and reliance on operational volume scan observations, the network calibration technique can be seamlessly implemented as part of standard quality-monitoring procedures [324]. In particular, this method can be used in synergy with other single-radar calibration monitoring techniques, such as relative calibration adjustment unsing clutter echoes (discussed in Section 7.4.3), to track deviations in reflectivity measurements.

10.5.4   Attenuation Correction

A network of radars viewing a common volume in space can experience different path integrated attenuation in their measurements. Recognizing this, Chandrasekar and Lim [314] developed a new approach to attenuation correction. This network-based solution is especially relevant for attenuating frequencies such as the X band. Consider a network of three radars operating at the same attenuating frequency, as shown in Figure 10.11. The shaded contours represent a hypothetical field of intrinsic (unattenuated) reflectivity. For example, for observations of the common volume V_3, the reflectivity seen by radar C is expected to be smaller than the reflectivity observed by the other two radars because of enhanced attenuation through the core of the precipitation system. For this example, it's likely that $Z'_{h,B} < Z'_{h,A} < Z'_{h,C}$, where the primed variables indicate observations (not corrected for attenuation), and the letters denote the radar nodes in the network. In the absence of attenuation and uncertainties, the reflectivity in a common volume for each radar should be the same, that is, $Z_{h,A}(V_N) = Z_{h,B}(V_N) = Z_{h,C}(V_N)$, where the un-primed variables indicate intrinsic values. However, the observed reflectivity in a common volume are different for each radar because of the different PIAs and measurement uncertainties.

Consider the expression for specific attenuation in the rain-profiling algorithm (eq. [9.51]). In that formulation, the attenuation profile could be retrieved thanks to a

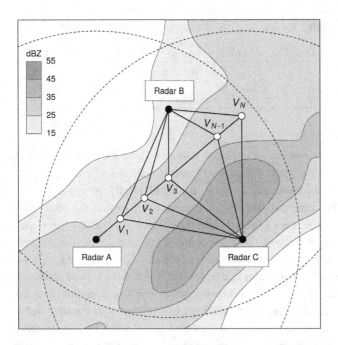

Figure 10.11 Schematic of the conceptual arrangement for attenuation retrieval in a network of three radars with overlap ratio $M = 2.4$. $V_1, V_2, V_3, \ldots, V_N$ indicate common volumes in the overlap region, and the dashed lines represent the maximum range of each radar.

constraint on the total path-integrated attenuation estimated from Φ_{dp} measurements. Now, suppose that an estimate of intrinsic reflectivity is available at the reference range r_{max}. Then, the specific attenuation can be estimated as

$$A_h(r) = \frac{[Z_h'(r)]^b \left(10^{0.1b\Delta Z_h(r_{max})} - 1\right)}{I(r_{min}, r_{max}) + \left(10^{0.1b\Delta Z_h(r_{max})} - 1\right) I(r, r_{max})}, \tag{10.31}$$

where $Z_h'(r)$ is the observed reflectivity at range r, $\Delta Z_h(r_{max})$ is the difference between unattenuated and attenuated reflectivity at range r_{max} (i.e., the two-way cumulative attenuation, and the integral terms $I(r_{min}, r_{max})$ and $I(r, r_{max})$ are given by eq. (9.48). In practice, the intrinsic reflectivity $Z_h(r_{max})$ is not known, but in a networked environment, the redundancy provided by multiple independent radar observations can be exploited. The basic idea is then to replace the constraint provided by $\Delta\Phi_{dp}$ in eq. (9.51) with the constraint that multiple radar observations should match at any common volume V_i (located at range $r_{max,A}$, $r_{max,B}$, $r_{max,C}$ from the radar nodes) after attenuation correction. This can be accomplished through an iterative procedure, starting from a first-guess reflectivity. The first estimated value of intrinsic reflectivity at a common volume V_N is set to the maximum value of observed reflectivities for that volume. Next, using eq. (10.31), the specific attenuation for radar A is estimated along the path between radar A and V_N, and the corrected reflectivity is evaluated using eq. (9.40). From these attenuation-corrected estimates, the specific attenuation profiles for radars B and C can also be calculated. Following this process, three independent estimates of specific attenuation, namely $\widehat{A}_{h,A}$, $\widehat{A}_{h,B}$, $\widehat{A}_{h,C}$, at the common volumes $V_1, V_2, \ldots, V_N$, are available to calculate a cost function:

$$\delta k = \frac{|\widehat{A}_{h,A}(V_i) - \overline{A_h}(V_i)| + |\widehat{A}_{h,B}(V_i) - \overline{A_h}(V_i)| + |\widehat{A}_{h,C}(V_i) - \overline{A_h}(V_i)|}{\overline{A_{h,A}}(V_i)};$$

$$\overline{A_h}(V_i) = \frac{\widehat{A}_{h,A}(V_i) + \widehat{A}_{h,B}(V_i) + \widehat{A}_{h,C}(V_i)}{3},$$

$$\tag{10.32}$$

where N is the number of common volumes along the path between the radar and V_N. δk is the weighted difference of specific attenuation along each radar path in the common volumes, and as such, it should ideally be equal to zero for perfect (unattenuated) observations. In practice, in order to find the value of unattenuated reflectivity at V_N, the estimated reflectivity value is perturbed from the initial guess until the value of the cost function reaches a minimum. Note that the only coefficient needed for the retrieval is the exponent of the $A_h(Z_h)$ relation in eq. (9.44) (coefficient b in eq. [10.31]), whose value can be set to $b \approx 0.8$ at the C band and the X band [279].

Analogous to the rain-profiling algorithm with Φ_{dp} constraint (Section 9.4.2), the retrieved estimate of specific attenuation is immune to possible radar calibration biases. However, the reflectivity retrieval is affected by the system bias of the individual radar nodes, with the retrieved reflectivity bias being nearly the same as the mean bias between the radars. Another important possible source of error is related to the geometrical mismatch between the radars' resolution bins at common volumes. The observations should match in altitude, which implies a preferential use of low-elevation scans. In general, the assumption that the reflectivity of the common volume is the same for all radars is only valid for small radar resolution volumes. Observations at large distances from the radars or in the presence of strong precipitation gradients should thus be dealt with care.

This method was implemented for real-time operation by Lim et al. [325] on the CASA IP1 network in Oklahoma. The results showed that the network-based attenuation-correction algorithm works well in real time, with the retrieved reflectivity in good agreement with the results of the conventional Φ_{dp}-based attenuation-correction method as well as the observations of the S-band WSR-88D radars.

10.6   Selected Problems

1. Consider a radar observation domain that covers a 200×200 km^2 region over a flat terrain.
    a) How many radars would be needed to cover the entire area using triangular network cells and imposing that the average minimum beam height over the network should not exceed 100 m above ground? See the definition of minimum beam height b_h in eq. (10.4).
    b) Consider three possible operating frequencies for the radars in the network. If the systems have pulse durations of $T_{tx} = 1.5$ μs (S band), $T_{tx} = 1.0$ μs (C band), $T_{tx} = 0.5$ μs (X band), and they all have a duty cycle (the fraction of the time a radar is transmitting, i.e. $T_{tx} \times$ PRF) of 0.1%, what would be the unambiguous velocity (in single PRF mode) for operating frequencies of 3 GHz (S band), 5 GHz (C band), and 10 GHz (X band)?
    c) What do you think would be the most cost-effective solution in terms of operating frequency for such a network and why?
2. Figure 10.12 shows the WSR-88D sites. There are about 155 WSR-88D radars in the US, with some of them deployed at US locations outside the contiguous states. For simplicity assume we have 150 radars in the continental United States (CONUS).
    a) Assuming a uniform distribution of the radars over the CONUS area, determine the area each radar covers.
    b) Visually, the deployment pattern of the state of Colorado is very simple. For simplicity, assume Colorado is a simple rectangle and the radars are deployed in the midpoint of the four sides and one in the middle of the state. Using the reflectivity and velocity coverage limits of WSR-88D and simple geometric calculations, determine the fraction of the state of Colorado which does not have weather radar coverage for this example.
    c) Most WSR-88D use 0.5 degrees as their lowest scanning elevation. Once again, using simple geometry and flat-Earth approximation, calculate the fraction of the atmosphere's volume below 1000 meters above ground level that is covered by a single weather radar. Use simple assumptions about the lowest beam height.

**NEXRAD WSR-88D Doppler Radar Sites
Over the Contiguous United States**

Figure 10.12 Map of the contiguous United States with locations of the NEXRAD WSR-88D Doppler radars.

    d) Colorado has the Rocky Mountains that block the radar beams at lower elevation angles. Consider a simple model for these mountains as a wall with a height of 2200 meters above the radar's altitude, running north-south at 3/5 of the way across the state starting from the west. Revise your calculations for the radar coverage below 1000 meter above ground level.

    e) Extended project: using an actual topographical map of Colorado, refine the coverage answers for the state.

3. The CASA network in Dallas/Fort Worth has deployed seven X-band radars to cover the urban environment. Compare the CASA network with the X-NET, a five-radar X-band urban radar networks for the city of Tokyo. Assuming flat terrain and the cities area as the coverage area, compare the two radar networks' performance metrics: coverage area, population coverage, volume revisit time, and multiple-Doppler capability. (Hint, additional details are available in the literature.)

4. While this chapter discussed the scientific aspects of radar networks, there are also additional aspects such as cost, ease of deployment, maintenance and site availability. New radar deployments in many countries have "organically evolved" as a hybrid of C-, X- and S-band radars. The advent of dual polarization and advanced techniques has extended the reach of C-band radars. A radar with extremely long range has been shown to have limitations due to Earth's curvature and terrain blockage. Urban applications, where the deployment is a challenge, have resorted to a proliferation of X-band radar networks to provide higher spatial and temporal coverage. As a thought exercise, if we were to replace the

S-band radars in the state of Colorado with a network of C-band radars, or a combination of network of C-band radars and X-band radars for metro region, develop a deployment map, of where the C-band radars will be deployed and how many. Please note that you need some simple assumptions, a factor to account for the ratio of the cost of C-band and S-band radars, including cost of land and maintenance. You may also limit coverage of each radar to 150 km. Note that this is a design problem and there is no unique answer.

5. There are many lakes and reservoirs in mountain regions with complex terrain around the world. These ultimately are caches that collect precipitation and the water in these lakes are a critical part of the societal resource. Examples include Lake Geneva in Switzerland and France, Lake Maggiore between Italy and Switzerland, lake Tanganyika in Africa. Many of these lakes and reservoirs are deep in mountainous terrain and therefore are poorly covered by traditional national weather radar networks. It is increasingly common to use smaller, higher-frequency radars (such as X-band radars) to provide gap-filling measurements. Observations in these mountainous region not only measure precipitation, but also measure atmospheric flow and the type of precipitation (snow versus rain), that falls in the region. Consider two radar applications, precipitation nowcasting and precipitation measurement, with focus on orographic effect (implying a need to document the microphysics from close to ground to storms' tops). Devise two different multi-radar deployment strategies, using the following scenario assumptions, and describe the end-user products each strategy will produce:

- A lake is approximated as a rectangle with a width of 10 km and length of 60 km.
- The terrain around the lake rises up to between 1000 m and 2000 m compared to the lake level.
- Develop a set of requirements from higher-level application requirements, to include vector Doppler measurements, accurate polarization measurements, and coverage at low level up to 100 m above local ground level.
- Often infrastructure costs control the decision process, as an example, in long lakes, cities have developed at the end of the lakes, due to historic and topographic reasons. It may be cheaper, to install a radar at the end cities, rather than run roads, power and IT infrastructure to an ideal midpoint of the lake (this infrastructure can often cost 10 times or more than the radar cost).
- Finally, for choice of radar wavelength. Again some assumptions are needed here, such as size of radars is directly proportional to wavelength, and volume/weight and other infrastructure can be considered proportional to size-ratio cubed. To account for power demand etc. let us assume doubling the wavelength will introduce a factor of 10 cost, which includes, power, deployment real-estate costs, and societal footprint such as ability to deploy on existing infrastructure like existing building tops, as it may be nearly impossible to build new infrastructure in historic locations or expensive tourist locations.

Appendix   Complex Numbers

Complex numbers provide a convenient mathematical representation to describe radar signals and they are used extensively in weather radar signal processing. The complex-valued number is defined by two values: a real and an imaginary part, or an amplitude and a phase. In radar systems, the in-phase (I) and quadrature (Q) voltages of the signal's envelope are the real and imaginary parts of a complex-valued signal, respectively. In particular for radars, the complex-valued IQ signal provides a measurement of the signal's amplitude and its phase. Complex-valued signal processing is used extensively in Chapters 5 and 6 to estimate the properties of precipitation from the received radar echoes.

A complex-valued number is given as $x = x_I + jx_Q$, with a real part (x_I) and an imaginary part (x_Q). The imaginary unit is $j = \sqrt{-1}$, and $j^2 = -1$, $j^3 = -j$, $j^4 = 1$. (Note that in engineering, j is typically used as the symbol for the imaginary number, and in physics, i is commonly used.) For mathematical operations involving complex numbers, the complex conjugate is also frequently used, which is defined as $x^* = x_I - jx_Q$. The simple transpose of a matrix is $[A]^T$. For complex-valued matrices, the complex transpose can also be used $[A]^H = ([A]^*)^T$. For the complex transpose, which is also referred to as the *Hermitian transpose*, both the complex conjugate and the simple transpose are performed on the matrix.

To graphically represent the complex-valued numbers, two axes (a complex plane) are required. The real part (I) is plotted along the x-axis, and the imaginary part (Q) is plotted along the y-axis. In panel (a) of Figure A.1, a complex number x and its complex conjugate x^* are shown on the complex plane. The IQ signal x can also be described in terms of its magnitude, a, and phase angle, φ, which is known as the *phasor representation*:

$$x = ae^{j\varphi}, \tag{A.1}$$

where the amplitude of the signal is

$$a = |x| = \sqrt{x_I^2 + x_Q^2}, \tag{A.2a}$$

and the phase of the signal is the angle between the signal, represented as a vector in the IQ plane, and the positive real axis (i.e., the in-phase axis):

$$\varphi = \angle x = \arg(x), \tag{A.2b}$$

To calculate $\angle x$, the function $\text{atan2}(x_Q, x_I)$ is typically used. The function uniquely identifies the angle within the four quadrants of the unit circle:

$$
\text{atan2}(x_Q, x_I) = \begin{cases} \text{atan}\left(\frac{x_Q}{x_I}\right) & x > 0 \\[2mm] \text{atan}\left(\frac{x_Q}{x_I}\right) + \pi & x < 0, y \geq 0 \\[2mm] \text{atan}\left(\frac{x_Q}{x_I}\right) - \pi & x < 0, y < 0 \\[2mm] \frac{\pi}{2} & x = 0, y > 0 \\[2mm] -\frac{\pi}{2} & x = 0, y < 0 \\[2mm] \text{undefined} & x = 0, y = 0 \end{cases} \tag{A.3}
$$

The magnitude a is a nonnegative number (i.e., $a \geq 0$). (In the IQ plane, a "negative number" is effectively implemented by a 180° phase shift.) The phase angle, φ, spans a range of 2π radians (or 360°). The selection of the transition point of this span is somewhat arbitrary, but typical (and convenient) selections are from $[0, 2\pi)$ or $(-\pi, \pi]$, as shown in panels (b) and (c) of Figure A.1, respectively. (The set notation "[" or "]" denotes inclusion of the endpoint, and "(" or ")" denotes up to, but not including, the endpoint.)

In weather radar signal processing, the received signal's power is proportional to the reflectivity factor of the precipitation. The radar signal's power is its most common measurement. In Doppler weather radar, the receiver's signals are sampled

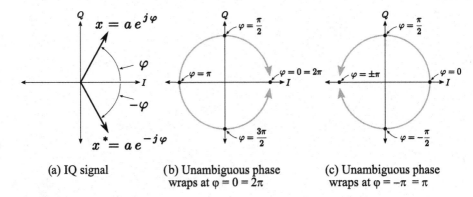

(a) IQ signal

(b) Unambiguous phase wraps at $\varphi = 0 = 2\pi$

(c) Unambiguous phase wraps at $\varphi = -\pi = \pi$

Figure A.1 A complex signal is defined by two dimensions. This can be I and Q in Cartesian coordinates or amplitude (a) and phase (φ) in polar coordinates. The signal's complex-conjugate (x^*) has a negative-phase or negative-quadrature component. Complex-valued variables are used extensively to describe radar signals and perform signal-processing tasks. The phase of the signal is unambiguous only over a range of 2π radians. The point at which the phase "wraps" is typically at $\varphi = 0 = 2\pi$ or $\varphi = -\pi = \pi$. The second representation is convenient for working with estimated Doppler velocities.

as complex-valued voltages. The power of the signal is the product of the complex voltage with its complex conjugate as

$$p = x^*x = xx^* = a^2 = x_I^2 + x_Q^2. \tag{A.4}$$

It is important to note that the result of a complex signal multiplied by its conjugate is always nonnegative and real-valued. Now consider the product of two complex-valued numbers, $x = x_I + jx_Q$ and $y = y_I + jy_Q$ as

$$x^*y = (xy^*)^* = (x_I - jx_Q)(y_I + jy_Q) = (x_Iy_I + x_Qy_Q) + j(x_Iy_Q - x_Qy_I). \tag{A.5}$$

The complex-valued multiplication can also be evaluated using the phasor form of the complex numbers, where $x = a_x e^{j\varphi_x}$ and $y = a_y e^{j\varphi_y}$, and their product is

$$x^*y = a_x e^{-j\varphi_x} a_y e^{j\varphi_y} = a_x a_y e^{j(\varphi_y - \varphi_x)}. \tag{A.6}$$

The difference in the phase between two complex numbers, x and y, is $\angle x^*y = \varphi_y - \varphi_x$. Also note the effect of the conjugation on the phase of their product: $\angle x^*y = -\angle xy^*$. The phase of the product of two samples is used to estimate the Doppler velocity, using two samples in time. It is also used to estimate the differential phase shift between the two polarization's signals.

References

[1] J. K. Petersen, *Handbook of Surveillance Technologies*, 3rd ed. New York: Routledge, 2012.

[2] H. Bluestein, *Severe Convective Storms and Tornadoes: Observations and Dynamics*, ser. Springer Praxis Books. Berlin: Springer, 2013.

[3] D. Mclaughlin, D. Pepyne, V. Chandrasekar, B. Philips, J. Kurose, M. Zink, K. Droege-meier, S. Cruz-Pol, F. Junyent, J. Brotzge, D. Westbrook, N. Bharadwaj, Y. Wang, E. Lyons, K. Hondl, Y. Liu, E. Knapp, M. Xue, A. Hopf, K. Kloesel, A. Defonzo, P. Kollias, K. Brewster, R. Contreras, B. Dolan, T. Djaferis, E. Insanic, S. Frasier, and F. Carr, "Short-wavelength technology and the potential for distributed networks of small radar systems," *Bulletin of the American Meteorological Society*, vol. 90, no. 12, pp. 1797–1817, 2009.

[4] International Telecommunication Union, "Radio Regulations," https://www.itu.int/en/myitu/Publications/2020/09/02/14/23/Radio-Regulations-2020, 2020.

[5] "IEEE standard letter designations for radar-frequency bands," IEEE Std. 521-2002 (Revision of IEEE Std. 521-1984), 2003.

[6] Y. Liu, B. Geerts, M. Miller, P. Daum, and R. McGraw, "Threshold radar reflectivity for drizzling clouds," *Geophysical Research Letters*, vol. 35, no. 3, pp. 1–5, 2008.

[7] J. R. Probert-Jones, "The radar equation in meteorology," *Quarterly Journal of the Royal Meteorological Society*, vol. 88, no. 378, pp. 485–495, 1962.

[8] E. K. Smith and S. Weintraub, "The constants in the equation for atmospheric refractive index at radio frequencies," *Proceedings of the IRE*, vol. 41, no. 8, pp. 1035–1037, 1953.

[9] R. J. Doviak, V. Bringi, A. Ryzhkov, A. Zahrai, and D. Zrnić, "Considerations for polarimetric upgrades to operational WSR-88D radars," *Journal of Atmospheric and Oceanic Technology*, vol. 17, no. 3, pp. 257–278, 2000. [Online]. Available: https://doi.org/10.1175/1520-0426(2000)017<0257:CFPUTO>2.0.CO;2

[10] R. G. Carter, *Microwave and RF Vacuum Electronic Power Sources*, ser. Cambridge RF and Microwave Engineering Series. Cambridge, UK: Cambridge University Press, 2018.

[11] U. S. Navy, *Navy Electricity and Electronics Training Series (NEETS): Module 18—Radar Principles*. Pensacola, FL: Naval Education and Training Professional Development and Technology Center, 1998.

[12] F. Loomis, "Bell System Plans for Broadband Network Facilities," *Tele-Tech*, vol. 12, no. 4, p. 80, 1953.

[13] C. Cook, "Pulse Compression—Key to More Efficient Radar Transmission," *Proceedings of the IRE*, vol. 48, no. 3, pp. 310–316, 1960.

[14] G. A. Sadowy, A. C. Berkun, W. Chun, E. Im, and S. L. Durden, "Development of an advanced airborne precipitation radar," *Microwave Journal*, vol. 46, pp. 84–93, 2003.

[15] M. Vega, V. Chandrasekar, J. Carswell, R. Beauchamp, M. Schwaller, and C. Nguyen, "Salient features of the dual-frequency, dual-polarized, Doppler radar for remote sensing of precipitation," *Radio Science*, vol. 49, no. 11, pp. 1087–1105, 2014.

[16] H. Dai, X. Wang, H. Xie, S. Xiao, and J. Luo, *Spatial Polarization Characteristics of Radar Antenna: Analysis, Measurement and Anti-jamming Application*. New York: Springer, 2018.

[17] J. D. Kraus and R. J. Marhefka, *Antennas for All Applications*, 3rd ed., ser. Electrical Engineering Series. New York: McGraw-Hill, 2002.

[18] W. L. Stutzman and G. A. Thiele, *Antenna Theory and Design*, 3rd ed. Hoboken, NJ: John Wiley & Sons, 2012.

[19] V. Chandrasekar and R. J. Keeler, "Antenna pattern analysis and measurements for multiparameter radars," *Journal of Atmospheric and Oceanic Technology*, vol. 10, no. 5, pp. 674–683, 1993. [Online]. Available: https://doi.org/10.1175/1520-0426(1993)010<0674:APAAMF>2.0.CO;2

[20] Y. Wang and V. Chandrasekar, "Polarization isolation requirements for linear dual-polarization weather radar in simultaneous transmission mode of operation," *IEEE Transactions on Geoscience and Remote Sensing*, vol. 44, no. 8, pp. 2019–2028, 2006.

[21] V. N. Bringi and V. Chandrasekar, *Polarimetric Doppler Weather Radar: Principles and applications*, 1st ed. Cambridge, UK: Cambridge University Press, 2001.

[22] F. Junyent, V. Chandrasekar, D. Mclaughlin, E. Insanic, and N. Bharadwaj, "The CASA integrated project 1 networked radar system," *Journal of Atmospheric and Oceanic Technology*, vol. 27, no. 1, pp. 61–78, 2010.

[23] J. Marshall and W. M. Palmer, "The distribution of raindrops with size," *Journal of Meteorology*, vol. 5, pp. 165–166, 1948.

[24] K. K. Lo and R. E. Passarelli Jr., "The growth of snow in winter storms: An airborne observational study," *Journal of the Atmospheric Sciences*, vol. 39, no. 4, pp. 697–706, 1982.

[25] P. R. Field, "Aircraft observations of ice crystal evolution in an altostratus cloud," *Journal of the Atmospheric Sciences*, vol. 56, no. 12, pp. 1925–1941, 1999.

[26] P. R. Field and A. J. Heymsfield, "Aggregation and scaling of ice crystal size distributions," *Journal of the Atmospheric Sciences*, vol. 60, no. 3, pp. 544–560, 2003.

[27] A. H. Auer, "Distribution of graupel and hail with size," *Monthly Weather Review*, vol. 100, no. 5, pp. 325–328, 1972.

[28] M. Schönhuber, G. Lammer, and W. Randeu, "The 2D-video-distrometer," in *Precipitation: Advances in Measurement, Estimation and Prediction*, S. Michaelides, ed. Berlin: Springer, 2008, ch. 1, pp. 3–31.

[29] S.-G. Park, H.-L. Kim, Y.-W. Ham, and S.-H. Jung, "Comparative evaluation of the OTT PARSIVEL^2 using a collocated two-dimensional video disdrometer," *Journal of Atmospheric and Oceanic Technology*, vol. 34, no. 9, pp. 2059–2082, 2017.

[30] P. L. Smith, "Sampling issues in estimating radar variables from disdrometer data," *Journal of Atmospheric and Oceanic Technology*, vol. 33, no. 11, pp. 2305–2313, 2016.

[31] C. W. Ulbrich, "Natural variations in the analytical form of the raindrop size distribution," *Journal of Climate and Applied Meteorology*, vol. 22, no. 10, pp. 1764–1775, 1983. [Online]. Available: https://doi.org/10.1175/1520-0450(1983)022<1764:NVITAF>2.0.CO;2

[32] R. S. Sekhon and R. C. Srivastava, "Doppler radar observations of drop-size distributions in a thunderstorm," *Journal of the Atmospheric Sciences*, vol. 28, no. 6, pp. 983–994, 1971. [Online]. Available: https://doi.org/10.1175/1520-0469(1971)028<0983:DROODS>2.0.CO;2

[33] P. T. Willis, "Functional fits to some observed drop size distributions and parameterization of rain," *Journal of the Atmospheric Sciences*, vol. 41, no. 9, pp. 1648–1661, 1984. [Online]. Available: https://doi.org/10.1175/1520-0469(1984)041<1648:FFTSOD>2.0.CO;2

[34] J. Testud, S. Oury, R. A. Black, P. Amayenc, and X. Dou, "The concept of 'normalized' distribution to describe raindrop spectra: A tool for cloud physics and cloud remote sensing," *Journal of Applied Meteorology*, vol. 40, no. 6, pp. 1118–1140, 2001.

[35] A. W. Green, "An approximation for the shapes of large raindrops," *Journal of Applied Meteorology*, vol. 14, no. 8, pp. 1578–1583, 1975.

[36] V. Chandrasekar, W. A. Cooper, and V. N. Bringi, "Axis ratios and oscillations of raindrops," *Journal of the Atmospheric Sciences*, vol. 45, no. 8, pp. 1323–1333, 1988.

[37] H. R. Pruppacher and K. V. Beard, "A wind tunnel investigation of the internal circulation and shape of water drops falling at terminal velocity in air," *Quarterly Journal of the Royal Meteorological Society*, vol. 96, no. 408, pp. 247–256, 1970.

[38] K. V. Beard and C. Chuang, "A New Model for the Equilibrium Shape of Raindrops," *Journal of the Atmospheric Sciences*, vol. 44, no. 11, pp. 1509–1524, 1987.

[39] R. J. Kubesh and K. V. Beard, "Laboratory measurements of spontaneous oscillations for moderate-size raindrops," *Journal of the Atmospheric Sciences*, vol. 50, no. 8, pp. 1089–1098, 1993.

[40] K. V. Beard, V. Bringi, and M. Thurai, "A new understanding of raindrop shape," *Atmospheric Research*, vol. 97, no. 4, pp. 396–415, 2010. [Online]. Available: http://www.sciencedirect.com/science/article/pii/S0169809510000268

[41] M. Szakáll, S. K. Mitra, K. Diehl, and S. Borrmann, "Shapes and oscillations of falling raindrops—a review," *Atmospheric Research*, vol. 97, no. 4, pp. 416–425, 2010. [Online]. Available: http://www.sciencedirect.com/science/article/pii/S0169809510000736

[42] K. Andsager, K. V. Beard, and N. F. Laird, "Laboratory measurements of axis ratios for large raindrops," *Journal of the Atmospheric Sciences*, vol. 56, no. 15, pp. 2673–2683, 1999.

[43] E. A. Brandes, G. Zhang, and J. Vivekanandan, "Experiments in rainfall estimation with a polarimetric radar in a subtropical environment," *Journal of Applied Meteorology*, vol. 41, no. 6, pp. 674–685, 2002.

[44] H. R. Pruppacher and R. L. Pitter, "A semi-empirical determination of the shape of cloud and rain drops," *Journal of the Atmospheric Sciences*, vol. 28, no. 1, pp. 86–94, 1971.

[45] K. V. Beard and R. J. Kubesh, "Laboratory measurements of small raindrop distortion. Part 2: Oscillation frequencies and modes," *Journal of the Atmospheric Sciences*, vol. 48, no. 20, pp. 2245–2264, 1991.

[46] R. Gunn and G. D. Kinzer, "The terminal velocity of fall for water droplets in stagnant air," *Journal of Meteorology*, vol. 6, no. 4, pp. 243–248, 1949.

[47] D. Atlas and C. W. Ulbrich, "Path- and area-integrated rainfall measurement by microwave attenuation in the 1–3 cm band," *Journal of Applied Meteorology*, vol. 16, no. 12, pp. 1322–1331, 1977.

[48] R. S. Sekhon and R. C. Srivastava, "Doppler radar observations of drop-size distributions in a thunderstorm," *Journal of the Atmospheric Sciences*, vol. 28, no. 6, pp. 983–994, 1971.

[49] D. Atlas, R. C. Srivastava, and R. S. Sekhon, "Doppler radar characteristics of precipitation at vertical incidence," *Reviews of Geophysics*, vol. 11, no. 1, pp. 1–35, 1973.

[50] G. C. McCormick, A. Hendry, and B. L. Barge, "The anisotropy of precipitation media," *Nature*, vol. 238, pp. 214–216, 1972.

[51] M. Saunders, "Cross polarization at 18 and 30 GHz due to rain," *IEEE Transactions on Antennas and Propagation*, vol. 19, no. 2, pp. 273–277, March 1971.

[52] G. Brussaard, "Rain-induced crosspolarisation and raindrop canting," *Electronics Letters*, vol. 10, no. 20, pp. 411–412, October 1974.

[53] G. Brussaard, "A meteorological model for rain-induced cross polarization," *IEEE Transactions on Antennas and Propagation*, vol. 24, no. 1, pp. 5–11, January 1976.

[54] K. V. Beard, H. T. Ochs III, and R. J. Kubesh, "Natural oscillations of small raindrops," *Nature*, vol. 342, pp. 408–410, 1989.

[55] F. Y. Testik and A. P. Barros, "Toward elucidating the microstructure of warm rainfall: A survey," *Reviews of Geophysics*, vol. 45, no. 2, 2007. [Online]. Available: https://agupubs .onlinelibrary.wiley.com/doi/abs/10.1029/2005RG000182

[56] G.-J. Huang, V. N. Bringi, and M. Thurai, "Orientation angle distributions of drops after an 80-m fall using a 2D video disdrometer," *Journal of Atmospheric and Oceanic Technology*, vol. 25, no. 9, pp. 1717–1723, 2008.

[57] M. Thurai and V. N. Bringi, "Drop axis ratios from a 2D video disdrometer," *Journal of Atmospheric and Oceanic Technology*, vol. 22, no. 7, pp. 966–978, 2005.

[58] V. N. Bringi, M. Thurai, and D. A. Brunkow, "Measurements and inferences of raindrop canting angles," *Electronics Letters*, vol. 44, no. 24, pp. 1425–1426, November 2008.

[59] R. R. Rogers and M. K. Yau, *A Short Course in Cloud Physics*, 3rd ed. Oxford: Pergamon Press Oxford, 1989.

[60] H. R. Pruppacher and J. D. Klett, *Microphysics of Clouds and Precipitation*. Dordrecht, Holland: D. Reidel, 1978.

[61] E. G. Bowen, "The formation of rain by coalescence," *Australian Journal of Scientific Research. Series A: Physical Sciences*, vol. 3, p. 193, Jun. 1950.

[62] J. W. Telford, "A new aspect of coalescence theory," *Journal of Meteorology*, vol. 12, no. 5, pp. 436–444, 1955.

[63] O. P. Prat and A. P. Barros, "A robust numerical solution of the stochastic collection-breakup equation for warm rain," *Journal of Applied Meteorology and Climatology*, vol. 46, no. 9, pp. 1480–1497, 2007. [Online]. Available: https://doi.org/ 10.1175/JAM2544.1

[64] L. P. D'Adderio, F. Porcù, and A. Tokay, "Identification and analysis of collisional breakup in natural rain," *Journal of the Atmospheric Sciences*, vol. 72, no. 9, pp. 3404– 3416, 2015. [Online]. Available: https://doi.org/10.1175/JAS-D-14-0304.1

[65] T. B. Low and R. List, "Collision, coalescence and breakup of raindrops. Part I: Experimentally established coalescence efficiencies and fragment size distributions in breakup," *Journal of the Atmospheric Sciences*, vol. 39, no. 7, pp. 1591– 1606, 1982. [Online]. Available: https://doi.org/10.1175/1520-0469(1982)039<1591: CCABOR>2.0.CO;2

[66] T. B. Low and R. List, "Collision, coalescence and breakup of raindrops. Part II: Parameterization of fragment size distributions," *Journal of the Atmospheric Sciences*, vol. 39, no. 7, pp. 1607–1619, 1982. [Online]. Available: https://doi.org/10.1175/1520-0469(1982)039<1607:CCABOR>2.0.CO;2

[67] A. P. Barros, O. P. Prat, F. Shrestha, F. Y. Testik, and L. F. Bliven, "Revisiting Low and List (1982): Evaluation of raindrop collision parameterizations using laboratory observations and modeling," *Journal of the Atmospheric Sciences*, vol. 65, no. 9, pp. 2983–2993, 2008. [Online]. Available: https://doi.org/10.1175/2008JAS2630.1

[68] R. J. Houze, *Cloud Dynamics*, vol. 104. Cambridge, MA: Academic Press, 2014.

[69] M. R. Kumjian and O. P. Prat, "The impact of raindrop collisional processes on the polarimetric radar variables," *Journal of the Atmospheric Sciences*, vol. 71, no. 8, pp. 3052–3067, 2014.

[70] X. Li and R. C. Srivastava, "An analytical solution for raindrop evaporation and its application to radar rainfall measurements," *Journal of Applied Meteorology*, vol. 40, no. 9, pp. 1607–1616, 2001.

[71] M. R. Kumjian and A. V. Ryzhkov, "The impact of evaporation on polarimetric characteristics of rain: Theoretical model and practical implications," *Journal of Applied Meteorology and Climatology*, vol. 49, no. 6, pp. 1247–1267, 2010.

[72] J. M. Straka, D. S. Zrnić, and A. V. Ryzhkov, "Bulk hydrometeor classification and quantification using polarimetric radar data: Synthesis of relations," *Journal of Applied Meteorology and Climatology*, vol. 29, no. 8, pp. 1341–1372, 2000.

[73] J.-P. Chen and D. Lamb, "The theoretical basis for the parameterization of ice crystal habits: Growth by vapor deposition," *Journal of the Atmospheric Sciences*, vol. 51, no. 9, pp. 1206–1222, 1994.

[74] M. Chiruta and P. K. Wang, "The capacitance of rosette ice crystals," *Journal of the Atmospheric Sciences*, vol. 60, no. 6, pp. 836–846, 2003. [Online]. Available: https://doi.org/10.1175/1520-0469(2003)060<0836:TCORIC>2.0.CO;2

[75] C. D. Westbrook, R. J. Hogan, and A. J. Illingworth, "The capacitance of pristine ice crystals and aggregate snowflakes," *Journal of the Atmospheric Sciences*, vol. 65, no. 1, pp. 206–219, 2008. [Online]. Available: https://doi.org/10.1175/2007JAS2315.1

[76] M. P. Bailey and J. Hallett, "A comprehensive habit diagram for atmospheric ice crystals: Confirmation from the laboratory, AIRS II, and other field studies," *Journal of the Atmospheric Sciences*, vol. 66, no. 9, pp. 2888–2899, 2009.

[77] C. Magono and C. W. Lee, "Meteorological classification of natural snow crystals," *Journal of the Faculty of Science, Hokkaido University*, vol. 2, no. 4, pp. 321–335, 1968.

[78] P. Kennedy, M. Thurai, C. Praz, V. N. Bringi, A. Berne, and B. M. Notaroš, "Variations in snow crystal riming and Z_{DR}: A case analysis," *Journal of Applied Meteorology and Climatology*, vol. 57, no. 3, pp. 695–707, 2018. [Online]. Available: https://doi.org/10.1175/JAMC-D-17-0068.1

[79] K. Kikuchi, T. Kameda, K. Higuchi, and A. Yamashita, "A global classification of snow crystals, ice crystals, and solid precipitation based on observations from middle latitudes to polar regions," *Atmospheric Research*, vol. 132–133, pp. 460–472, 2013. [Online]. Available: http://www.sciencedirect.com/science/article/pii/S0169809513001841

[80] A. V. Korolev and I. P. Mazin, "Supersaturation of water vapor in clouds," *Journal of the Atmospheric Sciences*, vol. 60, no. 24, pp. 2957–2974, 2003. [Online]. Available: https://doi.org/10.1175/1520-0469(2003)060<2957:SOWVIC>2.0.CO;2

[81] A. Korolev, "Limitations of the Wegener–Bergeron–Findeisen mechanism in the evolution of mixed-phase clouds," *Journal of the Atmospheric Sciences*, vol. 64, no. 9, pp. 3372–3375, 2007.

[82] R. R. Weiss and P. V. Hobbs, "The use of a vertically pointing pulsed Doppler radar in cloud physics and weather modification studies," *Journal of Applied Meteorology*, vol. 14, no. 2, pp. 222–231, 1975.

[83] R. Bechini, L. Baldini, and V. Chandrasekar, "Polarimetric radar observations in the ice region of precipitating clouds at C-band and X-band radar frequencies," *Journal of Applied Meteorology and Climatology*, vol. 52, pp. 1147–1169, 2013.

[84] J. Hallet and S. C. Mossop, "Production of secondary ice particles during the riming process," *Nature*, vol. 249, pp. 26–28, 1974. [Online]. Available: http://dx.doi.org/10.1038/249026a0

[85] J. M. Vogel and F. Fabry, "Contrasting polarimetric observations of stratiform riming and nonriming events," *Journal of Applied Meteorology and Climatology*, vol. 57, no. 2, pp. 457–476, 2018. [Online]. Available: https://doi.org/10.1175/JAMC-D-16-0370.1

[86] V. A. Sinclair, D. Moisseev, and A. Lerber, "How dual-polarization radar observations can be used to verify model representation of secondary ice," *Journal of Geophysical Research: Atmospheres*, vol. 121, no. 18, pp. 10954–10970, 2016. [Online]. Available: https://agupubs.onlinelibrary.wiley.com/doi/abs/10.1002/2016JD025381

[87] P. R. Field, R. P. Lawson, P. R. A. Brown, G. Lloyd, C. Westbrook, D. Moisseev, A. Miltenberger, A. Nenes, A. Blyth, T. Choularton, P. Connolly, J. Buehl, J. Crosier, Z. Cui, C. Dearden, P. DeMott, A. Flossmann, A. Heymsfield, Y. Huang, H. Kalesse, Z. A. Kanji, A. Korolev, A. Kirchgaessner, S. Lasher-Trapp, T. Leisner, G. McFarquhar, V. Phillips, J. Stith, and S. Sullivan, "Secondary ice production: Current state of the science and recommendations for the future," *Meteorological Monographs*, vol. 58, pp. 7.1–7.20, 2017. [Online]. Available: https://doi.org/10.1175/AMSMONOGRAPHS-D-16-0014.1

[88] R. A. Houze, L. A. McMurdie, W. A. Petersen, M. R. Schwaller, W. Baccus, J. D. Lundquist, C. F. Mass, B. Nijssen, S. A. Rutledge, D. R. Hudak, S. Tanelli, G. G. Mace, M. R. Poellot, D. P. Lettenmaier, J. P. Zagrodnik, A. K. Rowe, J. C. DeHart, L. E. Madaus, H. C. Barnes, and V. Chandrasekar, "The Olympic Mountains experiment (Olympex)," *Bulletin of the American Meteorological Society*, vol. 98, no. 10, pp. 2167–2188, 2017. [Online]. Available: https://doi.org/10.1175/BAMS-D-16-0182.1

[89] J. Leinonen and A. von Lerber, "Snowflake melting simulation using smoothed particle hydrodynamics," *Journal of Geophysical Research: Atmospheres*, vol. 123, no. 3, pp. 1811–1825, 2018. [Online]. Available: https://agupubs.onlinelibrary.wiley.com/doi/abs/10.1002/2017JD027909

[90] M. P. Langleben, "The terminal velocity of snowflakes," *Quarterly Journal of the Royal Meteorological Society*, vol. 80, no. 346, pp. 640–642, 1954. [Online]. Available: https://rmets.onlinelibrary.wiley.com/doi/abs/10.1002/qj.49708034619

[91] M. Ishizaka, H. Motoyoshi, S. Nakai, T. Shiina, T. Kumakura, and K. ichiro Muramoto, "A new method for identifying the main type of solid hydrometeors contributing to snowfall from measured size-fall speed relationship," *Journal of the Meteorological Society of Japan. Ser. II*, vol. 91, no. 6, pp. 747–762, 2013.

[92] A. J. Heymsfield, A. Bansemer, C. Schmitt, C. Twohy, and M. R. Poellot, "Effective ice particle densities derived from aircraft data," *Journal of the Atmospheric Sciences*, vol. 61, no. 9, pp. 982–1003, 2004. [Online]. Available: https://doi.org/10.1175/1520-0469(2004)061<0982:EIPDDF>2.0.CO;2

[93] F. Fabry and W. Szyrmer, "Modeling of the melting layer. Part II: Electromagnetic," *Journal of the Atmospheric Sciences*, vol. 56, no. 20, pp. 3593–3600, 1999. [Online]. Available: https://doi.org/10.1175/1520-0469(1999)056<3593:MOTMLP>2.0.CO;2

[94] W. Szyrmer and I. Zawadzki, "Modeling of the melting layer. Part I: Dynamics and microphysics," *Journal of the Atmospheric Sciences*, vol. 56, no. 20, pp. 3573–3592, 1999. [Online]. Available: https://doi.org/10.1175/1520-0469(1999)056<3573:MOTMLP>2.0.CO;2

[95] A. J. Heymsfield, A. Bansemer, M. R. Poellot, and N. Wood, "Observations of ice microphysics through the melting layer," *Journal of the Atmospheric Sciences*, vol. 72, no. 8, pp. 2902–2928, 2015. [Online]. Available: https://doi.org/10.1175/JAS-D-14-0363.1

[96] E. Barthazy, W. Henrich, and A. Waldvogel, "Size distribution of hydrometeors through the melting layer," *Atmospheric Research*, vol. 47–48, pp. 193–208, 1998. [Online]. Available: http://www.sciencedirect.com/science/article/pii/S0169809598000659

[97] S. Troemel, A. V. Ryzhkov, P. Zhang, and C. Simmer, "Investigations of backscatter differential phase in the melting layer," *Journal of Applied Meteorology and Climatology*, vol. 53, no. 10, pp. 2344–2359, 2014. [Online]. Available: https://doi.org/10.1175/JAMC-D-14-0050.1

[98] A. J. Illingworth, J. W. F. Goddard, and S. M. Cherry, "Polarization radar studies of precipitation development in convective storms," *Quarterly Journal of the Royal Meteorological Society*, vol. 113, no. 476, pp. 469–489, 1987. [Online]. Available: https://rmets.onlinelibrary.wiley.com/doi/abs/10.1002/qj.49711347604

[99] V. N. Bringi, K. Knupp, A. Detwiler, L. Liu, I. J. Caylor, and R. A. Black, "Evolution of a Florida thunderstorm during the convection and precipitation/electrification experiment: The case of 9 August 1991," *Monthly Weather Review*, vol. 125, no. 9, pp. 2131–2160, 1997. [Online]. Available: https://doi.org/10.1175/1520-0493(1997)125<2131:EOAFTD>2.0.CO;2

[100] K. A. Scharfenberg, D. J. Miller, T. J. Schuur, P. T. Schlatter, S. E. Giangrande, V. M. Melnikov, D. W. Burgess, D. L. Andra, M. P. Foster, and J. M. Krause, "The joint polarization experiment: Polarimetric radar in forecasting and warning decision making," *Weather and Forecasting*, vol. 20, no. 5, pp. 775–788, 2005. [Online]. Available: https://doi.org/10.1175/WAF881.1

[101] M. R. Kumjian, A. P. Khain, N. Benmoshe, E. Ilotoviz, A. V. Ryzhkov, and V. T. J. Phillips, "The anatomy and physics of Z_{DR} columns: Investigating a polarimetric radar signature with a spectral bin microphysical model," *Journal of Applied Meteorology and Climatology*, vol. 53, no. 7, pp. 1820–1843, 2014. [Online]. Available: https://doi.org/10.1175/JAMC-D-13-0354.1

[102] D. M. Plummer, J. R. French, D. C. Leon, A. M. Blyth, S. Lasher-Trapp, L. J. Bennett, D. R. L. Dufton, R. C. Jackson, and R. R. Neely, "Radar-derived structural and precipitation characteristics of Z_{DR} columns within warm-season convection over the United Kingdom," *Journal of Applied Meteorology and Climatology*, vol. 57, no. 11, pp. 2485–2505, 2018. [Online]. Available: https://doi.org/10.1175/JAMC-D-17-0134.1

[103] J. Hubbert, V. N. Bringi, L. D. Carey, and S. Bolen, "CSU-CHILL polarimetric radar measurements from a severe hail storm in eastern Colorado," *Journal of Applied Meteorology*, vol. 37, no. 8, pp. 749–775, 1998. [Online]. Available: https://doi.org/10.1175/1520-0450(1998)037<0749:CCPRMF>2.0.CO;2

[104] M. Picca, R. Kumjian, and A. V. Ryzhkov, "Z_{DR} columns as a predictive tool for hail growth and storm evolution," presentation at the 25th Conference on Severe Local Storms, Denver, CO, American Meteorological Society, 2010.

[105] K. A. Browning and G. B. Foote, "Airflow and hail growth in supercell storms and some implications for hail suppression," *Quarterly Journal of the Royal Meteorological Society*, vol. 102, no. 433, pp. 499–533, 1976. [Online]. Available: https://rmets.onlinelibrary .wiley.com/doi/abs/10.1002/qj.49710243303

[106] G. B. Foote and C. G. Wade, "Case study of a hailstorm in Colorado. Part I: Radar echo structure and evolution," *Journal of the Atmospheric Sciences*, vol. 39, no. 12, pp. 2828–2846, 1982. [Online]. Available: https://doi.org/10.1175/1520-0469(1982)039<2828: CSOAHI>2.0.CO;2

[107] L. R. Lemon and C. A. Doswell, "Severe thunderstorm evolution and mesocyclone structure as related to tornadogenesis," *Monthly Weather Review*, vol. 107, no. 9, pp. 1184–1197, 1979. [Online]. Available: https://doi.org/10.1175/1520-0493(1979)107<1184: STEAMS>2.0.CO;2

[108] L. J. Battan, "Doppler radar observations of a hailstorm," *Journal of Applied Meteorology*, vol. 14, no. 1, pp. 98–108, 1975. [Online]. Available: https://doi.org/10.1175/1520-0450(1975)014<0098:DROOAH>2.0.CO;2

[109] K. Browning, J. Frankhauser, J.-P. Chalon, P. Eccles, R. Strauch, F. Merrem, D. Musil, E. May, and W. Sand, "Structure of an evolving hailstorm part v: Synthesis and implications for hail growth and hail suppression," *Monthly Weather Review*, vol. 104, no. 5, pp. 603–610, 1976. [Online]. Available: https://doi.org/10.1175/1520-0493(1976)104<0603:SOAEHP>2.0.CO;2

[110] P. C. Kennedy and A. G. Detwiler, "A case study of the origin of hail in a multicell thunderstorm using in situ aircraft and polarimetric radar data," *Journal of Applied Meteorology*, vol. 42, no. 11, pp. 1679–1690, 2003. [Online]. Available: https://doi.org/ 10.1175/1520-0450(2003)042<1679:ACSOTO>2.0.CO;2

[111] G. B. Foote, "A study of hail growth utilizing observed storm conditions," *Journal of Climate and Applied Meteorology*, vol. 23, no. 1, pp. 84–101, 1984. [Online]. Available: https://doi.org/10.1175/1520-0450(1984)023<0084:ASOHGU>2.0.CO;2

[112] R. M. Wakimoto and V. N. Bringi, "Dual-polarization observations of microbursts associated with intense convection: The 20 July storm during the MIST Project," *Monthly Weather Review*, vol. 116, no. 8, pp. 1521–1539, 1988. [Online]. Available: https://doi .org/10.1175/1520-0493(1988)116<1521:DPOOMA>2.0.CO;2

[113] K. Aydin, T. A. Seliga, and V. Balaji, "Remote sensing of hail with a dual linear polarization radar," *Journal of Climate and Applied Meteorology*, vol. 25, no. 10, pp. 1475–1484, 1986. [Online]. Available: https://doi.org/10.1175/1520-0450(1986)025<1475: RSOHWA>2.0.CO;2

[114] T. K. Depue, P. C. Kennedy, and S. A. Rutledge, "Performance of the hail differential reflectivity (H_{DR}) polarimetric radar hail indicator," *Journal of Applied Meteorology and Climatology*, vol. 46, no. 8, pp. 1290–1301, 2007. [Online]. Available: https://doi.org/10 .1175/JAM2529.1

[115] R. E. Rinehart and J. D. Tuttle, "Antenna beam patterns and dual-wavelength processing," *Journal of Applied Meteorology*, vol. 21, no. 12, pp. 1865–1880, 1982. [Online]. Available: https://doi.org/10.1175/1520-0450(1982)021<1865:ABPADW>2.0.CO;2

[116] F. Junyent, V. Chandrasekar, V. N. Bringi, S. A. Rutledge, P. C. Kennedy, D. Brunkow, J. George, and R. Bowie, "Transformation of the CSU-CHILL radar facility to a dual-frequency, dual-polarization Doppler system," *Bulletin of the American Meteorological Society*, vol. 96, no. 6, pp. 975–996, 2015. [Online]. Available: https://doi.org/10.1175/ BAMS-D-13-00150.1

[117] F. Junyent and V. Chandrasekar, "An examination of precipitation using csu-chill dual-wavelength, dual-polarization radar observations," *Journal of Atmospheric and Oceanic Technology*, vol. 33, no. 2, pp. 313–329, 2016. [Online]. Available: https://doi.org/10.1175/JTECH-D-14-00229.1

[118] M. R. Kumjian, Y. P. Richardson, T. Meyer, K. A. Kosiba, and J. Wurman, "Resonance scattering effects in wet hail observed with a dual-X-band-frequency, dual-polarization Doppler on wheels radar," *Journal of Applied Meteorology and Climatology*, vol. 57, no. 12, pp. 2713–2731, 2018. [Online]. Available: https://doi.org/10.1175/JAMC-D-17-0362.1

[119] N. C. Knight, "Hailstone shape factor and its relation to radar interpretation of hail," *Journal of Climate and Applied Meteorology*, vol. 25, no. 12, pp. 1956–1958, 1986. [Online]. Available: https://doi.org/10.1175/1520-0450(1986)025<1956:HSFAIR>2.0.CO;2

[120] P. R. Kry and R. List, "Angular motions of freely falling spheroidal hailstone models," *The Physics of Fluids*, vol. 17, no. 6, pp. 1093–1102, 1974. [Online]. Available: https://aip.scitation.org/doi/abs/10.1063/1.1694848

[121] V. N. Bringi, T. A. Seliga, and K. Aydin, "Hail detection with a differential reflectivity radar," *Science*, vol. 225, no. 4667, pp. 1145–1147, 1984. [Online]. Available: http://science.sciencemag.org/content/225/4667/1145

[122] R. M. Rasmussen and A. J. Heymsfield, "Melting and shedding of graupel and hail. Part I: Model physics," *Journal of the Atmospheric Sciences*, vol. 44, no. 19, pp. 2754–2763, 1987. [Online]. Available: https://doi.org/10.1175/1520-0469(1987)044<2754:MASOGA>2.0.CO;2

[123] R. M. Rasmussen, V. Levizzani, and H. R. Pruppacher, "A wind tunnel and theoretical study on the melting behavior of atmospheric ice particles: III. Experiment and theory for spherical ice particles of radius > 500 μm," *Journal of the Atmospheric Sciences*, vol. 41, no. 3, pp. 381–388, 1984. [Online]. Available: https://doi.org/10.1175/1520-0469(1984)041<0381:AWTATS>2.0.CO;2

[124] R. M. Rasmussen and A. J. Heymsfield, "Melting and shedding of graupel and hail. Part II: Sensitivity study," *Journal of the Atmospheric Sciences*, vol. 44, no. 19, pp. 2764–2782, 1987. [Online]. Available: https://doi.org/10.1175/1520-0469(1987)044<2764:MASOGA>2.0.CO;2

[125] A. V. Ryzhkov, M. R. Kumjian, S. M. Ganson, and A. P. Khain, "Polarimetric radar characteristics of melting hail. Part I: Theoretical simulations using spectral microphysical modeling," *Journal of Applied Meteorology and Climatology*, vol. 52, no. 12, pp. 2849–2870, 2013. [Online]. Available: https://doi.org/10.1175/JAMC-D-13-073.1

[126] A. V. Ryzhkov, M. R. Kumjian, S. M. Ganson, and P. Zhang, "Polarimetric radar characteristics of melting hail. Part II: Practical implications," *Journal of Applied Meteorology and Climatology*, vol. 52, no. 12, pp. 2871–2886, 2013. [Online]. Available: https://doi.org/10.1175/JAMC-D-13-074.1

[127] N. Balakrishnan and D. S. Zrnić, "Use of polarization to characterize precipitation and discriminate large hail," *Journal of the Atmospheric Sciences*, vol. 47, no. 13, pp. 1525–1540, 1990. [Online]. Available: https://doi.org/10.1175/1520-0469(1990)047<1525:UOPTCP>2.0.CO;2

[128] B. Dolan, B. Fuchs, S. A. Rutledge, E. A. Barnes, and E. J. Thompson, "Primary modes of global drop size distributions," *Journal of the Atmospheric Sciences*, vol. 75, no. 5, pp. 1453–1476, 2018. [Online]. Available: https://doi.org/10.1175/JAS-D-17-0242.1

[129] V. N. Bringi, V. Chandrasekar, J. Hubbert, E. Gorgucci, W. L. Randeu, and M. Schoenhuber, "Raindrop size distribution in different climatic regimes from disdrometer and dual-polarized radar analysis," *Journal of the Atmospheric Sciences*, vol. 60, no. 2, pp. 354–365, 2003. [Online]. Available: https://doi.org/10.1175/1520-0469(2003)060<0354:RSDIDC>2.0.CO;2

[130] F. W. Murray, "On the computation of saturation vapor pressure," *Journal of Applied Meteorology*, vol. 6, no. 1, pp. 203–204, 1967. [Online]. Available: https://doi.org/10.1175/1520-0450(1967)006<0203:OTCOSV>2.0.CO;2

[131] C. A. Balanis, *Advanced Engineering Electromagnetics*, 2nd ed. Hoboken, NJ: John Wiley & Sons, 2012.

[132] L. Tsang and J. A. Kong, *Scattering of Electromagnetic Waves: Advanced Topics*, 1st ed. Hoboken, NJ: John Wiley & Sons, 2001.

[133] S. Cloude, *Polarisation: Applications in Remote Sensing*, 1st ed. Oxford: Oxford University Press, 2009.

[134] M. I. Mishchenko, L. D. Travis, and D. W. Mackowski, "T-matrix computations of light scattering by nonspherical particles: A review," *Journal of Quantitative Spectroscopy and Radiative Transfer*, vol. 55, no. 5, pp. 535–575, 1996. [Online]. Available: http://www.sciencedirect.com/science/article/pii/0022407396000027

[135] J. Tyynelä and V. Chandrasekar, "Characterizing falling snow using multifrequency dual-polarization measurements," *Journal of Geophysical Research: Atmospheres*, vol. 119, no. 13, pp. 8268–8283, 2014.

[136] W. Ellison, "Permittivity of pure water, at standard atmospheric pressure, over the frequency range 0–25 THz and the temperature range 0–100°C," *Journal of Physical and Chemical Reference Data*, vol. 36, no. 1, pp. 1–18, 2007.

[137] D. D. Turner, S. Kneifel, and M. P. Cadeddu, "An improved liquid water absorption model at microwave frequencies for supercooled liquid water clouds," *Journal of Atmospheric and Oceanic Technology*, vol. 33, no. 1, pp. 33–44, 2016.

[138] C. Mäetzler, "Microwave dielectric properties of ice," in *Thermal Microwave Radiation: Applications for Remote Sensing*, vol. 52, C. Mäetzler, ed. London: Institution of Engineering and Technology, 2006, ch. 5.

[139] A. Sihvola, "Dielectric Polarization and Particle Shape Effects," *Journal of Nanomaterials*, vol. 2007, 2007.

[140] P. Chýlek, J. T. Kiehl, and M. K. W. Ko, "Narrow resonance structure in the Mie scattering characteristics," *Applied Optics*, vol. 17, no. 19, pp. 3019–3021, Oct. 1978. [Online]. Available: http://ao.osa.org/abstract.cfm?URI=ao-17-19-3019

[141] A. Sihvola, *Electromagnetic Mixing Formulas and Applications*, ser. Electromagnetic Waves. London: Institution of Engineering and Technology, 1999.

[142] R. D. Scott, P. R. Krehbiel, and W. Rison, "The use of simultaneous horizontal and vertical transmissions for dual-polarization radar meteorological observations," *Journal of Atmospheric and Oceanic Technology*, vol. 18, no. 4, pp. 629–648, 2001. [Online]. Available: https://doi.org/10.1175/1520-0426(2001)018<0629:TUOSHA>2.0.CO;2

[143] C. Tang and K. Aydin, "Scattering from ice crystals at 94 and 220 GHz millimeter wave frequencies," *IEEE Transactions on Geoscience and Remote Sensing*, vol. 33, no. 1, pp. 93–99, Jan. 1995.

[144] R. L. Olsen, "A review of theories of coherent radio wave propagation through precipitation media of randomly oriented scatterers, and the role of multiple scattering,"

Radio Science, vol. 17, no. 5, pp. 913–928, 1982. [Online]. Available: https://agupubs .onlinelibrary.wiley.com/doi/abs/10.1029/RS017i005p00913

[145] V. Chandrasekar, E. Gorgucci, and L. Baldini, "Evaluation of polarimetric radar rainfall algorithms at X-band," presentation at the Second European Conference on Radar in Meteorology and Hydrology, Delft, Netherlands, 2002.

[146] S. Lim and V. Chandrasekar, "A dual-polarization rain profiling algorithm," *IEEE Transactions on Geoscience and Remote Sensing*, vol. 44, no. 4, pp. 1011–1021, April 2006.

[147] T. Otto and H. W. J. Russchenberg, "Estimation of specific differential phase and differential backscatter phase from polarimetric weather radar measurements of rain," *IEEE Geoscience and Remote Sensing Letters*, vol. 8, no. 5, pp. 988–992, Sep. 2011.

[148] A. Sihvola, "Mixing rules with complex dielectric coefficient," *Subsurface Sensing Technologies and Applications*, vol. 1, pp. 393–415, 01 2000.

[149] J. C. M. Garnett and J. Larmor, "XII. Colours in metal glasses and in metallic films," *Philosophical Transactions of the Royal Society of London. Series A, Containing Papers of a Mathematical or Physical Character*, vol. 203, no. 359–371, pp. 385–420, 1904. [Online]. Available: https://royalsocietypublishing.org/doi/abs/10.1098/rsta.1904.0024

[150] V. A. Markel, "Introduction to the Maxwell Garnett approximation: Tutorial," *Journal of the Optical Society of America A*, vol. 33, no. 7, pp. 1244–1256, Jul. 2016. [Online]. Available: http://josaa.osa.org/abstract.cfm?URI=josaa-33-7-1244

[151] R. J. Hogan, P. R. Field, A. J. Illingworth, R. J. Cotton, and T. W. Choularton, "Properties of embedded convection in warm-frontal mixed-phase cloud from aircraft and polarimetric radar," *Quarterly Journal of the Royal Meteorological Society*, vol. 128, no. 580, pp. 451–476, 2002. [Online]. Available: https://rmets.onlinelibrary.wiley.com/doi/abs/10.1256/003590002321042054

[152] M. Ishizaka, "An accurate measurement of densities of snowflakes using 3-D micropho-tographs," *Annals of Glaciology*, vol. 18, p. 92–96, 1993.

[153] R. Meneghini and L. Liao, "Comparisons of cross sections for melting hydrometeors as derived from dielectric mixing formulas and a numerical method," *Journal of Applied Meteorology*, vol. 35, no. 10, pp. 1658–1670, 1996. [Online]. Available: https://doi.org/10.1175/1520-0450(1996)035<1658:COCSFM>2.0.CO;2

[154] L. Liao and R. Meneghini, "Examination of effective dielectric constants of nonspherical mixed-phase hydrometeors," *Journal of Applied Meteorology and Climatology*, vol. 52, no. 1, pp. 197–212, 2013. [Online]. Available: https://doi.org/10.1175/JAMC-D-11-0244.1

[155] H. Pruppacher and J. Klett, *Microphysics of Clouds and Precipitation*, vol. 18, ser. Atmospheric and Oceanographic Sciences Library. Dordrecht, Netherlands: Springer, 2010.

[156] A. V. Ryzhkov, S. E. Giangrande, V. M. Melnikov, and T. J. Schuur, "Calibration issues of dual-polarization radar measurements," *Journal of Atmospheric and Oceanic Technology*, vol. 22, no. 8, pp. 1138–1155, 2005. [Online]. Available: https://doi.org/10.1175/JTECH1772.1

[157] E. Gorgucci, G. Scarchilli, and V. Chandrasekar, "A procedure to calibrate multiparame-ter weather radar using properties of the rain medium," *IEEE Transactions on Geoscience and Remote Sensing*, vol. 37, no. 1, pp. 269–276, Jan. 1999.

[158] R. Bechini, L. Baldini, R. Cremonini, and E. Gorgucci, "Differential reflectivity calibration for operational radars," *Journal of Atmospheric and Oceanic Technology*,

vol. 25, no. 9, pp. 1542–1555, 2008. [Online]. Available: https://doi.org/10.1175/2008JTECHA1037.1

[159] S. Y. Matrosov, "Theoretical study of radar polarization parameters obtained from cirrus clouds," *Journal of the Atmospheric Sciences*, vol. 48, no. 8, pp. 1062–1070, 1991. [Online]. Available: https://doi.org/10.1175/1520-0469(1991)048<1062:TSORPP>2.0.CO;2

[160] S. Y. Matrosov, R. F. Reinking, R. A. Kropfli, B. E. Martner, and B. W. Bartram, "On the use of radar depolarization ratios for estimating shapes of ice hydrometeors in winter clouds," *Journal of Applied Meteorology*, vol. 40, no. 3, pp. 479–490, 2001. [Online]. Available: https://doi.org/10.1175/1520-0450(2001)040<0479:OTUORD>2.0.CO;2

[161] S. Y. Matrosov, G. G. Mace, R. Marchand, M. D. Shupe, A. G. Hallar, and I. B. McCubbin, "Observations of ice crystal habits with a scanning polarimetric W-band radar at slant linear depolarization ratio mode," *Journal of Atmospheric and Oceanic Technology*, vol. 29, no. 8, pp. 989–1008, 2012. [Online]. Available: https://doi.org/10.1175/JTECH-D-11-00131.1

[162] A. Myagkov, P. Seifert, M. Bauer-Pfundstein, and U. Wandinger, "Cloud radar with hybrid mode towards estimation of shape and orientation of ice crystals," *Atmospheric Measurement Techniques*, vol. 9, no. 2, pp. 469–489, 2016. [Online]. Available: https://www.atmos-meas-tech.net/9/469/2016/

[163] A. Papoulis and S. U. Pillai, *Probability, Random Variables, and Stochastic Processes*. New York: McGraw-Hill Higher Education, 2002.

[164] J. W. Cooley and J. W. Tukey, "An algorithm for the machine calculation of complex Fourier series," *Mathematics of Computation*, vol. 19, pp. 297–301, 1965.

[165] R. Doviak and D. Zrnić, *Doppler Radar and Weather Observations*. Cambridge, MA: Academic Press, 1993.

[166] P. J. Brockwell and R. A. Davis, *Time Series: Theory and Methods*. Berlin: Springer-Verlag, 1986.

[167] P. R. Krehbiel and M. Brook, "A broad-band noise technique for fast-scanning radar observations of clouds and clutter targets," *IEEE Transactions on Geoscience Electronics*, vol. 17, no. 4, pp. 196–204, 1979.

[168] A. Zahrai and D. Zrnić, "The 10-cm-wavelength polarimetric weather radar at NOAA's National Severe Storms Laboratory," *Journal of Atmospheric and Oceanic Technology*, vol. 10, no. 5, pp. 649–662, 1993.

[169] F. Harris, "On the use of windows for harmonic analysis with the discrete Fourier transform," *Proceedings of the IEEE*, vol. 66, no. 1, pp. 51–83, 1978.

[170] N. Geckinli and D. Yavuz, "Some novel windows and a concise tutorial comparison of window families," *IEEE Transactions on Acoustics, Speech, and Signal Processing*, vol. 26, no. 6, pp. 501–507, 1978.

[171] J. B. Mead, "Comparison of meteorological radar signal detectability with noncoherent and spectral-based processing," *Journal of Atmospheric and Oceanic Technology*, vol. 33, no. 4, pp. 723–739, 2016.

[172] M. B. Priestley, *Spectral Analysis and Time Series*, 1st ed. Cambridge, MA: Academic Press, 1982.

[173] D. N. Moisseev and V. Chandrasekar, "Polarimetric Spectral Filter for Adaptive Clutter and Noise Suppression," *Journal of Atmospheric and Oceanic Technology*, vol. 26, no. 2, pp. 215–228, Feb. 2009. [Online]. Available: http://journals.ametsoc.org/doi/abs/10.1175/2008JTECHA1119.1

[174] J. Hubbert and V. N. Bringi, "An iterative filtering technique for the analysis of copolar differential phase and dual-frequency radar measurements," *Journal of Atmospheric and Oceanic Technology*, vol. 12, no. 3, pp. 643–648, 1995. [Online]. Available: https://doi.org/10.1175/1520-0426(1995)012<0643:AIFTFT>2.0.CO;2

[175] Y. Wang and V. Chandrasekar, "Algorithm for estimation of the specific differential phase," *Journal of Atmospheric and Oceanic Technology*, vol. 26, no. 12, pp. 2565–2578, 2009.

[176] T. Maesaka, K. Iwanami, and M. Maki, "Non-negative K_{DP} estimation by monotone increasing Φ_{dp} assumption below melting layer," presentation at the Seventh European Conference on Radar in Meteorology and Hydrology, Toulouse, France, 2012.

[177] S. E. Giangrande, R. McGraw, and L. Lei, "An application of linear programming to polarimetric radar differential phase processing," *Journal of Atmospheric and Oceanic Technology*, vol. 30, no. 8, pp. 1716–1729, 2013.

[178] M. Schneebeli and A. Berne, "An extended Kalman filter framework for polarimetric X-band weather radar data processing," *Journal of Atmospheric and Oceanic Technology*, vol. 29, no. 5, pp. 711–730, 2012.

[179] S. Lim and V. Chandrasekar, "A robust attenuation correction system for reflectivity and differential reflectivity in weather radars," *IEEE Transactions on Geoscience and Remote Sensing*, vol. 54, no. 3, pp. 1727–1737, 2016.

[180] E. Ruzanski, J. C. Hubbert, and V. Chandrasekar, "Evaluation of the simultaneous multiple pulse repetition frequency algorithm for weather radar," *Journal of Atmospheric and Oceanic Technology*, vol. 25, no. 7, pp. 1166–1181, 2008.

[181] M. Sachidananda and D. S. Zrnić, "Systematic phase codes for resolving range overlaid signals in a Doppler weather radar," *Journal of Atmospheric and Oceanic Technology*, vol. 16, no. 10, pp. 1351–1363, 1999.

[182] V. Chandrasekar and N. Bharadwaj, "Orthogonal channel coding for simultaneous co- and cross-polarization measurements," *Journal of Atmospheric and Oceanic Technology*, vol. 26, no. 1, pp. 45–56, 2009.

[183] A. S. Mudukutore, V. Chandrasekar, and R. Jeffrey Keeler, "Pulse compression for weather radars," *IEEE Transactions on Geoscience and Remote Sensing*, vol. 36, no. 1, pp. 125–142, 1998.

[184] H. D. Griffiths and L. Vinagre, "Design of low-sidelobe pulse compression waveforms," *Electronics Letters*, vol. 30, no. 12, pp. 1004–1005, 1994.

[185] N. Bharadwaj and V. Chandrasekar, "Wideband waveform design principles for solid-state weather radars," *Journal of Atmospheric and Oceanic Technology*, vol. 29, no. 1, pp. 14–31, Jan 2012. [Online]. Available: http://journals.ametsoc.org/doi/abs/10.1175/JTECH-D-11-00030.1

[186] J. M. Kurdzo, B. L. Cheong, R. D. Palmer, G. Zhang, and J. B. Meier, "A pulse compression waveform for improved-sensitivity weather radar observations," *Journal of Atmospheric and Oceanic Technology*, vol. 31, no. 12, pp. 2713–2731, 2014. [Online]. Available: http://journals.ametsoc.org/doi/abs/10.1175/JTECH-D-13-00021.1

[187] J. E. Cilliers and J. C. Smit, "Pulse compression sidelobe reduction by minimization of Lp-norms," *IEEE Transactions on Aerospace and Electronic Systems*, vol. 43, no. 3, pp. 1238–1247, 2007.

[188] R. M. Beauchamp, S. Tanelli, E. Peral, and V. Chandrasekar, "Pulse compression waveform and filter optimization for spaceborne cloud and precipitation radar," *IEEE Transactions on Geoscience and Remote Sensing*, vol. 55, no. 2, pp. 915–931, 2017.

[189] V. M. Melnikov and D. S. Zrnić, "Simultaneous transmission mode for the polarimetric WSR-88D: Statistical biases and standard deviations of polarimetric variables," Cooperative Institute for Mesoscale Meteorological Studies, University of Oklahoma, Norman, OK 2004. [Online]. Available: https://www.nssl.noaa.gov/publications/wsr88d_reports/SHV_statistics.pdf

[190] E. Gorgucci, G. Scarchilli, and V. Chandrasekar, "Specific differential phase estimation in the presence of nonuniform rainfall medium along the path," *Journal of Atmospheric and Oceanic Technology*, vol. 16, no. 11, pp. 1690–1697, 1999.

[191] E. Saltikoff, J. Y. N. Cho, P. Tristant, A. Huuskonen, R. Allmon, R. Cook, E. Becker, and P. Joe, "The threat to weather radars by wireless technology," *Bulletin of the American Meteorological Society*, vol. 97, no. 7, pp. 1159–1167, 2016. [Online]. Available: https://doi.org/10.1175/BAMS-D-15-00048.1

[192] M. Vaccarono, C. V. Chandrasekar, R. Bechini, and R. Cremonini, "Survey on electromagnetic interference in weather radars in northwestern Italy," *Environments*, vol. 6, no. 12, 2019. [Online]. Available: https://www.mdpi.com/2076-3298/6/12/126

[193] J. Y. N. Cho, "A new radio frequency interference filter for weather radars," *Journal of Atmospheric and Oceanic Technology*, vol. 34, no. 7, pp. 1393–1406, 06 2017. [Online]. Available: https://doi.org/10.1175/JTECH-D-17-0028.1

[194] V. N. Bringi, R. Hoferer, D. A. Brunkow, R. Schwerdtfeger, V. Chandrasekar, S. A. Rutledge, J. George, and P. C. Kennedy, "Design and performance characteristics of the new 8.5-m dual-offset Gregorian antenna for the CSU-CHILL radar," *Journal of Atmospheric and Oceanic Technology*, vol. 28, no. 7, pp. 907–920, 2011. [Online]. Available: https://doi.org/10.1175/2011JTECHA1493.1

[195] G. Scarchilli, E. Gorgucci, and V. Chandrasekar, "Detection and estimation of reflectivity gradients in the radar resolution volume using multiparameter radar measurements," *IEEE Transactions on Geoscience and Remote Sensing*, vol. 37, no. 2, pp. 1122–1127, March 1999.

[196] A. Ryzhkov and D. Zrnić, "Beamwidth effects on the differential phase measurements of rain," *Journal of Atmospheric and Oceanic Technology*, vol. 15, no. 3, pp. 624–634, 1998. [Online]. Available: https://doi.org/10.1175/1520-0426(1998)015<0624:BEOTDP>2.0.CO;2

[197] M. Gosset, "Effect of nonuniform beam filling on the propagation of radar signals at X-band frequencies. Part II: Examination of differential phase shift," *Journal of Atmospheric and Oceanic Technology*, vol. 21, no. 2, pp. 358–367, 2004. [Online]. Available: https://doi.org/10.1175/1520-0426(2004)021<0358:EONBFO>2.0.CO;2

[198] A. V. Ryzhkov, "The impact of beam broadening on the quality of radar polarimetric data," *Journal of Atmospheric and Oceanic Technology*, vol. 24, no. 5, pp. 729–744, 2007. [Online]. Available: https://doi.org/10.1175/JTECH2003.1

[199] R. M. Beauchamp and V. Chandrasekar, "Real-Time Noise Estimation and Correction in Dual-Polarization Radar Systems," *IEEE Transactions on Geoscience and Remote Sensing*, vol. 53, no. 11, pp. 6183–6195, 2015.

[200] I. R. Ivić, C. Curtis, and S. M. Torres, "Radial-Based Noise Power Estimation for Weather Radars," *Journal of Atmospheric and Oceanic Technology*, vol. 30, no. 12, pp. 2737–2753, Dec. 2013.

[201] S. A. Gauthreaux, D. S. Mizrahi, and C. G. Belser, "Bird migration and bias of WSR-88d wind estimates," *Weather and Forecasting*, vol. 13, no. 2, pp. 465–481, 1998. [Online]. Available: https://doi.org/10.1175/1520-0434(1998)013<0465:BMABOW>2.0.CO;2

[202] P. Jatau and V. Melnikov, "Classifying bird and insect radar echoes at S-band," presentation at the 35th Conference on Environmental Information Processing Technologies, Phoenix, AZ, American Meteorological Society, 2018.

[203] D. S. Zrnić and A. V. Ryzhkov, "Observations of insects and birds with a polarimetric radar," *IEEE Transactions on Geoscience and Remote Sensing*, vol. 36, no. 2, pp. 661–668, 1998.

[204] J. C. Hubbert, M. Dixon, and S. M. Ellis, "Weather radar ground clutter. Part II: Real-time identification and filtering," *Journal of Atmospheric and Oceanic Technology*, vol. 26, no. 1973, pp. 1181–1197, 2009.

[205] A. D. Siggia and R. E. Passarelli, "Gaussian model adaptive processing (GMAP) for improved ground clutter cancellation and moment calculation," in *Proceedings of ERAD (2004)*. Göttingen, Germany: Copernicus Publications, 2004, pp. 67–73.

[206] C. M. Nguyen and V. Chandrasekar, "Gaussian model adaptive processing in time domain (GMAP-TD) for weather radars," *Journal of Atmospheric and Oceanic Technology*, vol. 30, no. 11, pp. 2571–2584, 2013. [Online]. Available: http://journals.ametsoc.org/doi/abs/10.1175/JTECH-D-12-00215.1

[207] A. Battaglia, S. Tanelli, S. Kobayashi, D. Zrnić, R. J. Hogan, and C. Simmer, "Multiple-scattering in radar systems: A review," *Journal of Quantitative Spectroscopy and Radiative Transfer*, vol. 111, no. 6, pp. 917–947, 2010. [Online]. Available: http://www.sciencedirect.com/science/article/pii/S0022407309003677

[208] K. Aydin, S. H. Park, and T. M. Walsh, "Bistatic dual-polarization scattering from rain and hail at S- and C-band frequencies," *Journal of Atmospheric and Oceanic Technology*, vol. 15, no. 5, pp. 1110–1121, 1998. [Online]. Available: https://doi.org/10.1175/1520-0426(1998)015<1110:BDPSFR>2.0.CO;2

[209] J. Picca and A. Ryzhkov, "A dual-wavelength polarimetric analysis of the 16 May 2010 Oklahoma City extreme hailstorm," *Monthly Weather Review*, vol. 140, no. 4, pp. 1385–1403, 2012. [Online]. Available: https://doi.org/10.1175/MWR-D-11-00112.1

[210] J. C. Hubbert and V. N. Bringi, "The effects of three-body scattering on differential reflectivity signatures," *Journal of Atmospheric and Oceanic Technology*, vol. 17, no. 1, pp. 51–61, 2000. [Online]. Available: https://doi.org/10.1175/1520-0426(2000)017<0051:TEOTBS>2.0.CO;2

[211] V. Chandrasekar, L. Baldini, N. Bharadwaj, and P. L. Smith, "Calibration procedures for global precipitation-measurement ground-validation radars," *URSI Radio Science Bulletin*, vol. 2015, no. 355, pp. 45–73, Dec. 2015.

[212] A. Huuskonen and I. Holleman, "Determining weather radar antenna pointing using signals detected from the sun at low antenna elevations," *Journal of Atmospheric and Oceanic Technology*, vol. 24, no. 3, pp. 476–483, 2007. [Online]. Available: https://doi.org/10.1175/JTECH1978.1

[213] P. Altube, J. Bech, O. Argemí, and T. Rigo, "Quality control of antenna alignment and receiver calibration using the sun: Adaptation to midrange weather radar observations at low elevation angles," *Journal of Atmospheric and Oceanic Technology*, vol. 32, no. 5, pp. 927–942, 05 2015. [Online]. Available: https://doi.org/10.1175/JTECH-D-14-00116.1

[214] D. S. Silberstein, D. B. Wolff, D. A. Marks, D. Atlas, and J. L. Pippitt, "Ground clutter as a monitor of radar stability at Kwajalein, RMI," *Journal of Atmospheric and Oceanic Technology*, vol. 25, no. 11, pp. 2037–2045, 2008. [Online]. Available: https://doi.org/10.1175/2008JTECHA1063.1

[215] D. B. Wolff, D. A. Marks, and W. A. Petersen, "General application of the relative calibration adjustment (RCA) technique for monitoring and correcting radar reflectivity calibration," *Journal of Atmospheric and Oceanic Technology*, vol. 32, no. 3, pp. 496–506, 2015. [Online]. Available: https://doi.org/10.1175/JTECH-D-13-00185.1

[216] M. Kurri and A. Huuskonen, "Measurements of the transmission loss of a radome at different rain intensities," *Journal of Atmospheric and Oceanic Technology*, vol. 25, no. 9, pp. 1590–1599, 09 2008. [Online]. Available: https://doi.org/10.1175/2008JTECHA1056.1

[217] R. Bechini, V. Chandrasekar, R. Cremonini, and S. Lim, "Radome attenuation at X-band radar operations," presentation at the Sixth European Conference on Radar in Meteorology and Hydrology, Sibiu, Romania, 2010.

[218] E. Gorgucci, R. Bechini, L. Baldini, R. Cremonini, and V. Chandrasekar, "The influence of antenna radome on weather radar calibration and its real-time assessment," *Journal of Atmospheric and Oceanic Technology*, vol. 30, no. 4, pp. 676–689, 04 2013. [Online]. Available: https://doi.org/10.1175/JTECH-D-12-00071.1

[219] S. J. Frasier, F. Kabeche, J. Figueras i Ventura, H. Al-Sakka, P. Tabary, J. Beck, and O. Bousquet, "In-place estimation of wet radome attenuation at X band," *Journal of Atmospheric and Oceanic Technology*, vol. 30, no. 5, pp. 917–928, 2013. [Online]. Available: https://doi.org/10.1175/JTECH-D-12-00148.1

[220] J. Figueras i Ventura, A.-A. Boumahmoud, B. Fradon, P. Dupuy, and P. Tabary, "Long-term monitoring of French polarimetric radar data quality and evaluation of several polarimetric quantitative precipitation estimators in ideal conditions for operational implementation at C-band," *Quarterly Journal of the Royal Meteorological Society*, vol. 138, no. 669, pp. 2212–2228, 2012. [Online]. Available: https://rmets.onlinelibrary.wiley.com/doi/abs/10.1002/qj.1934

[221] E. A. Brandes and K. Ikeda, "Freezing-level estimation with polarimetric radar," *Journal of Applied Meteorology*, vol. 43, no. 11, pp. 1541–1553, 2004. [Online]. Available: https://doi.org/10.1175/JAM2155.1

[222] S. E. Giangrande, J. M. Krause, and A. V. Ryzhkov, "Automatic designation of the melting layer with a polarimetric prototype of the WSR-88d radar," *Journal of Applied Meteorology and Climatology*, vol. 47, no. 5, pp. 1354–1364, 2008. [Online]. Available: https://doi.org/10.1175/2007JAMC1634.1

[223] S. Boodoo, D. Hudak, N. Donaldson, and M. Leduc, "Application of dual-polarization radar melting-layer detection algorithm," *Journal of Applied Meteorology and Climatology*, vol. 49, no. 8, pp. 1779–1793, 2010. [Online]. Available: https://doi.org/10.1175/2010JAMC2421.1

[224] J. S. Kain, S. M. Goss, and M. E. Baldwin, "The melting effect as a factor in precipitation-type forecasting," *Weather and Forecasting*, vol. 15, no. 6, pp. 700–714, 2000. [Online]. Available: https://doi.org/10.1175/1520-0434(2000)015<0700:TMEAAF>2.0.CO;2

[225] L. Baldini and E. Gorgucci, "Identification of the melting layer through dual-polarization radar measurements at vertical incidence," *Journal of Atmospheric and Oceanic Technology*, vol. 23, no. 6, pp. 829–839, 2006. [Online]. Available: https://doi.org/10.1175/JTECH1884.1

[226] M. Skolnik, *Radar Handbook*, 3rd ed., ser. Electronics electrical engineering. New York: McGraw-Hill Education, 2008.

[227] J. P. Zagrodnik, L. A. McMurdie, and R. A. Houze, "Stratiform precipitation processes in cyclones passing over a coastal mountain range," *Journal of the Atmospheric Sciences*,

vol. 75, no. 3, pp. 983–1004, 2018. [Online]. Available: https://doi.org/10.1175/JAS-D-17-0168.1

[228] L. A. McMurdie, A. K. Rowe, R. A. Houze Jr, S. R. Brodzik, J. P. Zagrodnik, and T. M. Schuldt, "Terrain-enhanced precipitation processes above the melting layer: Results from Olympex," *Journal of Geophysical Research: Atmospheres*, vol. 123, no. 21, pp. 12,194–12,209, 2018. [Online]. Available: https://agupubs.onlinelibrary .wiley.com/doi/abs/10.1029/2018JD029161

[229] R. A. Houze and S. Medina, "Turbulence as a mechanism for orographic precipitation enhancement," *Journal of the Atmospheric Sciences*, vol. 62, no. 10, pp. 3599–3623, 2005. [Online]. Available: https://doi.org/10.1175/JAS3555.1

[230] D. N. Moisseev, S. Lautaportti, J. Tyynela, and S. Lim, "Dual-polarization radar signatures in snowstorms: Role of snowflake aggregation," *Journal of Geophysical Research: Atmospheres*, vol. 120, no. 24, pp. 12 644–12 655, 2015. [Online]. Available: https://agupubs.onlinelibrary.wiley.com/doi/abs/10.1002/2015JD023884

[231] J. Grazioli, G. Lloyd, L. Panziera, C. R. Hoyle, P. J. Connolly, J. Henneberger, and A. Berne, "Polarimetric radar and in situ observations of riming and snowfall microphysics during CLACE 2014," *Atmospheric Chemistry and Physics*, vol. 15, no. 23, pp. 13 787–13 802, 2015. [Online]. Available: https://www.atmos-chem-phys.net/15/ 13787/2015/

[232] D. Rosenfeld and C. W. Ulbrich, "Cloud microphysical properties, processes, and rainfall estimation opportunities," *Meteorological Monographs*, vol. 52, pp. 237–258, 2003. [Online]. Available: https://doi.org/10.1175/0065-9401(2003)030<0237: CMPPAR>2.0.CO;2

[233] L. R. Lemon and S. Parker, "The Lahoma storm deep convergence zone: Its characteristics and role in storm dynamics and severity," presentation at the 26th Conference on Radar Meteorology, Norman, OK, American Meteorological Society, 1996.

[234] J. C. Snyder, A. V. Ryzhkov, M. R. Kumjian, A. P. Khain, and J. Picca, "A Z_{DR} column detection algorithm to examine convective storm updrafts," *Weather and Forecasting*, vol. 30, no. 6, pp. 1819–1844, 2015. [Online]. Available: https://doi.org/10.1175/WAF-D-15-0068.1

[235] M. van Lier-Walqui, A. M. Fridlind, A. S. Ackerman, S. Collis, J. Helmus, D. R. MacGorman, K. North, P. Kollias, and D. J. Posselt, "On polarimetric radar signatures of deep convection for model evaluation: Columns of specific differential phase observed during MC3E," *Monthly Weather Review*, vol. 144, no. 2, pp. 737–758, 2016. [Online]. Available: https://doi.org/10.1175/MWR-D-15-0100.1

[236] M. P. M. Hall, J. W. F. Goddard, and S. M. Cherry, "Identification of hydrometeors and other targets by dual-polarization radar," *Radio Science*, vol. 19, no. 1, pp. 132–140, 1984. [Online]. Available: https://agupubs.onlinelibrary.wiley.com/doi/abs/10 .1029/RS019i001p00132

[237] H. Höller, M. Hagen, P. F. Meischner, V. N. Bringi, and J. Hubbert, "Life cycle and precipitation formation in a hybrid-type hailstorm revealed by polarimetric and Doppler radar measurements," *Journal of the Atmospheric Sciences*, vol. 51, no. 17, pp. 2500–2522, 1994. [Online]. Available: https://doi.org/10.1175/1520-0469(1994)051<2500: LCAPFI>2.0.CO;2

[238] H. Liu and V. Chandrasekar, "Classification of hydrometeors based on polarimetric radar measurements: Development of fuzzy logic and neuro-fuzzy systems, and in situ

verification," *Journal of Atmospheric and Oceanic Technologies*, vol. 17, pp. 140–164, 2000.

[239] D. S. Zrnić, A. Ryzhkov, J. Straka, Y. Liu, and J. Vivekanandan, "Testing a procedure for automatic classification of hydrometeor types," *Journal of Atmospheric and Oceanic Technology*, vol. 18, pp. 892–913, 2001.

[240] H. Liu and V. Chandrasekar, "Classification of hydrometeor type based on multiparameter radar measurements," in *International Conference on Cloud Physics*. Boston: American Meteorological Society, 1998, pp. 253–256.

[241] S. Lim, V. Chandrasekar, and V. N. Bringi, "Hydrometeor classification system using dual-polarization radar measurements: Model improvements and in situ verification," *IEEE Transactions on Geoscience and Remote Sensing*, vol. 43, pp. 792–801, 2005.

[242] R. Bechini and V. Chandrasekar, "A semi-supervised robust hydrometeor classification method for dual-polarization radar applications," *Journal of Atmospheric and Oceanic Technology*, vol. 32, pp. 22–47, 2015.

[243] H. Al-Sakka, A.-A. Boumahmoud, B. Fradon, S. J. Frasier, and P. Tabary, "A new fuzzy logic hydrometeor classification scheme applied to the French X-, C-, and S-band polarimetric radars," *Journal of Applied Meteorology and Climatology*, vol. 52, pp. 2328–2344, 2013.

[244] N. Roberto, L. Baldini, E. Adirosi, L. Facheris, F. Cuccoli, A. Lupidi, and A. Garzelli, "A support vector machine hydrometeor classification algorithm for dual-polarization radar," *Atmosphere*, vol. 8, no. 8, 2017. [Online]. Available: https://www.mdpi.com/2073-4433/8/8/134

[245] G. Wen, A. Protat, P. T. May, W. Moran, and M. Dixon, "A cluster-based method for hydrometeor classification using polarimetric variables. Part II: Classification," *Journal of Atmospheric and Oceanic Technology*, vol. 33, no. 1, pp. 45–60, 2016. [Online]. Available: https://doi.org/10.1175/JTECH-D-14-00084.1

[246] N. Besic, J. Figueras i Ventura, J. Grazioli, M. Gabella, U. Germann, and A. Berne, "Hydrometeor classification through statistical clustering of polarimetric radar measurements: A semi-supervised approach," *Atmospheric Measurement Techniques*, vol. 9, no. 9, pp. 4425–4445, 2016. [Online]. Available: https://www.atmos-meas-tech.net/9/4425/2016/

[247] J. Grazioli, D. Tuia, and A. Berne, "Hydrometeor classification from polarimetric radar measurements: A clustering approach," *Atmospheric Measurement Techniques*, vol. 8, no. 1, pp. 149–170, 2015. [Online]. Available: https://www.atmos-meas-tech.net/8/149/2015/

[248] J.-F. Ribaud, L. A. T. Machado, and T. Biscaro, "X-band dual-polarization radar-based hydrometeor classification for Brazilian tropical precipitation systems," *Atmospheric Measurement Techniques*, vol. 12, no. 2, pp. 811–837, 2019. [Online]. Available: https://www.atmos-meas-tech.net/12/811/2019/

[249] M. Yurkin and A. Hoekstra, "The discrete dipole approximation: An overview and recent developments," *Journal of Quantitative Spectroscopy and Radiative Transfer*, vol. 106, no. 1, pp. 558–589, 2007. [Online]. Available: http://www.sciencedirect.com/science/article/pii/S0022407307000556

[250] A. H. Auer and D. L. Veal, "The dimension of ice crystals in natural clouds," *Journal of the Atmospheric Sciences*, vol. 27, no. 6, pp. 919–926, 1970. [Online]. Available: https://doi.org/10.1175/1520-0469(1970)027<0919:TDOICI>2.0.CO;2

[251] M. Wolde and G. Vali, "Polarimetric signatures from ice crystals observed at 95 GHz in winter clouds. Part I: Dependence on crystal form," *Journal of the Atmospheric Sciences*, vol. 58, no. 8, pp. 828–841, 2001. [Online]. Available: https://doi.org/10.1175/1520-0469(2001)058<0828:PSFICO>2.0.CO;2

[252] J. Andric, D. Zrnić, and V. Melnikov, "Two-layer patterns of enhanced Z_{DR} in clouds," presentation at the 34th Conference on Radar Meteorology, Wiilliamsburg, VA, American Meteorological Society, 2009. [Online]. Available: https://ams.confex.com/ams/pdfpapers/155481.pdf

[253] W. J. Keat and C. D. Westbrook, "Revealing layers of pristine oriented crystals embedded within deep ice clouds using differential reflectivity and the copolar correlation coefficient," *Journal of Geophysical Research: Atmospheres*, vol. 122, no. 21, pp. 11737–11759, 2017. [Online]. Available: https://agupubs.onlinelibrary.wiley.com/doi/abs/10.1002/2017JD026754

[254] P. C. Kennedy and S. A. Rutledge, "S-band dual-polarization radar observations of winter storms," *Journal of Applied Meteorology and Climatology*, vol. 50, no. 4, pp. 844–858, 2011. [Online]. Available: https://doi.org/10.1175/2010JAMC2558.1

[255] E. M. Griffin, T. J. Schuur, and A. V. Ryzhkov, "A polarimetric analysis of ice microphysical processes in snow, using quasi-vertical profiles," *Journal of Applied Meteorology and Climatology*, vol. 57, no. 1, pp. 31–50, 2018. [Online]. Available: https://doi.org/10.1175/JAMC-D-17-0033.1

[256] P. V. Hobbs, S. Chang, and J. D. Locatelli, "The dimensions and aggregation of ice crystals in natural clouds," *Journal of Geophysical Research*, vol. 79, no. 15, pp. 2199–2206, 1974. [Online]. Available: https://agupubs.onlinelibrary.wiley.com/doi/abs/10.1029/JC079i015p02199

[257] E. R. Williams, D. J. Smalley, M. F. Donovan, R. G. Hallowell, K. T. Hood, B. J. Bennett, R. Evaristo, A. Stepanek, T. Bals-Elsholz, J. Cobb, J. Ritzman, A. Korolev, and M. Wolde, "Measurements of differential reflectivity in snowstorms and warm season stratiform systems," *Journal of Applied Meteorology and Climatology*, vol. 54, no. 3, pp. 573–595, 2015. [Online]. Available: https://doi.org/10.1175/JAMC-D-14-0020.1

[258] P. Bukovčić, A. Ryzhkov, D. Zrnić, and G. Zhang, "Polarimetric radar relations for quantification of snow based on disdrometer data," *Journal of Applied Meteorology and Climatology*, vol. 57, no. 1, pp. 103–120, 2018. [Online]. Available: https://doi.org/10.1175/JAMC-D-17-0090.1

[259] J. Vivekanandan, V. N. Bringi, M. Hagen, and P. Meischner, "Polarimetric radar studies of atmospheric ice particles," *IEEE Transactions on Geoscience and Remote Sensing*, vol. 32, no. 1, pp. 1–10, Jan 1994.

[260] P. Bukovčić, A. Ryzhkov, and D. Zrnić, "Polarimetric relations for snow estimation—radar verification," *Journal of Applied Meteorology and Climatology*, vol. 59, no. 5, pp. 991–1009, 05 2020. [Online]. Available: https://doi.org/10.1175/JAMC-D-19-0140.1

[261] B. Vonnegut, "Orientation of ice crystals in the electric field of a thunderstorm," *Weather*, vol. 20, no. 10, pp. 310–312, 1965. [Online]. Available: https://rmets.onlinelibrary.wiley.com/doi/abs/10.1002/j.1477-8696.1965.tb02740.x

[262] A. J. Weinheimer and A. A. Few, "The electric field alignment of ice particles in thunderstorms," *Journal of Geophysical Research: Atmospheres*, vol. 92, no. D12, pp. 14833–14844, 1987. [Online]. Available: https://agupubs.onlinelibrary.wiley.com/doi/abs/10.1029/JD092iD12p14833

[263] I. J. Caylor and V. Chandrasekar, "Time-varying ice crystal orientation in thunderstorms observed with multiparameter radar," *IEEE Transactions on Geoscience and Remote Sensing*, vol. 34, no. 4, pp. 847–858, July 1996.

[264] S. A. Amburn and P. L. Wolf, "VIL density as a hail indicator," *Weather and Forecasting*, vol. 12, no. 3, pp. 473–478, 1997. [Online]. Available: https://doi.org/10.1175/1520-0434(1997)012<0473:VDAAHI>2.0.CO;2

[265] D. R. Greene and R. A. Clark, "Vertically integrated liquid water-a new analysis tool," *Monthly Weather Review*, vol. 100, no. 7, pp. 548–552, 1972. [Online]. Available: https://doi.org/10.1175/1520-0493(1972)100<0548:VILWNA>2.3.CO;2

[266] H. Höller, "Radar-derived mass-concentrations of hydrometeors for cloud model retrievals," presentation at the 27th Conference on Radar Meteorology, Vail, CO, Oct. 9–13 1995. [Online]. Available: https://elib.dlr.de/32103/

[267] A. R. Jameson, "Microphysical interpretation of multi-parameter radar measurements in rain. Part I: Interpretation of polarization measurements and estimation of raindrop shapes," *Journal of the Atmospheric Sciences*, vol. 40, no. 7, pp. 1792–1802, 1983.

[268] C. W. Ulbrich and D. Atlas, "Rainfall microphysics and radar properties: Analysis methods for drop size spectra," *Journal of Applied Meteorology*, vol. 37, no. 9, pp. 912–923, 1998. [Online]. Available: https://doi.org/10.1175/1520-0450(1998)037<0912:RMARPA>2.0.CO;2

[269] G. Scarchilli, V. Gorgucci, V. Chandrasekar, and A. Dobaie, "Self-consistency of polarization diversity measurement of rainfall," *IEEE Transactions on Geoscience and Remote Sensing*, vol. 34, no. 1, pp. 22–26, Jan. 1996.

[270] E. Gorgucci, L. Baldini, and V. Chandrasekar, "What is the shape of a raindrop? An answer from radar measurements," *Journal of the Atmospheric Sciences*, vol. 63, no. 11, pp. 3033–3044, 2006.

[271] E. Gorgucci, V. Chandrasekar, V. N. Bringi, and G. Scarchilli, "Estimation of raindrop size distribution parameters from polarimetric radar measurements," *Journal of the Atmospheric Sciences*, vol. 59, no. 15, pp. 2373–2384, 2002. [Online]. Available: https://doi.org/10.1175/1520-0469(2001)058<0828:PSFICO>2.0.CO;2 https://doi.org/10.1175/1520-0469(2002)059<2373:EORSDP>2.0.CO;2

[272] E. Gorgucci, G. Scarchilli, V. Chandrasekar, and V. N. Bringi, "Measurement of mean raindrop shape from polarimetric radar observations," *Journal of the Atmospheric Sciences*, vol. 57, no. 20, pp. 3406–3413, 2000. [Online]. Available: https://doi.org/10.1175/1520-0469(2000)057<3406:MOMRSF>2.0.CO;2

[273] J. Pippitt, D. Wolff, W. Petersen, and D. Marks, "Data and operational processing for NASA's GPM ground validation program," presentation at the 37th International Conference on Radar Meteorology, Norman, OK, American Meteorological Society, 2015. [Online]. Available: https://ams.confex.com/ams/37RADAR/webprogram/Manuscript/Paper275627/37radarmanuscript.pdf

[274] L. D. Carey and W. A. Petersen, "Sensitivity of C-band polarimetric radar-based drop size estimates to maximum diameter," *Journal of Applied Meteorology and Climatology*, vol. 54, no. 6, pp. 1352–1371, 2015.

[275] E. Gorgucci and L. Baldini, "Drop shape and DSD retrieval with an X-band dual polarization radar," presentation at the 33rd Conference on Radar Meteorology, Cairns, Australia, American Meteorological Society, 2007.

[276] W. Hitschfeld and J. Bordan, "Errors inherent in the radar measurement of rainfall at attenuating wavelengths," *Journal of Meteorology*, vol. 11, no. 1, pp. 58–67, 1954.

[277] M. Marzoug and P. Amayenc, "A class of single- and dual-frequency algorithms for rain-rate profiling from a spaceborne radar. Pad I: Principle and tests from numerical simulations," *Journal of Atmospheric and Oceanic Technology*, vol. 11, no. 6, pp. 1480–1506, 1994.

[278] T. J. Smyth and A. J. Illingworth, "Correction for attenuation of radar reflectivity using polarization data," *Quarterly Journal of the Royal Meteorological Society*, vol. 124, no. 551, pp. 2393–2415, 1998. [Online]. Available: https://rmets.onlinelibrary.wiley.com/doi/abs/10.1002/qj.49712455111

[279] J. Testud, E. L. Bouar, E. Obligis, and M. Ali-Mehenni, "The rain profiling algorithm applied to polarimetric weather radar," *Journal of Atmospheric and Oceanic Technology*, vol. 17, no. 3, pp. 332–356, 2000.

[280] A. Ryzhkov, M. Diederich, P. Zhang, and C. Simmer, "Potential utilization of specific attenuation for rainfall estimation, mitigation of partial beam blockage, and radar networking," *Journal of Atmospheric and Oceanic Technology*, vol. 31, no. 3, pp. 599–619, 2014. [Online]. Available: https://doi.org/10.1175/JTECH-D-13-00038.1

[281] G. Delrieu, S. Caoudal, and J. D. Creutin, "Feasibility of using mountain return for the correction of ground-based X-band weather radar data," *Journal of Atmospheric and Oceanic Technology*, vol. 14, no. 3, pp. 368–385, 1997.

[282] L. D. Carey, S. A. Rutledge, D. A. Ahijevych, and T. D. Keenan, "Correcting propagation effects in C-band polarimetric radar observations of tropical convection using differential propagation phase," *Journal of Applied Meteorology*, vol. 39, no. 9, pp. 1405–1433, 2000. [Online]. Available: https://doi.org/10.1175/1520-0450(2000)039<1405:CPEICB>2.0.CO;2

[283] S.-G. Park, V. N. Bringi, V. Chandrasekar, M. Maki, and K. Iwanami, "Correction of radar reflectivity and differential reflectivity for rain attenuation at X band. Part I: Theoretical and empirical basis," *Journal of Atmospheric and Oceanic Technology*, vol. 22, no. 11, pp. 1621–1632, 2005. [Online]. Available: https://doi.org/10.1175/JTECH1803.1

[284] V. N. Bringi, T. D. Keenan, and V. Chandrasekar, "Correcting C-band radar reflectivity and differential reflectivity data for rain attenuation: A self-consistent method with constraints," *IEEE Transactions on Geoscience and Remote Sensing*, vol. 39, no. 9, pp. 1906–1915, Sep. 2001.

[285] J. S. Marshall, R. C. Langille, and W. M. K. Palmer, "Measurement of rainfall by radar," *Journal of Meteorology*, vol. 4, no. 6, pp. 186–192, 1947.

[286] L. Battan, *Radar Observation of the Atmosphere*. Chicago: University of Chicago Press, 1973.

[287] R. Keränen and V. Chandrasekar, "Detection and estimation of radar reflectivity from weak echo of precipitation in dual-polarized weather radars," *Journal of Atmospheric and Oceanic Technology*, vol. 31, no. 8, pp. 1677–1693, 2014. [Online]. Available: https://doi.org/10.1175/JTECH-D-13-00155.1

[288] T. A. Seliga and V. N. Bringi, "Potential use of radar differential reflectivity measurements at orthogonal polarizations for measuring precipitation," *Journal of Applied Meteorology*, vol. 15, no. 1, pp. 69–76, 1976. [Online]. Available: https://doi.org/10.1175/1520-0450(1976)015<0069:PUORDR>2.0.CO;2

[289] E. Gorgucci, V. Chandrasekar, and G. Scarchilli, "Radar and surface measurement of rainfall during CaPE: 26 July 1991 case study," *Journal of Applied Meteorology*,

vol. 34, no. 7, pp. 1570–1577, 1995. [Online]. Available: https://doi.org/10.1175/1520-0450-34.7.1570

[290] R. Cifelli, V. Chandrasekar, S. Lim, P. C. Kennedy, Y. Wang, and S. A. Rutledge, "A new dual-polarization radar rainfall algorithm: Application in Colorado precipitation events," *Journal of Atmospheric and Oceanic Technology*, vol. 28, no. 3, pp. 352–364, 2011. [Online]. Available: https://doi.org/10.1175/2010JTECHA1488.1

[291] S. E. Giangrande and A. V. Ryzhkov, "Estimation of rainfall based on the results of polarimetric echo classification," *Journal of Applied Meteorology and Climatology*, vol. 47, no. 9, pp. 2445–2462, 2008. [Online]. Available: https://doi.org/10.1175/2008JAMC1753.1

[292] H. Chen, V. Chandrasekar, and R. Bechini, "An improved dual-polarization radar rainfall algorithm (DROPS2.0): Application in NASA IFloodS field campaign," *Journal of Hydrometeorology*, vol. 18, no. 4, pp. 917–937, 2017.

[293] R. R. Lee and R. M. Steadham, "WSR-88D algorithm comparisons of VCP 11 and new VCP 12," presentation at the 20th International Conference on Interactive Information and Processing Systems (IIPS) for Meteorology, Oceanography, and Hydrology, American Meteorological Society, 2004.

[294] B. Philips and V. Chandrasekar, "The Dallas Fort Worth urban remote sensing network," in *2012 IEEE International Geoscience and Remote Sensing Symposium*. Piscataway, NJ: Institute of Electrical and Electronics Engineers, 2012, pp. 6911–6913.

[295] B. Philips, D. Pepyne, D. Westbrook, E. Bass, J. Brotzge, W. Diaz, K. Kloesel, J. Kurose, D. Mclaughlin, H. Rodriguez, and M. Zink, "Integrating end user needs into system design and operation: The Center for Collaborative Adaptive Sensing of the Atmosphere (CASA)," presentation at the 16th Conference on Applied Climatology, San Antonio, TX, American Meteorological Society, 2007.

[296] D. J. Mclaughlin, V. Chandrasekar, K. Droegemeier, S. Frasier, J. Kurose, F. Junyent, B. Philips, S. Cruz-pol, and J. Colom, "Distributed collaborative adaptive sensing (DCAS) for improved detection, understanding, and predicting of atmospheric hazards," presentation at the Ninth Symposium on Integrated Observing and Assimilation Systems for the Atmosphere, Oceans, and Land Surface, American Meteorological Society, 2005.

[297] D. Pepyne, D. Westbrook, B. Philips, E. Lyons, M. Zink, and J. Kurose, "Distributed collaborative adaptive sensor networks for remote sensing applications," in *2008 American Control Conference*. Piscataway, NJ: Institute of Electrical and Electronics Engineers, pp. 4167–4172, 2008. [Online]. Available: http://ieeexplore.ieee.org/lpdocs/epic03/wrapper.htm?arnumber=4587147

[298] F. Junyent and V. Chandrasekar, "Theory and characterization of weather radar networks," *Journal of Atmospheric and Oceanic Technology*, vol. 26, no. 3, pp. 474–491, 2009.

[299] V. Chandrasekar, H. Chen, and B. Philips, "Principles of high-resolution radar network for hazard mitigation and disaster management in an urban environment," *Journal of the Meteorological Society of Japan. Series II*, vol. 96A, pp. 119–139, 2018.

[300] G. P. Cressman, "An operational objective analysis system," *Monthly Weather Review*, vol. 87, no. 10, pp. 367–374, 1959. [Online]. Available: http://journals.ametsoc.org/doi/abs/10.1175/1520-0493%281959%29087%3C0367%3AAOOAS%3E2.0.CO%3B2

[301] S. L. Barnes, "A technique for maximizing details in numerical weather map analysis," *Journal of Applied Meteorology*, vol. 3, no. 4, pp. 396–409, 1964.

[302] M. A. Askelson, J.-P. Aubagnac, and J. M. Straka, "An adaptation of the Barnes filter applied to the objective analysis of radar data," *Monthly Weather Review*, vol. 128, no. 9, pp. 3050–3082, 2000.

[303] V. Lakshmanan, T. Smith, K. Hondl, G. J. Stumpf, and A. Witt, "A real-time, three-dimensional, rapidly updating, heterogeneous radar merger technique for reflectivity, velocity, and derived products," *Weather and Forecasting*, vol. 21, no. 5, pp. 802–823, 2006.

[304] A. Augst and M. Hagen, "Interpolation of operational radar data to a regular Cartesian grid exemplified by Munich's airport radar configuration," *Journal of Atmospheric and Oceanic Technology*, vol. 34, no. 3, pp. 495–510, 2017.

[305] C. G. Mohr and R. L. Vaughan, "An economical procedure for Cartesian interpolation and display of reflectivity factor data in three-dimensional space," *Journal of Applied Meteorology*, vol. 18, no. 5, pp. 661–670, 1979. [Online]. Available: https://doi.org/10.1175/1520-0450(1979)018<0661:AEPFCI>2.0.CO;2

[306] L. Jay Miller, C. G. Mohr, and A. J. Weinheimer, "The simple rectification to Cartesian space of folded radial velocities from Doppler radar sampling," *Journal of Atmospheric and Oceanic Technology*, vol. 3, no. 1, pp. 162–174, 1986. [Online]. Available: https://doi.org/10.1175/1520-0426(1986)003<0162:TSRTCS>2.0.CO;2

[307] S. S. Weygandt, A. Shapiro, and K. K. Droegemeier, "Retrieval of model initial fields from single-Doppler observations of a supercell thunderstorm. Part I: Single-Doppler velocity retrieval," *Monthly Weather Review*, vol. 130, no. 3, pp. 433–453, 2002. [Online]. Available: https://doi.org/10.1175/1520-0493(2002)130<0433:ROMIFF>2.0.CO;2

[308] A. Shapiro, P. Robinson, J. Wurman, and J. Gao, "Single-Doppler velocity retrieval with rapid-scan radar data," *Journal of Atmospheric and Oceanic Technology*, vol. 20, no. 12, pp. 1758–1775, 2003. [Online]. Available: https://doi.org/10.1175/1520-0426(2003)020<1758:SVRWRR>2.0.CO;2

[309] G. M. Heymsfield, "Statistical objective analysis of dual-Doppler radar data from a tornadic storm," *Journal of Applied Meteorology*, vol. 15, no. 1, pp. 59–68, 1976. [Online]. Available: https://doi.org/10.1175/1520-0450(1976)015<0059:SOAODD>2.0.CO;2

[310] J. Gao, M. Xue, A. Shapiro, and K. K. Droegemeier, "A variational method for the analysis of three-dimensional wind fields from two Doppler radars," *Monthly Weather Review*, vol. 127, no. 9, pp. 2128–2142, 1999. [Online]. Available: https://doi.org/10.1175/1520-0493(1999)127<2128:AVMFTA>2.0.CO;2

[311] J. Zhang, K. Howard, and J. J. Gourley, "Constructing three-dimensional multiple-radar reflectivity mosaics: Examples of convective storms and stratiform rain echoes," *Journal of Atmospheric and Oceanic Technology*, vol. 22, no. 1, pp. 30–42, 2005. [Online]. Available: https://doi.org/10.1175/JTECH-1689.1

[312] S. L. Barnes, "Mesoscale objective map analysis using weighted time-series observations," National Oceanic and Atmospheric Administration, Commerce Department, Technical Repport, 1973.

[313] R. C. Srivastava and L. Tian, "Measurement of attenuation by a dual-radar method: Concept and error analysis," *Journal of Atmospheric and Oceanic Technology*, vol. 13, no. 5, pp. 937–947, 1996. [Online]. Available: https://doi.org/10.1175/1520-0426(1996)013<0937:MOABAD>2.0.CO;2

[314] V. Chandrasekar and S. Lim, "Retrieval of reflectivity in a networked radar environment," *Journal of Atmospheric and Oceanic Technology*, vol. 25, no. 10, pp. 1755–1767, 2008. [Online]. Available: https://doi.org/10.1175/2008JTECHA1008.1

[315] E. Yoshikawa, V. Chandrasekar, T. Ushio, and T. Matsuda, "A Bayesian approach for integrated raindrop size distribution (DSD) retrieval on an X-band dual-polarization radar network," *Journal of Atmospheric and Oceanic Technology*, vol. 33, no. 2, pp. 377–389, 2016.

[316] Y. Wang and V. Chandrasekar, "Quantitative precipitation estimation in the CASA X-band dual-polarization radar network," *Journal of Atmospheric and Oceanic Technology*, vol. 27, no. 10, pp. 1665–1676, 2010.

[317] R. P. Davies-Jones, "Dual-Doppler radar coverage area as a function of measurement accuracy and spatial resolution," *Journal of Applied Meteorology*, vol. 18, no. 9, pp. 1229–1233, 1979. [Online]. Available: https://doi.org/10.1175/1520-0450-18.9.1229

[318] Y. Wang, V. Chandrasekar, and B. Dolan, "Development of scan strategy for dual Doppler retrieval in a networked radar system," in *International Geoscience and Remote Sensing Symposium (IGARSS)*, vol. 5. Piscataway, NJ: Institute of Electrical and Electronics Engineers, 2008, pp. 322–325.

[319] H. Chen and V. Chandrasekar, "Real-time wind velocity retrieval in the precipitation system using high-resolution operational multi-radar network," in *Remote Sensing of Aerosols, Clouds, and Precipitation*, T. Islam, Y. Hu, A. Kokhanovsky, and J. Wang, eds. Amsterdam:. Elsevier, 2017, pp. 315–339. [Online]. Available: http://dx.doi.org/10.1016/B978-0-12-810437-8.00015-3

[320] D. C. Dowell and A. Shapiro, "Stability of an iterative dual-Doppler wind synthesis in Cartesian coordinates," *Journal of Atmospheric and Oceanic Technology*, vol. 20, no. 11, pp. 1552–1559, 2003. [Online]. Available: https://doi.org/10.1175/1520-0426(2003)020<1552:SOAIDW>2.0.CO;2

[321] O. Bousquet and M. Chong, A multiple-Doppler synthesis and continuity adjustment technique (MUSCAT) to recover wind components from Doppler radar measurements," *Journal of Atmospheric and Oceanic Technology*, vol. 15, no. 2, pp. 343–359, 1998. [Online]. Available: https://doi.org/10.1175/1520-0426(1998)015<0343:AMDSAC>2.0.CO;2

[322] A. Shapiro and J. J. Mewes, "New formulations of dual-Doppler wind analysis," *Journal of Atmospheric and Oceanic Technology*, vol. 16, no. 6, pp. 782–792, 1999. [Online]. Available: https://doi.org/10.1175/1520-0426(1999)016<0782:NFODDW>2.0.CO;2

[323] J. Sun and N. A. Crook, "Dynamical and microphysical retrieval from Doppler radar observations using a cloud model and its adjoint. Part I: Model development and simulated data experiments," *Journal of the Atmospheric Sciences*, vol. 54, no. 12, pp. 1642–1661, 1997. [Online]. Available: https://doi.org/10.1175/1520-0469(1997)054<1642:DAMRFD>2.0.CO;2

[324] M. Vaccarono, R. Bechini, C. V. Chandrasekar, R. Cremonini, and C. Cassardo, "An integrated approach to monitoring the calibration stability of operational dual-polarization radars," *Atmospheric Measurement Techniques*, vol. 9, no. 11, pp. 5367–5383, 2016. [Online]. Available: https://www.atmos-meas-tech.net/9/5367/2016/

[325] S. Lim, V. Chandrasekar, P. Lee, and A. P. Jayasumana, "Real-time implementation of a network-based attenuation correction in the CASA IP1 testbed," *Journal of Atmospheric and Oceanic Technology*, vol. 28, no. 2, pp. 197–209, 2011. [Online]. Available: https://doi.org/10.1175/2010JTECHA1441.1

Index